A SYSTEMS APPROACH TO CIVIL ENGINEERING PLANNING AND DESIGN

A SYSTEMS APPROACH TO CIVIL ENGINEERING PLANNING AND DESIGN

Thomas K. Jewell
Union College
Schenectady, New York

Harper & Row, Publishers, New York

Cambridge, Philadelphia, San Francisco,

1817   London, Mexico City, São Paulo, Singapore, Sydney

Sponsoring Editor: Cliff Robichaud
Project Editor: Steven Pisano
Cover Design: Karen Salsgiver
Text Art: Reproduction Drawing Ltd.
Production: Delia Tedoff
Compositor: TAPSCO, Inc.
Printer and Binder: Maple Press

A SYSTEMS APPROACH TO CIVIL ENGINEERING PLANNING AND DESIGN

Library of Congress Cataloging in Publication Data

Jewell, Thomas K.
    A systems approach to civil engineering planning and
design.
    Bibliography: p.
    Includes index.
    1. Civil engineering. 2. System analysis. I. Title.
TA153.J49   1986   624   85-16453
ISBN 0-06-043317-5

85 86 87 88 9 8 7 6 5 4 3 2 1

Dedicated to Pam, Jim, Keith, and John;
without whose help and support
this project could never have been completed.

CONTENTS

PREFACE

In order to be competent and effective, engineers must be able to apply learned technical knowledge to real-world situations. The purpose of this text is to smooth the transition from structured, textbook-type problem solving to the solution of open-ended problems that may have a number of technically feasible solutions. By using the problem-solving techniques developed in the text, engineers will be better able to judge objectively which of the possible solutions are better, or, ideally, which is the best.

The systems approach as it is developed in this text provides a framework for systematic analysis of an engineering planning or design problem, development of alternative ways of solving the problem, and a choice of the best alternative for a particular set of circumstances. Imbedded within the systems approach are the tools of systems analysis, which can be utilized within the systems approach to make the analysis and choice of alternatives more objective. Using the methodology and techniques developed in the text, present as well as future engineers will better be able to advise the person with decision-making authority. The engineer will also learn how to choose the proper tool for use in a given situation and will learn to recognize those situations where an outside expert should be consulted.

ACKNOWLEDGMENTS

My wife, Pam, through her patience and skill as an editor and typist, has contributed a great deal to the preparation of this manuscript. Dr. R. Paul Mason, who taught the systems analysis course at Union College before my arrival, provided me with a great deal of background information which I have incorporated into my class notes, and thence into this text. Although they are too numerous to mention individually, many of my students over the past five years have helped in the development and checking of example and assignment problems. Professor Frank Griggs, former chairman of the Civil Engineering Department, and Dr. Thomas D'Andrea, Vice-president for Academic Affairs at Union College have lent considerable support through sabbatical and other release time.

Many excellent ideas for revisions were supplied by the reviewers: Dr. John C. Liebman of the University of Illinois at Champagne-Urbana, Dr. Jon D. Fricker of Purdue University, and Dr. John W. Labadie of Colorado State University. Guidance for marketing was provided by Dr. David Marks of MIT, Dr. James Male of the University of Massachusetts at Amherst, and Dr. Norman Bolyea of the University of Lowell.

I am grateful to the Literary Executor of the late Sir Ronald A. Fisher, F.R.S., to Dr. Frank Yates, F.R.S., and to Longman Group Ltd. London, for permission to reprint Tables III and VII from their book *Statistical Tables for Biological, Agricultural and Medical Research* (6th edition, 1974).

Ideas for this text have been gleaned from many sources. Particular references that have weighed heavily in the development of a particular section are cited in the text. General references that were utilized in developing concepts are given in the Bibliography. I would like to express my appreciation to all those who have helped, and from whom I have acquired information, and apologize to anyone whom I may have inadvertently neglected to mention.

Final responsibility for the accuracy and completeness of the text, of course, lies with the author.

THOMAS K. JEWELL

A NOTE TO THE INSTRUCTORS

In undergraduate civil engineering education, systems analysis techniques need to be adequately coupled with the teaching of the systems approach. The systems approach and systems analysis must be adequately integrated into the civil engineering curriculum so that they will provide unifying factors in the teaching of planning and design, and will represent a formalizing and reinforcing of the procedures inherent in any good design process. Too much emphasis in the past seems to have been placed on systems analysis techniques as isolated tools.

This text presents an integrated approach, putting all aspects of systems theory into proper perspective. It is intended to be used in an upper-level undergraduate or introductory graduate course in systems analysis for civil engineering students. The theories of the systems approach and systems analysis are developed and then applied to the civil engineering planning and design processes. Integration of concepts and applications to practical problems are emphasized. After mastering the material in this text, a student will be well prepared for a comprehensive design course. The text will also help students make the transition from textbook problem solving to real-world problem solving. Open-ended problems in which it is necessary to choose from among several alternatives are emphasized. The student will learn that most real-world engineering decisions cannot be made with high levels of precision and complete certainty as to the accuracy of the results.

Students using this text are assumed to have completed the engineering science sequence of courses through strength of materials (mechanics of deformable solids) and fluid mechanics. Completion of structural analysis, soil mechanics, and advanced fluids (hydraulics or water resources engineering) is also desirable, but not necessary, for comprehension of the concepts in the text. The text assumes that students have a background in the following areas:

Linear algebra

Engineering economics

Probability theory

Computer usage

Numerical analysis

However, since not all students will have taken courses in all of these areas, brief reviews of the concepts from each of these areas used in the text are contained in the appendixes. Lists of additional references are also given.

It is becoming increasingly important for engineering students to be familiar with computer usage. Although computer programming literacy is not necessary for a student to understand the theory of systems analysis and the systems approach, it is essential if many of the concepts are to be applied to real-world problems. Since computer centers have applications packages for implementing the procedures described in this text, computer programs are not presented. However, several programs that may be useful to instructors are included in the Solutions Manual.

The text is organized into four parts, plus appendixes. The first three parts develop the systems approach methodology, with particular systems analysis techniques developed at the applicable stage of the systems approach. The fourth section contains specialized systems analysis techniques that are useful for particular types of problems. One approach to teaching this material would be for an instructor to cover the material contained in Parts One through Three and any remedial material necessary from the appendixes, in a normal one-semester/term course. Additional material from Part Four may be presented as time permits, or as special needs dictate. An alternate approach would be to cover the whole book in a two-semester/term course sequence.

Detailed example problems are presented whenever new theory is introduced. Homework problems given at the end of each chapter emphasize the practical application of the theory learned, and require the student to use his or her previously gained technical expertise to develop system models, to identify interactions among system components, subsystems, and the problem environment, and to develop optimum feasible solutions to system problems using appropriate techniques.

At the completion of the course the student will have a firm understanding of the systems philosophy toward planning and design, and will also have mastered several new analysis tools to help in objective decision making.

PART ONE

THE PLANNING/DESIGN PROCESS: PRELIMINARY STEPS

Chapter 1

The Planning/Design Process

INTRODUCTION

This chapter will provide a general introduction to the planning/design processes and their relationship to the systems approach and systems analysis. Subsequent chapters will consider specific aspects of these concepts in more detail.

Planning and design are integral parts of the solution to most nontrivial engineering problems. Both of these topics should receive significant emphasis in civil engineering curricula. Students spend a great deal of time studying engineering science subjects and the theory of civil engineering problem solving. Frequently, however, the problems that they are required to solve are limited in scope and do not require application of problem-solving skills beyond what is contained in the textbook. Training in good planning/design procedures, referred to as the systems approach, will help students in their transition to practical, applied problem solving.

To be effective, the teaching of planning and design must be integrated throughout the curriculum. Even at the freshman level, students can be introduced to these processes through creative problems in courses such as graphics and basic mechanics. At the same time, they will be introduced to the systems approach. As students progress, some planning and design should be incorporated into each of their engineering courses.

Planning and design work represents the next step beyond textbook problem solving. Students should be expected to integrate and synthesize their acquired knowledge under a given set of restrictive conditions (constraints) to solve realistic problems that have no fixed solution (open-ended problems). These problems may have several distinct solutions, some more desirable than others, with one or more best, or optimal solutions. The student must learn to recognize these optimal solutions.

Students must also learn to deal with problems that are subjective, or nonquantifiable, in the usual engineering sense. This may be due to the complexity of the problem, or to its nature. For example, it is difficult to assign a specific value to the beauty of nature or to the damage done by sound intrusion. Numerical ratings can be assigned to a subjective evaluation scale for these entities; however, care must be taken to remember the basis for these numbers when combining them with quantifiable entities.

A course in systems concepts (the systems approach and systems analysis) will help the student to develop the tools necessary to undertake meaningful planning and design related to real world problems, to properly define these problems, and to find solutions to these problems that are both correct and efficient. Optimization is the term applied to the process undertaken to find the most efficient solution possible. "Most efficient," as the term is used in this text, could be defined as the maximum output per unit input, or minimum input per unit output.

Proper curricular integration of systems concepts will draw together and solidify the various problem-solving approaches that the student has learned. All of these concepts can then be utilized in subsequent upper-level design courses, and ideally in a final comprehensive planning and design course. Figure 1.1 is a schematic that shows how a course in systems concepts can be effectively integrated into the civil engineering curriculum.

Systems concepts can be applied to a wide variety of engineering problems, including the planning and design of structures and transportation systems, finding best solutions for water resources and environmental problems, and construction planning and management. It is becoming increasingly important that all educated people have some knowledge of systems analysis and the systems approach. At the same time, it is becoming absolutely necessary for engineers to have a working knowledge of both of these concepts.

Planning/Design, the Systems Approach, and Systems Analysis

Planning and design are undertaken with the objective of producing a workable product, whether it be an automobile, a building, a highway system, or whatever. However, it is not necessarily enough just to produce a workable product. There may be several workable designs, some of which are better, or more competitive, than others. The systems approach allows the designer/planner to identify and evaluate a range of alternate solutions, to develop information about trade-offs among these solutions, and to make recommendations concerning the best solutions to whomever is responsible for making final decisions. Application of the systems approach facilitates determining that a problem exists, refining the problem, generating possible solutions within a defined set of limiting conditions (constraints), and then determining which solution is best according to stated criteria. These criteria form what is called the objective function. The process used in applying the systems approach is commonly called optimization of the objective function; the purpose of which may be to minimize cost, minimize material used, minimize pollution potential entering a stream, or maximize benefits. Essentially it is finding the most efficient solution to a problem which will maximize or minimize either system input or system output for a given level of the

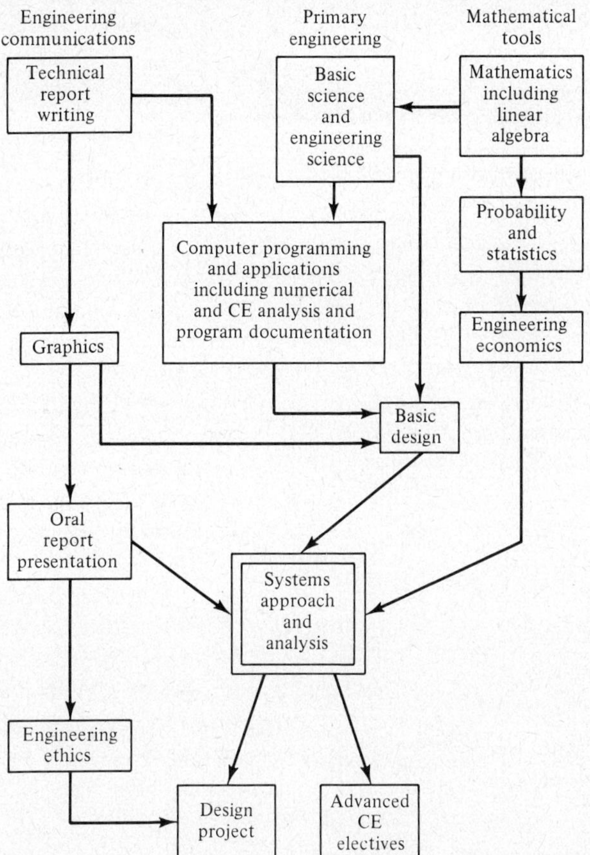

Figure 1.1 Integration of systems concepts into the civil engineering curriculum.

other. Systems analysis is primarily a body of mathematical techniques that can be used to assist in and strengthen the application of the systems approach by producing quantitative measures of performance. Some systems analysis techniques include mathematical optimization, economic analysis, and decision theory.

Figure 1.2 shows the author's conception of the relationship between systems analysis and the systems approach as applied to the planning/design process. It defines the various aspects of systems analysis as the tools and the systems approach as the integrating procedure that facilitates application of the tools to find the best solution to engineering problems. The systems approach should not be confused with systems analysis. The systems approach can be applied without the aid of systems analysis, but it will lose much of its objectivity. Conversely, systems analysis cannot take the place of effective application of the systems approach. Not all elements of systems analysis need be used in each situation; rather, it is up to the engineer/planner to apply those aspects of systems analysis that will effectively complement the systems approach in developing a specific problem solution.

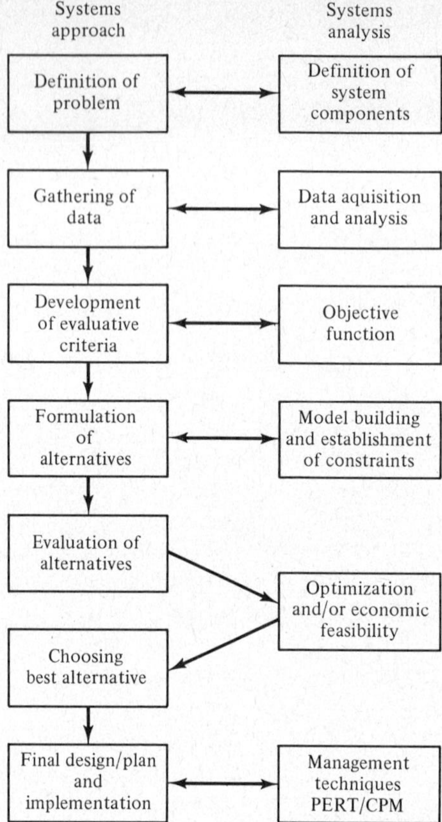

Figure 1.2 Planning and design process. Feedback may be necessary from any step to any previous step. This is illustrated in Figure 1.4.

The remainder of Chapter 1 will be devoted to an introduction to what systems are, and to how planning and design, the systems approach, and systems analysis are related to these systems. Detailed study of these concepts will be undertaken in subsequent chapters. A major objective of this text is to provide engineers with a thorough understanding of these concepts and to show how they can be applied to help solve real-world problems.

PLANNING/DESIGN

Historical Perspective of Planning/Design

Since before the beginning of recorded time, people have been innovative, if not systematic, in their approach to problem solving. Most innovation was in response to an easily recognizable need, such as:

Improvement in survival chances through improved hunting utensils, improved agricultural equipment, and improved housing.

Improvement in transportation.

Desire for dominance over neighbors.

Desire to be able to control and exploit the environment.

Desire for a better quality of life.

The basic activities of design were present. These included the mental activities of idea generation and evaluation, and the physical activity of communication of the workable design. Idea generation and evaluation were largely by trial and error. If an idea worked, it was retained and possibly improved upon through more trial-and-error development. If an idea did not work, it was usually discarded with little thought or analysis as to why it did not. Early communication was accomplished by passing on the product itself, where possible, or by word of mouth. Even after written communication became prevalent, distribution of engineering designs was not widespread because of the laborious task of reproducing plans, and because of transportation difficulties. Until recently, little thought was given to the secondary effects of engineering innovations.

As technical knowledge evolved, engineering innovation became more systematic. More emphasis was placed on why a particular idea might not work, and on how that idea might be modified or combined with other ideas to produce a workable solution to a problem.

With increasing knowledge came more complex solutions and a spiraling rise in innovation. One new innovative design often revealed several more designs that could be undertaken. The increasing complexity of design necessitated a more systematic approach, and in most cases a team design effort. No longer could one engineer direct all aspects of a large project. The chief engineer was forced to act as a manager for a hierarchy of engineering subordinates, each responsible for a particular aspect of the project and a subhierarchy of engineering and technical subordinates. Thus was born the need for the systems approach. It provided a means of coordinating the diverse activities of the planning/design team; assisted in carefully defining the problem in terms of the various levels of systems associated with it; and facilitated the identification of the complex interaction among these systems. The systems approach allowed a more rational approach to planning/design, which would lead to a more nearly optimal solution. Needs surfaced for quantitative methods to assist in the systems approach. Systems analysis techniques were developed; however, many of them had to wait for the development of large computers before they could be applied to practical problems. Development of systems analysis techniques also influenced further refinement of the systems approach.

Planning

Planning and design are so closely related that it is difficult to separate one from the other. Planning is involved to some extent in everything that an engineer does. A poorly planned engineering project will almost certainly not be executable. What will be discussed in this text is a systematic approach to planning, and how planning can help civil engineers achieve desirable goals.

The planning process closely follows the systems approach, and may involve the use of sophisticated analysis and computer tools. However, the scope of problems addressed by planning is different from the scope of design problems. Basically, planning is the formulation of goals and objectives that are consistent with political, social, environmental, economic, technological, and aesthetic constraints; and the general definition of procedures designed to meet those goals and objectives. Goals are the desirable end states that are sought. They may be influenced by actions or desires of government bodies, such as legislatures or courts; of special interest groups; or of administrators. Goals may change as the interests of the concerned groups change. Objectives relate ways in which the goals can be reached. Planning should be involved in all aspects of an engineering project, including preliminary investigations, feasibility studies, detailed analysis and specifications for implementation and/or construction, and monitoring and maintenance. A good plan will bring together diverse ideas, forces, or factors, and combine them into a coherent, consistent structure that when implemented, will improve target conditions without deprecating nontarget conditions. Effective use of the systems approach will help ensure that planning studies address the true problem at hand. Planning studies that do not do this could not, if implemented, produce useful and desirable changes.

Comprehensive transportation plans and construction (management) plans for a physical project are two specific examples of engineering planning studies. The latter plan is generally implementable, because the scope and definition of the problem can be readily established. Definition of the total problem environment for the former type of study may be much more difficult, if not impossible. Thus, the decision maker is left with the problem of how to fill in the blanks.

Design

The Accreditation Board for Engineering and Technology, ABET (1980), defines design as:

> the process of devising a system, component, or process to meet desired needs. It is a decision-making process (often iterative), in which the basic sciences, mathematics, and engineering sciences are applied to convert resources optimally to meet a stated objective. Among the fundamental elements of the design process are the establishment of objectives and criteria, synthesis, analysis, construction, testing, and evaluation.

Design is the process of determining the specific form of an end product. Design is the next step beyond text book analysis, and it is often difficult for students to fully comprehend when they are initially introduced to it. For the first time, not all of the relevant information is supplied, and open-ended type problems are encountered. Engineering judgment must be applied and previously learned technical expertise must be synthesized, recalled, and put together, to solve the new problem.

Design can also be defined with respect to systems. System design is the process of selecting, putting together, and/or developing the components (elements or procedures) that will produce a system that will optimally satisfy some rationally arrived at goals. Optimality in this sense indicates a system that achieves its finest level of performance.

The only way to learn design is through practice. The elements of design can be taught at the undergraduate level, but the process of becoming a true professional designer takes several years of practice under professional engineers. This is the main reason for the four-year design experience required before a new engineer can take the professional engineer's licensing examination. An understanding of the systems approach to design and an appreciation of the fact that design problems are generally open-ended, with several possible solutions, are concepts that a student should gain through undergraduate design training.

Design is an iterative, or cyclical, process, in that the initially completed design often points to improvements that could be made in some earlier stage of the process. By recycling to this step of the design process, the overall design is improved. Good design takes a certain amount of creativity and a lot of technical expertise. It takes a group effort by several professionals from differing areas of expertise to produce a workable and efficient design.

During an introduction to design at the undergraduate level, the student should learn that design often will have to be undertaken without adequate information (data), without adequate time to fully explore the problem, and/or without adequate funds to gather all the information or conduct all of the tests needed. As part of the design education, the student should practice working as part of a project team. Participation in the activities of the project team will demonstrate the value of interaction with other engineers, experts from other disciplines, and the lay public; the necessity of completing work on time, and the importance of preparing oral and written reports in a professional and informative manner.

Hill (1981) has stated that students who have mastered the fundamentals of design will habitually:

1. Get a correct answer to the correct problem.
2. Be counted on to:
   a. Make appropriate assumptions.
   b. Select reasonable values for missing data.
   c. Discard extraneous information.
3. Set up problems in a concise tabular form when iterative or repetitive calculations are required.
4. Establish an efficient procedure before attempting to solve a problem, and then use that procedure.
5. Relate mathematical models to their real-world counterparts.
6. Have an understanding of the limitations of design formulas and pertinent design procedures.
7. Have the ability to comfortably and confidently:
   a. Arrive at an appropriate solution.
   b. Seek the help of a consultant (specialist).

The author would like to add an eighth characteristic of competent students of design:

8. Determine the most efficient computational tool: mind, calculator, or computer.

These characteristics are not in any particular order of importance, with the possible exception of item 1. If a student cannot arrive at an acceptable solution to the correct problem, then the rest of the characteristics are meaningless. A thorough understanding of the systems approach will help the student in properly defining the problem to be solved and will assist in not only arriving at a workable solution, but in finding the best or optimal solution.

Interaction of Planning and Design

The planning and design processes complement each other. Essentially the same steps are required in both, and the systems approach can be applied equally as effectively to either process.

The systems approach is essentially a planning framework to improve the design process. After the designer determines the best design for the end product, be it a high-rise building or a section of highway, the orderly implementation (construction) of the design must be planned.

Generally a plan does not specify the final form of a physical object, and several design problems may be imbedded within it. Take for example, a transportation plan for an urban area which may require the design of highways, interchanges, and even possibly an entire rail rapid transit system to meet the objectives of the plan.

The planning and design processes are characterized by repetitive recycling. Additional information that becomes available, or changing conditions that become apparent, as a result of the planning/design process, may point to the need for replanning or redesigning at a previous step. Through this recycling within the planning/design process, the engineer is able to anticipate possible problems and adjust the process for their solution during the planning, engineering design, architectural design, or construction stages. In other words, the engineer can anticipate these problems and solve them before they become serious enough to threaten the financial viability or the safety of a project or the implementability of a plan.

SYSTEMS

There are many variations in the definition of what a system is, but all of the definitions share many common traits. Some kind of system is inherent in all but the most trivial civil engineering planning and design problems. To understand a problem, the engineer must be able to recognize and understand the system that surrounds and includes it.

Most engineering projects are systems themselves, or are a subsystem or component of a larger system. This is true today as well as in historical engineering projects. Failure to recognize the systems nature of projects and to properly define the system and its associated interactions with the surrounding environment have led to many notable engineering failures in the past, including the French efforts to construct the Panama Canal, the Tacoma Narrows Bridge and many earlier suspension bridges, and the urban highway transportation systems for many of our metropolitan areas. Some notable examples of adequate system planning and design are the Navy's Polaris Missile project, the U.S. air traffic control system, and the Apollo moon landing program. Each of these represents a complex system that would not function properly without adequate definition of what makes up the system, how the system works, and how it interacts with its environment. Some of the reasons for the poor system definition in the former projects include: poor communications, lack of knowledge of interrelationships, politics, limited objectives, and transportation difficulties.

What, then, is a system? A system is some collection of components that are connected by some type of interaction or interrelationship. These components collectively will respond to a given input and produce some form of output. The components, and thus the system, will be subject to certain constraints which set a boundary around the system. Definition of these system boundaries depends on the chosen objectives for the system. In short, a system is a collection of things that function together to fulfill some specific purpose or function. Some examples of systems could include an automobile, an urban mass transit system, or our system of state and federal governments. The systems approach can also be thought of as a system, with the tools of systems analysis comprising its components.

The components of the system can themselves be systems, or subsystems of the higher-order system. Thus a hierarchy of systems can be defined. The subsystems interact with various other subsystems at the same level in the hierarchy, as well as with systems above and below them in the hierarchy. Such systems are pyramidal, with the number of subsystems at each higher level becoming smaller until there is one system at the top that encompasses all of the subsystems. An engineer must be able not only to identify the overall system, but also, to identify the various subsystems and the interactions involved, at least down to the minimum level that will have to be analyzed for the problem at hand. The engineer must also be able to determine at what level within the hierarchical structure each particular problem lies. Identifying the proper level may have a great deal to do with the outcome of any analysis that may be undertaken.

Figure 1.3 depicts the pyramidal structure of a typical hierarchy of systems. Generally the types of systems in such a hierarchy progress from the more general at the top to the more specific at the bottom. The degree of technical definition also has a tendency to increase for the lower-order subsystems. Thus, it often may be possible to optimize subsystems, while the overall system is too complex to optimize. Also, some factors affecting the overall system may not be known, or some of the factors affecting the system may not be amenable to evaluation and/or quantification. If the interactions between the system levels are understood, higher-level systems can sometimes be made near optimal even though they are themselves too complex to be analyzed completely.

The concept of a hierarchy of systems is not new. Aristotle wrote of the hierarchy of biological systems, placing man at the top. However, Churchman (1979) has written that Aristotle also believed that the entire conceptual scheme of nature needed to be tied together and could not be regarded in terms of separable components. Churchman used this to ask the question: "Can nonhierarchical system thinking be more fruitful than hierarchical system thinking?" His premise was that nonhierarchical thinking might be more all-encompassing and less compartmentalized. While this may seem reasonable, the fact remains that the proper definition of a system must also include

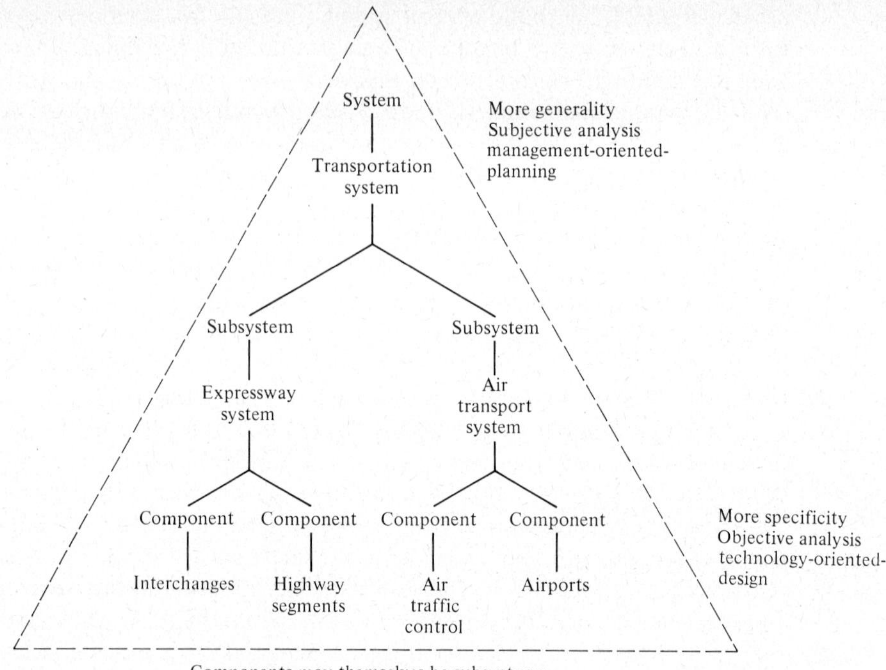

Figure 1.3 Pyramidal structure of system hierarchy.

the definitions of the various subsystems and all of their interactions. Furthermore, subsystems have to be defined if the decision maker is to be able to discern the proper level at which efforts should be concentrated.

SYSTEMS APPROACH

The systems approach is a general problem-solving technique that brings more objectivity to the planning/design processes. It is, in essence, good design; a logical and systematic approach to problem solution in which assumptions, goals, objectives, and criteria are clearly defined and specified. Emphasis is placed on relating system performance to these goals. A hierarchy of systems, which allows handling of a complex system by looking at its component parts or subsystems, is identified. Quantifiable and nonquantifiable aspects of the problem are identified, and immediate- and long-range implications of suggested alternatives are evaluated.

The systems approach establishes the proper mood of inquiry and helps in the selection of the best course of action that will accomplish a prescribed goal by broadening the information base of the decision maker; by providing a better understanding of the system and the interrelatedness of the system and its component subsystems; and by facilitating the prediction of the consequences of several alternative courses of action. Quantitative and/or qualitative analysis may be used in the application of the systems approach. The methods of systems analysis frequently assist the systems approach in the objective analysis of quantifiable factors. Subjective analysis must be used to evaluate the nonquantifiable factors.

The systems approach is a framework for analysis and decision making. It does not solve problems, but does allow the decision maker to undertake resolution of a problem in a logical, rational manner. While there is some art involved in the efficient application of the systems approach, other factors play equally important roles. The magnitude and complexity of decision processes, even in simple applications, require the most effective use possible of the scientific (quantitative) methods of systems analysis. However, one has to be careful not to rely too heavily on the methods of systems analysis, if to do so requires use of overly simplified models that do not adequately reflect physical and social reality. Outputs from such analyses have a tendency to take on a false validity because of their complexity and technical elegance.

One object of the systems approach is to plan for and redesign the future, rather than waiting passively and trying to react on a problem-solving basis when the future occurs. In applying the systems approach, decision makers look at a problem in its historical perspective, try to predict future implications of solutions, and look at lateral effects of solutions. Use of the systems approach allows the engineer to anticipate a variety of viewpoints and requirements, to plan for accommodating these viewpoints and requirements, and to minimize criticism that may arise from all sectors of society.

The systems approach can provide students with a suitable background for meaningful applied civil engineering planning and design courses. It also ties together many problem-solving techniques that students have learned in earlier technical courses.

The steps in the systems approach include:

1. Definition of the problem.
2. Gathering of data.
3. Development of criteria for evaluating alternatives.
4. Formulation of alternatives.
5. Evaluation of alternatives.
6. Choosing the best alternative.
7. Final design/plan implementation.

The relationship of each of these steps to various aspects of systems analysis was illustrated in Figure 1.2. Each of these steps will be discussed in more detail later in this chapter and in subsequent chapters. Often several steps in the systems approach are considered simultaneously, facilitating feedback and allowing a natural progression in the problem-solving process. Figure 1.4 depicts this overlap in the steps of the systems approach and shows the most common paths of feed forward and feedback between steps.

The systems approach has several defining characteristics. It is a repetitive process, with feedback allowed from any step to any previous step. Without this feedback ability, the systems approach is not being applied. Frequently, because systems analysis takes such a broad approach to problem solving, interdisciplinary teams must be called in. Coordination and commonality of technique among the disciplines is some-

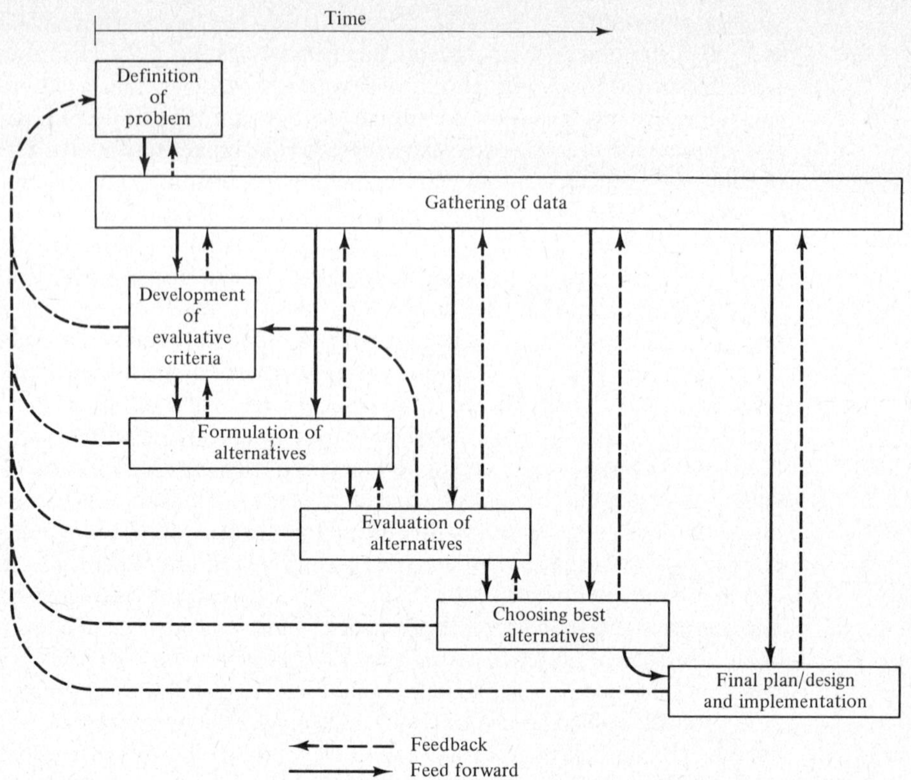

Figure 1.4 Activity overlap and feedback within systems approach.

times hard to achieve. However, if applied with ingenuity and flexibility, the systems approach can provide a common basis for understanding among specialists from seemingly unrelated fields and disciplines. Close communication among the parties involved in applying the systems approach is essential if this understanding is to be achieved.

Efficient control is a central theme of the systems approach. Through analysis and design, the systems approach aims to efficiently control systems of objects, phenomena, and/or agencies in order to focus their contributions toward a common goal. The systems approach will be most beneficial when the goals and objectives of the systems involved can be reasonably well defined. The systems approach also aims for optimization of the overall system rather than piecemeal suboptimization of its components or subsystems. This is important because the larger and more complex a system becomes, the less likely it is that an optimization of all subsystems will give an overall optimum for the total system. Also, it may be difficult to define the optimum when systems have multiple objectives. Without efficient control, overall optimization, or the most nearly optimal solution in the case of multiobjective problems, would not likely occur.

One principle that is useful in the application of the systems approach is the principle of least commitment. This states that when progressing through the decision-making process (the steps of the systems approach), no irreversible decision should be made until it must be made. This maintains maximum flexibility and keeps more possible alternatives available to the planner or designer. Elimination of alternatives that in the end may prove to be optimal is minimized. Application of least commitment should reduce the requirement for iteration in the systems approach, and should result in more efficient utilization of the total resources available.

The systems approach has become more important since the mid-1940s because of the increasing complexity and magnitude of contemporary problems. The most rational and systematic approach possible is required. Problems are typified by a large number of interacting variables, many of which defy quantification. Many alternative decisions may satisfy the stated objectives; however, the best decision is not obvious.

Some of the obstacles to proper application of the systems approach include perceptual, cultural, and habitual obstacles. Perceptual obstacles are related to peoples' inability to visualize remote relationships or to make the link between cause and effect. Frequently people do not use all of their senses for observation, but rather choose the data that conforms to their preconceived notion of how things should be. Cultural obstacles develop from local customs and procedures. Engineers are noted for their reliance on reason, logic, and established procedures, sometimes to the detriment of creativity and innovation. This leads to the third obstacle, habit. It is much easier to use established procedures than to develop new and innovative procedures. A final obstacle in many problems is the lack of time to fully develop the data and analyses required for a proper application of the systems approach.

Proper application of the systems approach places considerable responsibility on the shoulders of engineers. They must recognize the complexity of the problem, and provide a rational framework for addressing all aspects of the problem that are relevent, while blocking out aspects that are not. They must work to understand the nature of the environment surrounding any problem and the response phenomena associated with the problem. All components involved in the problem and the interactions among these components and with the environment must be recognized and treated. The engineers must recognize the system level at which analysis should be undertaken, and must accommodate and facilitate interdisciplinary interaction. Engineers will also generally be responsible for constructing a theory of reality (model) which will guide all interested parties in their observations of the problem, which in turn will help in the revision of the theory of reality.

Definition of Problem

If the use of the systems approach is to be effective, an engineer must be able to correctly identify the problem that requires investigation. Proper problem definition is analogous to proper definition of the free body in solid mechanics or control volume in fluid mechanics. Without proper problem definition at this stage, any later analysis and decision making will be flawed because the incorrect problem will be investigated.

Problem definition may require iteration and careful investigation, because problem symptoms may mask the true cause of the problem. A number of motor vehicle accidents at an intersection is a symptom of a problem, but the actual problem that generates the symptoms may be high traffic volumes, poor geometry, inadequate signal control, inadequate sight distance, or a combination of problems.

A key step of problem definition requires the identification of any systems and subsystems that are part of the problem, or related in some way to it. This set of systems and interrelationships is called the environment of the problem. This environment sets the limit on factors that will be considered when analyzing the problem. Any factors that cannot be included in the problem environment must be included as inputs to, or outputs from, the problem environment.

When defining the problem and its environment, the best approach is to make the definition as general as possible. The largest problem over which there is a reasonable chance of maintaining control should be the problem defined. This definition can either be expanded or restricted as more information becomes available. The greater the generality in the problem definition, particularly at the beginning of the study, the greater is the probability that desirable solutions will not be eliminated from consideration before examination or analysis. Also, generality in problem definition minimizes presupposing and prejudicing solutions.

Many problems are so complex that it is virtually impossible to isolate and identify all interacting components or subsystems. Also, some components/subsystems may have to be excluded from consideration if analysis is to be practical. In these cases, the principal interacting components/subsystems that can be identified are considered to be the problem environment or system. Uncertainty as to additional components/subsystems must be evaluated at least subjectively, and a statement as to their possible implications included with any final solution. Interacting components/subsystems that are identified but not included in analysis for practical reasons must be considered through the specification of the interactions with the environment in the form of inputs to and outputs from the system.

Gathering of Data

Gathering of data to assist in planning and design decision making through the systems approach will generally be done in conjunction with several steps. Some background data will have to be gathered at the problem definition stage and data gathering and analysis will continue through the final plan/design and implementation stage. Data that are gathered as the approach continues will help to identify when feedback to a previous step is required. If adequate data are not gathered throughout the process, the necessity for feedback may not be recognized until critical problems arise that will be expensive to remedy, until the product of the planning/design process will be rendered useless for all practical purposes, or until, in the worst scenario, some catastrophic failure will occur.

Data will be required at the problem definition stage to evaluate if a problem really exists; to establish what components, subsystems, and elements can be reasonably included in the delineation of the problem environment; and to define interactions between components and subsystems. Data will be needed during later steps to establish constraints on the problem and systems involved in it, to increase the set of quantifiable variables and parameters (constants) through statistical observation or development of measuring techniques, to suggest what mathematical models might contribute effectively to the analysis, to estimate values for coefficients and parameters used in any mathematical models of the system, and to check the validity of any estimated system outputs. When feedback is required, the data previously acquired can assist in redefining the problem, systems, or system models.

How much data is enough can only be answered in terms of the goals and objectives defined for the problem environment. There are certain statistical tools available that can assist in estimating how much confidence can be placed in assertions derived from the data. Often the data needed to assess a problem may encompass many fields outside of engineering; for example, sociology, geography, economics, meteorology, or medicine. The engineer must use these data, with the help of qualified specialists, to develop constraints and criteria in unfamiliar areas if a successful plan or design is to result.

Development of Evaluative Criteria

Evaluative criteria must be developed to measure the degree of attainment of system objectives. These evaluative criteria will facilitate a rational choice of a particular set of actions (from among a large number of feasible alternatives) which will best accomplish the established objectives. Some evaluative criteria will provide an absolute value of how good the solution is, such as the cost of producing one unit of some product. Other evaluative criteria will only produce relative values that can be compared among the alternatives to rank them in order of preference, as in economic comparisons such as benefit/cost analysis.

In most complex real world problems, more than one objective can be identified. A quantitative or qualitative analysis of the trade-offs between the objectives must be made. It may be possible to specify all but one objective as set levels of performance, which become system constraints. Then the system can be designed to perform optimally in terms of the remaining objective. For many problems, cost effectiveness would be the primary objective. Cost effectiveness can be defined as the lowest possible cost for a set level of control of a system, or the highest level of system control for a set cost.

An example of competing objectives can be seen in corporate objectives. The objective of the corporation is to maximize return of investment. Government regulations may impose on the corporation the objective of maintenance of environmental quality. Some specific objectives that the corporation might consider would be:

1. Maximization of corporate profits.
2. Minimization of waste products discharge.
3. Minimization of treatment costs.

These objectives may seem mutually exclusive; however, if the problem is properly formulated, they can all be accommodated. The objective could be maximization of corporate profits subject to the specified constraint of allowable waste discharge. Maximizing corporate profits should implicitly accomplish the objective of minimizing waste treatment costs for the allowable waste discharge. The objective of maximizing corporate profits would probably not be achieved if either of the other objectives was chosen as the primary objective. The choice of the primary objective will have a great deal of influence on the overall definition of the problem and on the evaluative criteria developed.

In other problems, it may not be appropriate to identify a primary objective, so multiobjective analysis techniques, which will be discussed in Chapter 13, will have to be used. For all problems, engineers must avoid proposing mutually exclusive or unrealistic objectives.

After establishment of reasonable objectives, the formulation of evaluative criteria is crucial. They will determine to a large extent the form of the final plan or design through the selection of a preferred alternative. Certain alternatives will make them appear infeasible. Therefore, realistic objectives and criteria will help simplify analysis and decision making, while unrealistic objectives and criteria may eliminate viable alternatives. The choice of evaluative criteria is itself a design decision, and should be treated as such.

Formulation of Alternatives

Formulation of alternatives is essentially the development of system models that in conjunction with evaluative criteria will be used in later analysis and decision making. If at all possible, these models should be mathematical in nature. However, it should not be assumed that mathematical model building and optimization techniques are either required or sufficient for application of the systems approach. Many problems contain unquantifiable variables and parameters which would render results generated by even the most elegant mathematical model meaningless. If it is not practical to develop mathematical models, subjective models that describe the problem environment and systems included can be constructed. Models allow a more explicit description of the problem and its systems and facilitate the rapid examination of alternatives. The primary emphasis in this text will be on problems that can at least partially be represented by mathematical models. However, limitations of these models and appropriate applications of subjective models will be indicated.

Effective model building is a combination of art and science. The science includes the technical principles of mathematics, physics, and engineering science. The art is the creative application of these principles to describe physical or social phenomena. Practice is the best way to learn the art of model building, but this practice must be predicated on a thorough understanding of the science. Although the art of modeling cannot be taught in a single course, a course that introduces the student to the systems approach will lay the foundation for further development.

In learning the art of modeling, students must be willing to probe the limits of their present level of expertise and be able to examine problems in new ways. If a certain process is considered to be deterministic (a particular input would produce a predictable output), what would happen if it were assumed to be stochastic (a particular input would produce a range of outputs based on some probability distribution)? If a phenomena or process is modeled as a first-order ordinary differential equation, what would happen if it were modeled as a second- or third-order process or if it were modeled using a partial differential equation? If an hypothesized objective function is linear, what would happen if it were made nonlinear? These are the types of questions that the modeler should learn to ask if a model is to represent the needs of a given problem realistically. An engineer who learns to develop good realistic models has successfully progressed beyond the problem solution by formulas stage.

The modeler must also realize that any model, whether developed locally or acquired from outside sources, must be calibrated and verified before it can be used to produce information of value. Calibration involves adjustment of model parameters (constants) so that model output will agree (within acceptable bounds) with measured system output for a given input. Verification is the further confirmation of model accuracy by testing it against another set of measured system data. Both calibration and verification involve gathering and proper analysis of system data, and will be discussed in detail in Chapter 3.

In order to be able to develop a realistic model, the decision maker should have a firm grasp on all of the underlying physical concepts inherent in the problem under study. Control over the modeling effort must be maintained; although, the decision maker may feel that outside qualified technical expertise is necessary for specific aspects of the modeling process. The modeler must, however, understand all material that is provided by an outside source before it is incorporated into the model. In this way, the modeler will have the necessary knowledge to have confidence in the model results, and will be able to recognize when updated information requires model refinements.

Before beginning any modeling effort, students must have a basic understanding of what a model is and what it can realistically be expected to accomplish. In one sense, a model is a simplified representation of reality which can be used by the decision maker to estimate the response of the actual system to a certain input. In another sense, a model is a method of trying to measure or approximate reality. A model may provide structure for a method of inquiry or observation needed to assist in planning and design decision making. A mathematical model must be sufficiently analogous to the real problem to be meaningful; but at the same time, it must be simple enough to be amenable to quantitative analysis.

A mathematical model is a set of independent and dependent variables that are related through mathematical functions. The modeler has control over the independent variables and can manipulate them at will. Therefore, they are referred to as decision variables. Dependent variables, as their name implies, are dependent on the values assigned to the decision variables, and represent the output of the model. Mathematical models can also be defined as a set of state variables and transformation functions. The state variables define the status of the system being modeled at any particular time. If values of one or more of the state variables is (are) changed, the transformation functions are used to define new values for the other state variables, and thus a new state for the modeled system.

In any type of modeling, necessary and sufficient data must be available to decide which variables should be included in the model, and what form of model is appropriate.

One type of modeling may involve systems synthesis. This involves the initial conception of an engineering system for the purpose of satisfying certain established needs. A model needs to be developed to analyze the interactions and output of the system to certain inputs. Results of this model analysis can be used to complete the final design of the system.

Descriptive and prescriptive models are two other types of models. Descriptive models are the classic physical science and engineering models. They describe the way the system works to the degree of precision necessary for the analysis being undertaken. Descriptive models are analysis models which leave the actual decision making to the modeler, whereas, prescriptive models actually suggest which decisions should be made so that the system works most efficiently to achieve the given objective or objectives.

The optimization model is one type of prescriptive model that is often used in systems analysis to assist with the systems approach. An optimization model is a conceptual model consisting of some objective function (evaluative criterion) that is to be maximized or minimized subject to some set of constraints. Constraints represent either absolute upper limits on certain system conditions, such as an amount of a certain resource available, or conditions, which if exceeded, will not contribute to the objective function for the excess amount, such as specifying that at least a certain level of a constituent must be provided by the model solution, but not limiting the content to that level. Constraints can be dictated by a number of conditions, including:

Laws, patents, standards, or regulation

Economics

Resource limitations

Political and social pressures

Morality and ethical and professional responsibilities

Physical principles

For a problem to be properly constrained, the constraints must form a bounded space. Solutions to the problem that lie within the boundary of this space are called feasible solutions. If both the objective function and constraints are linear, the model is called a linear programming model. If either or both are nonlinear, the model is a nonlinear programming model.

Evaluation of Alternatives

To evaluate the alternatives that have been developed, some form of analysis procedure must be used. Numerous mathematical techniques are available, including the simplex method for linear programming models, analytical and search techniques for solving nonlinear programming models, the various methods for solving ordinary and partial differential equations or systems of differential equations, matrix algebra, various economic analyses, and deterministic or stochastic computer simulation. Subjective analysis techniques may be used for multiobjective analysis, or subjective analysis of intangibles.

The appropriate analysis procedures for a particular problem will generate a set of solutions for the alternatives that can be tested according to the established evaluative criteria. In addition, these solution procedures should allow efficient utilization of manpower and computational resources.

More complex models, such as many nonlinear programming models, may defy rigorous analysis and optimization techniques. In such cases, informal search, or trial-

and-error exploration, of feasible combinations of decision variable values (values that satisfy all constraints simultaneously) can be utilized to arrive at an optimal or near optimal solution to a problem. Judicious use of computers can greatly increase the number of alternative solutions that are investigated. Pattern search and/or gradient search techniques may be helpful in this respect. Essential elements of these techniques are presented in Appendix E. Even with the help of a computer, however, search techniques often give approximate results, since only a fraction of possible solutions can be evaluated, and it can not always be proved that the best of the alternative solutions investigated is the overall (or global) optimum solution. The success of this approach depends on the selection of the search strategy and the effective implementation and interpretation of this strategy.

As part of the analysis stage, the importance of each variable should be checked. This is called sensitivity analysis, and it involves testing how much the model output will change for given changes in the values of the decision variables and model parameters.

Choosing the Best Alternative

Choice of the best alternative from among those analyzed must be made in the context of the objectives and evaluative criteria previously established, but also must take into account nonquantifiable aspects of the problem such as aesthetic and political considerations. The chosen alternative will greatly influence the development of the final plan/design and will determine in large part the implementability of the suggested solution.

Preferably the best alternative can be chosen from the mathematical optimization within feasibility constraints. Frequently, however, a system cannot be completely optimized. Near optimum solutions can still be useful, especially if sensitivity analysis has shown that the solution (and thus the objective function) is not sensitive to changes in the decision variables near the optimum point.

Final Planning/Design and Implementation

Final planning/design is the culmination of the systems approach. It is the stage at which concrete results should show. A lack of concrete results at this stage points to a serious flaw in the systems approach: the engineer should have sensed that major problems were arising and should have corrected these problems at an earlier stage in the process.

At this stage, there should be minimal need for feedback to earlier steps. If it becomes apparent that feedback is necessary, however, the engineer must not hesitate to retrace the steps in the decision-making process. It is also at this stage that the engineer should check that the original objectives have not changed during the course of the application of the systems approach.

Even after completion of the final plan/design, the engineer must still be prepared to expend considerable effort toward the successful implementation of the plan/design. Without the active participation of the planner/designer at the implementation stage, problems may arise which are beyond the abilities of the implementors to solve, and which may threaten the viability of the whole effort.

Actual final planning/design is primarily a technical matter that is conducted within the constraints and specifications developed in earlier stages of the systems approach. One of the end products of final planning or design is a report which describes the recommendations made. To be effective, this report must also include information on what has gone into the application of the systems approach to the problem. The report should be written in the context of the audience for which it is intended. A well-written nontechnical report can go a long way toward developing public support for the recommended problem solution, whereas, a well-written technical report given to the same audience may be intimidating and actually reduce support for the recommendations.

The report should contain some reference to all aspects of the systems approach. The problem should be defined and, where appropriate, excluded components and subsystems should be identified and reasons given for their exclusion. Data-gathering efforts and the relevance of the acquired data to the problem analysis should be described. Criteria that were used for evaluating alternatives should be presented along with discussion of why these were chosen and others excluded. Reasons for choice of alternatives and constraints should be justified. Pertinent literature or experimental verification should be cited. Enough of the underlying problem physics should be presented to show that any mathematical models utilized are realistic. Methods of solution should be described, and where practical, a sample solution given. Strengths and weaknesses of the model should be indicated and some indication given as to the sensitivity of the solution to changes in decision variable values. On this basis, the final plan/design can be presented along with the projected requirements for implementation.

Planners and designers have to be concerned with the difficulty associated with implementation of the conceived solution. They must be able to identify individuals and organizations which have the necessary authority and/or power to either promote or block implementation of the recommended solution. Frequently, the results of a systems study, even though good systems approach procedures are used, are ignored by the very group for which it was prepared. This may result from a general mistrust of the so-called systems experts, or by political decisions which may help an individual remain politically powerful but which have little rational basis. Engineers must be able to deal with this type of opposition in a rational and systematic manner.

In implementation of planning policy, the engineer must remain sensitive to the dynamic nature of any plan. Conditions may change or additional information be acquired. Implementation requires an adaptive learning process and possible recycling to previous steps. Newly acquired data allow refinement of the system and adaptation to a changing environment.

In implementation of designs, careful planning for and management of the construction process can often mean the difference between economic and technical success or failure of the overall project. This is especially important for projects involving numerous and complex subsystems and large work forces.

The following two examples will illustrate use of the systems approach. Example 1.1 shows some of the ways that the systems approach can contribute to the successful completion of a project, while Example 1.2 demonstrates how problems might arise when the systems approach is not used.

EXAMPLE 1.1

A shopping mall development corporation is contemplating expanding into a new market area. This example will outline how the corporation management might use the systems approach throughout the planning and design stages of the project to increase efficiency and to help ensure successful completion of the project.

First of all, it must be emphasized that the mall developers will have different goals from nearly any other group or agency that may have an interest in or be involved in the project. The overall goal of the development corporation will be to maximize return on investment. Their primary concern will be to develop objectives to meet this goal. The goals of other interest groups, such as conservation organizations or downtown chambers of commerce may be in conflict with the goals of the developers, or they may impose constraints on the plans of the developers.

Some of the objectives of the development corporation may be to optimize the location of the mall with respect to possible customers, to maximize the rental space possible for any chosen site, and to minimize construction costs. Secondary objectives might include minimization of required site preparation and integration of the mall into established traffic networks.

The project could be divided into four stages; feasibility planning, preliminary engineering, final engineering, and construction. Proper application of the systems approach is important to the successful completion of each of these phases.

Feasibility planning requires proper definition of the overall environment within which the developer will have to operate. Definition will involve factors such as the economic climate in the area, the state of development and the outlook for future development in the area, transportation facilities available and projected, availability of utility services, and the social and political climate of the area. Failure to consider all of these factors in the feasibility planning could easily result in a costly termination of the project at a more advanced stage. As an example, an otherwise viable project could be jeopardized by the failure to anticipate resistance by several organized citizens' groups. Such resistance can often be defused by anticipation and incorporation of the groups into the planning process, provided that all parties are willing to compromise to a certain extent. Also, resistance may arise if the mall site is chosen so that it will siphon off business from already established shopping districts, whereas, if it is proposed for an area where there is no well-defined shopping area it may be received favorably by all concerned groups.

Whatever the conditions, the development firm is best served by not being surprised by unexpected developments that may cause costly delays and reduced credibility.

Data gathering is essential to the feasibility planning stage. The development firm must have information on demographic distribution, home–work–shopping travel patterns, and probable future changes if they are to make an informed decision on location of the mall. Information would also have to be acquired concerning regulations and zoning ordinances that will have to be met. Unusual site or local conditions and possible lateral effects of the development will have to be defined so that mitigation measures can be planned for.

While gathering this information, the development firm will also be developing a hierarchy of systems that will assist in the decision-making process and the later detailed design and construction. Figure 1.5 shows a portion of this hierarchy. Several additional levels of subsystems and components could be defined to assist in detailed planning and design. Care must be taken to identify all possible interactions among subsystems and components.

Evaluative criteria will be the expected cash volume of business minus development costs. Alternatives at this stage will involve alternate market areas and then alternative sites within the chosen market areas. Linear programming or some other type of mathematical modeling may be employed to assist in evaluating alternatives. The alternative that maximizes the net return on investment would be evaluated for possible institutional, physical, or environmental constraints that might prevent completion of the project. It might be possible that a less than optimal location would have to be chosen because of other uncontrollable factors.

At the completion of the feasibility planning stage, a site should have been chosen, all possible future problems identified, and plans formulated for mitigation of any problems that cannot be immediately addressed. Feedback to earlier steps in the planning process is of minimal risk throughout the feasibility planning stage because no large-scale commitment of resources has taken place. Planners for the developers should ensure that all possible feedback routes have been investigated before committing to more detailed planning and design.

The preliminary engineering stage overlaps the feasibility planning stage somewhat, because some engineering input will be required to evaluate feasibility. An ex-

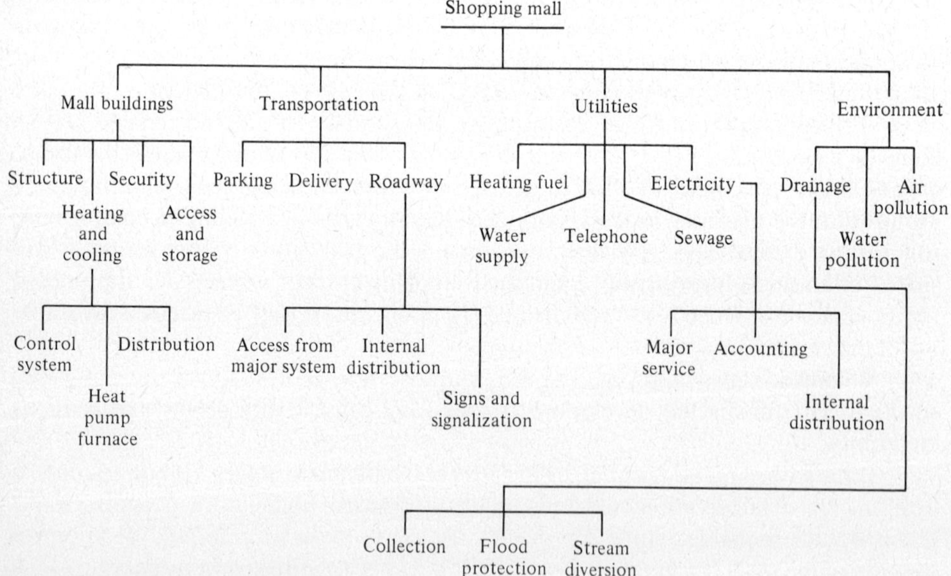

Figure 1.5   Hierarchy of systems.

ample would be the determination of foundation and drainage conditions existent on possible development sites. Therefore, it is necessary that a coordinated planning and design team be put together early in the planning process to facilitate communication and thorough consideration of alternatives and constraints.

As many alternatives as can practically be evaluated should be developed for each subsystem or component included in the initial engineering phase. Wherever possible, some formal optimization technique should be employed. When this is not possible, the design team should carefully evaluate the alternatives to arrive at a best estimate of the optimal solution. In addition, the design team would have to coordinate overall optimization of the mall project, which could not realistically be done analytically.

Collection and evaluation of data will continue in this and subsequent phases. Preliminary engineering will include consideration of coordination of mall traffic patterns with the existing road network and any necessary modifications in the existing system. Parking area requirements and preliminary layout will need to be specified. Access will have to be provided for delivery of goods. Initial planning will have to be undertaken for connection of the mall to public utilities, including necessary extensions of or modifications to the existing system. Environmental protection, including drainage, water pollution, and air pollution will have to be accommodated. Special problems, such as a water supply aquifer underlying the proposed site, will have to be given close attention. Mall layout and area requirements will have to be developed and architectural renderings produced so interested groups can visualize what the completed project will look like. Feedback during the preliminary engineering stage will still be fairly easy, because the financial commitment at this point should be limited to an option to purchase the site, and the planning and engineering staff time invested.

All questionable permits that have requirements that may be hard to meet should be obtained before making a final commitment to purchase and develop the property. This may require that some final design be completed. An example would be a detailed drainage design when there is a question of possible downstream damage that could be caused by stormwater runoff from the mall. The remainder of final design would involve the detailed specifications and drawings for site preparation, foundation, structural framework, utilities, exterior finishing, interior finishing, parking, and transportation. Feedback is still possible at this stage, and it should be implemented if changing conditions warrant it. However, at this point considerable investment has been made, and proper application of the systems approach to earlier stages should minimize the necessity for feedback. Evaluative criteria for the final design stage will be to minimize the costs of material and labor required to produce the finished product. Formulation and evaluation of alternatives would take place in all of the subsystem designs. Alternative structural systems, utility systems, fire control and warning systems, and traffic control systems are a few among the many that would have to be considered.

After initiation of the construction phase of the project, feedback becomes increasingly more difficult and expensive. Therefore, the need for it should be prevented beforehand. Management tools, such as the critical path method (which will be discussed in detail in Chapter 9), can be used to help control scheduling of construction activities. A large development firm may act as its own contractor; however, even if the work is contracted out, the developer must conduct an inspection program to make sure that the work is going according to plan, and that the correct materials are being used.

Proper application of the systems approach and its associated tools of systems analysis at all stages of this project will help ensure maximum return on investment for the development corporation.

EXAMPLE 1.2 ───

This example demonstrates how problems can arise when the systems approach is not used.

At one time, all three cities shown in Figure 1.6 drew their drinking water from the river and could use it without additional treatment.

Eventually, City C found that its water supply was being degraded by Cities A and B, so its water supply was moved to the well field shown. Now it finds that its well field is being contaminated by leachate from its landfill.

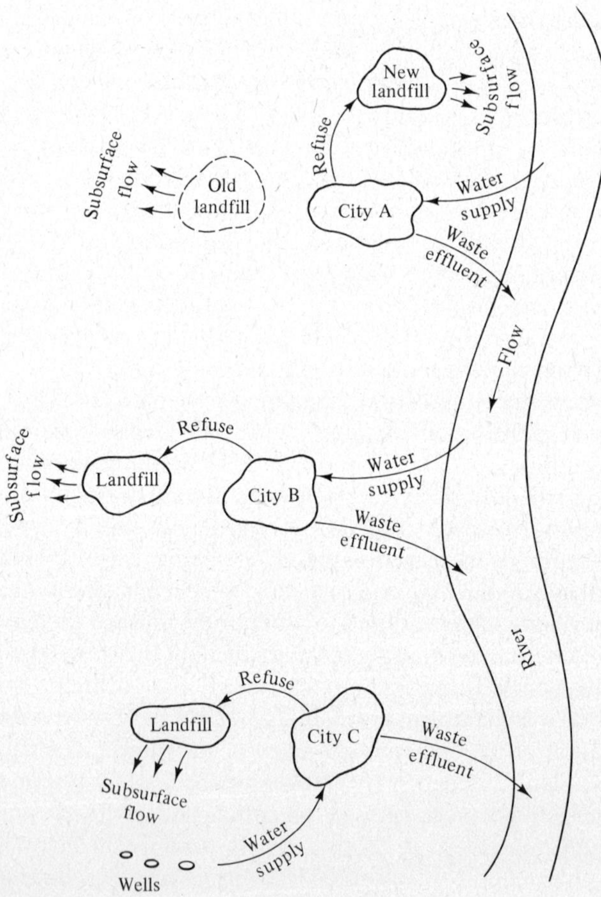

Figure 1.6 Absence of systems approach.

City B has been experiencing increased degradation of its water supply and has responded by implementing increasingly more stringent treatment.

City A filled its old landfill a few years ago, and subsequently opened the new landfill shown. Recently it has been experiencing degradation of its river water supply, even though there are no upstream urban or industrial areas.

Examples of how proper application of the systems approach could have helped to prevent some or all of the problems encountered by Cities A, B, and C, or could help solve the present problems and possibly prevent future problems will be developed in subsequent chapters.

SYSTEMS ANALYSIS

Systems analysis provides tools that, if used effectively, can greatly assist in implementation of the systems approach. However, when used improperly, these tools can produce misleading, and sometimes dangerous, information. As with any tool, they have to be understood to be used properly. Systems analysis involves the use of formal mathematical models and solution techniques, as opposed to verbal reasoning, to arrive at a problem solution. It provides for systematic quantitative analysis in the comparison of alternative courses of action involving design, construction, operation, maintenance, and management of engineering systems in order to increase overall efficiency. Meta Systems (1975) has described the optimal situation for application of systems analysis as being when conditions are changing too rapidly for new information to be assimilated by informal, unsystematic methods, but slowly enough to permit the development of a model that reflects the recent past and can predict the relevant future. Sufficient time must also be available to permit the accumulation and processing of data necessary for validating the model and estimating its parameters.

Unfortunately, systems analysis is sometimes mistaken for the systems approach, and solutions to problems are proposed that are based purely on systems analysis techniques without examining the overall context of the problem. Such solutions have little chance of successful implementation.

Many of the techniques that are now used for systems analysis were developed during the military operations concerned with World War II. During this period there was a dramatic growth in the complexity and uncertainty in military decision making that accompanied the technological revolution in weapons and the complex logistical problems of the allied military effort. Systems analysis helped keep control over these complex situations. Since the end of World War II, these techniques have been further developed and applied to many management and decision problems, sometimes with great success, sometimes with limited success.

In order to apply systems analysis it is necessary to be able to identify the systems, subsystems, components, and interactions involved in the problem. Systems analysis can provide a detailed examination of a system or subsystem which will help in providing a better understanding of its nature and its essential features. Systems analysis can be used to predict the response of a system to changing values of input variables.

Systems analysis cannot replace the systems approach, but it is often an integral part of it, especially if mathematical optimization is to be undertaken. Systems analysis also does not replace experience or technical competence; it augments it. Proper application of systems analysis to engineering problems requires a person who is expert in both the technical aspects of engineering and the techniques of systems analysis.

Some of the techniques that can be used in systems analysis are as follows.

Linear Programming   If a problem can be modeled as a linear objective function subject to a set of linear constraints, linear programming can be used to develop an optimum operating strategy for the system described by the model.

Dynamic Programming   Dynamic programming is used to optimize systems possessing a serial structure. A serial structure consists of components connected head to toe with no recycle.

Multiobjective Analysis   For some problems, no one dominant objective can be identified. Multiobjective analysis is a framework for optimization of systems to meet these multiple objectives. An example of multiple objectives might be the maximization of net national income benefits and the preservation of environmental quality.

Economic Analysis   Economic feasibility of proposed solutions is a primary consideration. Economic analysis can be divided into several types including profit/loss, benefit/cost, and cost effectiveness analysis. Each of these are different applications, but they share a common core of economic theory.

Decision Analysis   Decision analysis allows the decision maker to choose what is probably the best course of action under conditions of inadequate data and probabilistic events.

Management Techniques   These techniques can be used to efficiently schedule activities and resources during the execution of engineering projects. Critical path method (CPM) and program evaluation and review technique (PERT) are the most widely used examples.

Network and Linear Graph Analysis   Many systems can be modeled as a set of links and nodes connected together to form a network. Highway networks and surveying level networks are two examples. Network and linear graph analysis allow use of the network defined to assist in optimization of important aspects of the problem, such as flow capacity or routing.

SUMMARY

The methods and techniques of the systems approach and systems analysis, if properly developed and used, can significantly rationalize the planning/design process and streamline the art of decision making. The systems approach can be used to advantage on almost all problems. The degree of utilization of the systems approach should be based on a study of whether the cost of doing the analysis is less than the savings or benefits to be achieved. Generally the benefits will far outweigh the costs, except for the smallest projects.

Use of the systems approach, with help from systems analysis, adds an important structure to the design process and allows development of an optimal or more nearly optimal design. Answers to the following questions can give some measure of the success of the systems approach as applied to a particular problem:

1. Did the systems approach make the debate over the best choice of alternatives more rational and informed?
2. Did the systems approach introduce competitive alternatives that might not otherwise have been considered?

As an adjunct to the systems approach, systems analysis is a tool that should be in the inventory of every civil engineer. However, as with any tool, simply knowing the mechanics of the tool is not enough. The engineer must also know how to apply it to practical problems. Thus, it is essential that the systems approach be integrated into the teaching of systems analysis, and both systems analysis and the systems approach be utilized when the student is gaining proficiency in the design process.

APPLICATIONS EXERCISES

1-1. Reconsider the various aspects of Example 1.1 from the perspective of each of the following groups. How might each approach the proposal for the new mall, and what conflicts might arise?
    (a) The city council for the city that will house the mall.
    (b) The city council for a nearby city that has been redeveloping its own downtown shopping area.
    (c) A citizens group concerned with environmental protection.
    (d) The county chamber of commerce.
    (e) A large industrial employer in the area.
    (f) The zoning board for the host city.

1-2. Describe what might be involved in applying each step of the systems approach to each of the projects listed below. Approach the problem from the perspective of an engineer working for the primary group responsible for sponsoring the project.
    (a) A 100-mile segment of the interstate highway system.
    (b) A new interchange linking an interstate highway with a new sports complex.
    (c) A wastewater treatment plant for a city.
    (d) A regional wastewater treatment plant serving several cities.
    (e) A plan for downtown rehabilitation and redevelopment.
    (f) A high-rise apartment building.
    (g) A rail rapid transit system for a metropolitan area.
    (h) A downtown rail rapid transit station.
    (i) A suburban rail rapid transit station.
    (j) Rehabilitation of the New York City subway system.
    (k) The Olympic summer games sports complex.
    (l) A college sports stadium for football, track and field, and baseball.
    (m) A river basin comprehensive management plan.

    (n) A management plan for a popular national park.

    (o) A management plan for a national forest wilderness area.

    (p) A national toxic waste management plan.

    (q) A state toxic waste management plan.

    (r) A local toxic waste management plan.

1-3. What other groups might also be concerned with each of the projects described in Exercise 1-2? How would their perception of the proper application of the systems approach differ from that of the primary group?

1-4. Describe a project, from your own experience or observation, that was less than successful. Why was the project not successful? Cite examples of where the systems approach might not have been properly applied, and how specific steps in the systems approach might have been applied effectively to improve the chances for success.

1-5. Urban planners hope to improve the quality of life for residents in a rundown section of a city.

    (a) What factors might be appropriate for measuring the quality of life in the area?

    (b) How might the systems approach be utilized to assist in accomplishing the end goal?

    (c) What groups should be included in the planning process?

1-6. A major metropolitan area is planning for water supply and wastewater disposal requirements in the future. Explain how each step in the systems approach might be used to assist in this planning.

1-7. Five communities in fairly close proximity to one another want to develop a regional water supply system. At the present time, all homes and businesses in the region have individual systems; however, increased development has caused many wells to run dry during low rainfall periods. Three reservoirs could be developed at locations A, B, and C. The communities wish to develop a water supply system that will meet their needs, while minimizing the delivered cost of water. Relative locations of the communities and possible reservoir sites, with approximate separation distances, are given in the figure for Exercise 1-7.

    Describe how the systems approach could be applied to this problem. Be careful to identify what types of data might be required, and what groups should be included in the planning process.

1-8. An urban planning commission is trying to dissuade commuters from driving their personal vehicles into the central business district by promoting development of a mass transit system. It is hoped that the transit system will reduce traffic congestion and make everyone's commute less difficult. At the same time, the bridge authority, which is in charge of financing, building, and maintaining all the bridges and tunnels that carry traffic into the central business district, has a policy of promoting usage of their facilities by offering frequent users monthly passes at a reduced price. This tends to increase the volume of traffic into, and out of, the central business district.

    Describe the conflicts that exist in this problem, and how they might be alleviated. How could the systems approach be applied to this problem, to assist in finding an acceptable solution? What other factors should also be considered?

1-9. The river and bay, shown in the figure for Exercise 1-9, are highly polluted, unfit for anything but commercial boat traffic. Describe some pitfalls that may be encountered if attempts are made to alleviate the problem without applying the systems approach. Describe all factors that may have an influence on the problem, and how they could be incorporated into the systems approach.

1-10. You are the site engineer for the design consulting engineering firm at a construction project that includes earth moving and building of a loading dock and an access road

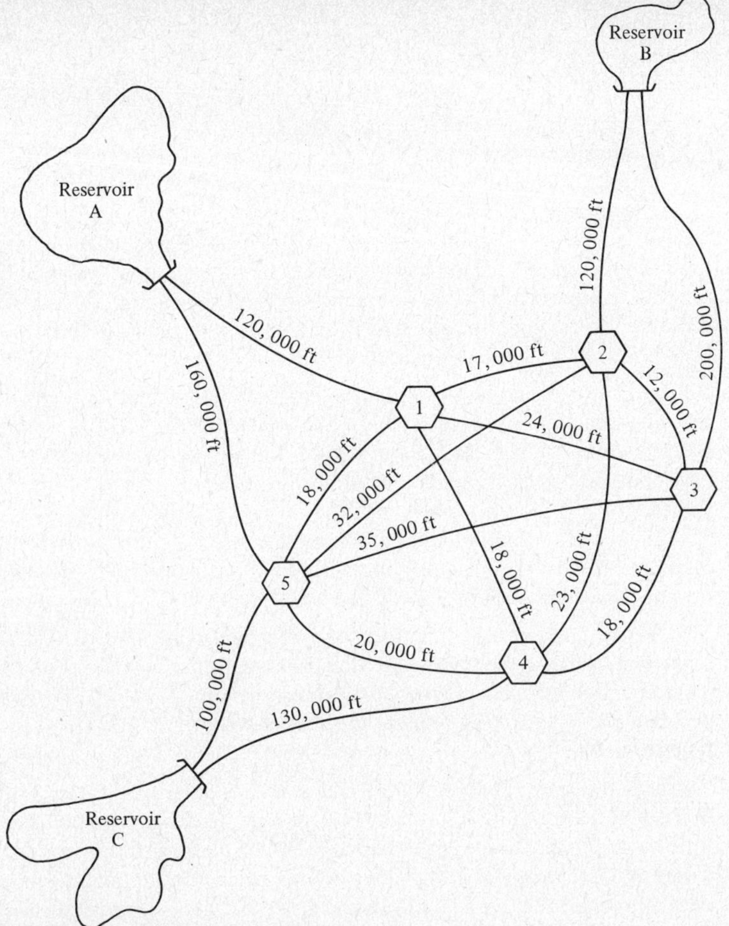

Figure for Exercise 1-7

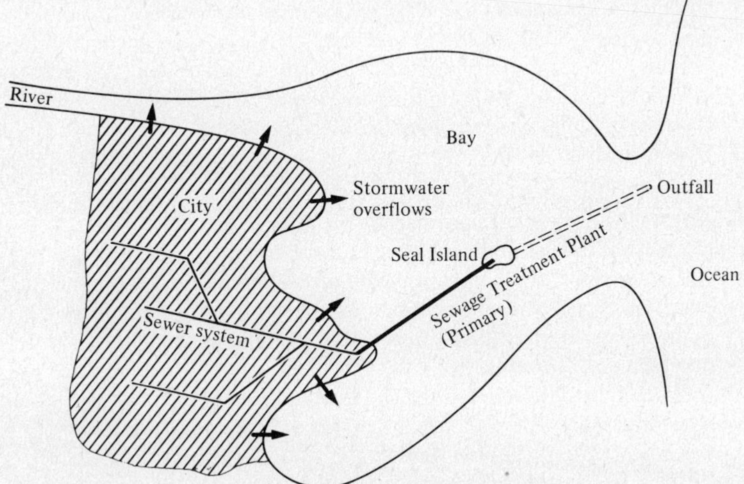

Figure for Exercise 1-9

system. A slope failure has occurred that threatens the stability of the loading dock foundations. Describe how the systems approach could be used to develop a plan of action for dealing with the problem.

1-11. A new interstate is scheduled to be built parallel to the old main road (U.S. Highway) that travels through rural areas and medium-sized town centers. The town centers have many businesses, such as restaurants and motels, that are oriented toward serving the highway travelers. The rural areas surrounding the town centers are almost totally devoted to farming. You are an engineer employed by the county planning commission, which is trying to assess the impacts of the interstate on the region. Discuss how these impacts could be identified and assessed in the framework of the systems approach. What steps would have to be taken to quantify the various impacts?

1-12. You are an engineer employed by the city planning department of the major city of a metropolitan area. An area of your city populated mostly by low income minorities has been troubled by riots and urban blight. The mayor has asked that your department develop a plan for implementation of transportation system improvements (both passenger and freight) that would help to alleviate the situation. Discuss what such a plan would have to take into account, and how it might help the residents. Organize your answer in the framework of the systems approach. What types of data might you need, and what outside assistance might you need to seek?

1-13. You are the transportation planner for a large metropolitan area with about 1.5 million inhabitants. It has a concentrated central business district surrounded by sprawling suburbs. There is essentially no existing mass transit system. What should you consider when deciding if a mass transit system should be developed? If you feel that such a system is necessary, what type would you recommend?

1-14. You are a transportation planner for a large metropolitan area. There is an existing airport in the urban area; however, it has not kept pace with the growth of the city, and there is little room for expansion in the area adjacent to the airport. Discuss the factors that you would have to consider in deciding where a new airport might be placed; that is, in determining the relative advantages and disadvantages of locating the airport in the urban center, in the urban fringe (suburban area), or beyond.

1-15. Your city has an air pollution problem that is generally believed to be caused by excessive automobile emissions. The U.S. Environmental Protection Agency has ordered that a plan to improve air quality to acceptable levels be developed. You have been assigned responsibility for the development. Your city has an existing rail rapid transit system that is working at capacity during rush hours, and carries about 10% of the commuter traffic. Describe how the systems approach could be used to develop and evaluate alternatives. Discuss several possible alternatives, and the secondary effects that these alternatives might have.

1-16. An existing expressway passes near a medium density residential development. Presently, residents of the development have to travel 5 miles in either direction to reach an interchange to enter the expressway. You are a planner for the State Department of Transportation. The Board of Selectmen for the community containing the residential development has requested that you consider building an interchange, ostensibly to serve the development. However, you know that open land near the proposed interchange is being considered by developers as the site for a large shopping mall. Discuss all the factors you should consider in evaluating the proposed interchange, and identify possible conflicts that might arise if the plan were to go forward.

Chapter 2

Definition of Problem

INTRODUCTION

Problem definition is probably the most difficult step in the application of the systems approach. It is also one of the most difficult steps to describe. The whole process of problem definition is evolutionary. An engineer with experience is able to recognize more subtle aspects of a problem, but with each new problem confronted, additional information is stored away in memory to help in future problem definitions.

Problem definition is an investigative stage in which the engineer has to decide if a problem does indeed exist. It is a preliminary screening stage during which boundaries are set on the problem, extraneous factors are screened out, and the salient aspects of the true problem at hand are identified. At the same time, the engineer has to remain within the constraints of jurisdiction, knowledge, time, and money, which may be working against an adequate problem definition.

Chapter 1 described briefly what we mean by problem definition. The present chapter will deal more with how to identify the various aspects of a problem; how to establish the environment surrounding a problem; and how to define the components, subsystems, and systems hierarchy that influence a problem. Although these various actions are divided into sections in the chapter, in an actual application they will be occurring simultaneously.

PROBLEM ENVIRONMENT

In defining the problem environment, and thus determining the correct problem, or question, to study, it is important to distinguish between the true problem and the symptoms of the problem. First of all, the needs associated with the problem have to be established. A need is something which is presently missing from the environment, and which, if satisfied, will improve conditions. If it turns out that there really are not any needs, then there is no reason to try to analyze the problem. It is, in essence, a nonproblem and should be set aside until and if it becomes a problem in the future. If needs can be established, they will have considerable influence on the definition of the problem. Well-established needs will lead to the definition of appropriate system goals.

A careful investigation of the environment surrounding the needs, and the problem, must be undertaken. Preconceived notions about the form of the problem or possible methods of solution must be avoided. All possible components and subsystems associated with the problems must be identified. These must then be organized into a hierarchy of systems. The interactions of components and subsystems must be identified and used to modify the systems hierarchy wherever appropriate. In conjunction with the established problem needs and in accordance with the hierarchy of systems, goals and objectives for the systems and subsystems must be identified. Objectives must be specific enough so that criteria can be established to measure how well the objectives have been met. Example 2.1 illustrates these considerations.

When establishing the problem environment, thought must be given to which aspects of the system are subject to control. Permissible ranges of these controllable aspects (variables) will later comprise the constraints on acceptable problem solutions. In determining which aspects are controllable, the planner/designer must be cognizant of the resources available and any special difficulties that must be overcome. Efforts must be undertaken to estimate how the uncontrollable aspects (variables) affect the system outputs. These aspects of the problem, components, or subsystems, may be beyond the planner/designer's control because they lie outside the constrained or limited problem environment, or because there is a lack of knowledge concerning a particular subsystem and its functioning. Pollution from upstream sources in a river when it enters a jurisdictional area would be an example of the former situation, while lack of knowledge of how improved highways will affect transportation demand and land development would be an example of the latter situation. It is possible that the problem definition should be expanded to include these aspects of the environment, and/or additional research be undertaken to acquire the knowledge necessary to include particular subsystems in the problem definition and subsequent analysis. The relative benefits to be gained by the increased understanding of the problem environment must be weighed against the increased costs (in both time and money) required.

Time, financial, labor, and material constraints may have considerable influence on all aspects of problem definition. Overly strict constraints in these areas may prevent adequate problem definition. Either means must be found to ease the constraints, or consideration should be given to not studying the problem until adequate time, financial, labor, and/or material resources become available. Otherwise, the results of any study undertaken may be irrelevant and unimplementable.

Definition of the problem should also include the development of preliminary criteria which will be used to judge the validity of a possible solution. These will be refined prior to evaluation of alternatives, but they may influence the problem environment definition. Problem definition is a compromise between including all factors influencing the problem and keeping the problem small enough to be solved with available time, budget, and manpower resources.

HIERARCHY OF SYSTEMS

As was discussed in Chapter 1, most systems are hierarchies of subsystems and components, arranged in a pyramidal structure (inverted tree), with the main system as the main trunk, and the components at the bottom of the branches. Higher-order subsystems would encompass subsystems and components of the network below it. In determining the hierarchical structure, it is imperative to examine everything contained within the system environment, and to identify aspects of the problem environment that could be considered system components or subsystems. Various approaches to system definition will be discussed, then applied in Example 2.1.

There is no set method that will always provide the best approach to defining a system. A group effort will provide the best chance of including all of the relevant subsystems. A good initial step would be to define the overall system, then work downward through the hierarchy of subsystems. When the hierarchy is established, then lateral interactions among subsystems can be identified. The hierarchical structure should be refined and updated until it is in its most appropriate form. Later analysis may point out additional changes that should be made in the system structure. The end product will be a model of the system structure that requires updating as more information becomes available.

There are several advantages to treating problems as systems, when such treatment is appropriate. Problems can be considered in their totality. Definition of the relevant systems will aid greatly in definition of the complete problem environment. Furthermore, definition of the system will point out the most effective points of control in the system which, in turn, can lead to the best solution to meet the system objectives. The range of possible control options will be increased, as will be the opportunities for efficient, integrated problem solutions.

Systems can be subdivided into two types: physical and organizational. Physical systems contain primarily physical objects as subsystems and components, but may also have organizational subsystems. The transportation system for a metropolitan area is an example of a physical system. Physical subsystems within this system could include the expressway system, the rapid transit system, and the air transport system. The airport authority (management structure) would be classified as an organizational subsystem within the air transport system.

Organizational systems are typically management or governance oriented, but they can also have physical systems as subsystems. An example of this is the management structure for a construction firm with a computer subsystem helping in project scheduling and accounting.

Structure of systems may be divided into deterministic and stochastic. In purely deterministic systems, every particular input will produce a predictable response. If the same input is repeated, the output will also be repeated. In stochastic systems, one cannot expect identical output (system behavior) when the system is subjected to identical inputs. In other words, the system response is somewhat random and is not reproducible at will for a specified input. A set of observations of the output for a specified input will comprise a sample taken from a probability distribution. Certain characteristics of this probability distribution can sometimes be inferred from a sample or a series of samples.

SUBSYSTEM AND COMPONENT INTERACTIONS

To assist in defining the interactions among various levels of the system hierarchy, systems analysts generally think in terms of goals, objectives, and criteria. These terms will also be useful when development of evaluative criteria and formulation of alternatives are discussed. A goal is the specific purpose or function of a particular system or subsystem. Goals are established by the needs that must be met by that system. Objectives reflect the ways in which the goal or goals of the system can be met. Objectives are essentially requirements that can be violated, but only at some cost or penalty. Criteria indicate how the degree of objective attainment can be measured. Normally the goal (purpose or function) of a particular level of a system is related to the objective of the next higher level of system. This can be used to establish the proper hierarchy of systems.

Interactions among subsystems at the same level of the hierarchy, or among subsystems at different levels and on different branches of the hierarchy, can be identified through logic and a knowledge of the relevant system behavior. Secondary and tertiary effects of changes in one subsystem or component on other subsystems or components must be considered. These secondary and tertiary effects may be desirable or undesirable, but they should not be unexpected. Some interactions may be in the form of conflicts that have to be resolved before the overall system can function properly. Other interactions may be delayed in time.

The engineer/planner must decide for what level in the system hierarchy a solution should or could be proposed. Interactions across that level and with higher and lower levels will influence this decision. Addressing too narrow a subset of the overall system may increase problems in other areas of the hierarchy. Such possible problems should be identified; and if they are critical, they should be included in the proposed solution.

Several examples of system hierarchy development and subsystem and component interaction identification will follow.

EXAMPLE 2.1 ——————————————————————————————————————

You are in charge of studying the feasibility of a hydroelectric power project for an electrical utility. As a preliminary step in applying the systems approach to this problem, you want to identify the hierarchy of systems, possible system goals, objectives, and criteria, and system/subsystem interactions. Figure 2.1(a) shows a portion of a possible hierarchy for a hydroelectric power project. Note that many of the subsystems would themselves be parent systems under different definitions of the problem to be studied. Other readily identifiable subsystems have been left out of Figure 2.1(a) because of space limitations.

After identifying a possible systems hierarchy, it is necessary to define preliminary goals, objectives, and criteria for the subsystems to see if the system hierarchy proposed is a logical one. Figure 2.1(b) shows a portion of this development for the hydroelectric project. The diagram does not explicitly identify the objectives for the various levels of systems. It has already been stated that the goals for one level of system are generally the objectives for the next higher level of system. Thus, the objectives for the dam portion of the project would be to minimize the size and complexity of the spillway structure, to minimize the volume required for the dam, and to choose the best site for the dam. Each of these objectives should contribute toward the goal of minimizing the construction costs for the dam. If the goals of lower subsystems do not serve as objectives for the next higher level of system, then either the goals and objectives have been formulated incorrectly or the systems hierarchy needs modifying. Since the goals and objectives depend primarily on the problem definition and the requirements of the client, they should be arrived at through consultation with all concerned parties. Additional goals, objectives, and criteria could be established to accompany the remainder of the subsystems of Figure 2.1(a).

After developing the system hierarchy, the engineer/planner should identify system and subsystem interactions, both laterally within the same layer of the hierarchy, and vertically between layers and between subsystems not on the same branch of the hierarchy. Figure 2.1(c) shows some of the interactions that may be encountered for a portion of the hydroelectric project. There would be lateral interaction between the transformer subsystem of the transmission system and the transformer subsystem of the generator system. There is also an obvious interaction between the turbine runner and the generator armature. Vertical interaction can be identified between the turbine housing subsystem and the support structures subsystem of the power plant. It can also be seen that one system may have several subsystems that are grouped into different branches of the systems hierarchy. The control subsystem of the power plant has its own subsystems for controlling the generators, turbines, transmission system, and other systems of the plant. The environment subsystem associated with the transmission lines is also a subsystem of the environmental subsystem of the project.

EXAMPLE 2.2

Development had been taking place for several years in areas A, B, and C shown in Figure 2.2(a). Storm drainage systems for these areas were constructed and then joined into one conduit above section D, an older residential area. Before development of A, B, and C, the drainage system through D was adequate to handle both the natural flow from upstream and the storm drainage from area D. After development, however, area D began experiencing frequent flooding problems during storm events.

Because of this, angry residents from area D began attending zoning board meetings regarding proposed further development in areas A, B, and C. They complained vigorously about the existing flooding conditions and voiced their fears that additional development would only make matters worse.

The town did not have an engineer on its staff; however, a member of the planning board who had also been attending the zoning board meetings knew that the town owned undeveloped parcels of land on both branches A and C of the drainage system. He suggested that the town public works department could construct a detention pond on each of these branches to reduce the flooding problem. The rationale was that by temporarily storing some of the stormwater flow, the stormwater would drain down into section D more slowly, reducing the peak flow in that area. All of the interested parties agreed that this seemed like a reasonable solution, and the developers agreed to delay starting their projects until after the detention ponds were finished.

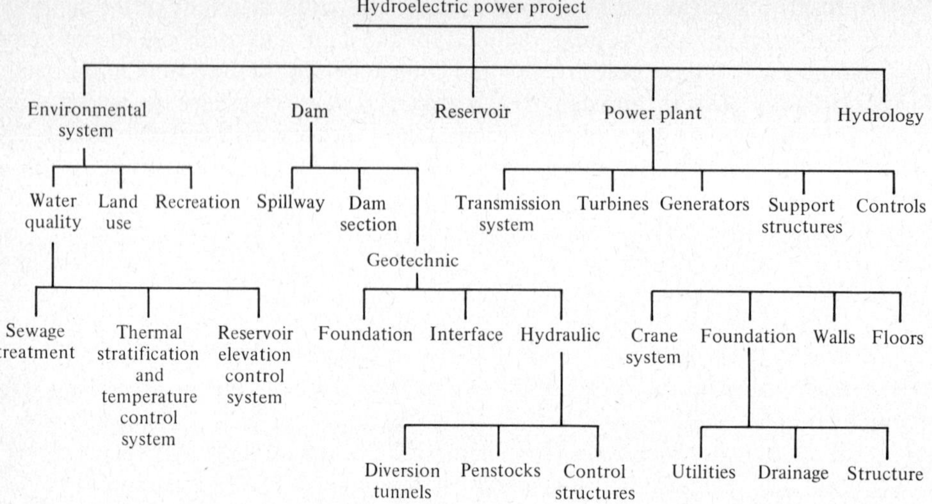

(a) Definition of hierarchy of systems.

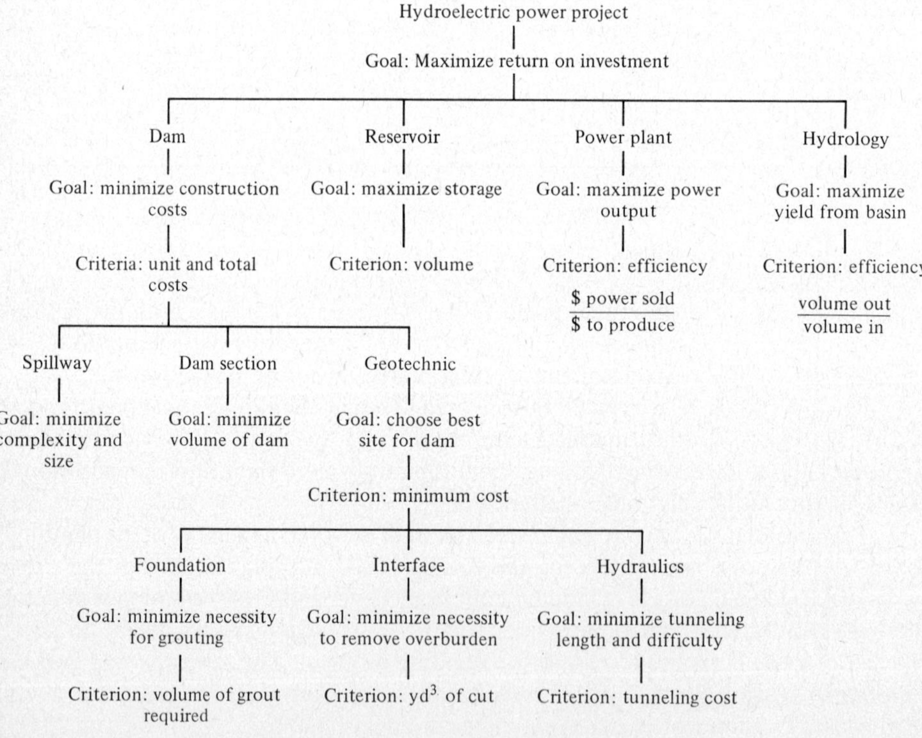

(b) Definition of goals, objectives, and criteria.

Figure 2.1   Hierarchy of systems.

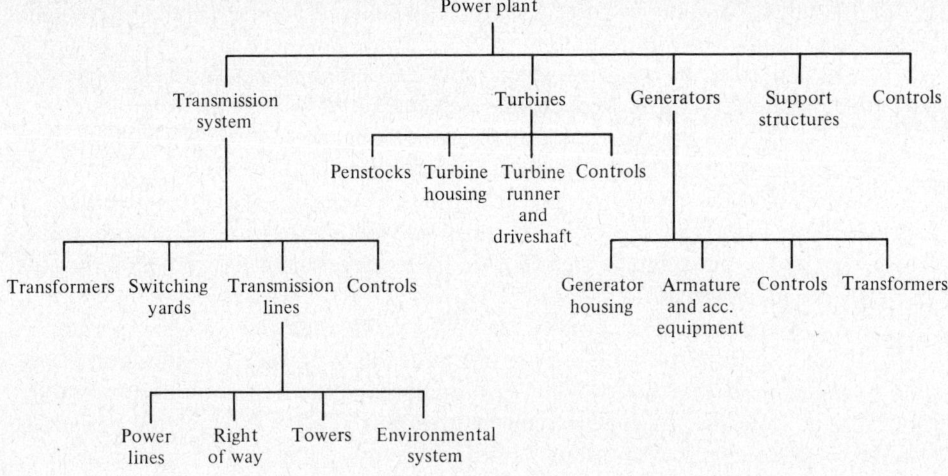

(c) Example of interactions between systems and subsystems.

Figure 2.1 *(Continued)*

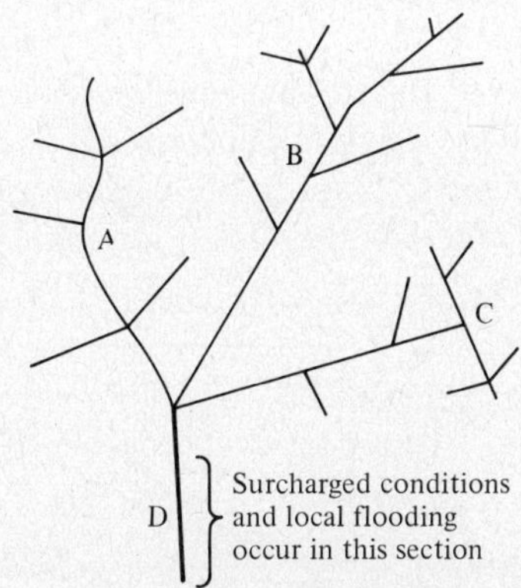

(a) Storm drainage system before adding detention storage.

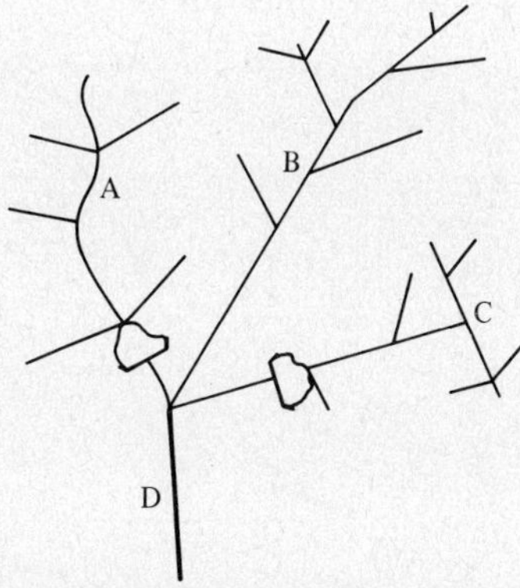

(b) System after adding detention storage.

Figure 2.2 Storm sewer system.

The public works department constructed one detention pond each on branches A and C, as shown in Figure 2.2(b). Orifice outlets were installed in the ponds to control the rate of release of stored water. After the detention ponds were constructed, but before any further development took place, the public works superintendent began to receive complaints that the flooding in area D was worse than before. Perplexed, the superintendent consulted with the planning board and the board of supervisors of the town, and they decided to hire a drainage engineer to study the problem. In the meantime, work had started on two new developments in the drainage area of branch A, and one in drainage area C. Two more proposals had also been submitted for developments in basin B.

The engineer studied the problem and reported that the new problem of increased flooding was a result of treating the symptoms of the old problem without taking the time to understand the true problem. An analysis of the hydraulic and hydrologic conditions in the basins revealed that the hydrographs (plots of flow vs. time) generated by a typical storm, prior to installation of the detention ponds, would appear as shown in Figure 2.3(a) when they reached the junction of the three branches. After these hydrographs combined and entered section D, the hydrograph would be as shown in Figure 2.3(b). The peak flowrate for predetention conditions was approximately 13 units.

The two detention ponds served to attenuate the peak flowrate from basins A and C, but in so doing, also delayed the new peak flowrates. The hydrograph for basin B remained unchanged, as did the total volume of flow passing through the system. The individual hydrographs at the junction after detention pond installation revealed

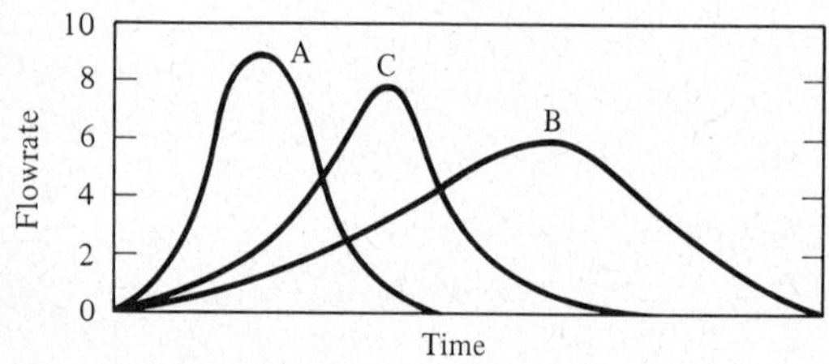

(a) Individual hydrographs prior to installation of detention basins.

(c) Individual hydrographs after installation of detention basins.

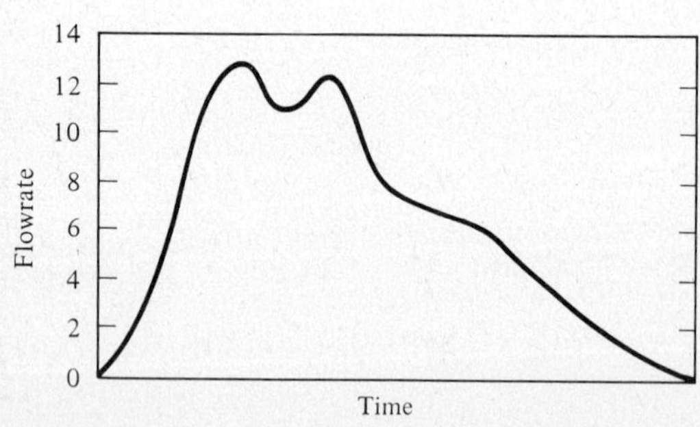

(b) Combined hydrograph prior to installation of detention basins.

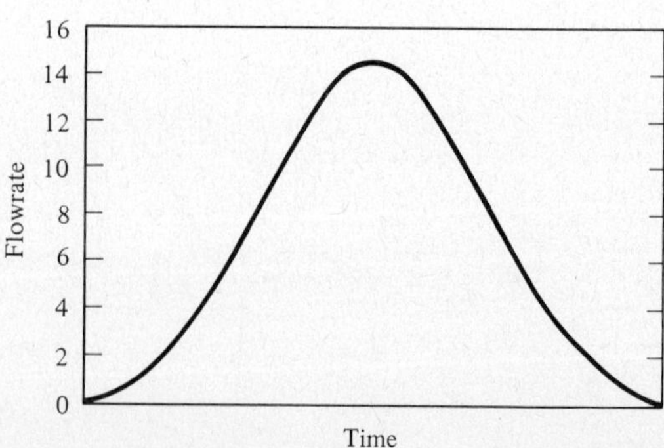

(d) Combined hydrograph after installation of detention basins.

Figure 2.3 Routed hydrograph at section *D*.

the results shown in Figure 2.3(c). These hydrographs combined to give the hydrograph shown in Figure 2.3(d), which had a peak flowrate of about 14.5 units, 1.5 units more than the predetention conditions.

Once the engineer had defined the actual problem, proper remedial measures could be taken to reduce flooding problems caused by past development in areas A, B, and C, as well as any new problems that could be caused by the developments now under construction. The town now has reason to delay approval of any further development until proper measures can be defined which will prevent future problems.

EXAMPLE 2.3 ———————————————————————————————————————

A highway bridge is to be designed to cross a river. This is a problem that is generally well defined with quantifiable variables and parameters. Problem definition would proceed as follows.

The goals for this bridge system would be the safe and efficient movement of the required volume of traffic. Objectives for the bridge system can be identified from the goals of the next level of subsystem, shown in Figure 2.4. These would include:

Pedestrian. Safe passage of pedestrians across the river.

Roadway. Sufficient carrying capacity at minimum cost.

Approach. Smooth transition from approach highways to bridge proper.

Support Structure. Support bridge live and dead loads at minimum cost.

Control System. Maintain safe conditions while maximizing throughput.

The environment surrounding the problem would include the approach highways, the river, topography surrounding the site, and any aesthetic considerations. Approach highways would dictate approximately where the endpoints of the bridge would have to be and would provide the traffic input to the bridge. The river conditions would control how long the bridge must be, and what type of supports might be required. Surrounding topography would have to be considered when deciding alignment of approach interchanges and ramps and placement of end abutments of the bridge.

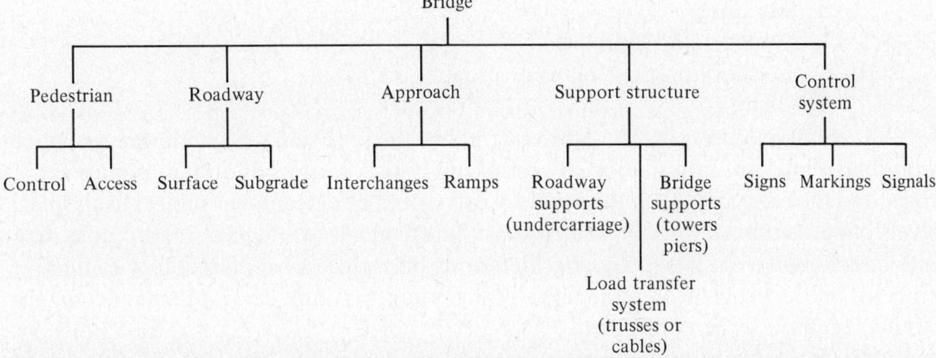

Figure 2.4 Definition of system hierarchy for bridge project.

Aesthetic considerations may dictate the form of the bridge structure if it is to be placed at a site that will be under public scrutiny.

In defining the problem environment, some insight must be gained into how the remainder of the steps of the systems approach might be undertaken. Data that should be gathered and analyzed can usually be ascertained by asking a series of questions about the problem.

Is the river navigable?

How many traffic lanes will be required?

Should pedestrian traffic be accommodated?

Will intermediate support be required for the bridge structure?

What aesthetic conditions should be considered?

What are the future development plans for the area?

Some of the data that could be gathered and analyzed to answer these questions would be:

1. Traffic counts.
    a. Volume (vehicle).
    b. Type (vehicle).
    c. Pedestrian.
2. Structural material strength and cost data.
3. Width, depth, and other river data.
4. Foundation conditions in river and at both ends of bridge site.
5. Legislated requirements.
6. Land availability and cost.
7. Seismic data.
8. Navigation data.
    a. Type and size of boats.
    b. Volume of traffic.
    c. Clearance requirements.
9. Aesthetic data.
    a. Site lines.
    b. Surrounding land use.
10. Future development plans or projections.

Evaluative criteria for the project would be the life cycle costs of the bridge which would take into account the initial cost of the bridge, estimated annual maintenance costs, and the expected life of the bridge. Costs would be determined subject to specified levels of performance for the various subsystems, which would each incorporate their individual objectives. Aesthetic considerations may place a supplementary evaluative criterion on the problem. Adaptiveness of the bridge to future development and change would also have to be considered.

The formulation of alternatives would include the different types of structure to be considered, length of span required, and number and type of supports. Alternate traffic projections might be used to estimate the number of lanes required. Constraints on the alternatives would include shipping restrictions and any highway regulation requirements. Even in a project of this size, it would be impractical to try to develop a constrained optimization model of the whole project. Mathematical models could be used effectively, however, to analyze, and in some cases optimize, subsystems of the bridge. Overall optimization would be left up to the judgment of the planners and designers. Not all alternatives could be considered, but experience will dictate which alternatives would probably have a chance of being feasible. Economic analysis of the overall project and design analysis of the various components will assist in the evaluation of the alternatives and the choice of the best alternative.

After a decision has been made as to the preferred alternative, final design would commence. This would include detailed specification of all structural and physical components of the system and development of fabrication and construction plans. Implementation would include management of the construction phase of the project using management aids such as PERT and CPM, and identification and interpretation of needs for changes in the design as the project proceeds toward completion.

Consideration of these subsequent steps will help in identifying other systems and components that should be included in the definition of the problem environment. This will reduce the need for feedback during subsequent steps, but not completely eliminate it. Engineer/planners must remain continually alert for new information that may become available and may indicate the need for feedback to a previous step in the process.

EXAMPLE 2.4

You are responsible for developing a solid-waste management plan for a large metropolitan region that encompasses area from two states. This is a problem that may be less definable than the previous problem because of the numerous competing interests and the many unknown factors involved in the generation and disposal stages.

The environment surrounding the problem would include the land area of the metropolitan area and the surrounding area that might be impacted as well as the government structures responsible for the various jurisdictions. Waste sources, disposal sites, and transportation systems would also be included in the problem definition, and thus, the problem environment. A first level of subsystems might be defined as is shown in Figure 2.5.

On a purely mathematical basis, this problem is definable. The engineer/planner could develop a constrained optimization model that could be used to estimate the most economic way to transport and dispose of solid wastes. Problems arise, however,

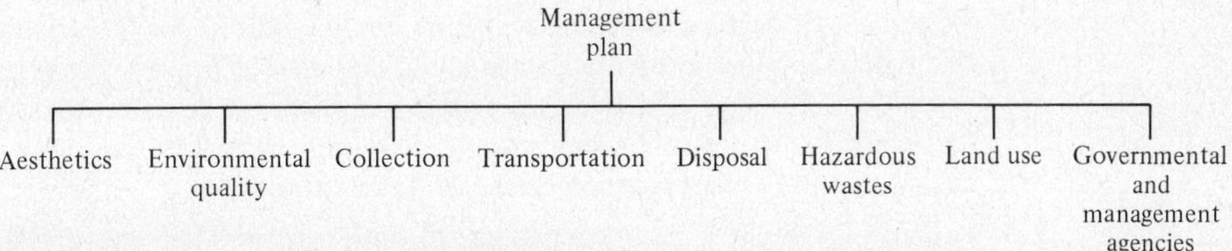

Figure 2.5 Solid-waste management plan.

in developing realistic constraints that would be acceptable to all concerned parties. Jurisdictional disputes would likely arise among the different municipal entities contained in the metropolitan area, and between the two responsible state governments.

Also, objections may be voiced by people living or working in a particular sector. Although these people realize that they are producing solid wastes, they will not be likely to support a plan to develop a nearby site for the disposal of these wastes, especially if the plan also calls for importation of wastes from other sectors. Citizen opposition has become increasingly strong as public awareness of waste disposal (particularly hazardous wastes) has grown. Thus, although a mathematical model could be developed, chances for implementation of its recommendations would be slight because of the effects of these exogenous and largely unquantifiable factors. Nonetheless, such a model could be extremely useful in developing an understanding of the system and in determining a set of alternative solutions for further study and refinement.

SUMMARY

Problem definition is a key element in successful application of the systems approach. Improper problem definition will lead to the wrong problem being studied. Incomplete problem definition can fail to identify vital components of the systems associated with the problem. In either case, there is little chance that any proposed problem solution can improve on existing conditions.

Problem definition is not a simple task. It requires considerable attention to detail and a thorough knowledge of the relevant processes, phenomena, and systems that comprise the problem environment. During later stages of the systems approach, problem definition also requires continued alertness for additional information which may help refine problem definition or indicate changing conditions that may require problem redefinition. It is easy to become sidetracked as the engineer becomes engrossed in the application of the various systems analysis tools. It is a good idea consistently to return to the original problem statement to ensure that the correct problem is still being solved.

APPLICATIONS EXERCISES

2-1. Identify further subsystems at the fourth and fifth levels that will contribute to the systems hierarchy developed for Example 2.1. Develop goals, objectives, and criteria for these subsystems, and identify additional lateral and vertical interactions that would exist among the subsystems of the systems hierarchy for the hydroelectric project.

2-2. Describe how detailed problem definition could continue for the situation developed in Example 2.2. What immediate steps could be taken to prevent further damage or deterioration of conditions? Develop a systems hierarchy with goals, objectives, and criteria that could assist in further analysis and decision making. What steps could be taken to ensure that the appropriate problem was being studied?

2-3. Restudy Example 1.2, concentrating on how to define the appropriate problem, and developing a hierarchy of systems for the problem.

2.4. Develop a third layer of subsystems for Figure 2.4. Also, indicate appropriate goals, objectives, and criteria for the subsystems of the hierarchy.

2-5. Further develop the systems hierarchy for Example 2.4, including goals, objectives, and criteria. Also, indicate where conflicts may arise among system and subsystem goals because of the nature of the systems themselves, or because of different groups that might be participating in the planning process. How might these conflicts be alleviated or accommodated?

2-6. Perform each of the activities listed at the end of this problem for each of the following situations:

(a) You are the city engineer put in charge of reorganizing the municipal public works system so that it is more responsive to the needs of the community and performs its responsibilities at minimum cost.

(b) You are an engineer overseeing the study of the entire highway system for a large metropolitan area.

(c) You are responsible for conducting a study of the wastewater treatment and disposal for a medium-sized urban area.

(d) You are responsible for developing a plan for a new commercial airport that may be built in your region.

(e) You are responsible for designing a water distribution system for a new planned community. The community will be developed as a suburb to an older metropolitan area.

(f) Your client wants to build a high rise building (with both office and apartment space) on the site of an existing theater.

(g) You are conducting a study of water diversion from one watershed to another for the purpose of water supply.

The activities for this problem are:

(1) Develop a systems hierarchy for each of the given situations. Include several levels of subsystems and components. Put yourself in the place of the engineer who will be responsible for the analysis and decision making.

(2) Describe what would be required to define the problem for each situation given. Include what would be necessary to define the problem environment and possible goals, objectives, and criteria for the systems you have defined. Indicate what interaction will probably occur among subsystems and components.

(3) Describe what types of data would have to be gathered throughout application of the systems approach to each situation. Indicate how difficult it may be to acquire the necessary data in each case, and why it is necessary to gather the particular data.

(4) Develop possible evaluative criteria for the situations described. Describe how your evaluative criteria might change depending on your relationship to the problem. How might your relationship to the problem affect your whole approach to the problem, and how can you as an engineer take this into account in your application of the systems approach? Put yourself in the place of several different groups that may have an interest in each situation.

(5) Describe qualitatively what would be involved in formulation of alternatives for the situations described. What elements may cause physical or organizational constraints on the problem. What types of models might be developed to assist in analysis? What methods might prove to be helpful in evaluating the alternatives? Why do you feel that each might be helpful? How would you approach the choice of the best alternative?

(6) List and describe the steps and procedures that might be necessary for final planning/design and implementation of the recommended solution for each situation.

2-7. In 1960 a major north–south expressway opened, passing through a suburban community of 4500 residents. The southern end of the expressway served a metropolitan area about 15 miles below the area in question. Further north the expressway passed near a major recreational area and then continued on to other metropolitan areas. Between 1960 and 1980, the population of the town swelled by over 400%, to almost 24,000. The town is situated in a county, part of which has experienced similar growth pressures while the remainder has remained predominantly rural.

In 1960 there were no regional planners in the area to foresee and plan for such growth. As a result, services in the town have failed to keep pace with the growth. Some private water systems are strained by the extra demand. There is no public system. The landfill does not meet state requirements to guard against pollution. The town has not bought a truck for plowing snow since 1973, though four more are needed. There are not enough parks. Although developers were required to donate 5% of their land for parks, they donated land that was unusable for homes or for any other purpose.

In order to maintain a balanced budget and avoid raising property taxes too much, the town has put off making improvements in the landfill, purchasing highway trucks, or making improvements in the infrastructure.

What is the pertinent problem for the situation described? How do the symptomatic problems relate to the overall problem? Develop a systems hierarchy with appropriate goals and objectives that could be helpful in trying to solve the existing problems. Suggest evaluative criteria that might be used to evaluate alternatives and measure the degree of success of solutions.

2-8. New York State has been mandated to provide high-technology hazardous waste treatment facilities, for the 1.3 million tons of potentially dangerous wastes generated within the state each year. Develop a hierarchy of systems that will help to define the environment for this problem. Indicate where interactions among subsystems can be expected to occur. Identify potential areas of conflict, and possible ways they might be avoided or mitigated.

2-9. The automobile has been blamed to some extent for all of the following conditions:
(a) The U.S. dependence on liquid hydrocarbon fuels.
(b) Air pollution problems in urban areas.
(c) The necessity for large highway systems.
(d) The large number of fatalities experienced on highways each year.
(e) The growth of the suburbs and the deterioration of cities.
(f) The deterioration of public transportation systems.

Discuss the definition of a problem or problems, plus the hierarchy of systems that could then conceivably be used to change some or all of the above listed conditions.

2-10. It has been proposed that one way to relieve downtown traffic congestion in a certain city would be to limit the number of parking places available, and vigorously enforce parking regulations. The city does have a surface bus transportation system, but it does not have a rail rapid transit system. A feasibility study for a rail rapid transit system is now being undertaken. You are the city engineer and the task of evaluating this proposal has been delegated to you. Discuss what should be considered in the definition of this problem, with particular emphasis on identifying where conflicts might arise.

2-11. An historic theater is located on 1 acre of prime downtown property. The theater is in good structural condition; however, it is in need of considerable cosmetic maintenance. The theater is owned by the city and is used for many cultural activities. It will seat about 2500 people. A development firm has proposed to purchase the property from the city, raze the theater, and construct a 50-story office and convention center that would attract up to 10,000 occupants at one time. You are an engineer for the city and have been asked to review this proposal. How would you go about conducting this review?

2-12. You are an engineer in charge of planning for a State Department of Transportation. The overall goal for the transportation system in your state is to move people and goods safely and efficiently. Develop a five-level systems hierarchy with goals, objectives, and criteria that could be used to help evaluate ways of meeting the overall goal.

2-13. Develop a hierarchy of systems that might be useful in defining the problem described in Exercise 1-5. Show how the hierarchy of systems might differ when developed from the perspective of the different groups involved in the planning process. Also, develop appropriate goals, objectives, and criteria, showing how these will differ for the various groups. What factors may have to be excluded from the definition of the problem environment, and how could their influence be weighed in the decision-making process?

2-14. The present water supply for the metropolitan area considered in Exercise 1-6 comes from both surface water and groundwater sources. Some of the area is sewered, with treatment at several locations prior to discharge into a fairly small river. The water quality in the river is marginal. Develop a hierarchy of systems that will assist in problem definition for trying to plan for future requirements. Include possible goals, objectives, and criteria. Show where system and subsystem interactions will take place. How could outside influences be included in the problem definition?

2-15. Develop a hierarchy of systems that would apply to the water supply system developed in Exercise 1-7. Include goals, objectives, and criteria for each level of subsystem identified. Define the environment which encompasses the problem. Describe factors, or systems, that are not included in your definition of the problem environment, and include how each might be taken into account.

2-16. Develop a systems hierarchy for the central business district traffic problem described in Exercise 1-8. Be sure to include all relevant systems that could influence the problem. What is the problem? Indicate where interactions will occur, and discuss possible remedies for conflicts.

2-17. Indicate possible solutions to the river and bay pollution situation described in Exercise 1-9. What interactions might take place between these solutions, and where might conflicts occur? Develop a systems hierarchy with goals and objectives that will help in defining the correct problem for which a proper mix of solutions can be found.

2-18. A proposal has been made to route a major interstate highway through a large metropolitan area. Put yourself in the place of the engineer responsible for planning this project. What information would you have to have to adequately define the problem, and what conflicts are likely to occur? Develop a three-level systems hierarchy for the project.

2-19. Describe the systems that might be involved in the interstate highway impact study of Exercise 1-11. Arrange these systems in a systems hierarchy that would be useful in assessing impacts. Identify subsystem interactions that are not obvious from the hierarchy.

2-20. Discuss how you would go about defining the problem for the situation described in Exercise 1-12. How might a boundary be drawn around the problem environment? Indicate what problem elements might be outside the environment as defined, and how the influence of these elements could be included in the evaluation process.

2-21. Discuss what steps would have to be taken to properly define the problem of air pollution posed in Exercise 1-15. What data need to be gathered to assist in this? What factors outside the problem environment may influence the problem, and how might they be given proper attention?

Chapter 3

Acquisition and Analysis of Data

INTRODUCTION

Gathering of the data necessary to assist in applying the systems approach is an ongoing process that starts with problem definition and continues through implementation. Data gathering includes acquisition (collection of data) and analysis (using that data to estimate parameter values or make inferences about systems conditions).

While the process of data acquisition and analysis is a valuable adjunct to reasoning, it cannot replace it. Reasoned assertions can be verified by observed data; but then, reasoning must again be employed to make further assertions. Observation can establish past and present conditions; then (if the assumption is made that the future will behave similarly to the past) reasoning can be used to forecast future conditions.

The person responsible for data acquisition and analysis is faced with a somewhat circular problem. The observational process has to be defined in terms of the problem and systems to be analyzed; while at the same time, definition of the problem and systems has to rely on observations that have already been made. That is why feedback and continued data gathering are necessary throughout the systems approach.

The use to which data will be put and the type of analysis to be undertaken will determine to a large extent how the acquisition program will be undertaken and how much data must be gathered. If the objective is to make estimates of parameter values for a descriptive model, then enough data would have to be gathered to make parameter estimates that would provide accurate model output. However, if the objective is to achieve optimal operating conditions for some system subject to established constraints, then data should be gathered until the marginal cost of gathering more data equals the marginal value of the information they yield.

Any data acquisition and analysis program must be carefully planned if worthwhile data are to be acquired without expenditure of excessive amounts of time and money. Attention to detail and good engineering judgment are essential ingredients in any successful data acquisition program.

Carefully acquired data may be analyzed to accomplish:

1. Model calibration and verification.
2. Establishment of trends.
3. Calculating basic statistics for comparison (mean, mode, standard deviation, etc.).
4. Estimation of model constants (regression).
5. Statistical correlation (regression and analysis of variance).
6. Frequency analysis.

If faulty data are input to the analysis stage, then final results, whether they are in the form of an empirical model or are estimates of parameters for a deterministic model, cannot help but be flawed. By the same token, faulty analysis of good data will also produce flawed results. Since errors in either the acquisition or analysis stages are difficult to detect, great care is essential. It may also be useful to analyze sensitivity of results to errors in various data inputs. This analysis will identify areas in which even additional caution is advised.

The remainder of this chapter will deal with the elements of a good data acquisition program, will cover several commonly used types of data analysis, and will provide descriptions of applications.

ACQUISITION

Data acquisition is a tedious and exacting process that students need to become familiar with through the engineering education process. Students must gain an appreciation of the "garbage in–garbage out" syndrome: that without good data, even the most elegant analysis techniques are useless. It is preferable to gain this appreciation during the formal educational process, not through making costly mistakes in practice later on.

The intent of this section is to introduce the student to the concepts of good data acquisition. These concepts should be used in other courses where the student is actually required to take data. Two things that should remain with the student are an appreciation for the need for attention to detail in any data acquisition program, and the suspicion that any data or instrument are inaccurate and imprecise unless it can be proven otherwise.

Types of Data

The various types of data can be broken down into the broad categories of time independent (static), time independent (dynamic), and time dependent (dynamic) data.

Any characteristic of a system which does not change is said to be static. Static data quantitatively describe one or more of these static properties. Measurement of the deflection of a beam under a given load is an example of static data. As long as the load remains constant, the deflection should be constant. Actual data gathering involves measurement of the deflection several times, followed by application of some measure of data central tendency to estimate the best value for the deflection. It is necessary to do this because there is some error inherent in any measurement procedure. Each time some static property is measured, a discrete data point is produced. A series of discrete data points taken for some static property represents a sample of all possible discrete data points for the given static property. Statistical analysis techniques can be used to make inferences from such samples. For the above type of data, the timing of observations is not important as long as all other conditions remain constant.

The second type of data are time independent dynamic data. These measure a property of some component or object (dependent variable) that is expected to vary with some other property (independent variable) but not as a function of time. From the previous example, the deflection of a beam is expected to vary in some manner with changing loads. The expected variation of dissolved oxygen in a river with water temperature is another example of time independent dynamic data. Data gathering for time independent dynamic data may involve either discrete (dependent variable observed for some number of independent variable values) or continuous observation. In any case, corresponding observations have to be made of the dependent and the independent variables. These observations form a data set that can be used to derive approximate functional relationships between the dependent and independent variables utilizing regression techniques, which minimize the error between the proposed functional relationship and the observed data. Relationships developed through regression are called empirical: they are based solely on experiment and observation rather than on theory. However, engineers generally base empirical models on the best known theory and then utilize regression analysis to estimate the parameters for the proposed models. Application of regression techniques will be described in a subsequent section.

Time synchronization is often important in the gathering of time independent dynamic data. Even though the data are not functions of time, it may be important to compare several different types of data temporally. If this is to be done, then the different types of data have to be gathered with respect to a common time clock. Or, it may be important to gather several different types of data at the same instant in time to facilitate later data analysis. It is good practice to maintain time synchronization in data acquisition programs unless it is impossible to do so.

Data that describe a time dependent dynamic process are the third type. For example, the exertion of biochemical oxygen demand on a body of water by waste products and the daily or seasonal variation of natural phenomena such as photosynthetic activity or average daily temperature are time dependent dynamic processes. Such phenomena may be dependent on factors other than time, but time seems to have primary influence. Time dependent phenomena can sometimes be described by regression analysis; but at other times, they must be described by time series analysis, a subject that is covered in more advanced texts.

It is important that the data gatherer be able to identify the type or types of data that need to be acquired. The type of data will have considerable influence on the design and carrying out of the whole data-gathering program.

Methods of Data Gathering

All of the numerous methods for data gathering will have to be utilized by the engineer at one time or another. A literature search in a public or private library or through municipal or other governmental archives may reveal much of the background data concerning a particular problem. Information that relates to the problem at hand may be uncovered in journals or textbooks, or it may be located in earlier reports and plans that are on file.

A valuable method for data gathering that should never be ignored is discussing the problem with people who are familiar with its various aspects. These people could include engineers, administrators, and other employees of the agencies involved (public works employees, for example), and any member of the general public who might be able to contribute valuable insights. Often abutters to a proposed project site know a great deal about previous efforts concerning the site, and may reveal possible public resistance to the project that the decision maker can then plan to ameliorate. Such information can often save considerable time and money.

Nothing should ever be decided before a site visit is made to ascertain data on actual conditions. Conditions can change rapidly, and these changes are often not reported. Site reconnaissance can vary from an informal visual survey to formal instrument survey and inspection of any physical facilities included in the problem to be analyzed. Checking of surface or sewer invert elevations and visual or remote camera inspection of storm and sanitary sewers are examples of these methods.

If the above steps do not produce enough data to define a problem properly or to undertake analysis, a formal data acquisition program must be undertaken. Such a program may involve field work, laboratory work, or a combination of the two. Data acquisition programs are expensive and must be carefully planned if useful data are to be obtained. The choice of which methods of data acquisition should be used depends on the proposed use of the acquired data and on the conditions of the problem. Cost and time constraints must also be considered. Example 3.1 shows what types of methods might be used to gather data for a stormwater characterization study.

EXAMPLE 3.1

As part of a sewer upgrading and redesign project, it is necessary to characterize the quantity and quality of stormwater to assist in determining if stormwater treatment might be required. The following outline indicates what methods of data gathering might be utilized to characterize the quantity and quality of stormwater.

Literature Review
Previous studies of same system

Previous studies of similar systems

Plans of sewer system

Topographic and other maps

Personal Interviews
Municipal engineer

Public works supervisor

Residents along sewer right-of-way, especially where problems are suspected

Employees who have worked on the sewer

Site Reconnaisance
Condition and security of access points

Interior condition of sewers

Conditions at outfall

Data Acquisition Program
Flowrate
  Type of primary device (weir, flume, etc.)
  Method of secondary data acquisition (data recording)
    Manual
    Remote recording
    Remote transmitting
Quality sampling
  Method of sampling
    Manual
    Automatic
Type of sample to be taken
  Discrete: collected at selected intervals
  Simple composite: series of small samples of equal volume collected at regular time intervals, then combined into single container
  Flow proportional composite: volume of small sample is proportional to volume of flow during sampling time period
  Sequential composite samples: series of short-term composite samples
Methods of sample analysis
  In-house
    Manual
    Automated
  Contract

Quality Control in Data Acquisition

In order for any data acquisition program to yield accurate results, quality control in both the gathering and analysis processes must be maintained. Although errors will exist in any measurement program, quality control will ensure that these errors are minimized and are largely of the type that can be compensated for.

Error is the difference between the measured value of some quantity and the true or exact value. The amount of error introduced depends on both the sensitivity and accuracy of the measuring device and the ability and skill of the observer. Some amount of error is always present, regardless of how good the observer is or how precise and accurate the instrumentation might be.

The sensitivity of a measuring device is the minimum amount of change in the measured parameter that the device is capable of detecting and indicating. For example, if the sensitivity of a water level measuring device is found to be +2% of full-scale deflection, and full-scale deflection is 3 ft, then a change of less than 0.06 ft would not be reliably reported by the instrument.

Accuracy refers to the agreement between the measurement and the true value of the quantity being measured. Accuracy is related to the statistical measures of central tendency or location (mean, median, and mode). Errors in accuracy are termed systematic errors and are usually caused by measuring equipment misadjustment or some repeated error in taking data. Such errors can be either positive or negative, but usually have the same sign throughout a particular set of measurements. They also frequently have either the same magnitude or an establishable relationship between the magnitude of the measurement and the magnitude of the error.

Precision refers to the reproducibility of results from one test to another. Precision, however, does not infer anything about the accuracy of results. Very precise results may not be accurate at all if there is some instrument calibration error that causes it to consistently report erroneous results. Precision is related to the statistical measures of dispersion or variability (variance, standard deviation, and coefficient of variation). Errors in precision are referred to as random errors. They are due to irregular causes and are generally not completely controllable. Random errors are normally both positive and negative, and are not likely to have a significant influence on the mean value for a set of data.

Systematic errors can often be detected by zeroing an instrument or by checking measured values against known standards. Random errors in a measuring instrument can be evaluated by replicating tests on a known sample. If statistical tests reveal excessive random error, then the measurement instrument may need replacing. Both types of errors can be minimized by following established procedures for all data measuring.

The chief aims of a data quality assurance program are to account for or eliminate systematic errors, and to minimize random errors as much as possible. As a result, the investigator can estimate the most probable value of the object of observation, and can estimate the amount of confidence that should be placed in this determination.

Planning for Data Acquisition Program

Careful planning will mean the difference between success and failure in a data acquisition program. Planning will include determination of the type of data required and an estimate of the amount needed. The required accuracy and precision of data must be established. Also, the availability of data gathered by others must be ascertained and some determination made as to its transferability and suitability to the present study.

If laboratory investigations are to be carried out, space and equipment requirements must be determined. Special requirements such as temperature and humidity control must be accommodated. Plans must be made for acquisition of samples for testing. If samples are to be taken in the field and transported to the laboratory for analysis, preservation and transportation requirements must be planned for.

If field sampling sites are to be used, then a thorough reconnaissance must be completed. Alternative sites must be evaluated in terms of security of equipment, ease of installation, accessibility, and type of measuring equipment that could be installed.

Selection of primary, secondary, and tertiary measuring equipment must be made. Primary measuring equipment provides the environment for measurement. In flow measurement, for example, the primary measuring equipment would be a weir or flume. In the measurement of the deformation of a material sample in a tensile test, it would be the strain gage. An example of secondary equipment for flow measurement would be the stage recorder, whereas, for the tensile test, it would be the readout instrument for the strain gage. Tertiary equipment for either would be data transmission and storage equipment, as well as any computer equipment used for analysis.

Installation of equipment and any required protective facilities must be planned for. Power requirements must be identified, and provisions made for calibration, maintenance, and servicing of equipment. Means for transmission, reduction, and processing of data must be established.

Basic design goals for any measuring instrument can be identified. Some of these include the following:

Range. The instrument must be able to measure the full range of values to be expected from the characteristic being measured.

Accuracy and Precision. The instrument must be able to provide the accuracy and precision required for the expected use of the data. Conversely, it may not be cost effective to use equipment that is more accurate and precise than is necessary. Accuracy and precision are usually measured as ±% error.

Reaction to Changes. If the measured characteristic changes rapidly, the instrument must be able to react to these changes. Also, the instrument must be sensitive enough to react to the smallest change in the characteristic which it is essential to be able to detect.

Installation. The measuring equipment must have the capability of being installed where it will be needed. When considering this, power and accessory equipment requirements must be taken into account.

Simplicity and Reliability. Reliability is a major requirement of any measuring instrument. Generally, the simpler the instrument, the more reliable it will be and the easier it will be to repair if problems develop.

Unattended Operation. If the instrument must operate without an attendant, proper installation provisions must be made, and periodic inspections must be scheduled to check the instrument's continued proper functioning.

Maintenance Requirements. The ease and frequency of required maintenance must be considered.

Adverse Ambient Conditions. If the instrument has to operate in a hostile environment, it must have the necessary protection.

Calibration Requirements. It will be necessary to determine if the instrument will need calibration each time data are taken or whenever it is moved. The ease of calibration and the length of time that the instrument will hold calibration must be considered.

Adaptability. The measuring instrument should be adaptable to a variety of situations.

Cost. The instrument must be affordable.

Special Requirements. Special requirements such as solids-handling capability for wastewater samplers, or sample preservation requirements must be considered.

Additional Requirements. Other considerations may include determining how rugged and portable the instrument must be and whether the instrument should be self-contained.

These design requirements provide a framework for the engineer to follow when an instrument is under consideration. They may not apply to every measuring instrument, and their order of importance would vary according to their specific application.

Field site selection must also be carefully undertaken. Requirements will vary considerably depending on the type of data to be gathered, but some general guidelines can be established. The primary consideration is whether or not the site can provide the required data. Some sites may be biased by outside influences that will invalidate any data gathered. For example, taking rainfall data close to a building that could influence the local wind patterns could also influence the amount of rainfall sensed by the instrument. Accessibility and security of the site are also important considerations. Sometimes at the most accessible sites it is impossible to protect equipment from theft and vandalism. Less accessible sites may be more secure, but they may not provide adequate working conditions. The safety of both personnel and equipment are also prime concerns.

Conduct of Data Acquisition Program

The engineer who is responsible for running a data acquisition program must remain alert to problems caused by unexpected occurrences. Careful planning can minimize the risk of such problems; but even well-laid plans can be affected by an unexpected turn of events. Automatic equipment seems to fail at the most inopportune moment, for example. Batteries may wear down prematurely or power may be interrupted during data acquisition periods. Equipment may be damaged by vandalism, by accident, or by extreme values of the parameter being measured. Recording charts can rip and pens can run out of ink. Sensors and sampling ports can become disconnected or blocked. The best defense against all of these is the frequent checking and maintenance of equipment. Also, if the data-gathering effort is associated with discrete events, such as in stormwater characterization, an attendant should be monitoring equipment during these events to minimize equipment malfunctions and maximize usable data gathered. Even if the equipment being used is fully automatic, periodic checking and maintenance are still necessary to assure accurate results.

Any abnormal situations that may affect the quality of data gathered should be noted as a permanent part of the data record. During visits to a sampling site, conditions may be observed that would affect the validity of previously gathered data. Equipment may have failed or malfunctioned due to a power interruption, for example. Primary measurement devices may have been damaged or adversely influenced. In stormwater flow, for instance, a rag that has lodged in the notch of a V-notch weir would adversely influence the measured water flow. Such anomalies should be noted and, whenever possible, some determination made as to the approximate quality of the resulting data (good, fair, poor, etc.).

Any data of questionable quality should not be used to make judgments or inferences. If such data are in fact erroneous, using them is worse than not having any data at all because of the erroneous conclusions that could be reached.

ANALYSIS

Once the quality of data has been ensured through careful planning and adherence to established practices during the data acquisition program, the data can be used to calculate various statistics. These statistics can be utilized as important input to the decision making process. Statistical procedures ranging from calculation of basic statistics, such as measures of central tendency or variability, through advanced procedures such as sequential analysis, nonparametric statistics, and time series analysis, may be involved. This text will introduce the student to the practical application of statistics and will cover statistical procedures through bivariate linear regression with tests for significance and confidence intervals.

Measures of Central Tendency

Mean   The mean is the arithmetic average of data, and it is generally referred to as $\bar{X}$ in equations.

$$\bar{X} = \frac{\sum_{i=1}^{n} X_i}{n} \tag{3.1}$$

where X_i = data value i

   n = number of data values

The mean is generally understood to be the average of a set of data values. Since the other measures of central tendency are also types of averages, the mean should be used only when referring to the arithmetic average. The mean is a valuable statistic which is also used in many other statistical analyses, including measures of variability and regression. Two properties of the mean are important. The algebraic sum of the deviation of individual data points from the mean, $\sum_{i=1}^{n}(X_i - \bar{X})$, is zero for a set of data. Also, the sum of the squares of deviations of individual data values from the mean $\sum_{i=1}^{n}(X_i - \bar{X})^2$ is less than the sum of squared deviations taken with respect to any other value of X. If it is desirable to calculate the overall mean value of several data sets of different size when the means of each data set are known, the equation

$$\bar{X}_0 = \frac{k_1\bar{X}_1 + k_2\bar{X}_2 + \cdots + k_l\bar{X}_l}{k_1 + k_2 + \cdots + k_l} = \frac{\sum_{i=1}^{l} k_i\bar{X}_i}{\sum_{i=1}^{l} k_i} \tag{3.2}$$

where $\quad k_i$ = number of data values in set i

$\bar{X}_i$ = mean value for set i

l = number of data sets

$\bar{X}_0$ = overall mean value

can be used. When it can be determined that certain data values may have more importance than others to the estimate of the mean, it may be desirable to calculate a weighted mean. This can be accomplished by the formula

$$\bar{X} = \frac{W_1X_1 + W_2X_2 + \cdots + W_nX_n}{W_1 + W_2 + \cdots + W_n}$$

$$= \frac{\sum_{i=1}^{n} W_iX_i}{\sum_{i=1}^{n} W_i} \tag{3.3}$$

where W_i is the weight attached to data value X_i.

EXAMPLE 3.2

Several rain gages are located in a large basin, but each covers a different proportion of the overall basin. It is desirable to calculate the weighted estimate of the mean rainfall for this basin, using the area of coverage as the weighting factor for each rain gage.

Rain gage number	Rainfall rate (in./hr)	Area of coverage (acres)
1	1.0	200
2	0.8	300
3	1.2	150
4	1.1	250

The rainfall rate will be the data value for this problem, and the area of coverage will be the weighting factor.

$$\bar{R} = \frac{\sum_{i=1}^{4} W_iR_i}{\sum_{i=1}^{4} W_i} = \frac{200(1.00) + 300(0.80) + 150(1.20) + 250(1.10)}{200 + 300 + 150 + 250}$$

$$= 0.99 \text{ in./hr}$$

Median The median is the middle value of a set of data values arranged in order of magnitude. If there are an even number of data values, then the median is the arithmetic average of the middle two. The median is a useful measure of central tendency when extreme values may occur on one or both ends of the data value spectrum which could unrealistically skew the mean one way or the other.

Mode   The mode is the value in a set of data which is repeated the most times. It is most useful when used in conjunction with frequency distributions and is not useful for small data sets. If a frequency distribution has multiple peaks, then it is said to be multimodal. A mode is determined for each peak. A two-peaked frequency distribution would be referred to as bimodal.

EXAMPLE 3.3

Calculate the (a) mean, (b) median, and (c) mode of the following set of traffic counts. All data values are in terms of hundreds of vehicles per hour.

(a) Mean

Hour	Hundreds of vehicles/hr
1	1
2	8
3	6
4	5
5	11
6	7
7	9
8	7
9	5
10	5
11	4
12	6
13	5
14	3
15	9
16	3
17	5
18	8
19	8
20	5

$$n = 20 \qquad \text{mean } \bar{X} = \frac{\sum_{i=1}^{20} X_i}{20} = \frac{120}{20} = 6.0$$

$$\bar{X} = 600 \text{ vehicles/hr}$$

(b) Median

Order of magnitude
1
3
3
4
5
5
5
5
5

Order of magnitude (*Continued*)

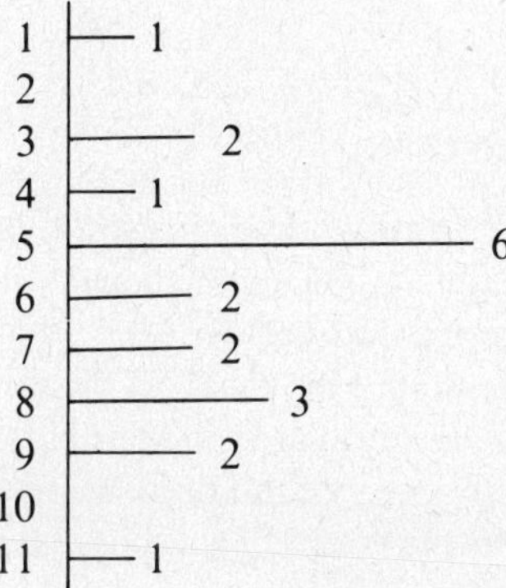

Since there are 20 values, the median is the arithmetic mean of the tenth and eleventh values, or 5.50.

$$\text{median} = 550 \text{ vehicles/hr}$$

(c) Mode

Frequency distribution

$$\text{mode} = 5 \text{ or } 500 \text{ vehicles/hr}$$

It could be argued that this distribution is multimodal, but the size of the data set is insufficient to support this.

This example also shows the effects of scaling. At times it is convenient to multiply or divide all data by some constant factor, usually a power of 10, to make the data more manageable. Measures of central tendency calculated from these data are in the same proportion as the original data.

Measures of Variability

Variance and Standard Deviation   The most commonly used measures of variability, the variance and standard deviation, provide ways of quantifying the scatter of a data set with respect to its mean value.

Each data point has a deviation from the mean given by

$$X_i - \bar{X} \tag{3.4}$$

The algebraic sum of the deviations would not be a useful measure of the scatter of the data about the mean because, as stated previously, the sum of deviations is zero. One measure that could be utilized is the mean deviation. It is the sum of the absolute values of the deviations divided by the number of data points. A more practical measure, however, utilizes the square of the deviation. This measure will be useful in further statistical testing. If each deviation is squared and all of the squares are summed, then

$$\text{SSD} = \sum_{i=1}^{n} (X_i - \bar{X})^2 \tag{3.5}$$

where SSD is the sum of the squared deviations. From this, the standard deviation of the sample data set is

$$S = \sqrt{\frac{\text{SSD}}{n}} = \sqrt{\frac{\sum_{i=1}^{n}(X_i - \bar{X})^2}{n}} \tag{3.6}$$

Units for S are the same as for the basic data variable. The variance of a sample is the square of the standard deviation, or S^2.

$$S^2 = \text{variance} = \frac{\sum_{i=1}^{n}(X_i - \bar{X})^2}{n} \tag{3.7}$$

Any data set taken is, in fact, a sample from the larger population, which includes all of the possible data values for the entity being measured. Definitions of population and sample are reviewed in Appendix C. The statistician's job is to infer something about the population from the sample data set.

If the mean of the sample is equal to the mean of the population, then estimates of parameters such as the standard deviation are called unbiased estimators. The mean of the sample, $\bar{X}$, is an unbiased estimator of the population mean, μ; however, if the mean of the sample is not equal to the mean of the population, which is usually the case, then the estimate of the standard deviation found from the sample is said to be biased. It can be shown that the sum of squared deviations from the sample mean, $\sum_{i=1}^{n}(X_i - \bar{X})^2$, is smaller than if any other value is substituted for $\bar{X}$. Thus, if $\bar{X} \neq \mu$, the estimate of the standard deviation from the sample will be less than the population standard deviation. To overcome this bias, the sum of squared deviations in Equation 3.6 is divided by $n - 1$ rather than n. Therefore, the estimate of the population standard deviation calculated from the sample, $S(X)$, is defined as

$$S(X) = \sqrt{\frac{\sum_{i=1}^{n}(X_i - \bar{X})^2}{n - 1}} \tag{3.8}$$

For large values of n ($n \gtrsim 30$), the two definitions are nearly identical. Also, if Equation 3.6 is used to calculate S, it can be converted to $S(X)$ by multiplying by $\sqrt{n/(n - 1)}$. This may be a more convenient method, since Equation 3.6 can be rearranged as follows:

$$S = \sqrt{\frac{\sum_{i=1}^{n}(X_i - \bar{X})^2}{n}}$$

$$= \sqrt{\frac{\sum_{i=1}^{n} X_i^2 - 2\sum_{i=1}^{n} \bar{X}X_i + \sum_{i=1}^{n} \bar{X}^2}{n}}$$

Since $\sum X_i/n = \bar{X}$, then

$$S = \sqrt{\frac{\sum_{i=1}^{n} X_i^2}{n} - 2\bar{X}^2 + \frac{n\bar{X}^2}{n}}$$

$$= \sqrt{\frac{\sum_{i=1}^{n} X_i^2}{n} - \bar{X}^2}$$

and
$$S(X) = \sqrt{\left(\frac{\sum_{i=1}^{n} X_i^2}{n} - \bar{X}^2\right) \cdot \sqrt{\left(\frac{n}{n-1}\right)}} \tag{3.9}$$

This form is easier to calculate in tabular calculations.

Statisticians refer to the actual standard deviation of a population as $\sigma(X)$. The following convention will be used in this text when referring to these various statistics:

Statistic	Sample	Estimate for population from sample	Population
Standard deviation	S	$S(X)$	$\sigma(X)$
Variance	S^2	$S^2(X)$	$\sigma^2(X)$

A large standard deviation indicates that there is a large amount of scatter in the data about the mean. A small standard deviation indicates that the data are tightly grouped about the mean. For a normally distributed variable (see Appendix C), the standard deviation is a measure of the fraction of the distribution that falls within any specified multiple of the standard deviation from the mean. For example, 95.4% of the distribution lies between $\pm 2.0 \; \sigma(X)$ from the mean. This characteristic will be utilized for estimating probabilities of occurrences in Chapter 10.

Coefficient of Variation The coefficient of variation is a measure of the relative dispersion of data, whereas, the standard deviation is a measure of the absolute dispersion.

$$\text{coefficient of variation} = V = \frac{S(X)}{\bar{X}} \tag{3.10}$$

V is a good measure of relative dispersion when the magnitudes of two data sets are different. Or, since V is dimensionless, it is useful for comparing two data sets when the units are different. The coefficient of variation fails to be useful when the mean value, $\bar{X}$, is close to zero.

One other measure of dispersion, the standard error of estimate, has many of the same properties as the standard deviation, but it is utilized to measure the scatter of data points about a regression line. It will be defined and utilized in the section of this book dealing with its use.

EXAMPLE 3.4

Using the traffic volume data set from Example 3.3, estimate the population standard deviation, variance, and coefficient of variation.

X_i	X_i^2
1	1
3	9
3	9
4	16
5	25
5	25
5	25
5	25
5	25
5	25
6	36
6	36
7	49
7	49
8	64
8	64
8	64
9	81
9	81
11	121
$\Sigma X_i = 120$	$\Sigma X_i^2 = 830$

$$\bar{X} = 6.00$$

$$S(X) = \sqrt{\left(\frac{830}{20} - (6.0)^2\right)} \cdot \sqrt{\left(\frac{20}{19}\right)} = \sqrt{5.5}\,\sqrt{1.05} = 2.41$$

$$\therefore S(X) = 241 \text{ vehicles/hr}$$

$$S^2(X) = 5.79 \times 10^4 \text{ (vehicles/hr)}^2$$

$$V = \frac{S(X)}{\bar{X}} = \frac{2.41}{6.00} = 0.401$$

This example shows the effects of scaling on measures of variability. Since the standard deviation has the same units as the original data, it is also scaled by the same factor. The actual magnitude has to be utilized, however, for calculating the variance.

Statistical Test for Difference Between Two Means   It is sometimes useful to be able to tell if the means of two sets of data are significantly different from each other with some estimable amount of statistical confidence in that determination. Two new concepts, degrees of freedom and hypothesis testing, are useful tools in learning the skills necessary to determine the significance of this difference. These concepts will also be used in later analyses.

Degrees of Freedom   The degrees of freedom (DF) is the number of independent data values that are available for estimating a particular statistical parameter. For example, if n data points are available, from which the mean, $\bar{X}$, is calculated, then $\bar{X}$ is estimated with n degrees of freedom. If the standard deviation is also estimated, there is now one less independent data value available for estimation. In calculating the deviation from the mean, the nth deviation from the mean could be found from the other $n - 1$ deviations and the mean. Therefore there are $n - 1$ degrees of freedom in calculating the standard deviation. If additional statistics are calculated, progressively fewer degrees of freedom are available.

Hypothesis Testing   Hypothesis testing involves setting up some hypothesis and then using statistical procedures either to reject or accept that hypothesis with some level of certainty. For the test of two means, the hypothesis that the population means of the two sets (μ_1 and μ_2) are equal could be formulated. In actually carrying out the computations in the following example, $\bar{X}_1$ and $\bar{X}_2$ are used as estimates of μ_1 and μ_2.

Two batches of steel are tested. The mean modulus of elasticity for the first is 30×10^6 psi. If the mean modulus of elasticity of the second set is 31×10^6 psi, is it significantly stronger than the first batch? How much confidence can be placed in the inference that has been drawn? These are the types of questions that will be answered through the test of two means.

The hypothesis is formulated

$$H_0 : \mu_1 = \mu_2 \tag{3.11}$$

The symbol H_0 is referred to as the null hypothesis because it is assumed that there is no difference between the parameters being evaluated. The null hypothesis will either be accepted or rejected after application of the appropriate statistical test.

The null hypothesis for the difference between two means is evaluated using a statistical procedure called the Student's t test. A value of t is calculated for the particular test to be done. In the case of the test of two means, t is calculated from

$$t = \frac{|\bar{X}_1 - \bar{X}_2|}{\sqrt{\dfrac{\Sigma' X_1^2 + \Sigma' X_2^2}{n_1 + n_2 - 2}} \cdot \sqrt{\dfrac{1}{n_1} + \dfrac{1}{n_2}}} \tag{3.12}$$

where $\Sigma' X_1^2 = \Sigma_{i=1}^{n_1}(X_{1i} - \bar{X}_1)^2$

$\Sigma' X_2^2 = \Sigma_{j=1}^{n_2}(X_{2j} - \bar{X}_2)^2$

and $\sqrt{\dfrac{\Sigma' X_1^2 + \Sigma' X_2^2}{n_1 + n_2 - 2}} =$ pooled estimate of standard deviation for combined data sets 1 and 2 (note that n_1 does not have to be the same as n_2)

After t is calculated for the pooled data sets, it is compared with t values in a standard table of the t distribution (see Table 3.1). If the calculated t is larger than the tabulated t value for a preselected confidence or significance level, α, then the hypothesis H_0: $\mu_1 = \mu_2$ is rejected, and the inference made that the means are significantly different. There is an α chance that this inference is wrong. This infers that there is an α chance that the difference in the means could have been caused purely by chance. When the calculated t value does not exceed the tabulated value for the preselected confidence level and number of degrees of freedom, then the inference can be drawn that there is no significant difference between the means. The reason for this result could be that there actually is no difference in the means, or that the data extracted in the samples will not support a rejection of the hypothesis. For example, samples with large standard deviations will result in small calculated t values, and therefore, a low level of rejection.

Engineering experience will help develop the judgment required to arrive at the appropriate value for the confidence level α. The choice should be based on the importance and significance of the decision to be made, and the potential cost of an incorrect decision. Values of α between 0.1 and 0.001 are common choices. At the 0.1 level, there is a 10% chance that a decision to reject the hypothesis is wrong. At the 0.001 level, there is only a 0.1% chance of erroneously rejecting the hypothesis. Most statisticians report that the hypothesis can be rejected at one level of significance, but not at the next more stringent level. This allows the engineer receiving the analysis to make the appropriate judgment concerning the results.

EXAMPLE 3.5 ____

Two batches of steel have mean values of modulus of elasticity of 30×10^6 psi and 31×10^6 psi for sample sizes of 8 and 6, respectively. If the calculated t is 2.85, can the hypothesis $H_0: \mu_1 = \mu_2$ be rejected, and if so, with what level of confidence?

$$t \text{ calculated} = 2.85$$
$$\text{DF} = n_1 + n_2 - 2 = 12$$

Check $t_{\alpha,\text{DF}}$ from Table 3.1.

$$t_{0.1,12} = 1.782$$

$$t_{0.05,12} = 2.179$$

$$t_{0.02,12} = 2.681 \qquad t \text{ calculated lies between } t_{0.02,12} \text{ and } t_{0.01,12}$$

$$t_{0.01,12} = 3.055$$

$$t_{0.001,12} = 4.318$$

Therefore, the hypothesis, $H_0: \mu_1 = \mu_2$, can be rejected at the 0.02 significance level, but not at the 0.01 significance level. In other words, we can infer that the two means are different with less than a 2% chance of being in error. We are 98% sure that the two means are different, but not 99% sure.

TABLE 3.1 t DISTRIBUTION α, PROBABILITY OF A LARGER ABSOLUTE VALUE OF t

DF	0.9	0.8	0.7	0.6	0.5	0.4	0.3	0.2	0.1	0.05	0.02	0.01	0.001
1	0.158	0.325	0.510	0.727	1.000	1.376	1.963	3.078	6.314	12.706	31.821	63.657	636.619
2	0.142	0.289	0.445	0.617	0.816	1.061	1.386	1.886	2.920	4.303	6.965	9.925	31.598
3	0.137	0.277	0.424	0.584	0.765	0.978	1.250	1.638	2.353	3.182	4.541	5.841	12.941
4	0.134	0.271	0.414	0.569	0.741	0.941	1.190	1.533	2.132	2.776	3.757	4.604	8.610
5	0.132	0.267	0.408	0.559	0.727	0.920	1.156	1.476	2.015	2.571	3.365	4.032	6.859
6	0.131	0.265	0.404	0.553	0.718	0.906	1.134	1.440	1.943	2.447	3.143	3.707	5.959
7	0.130	0.263	0.402	0.549	0.711	0.896	1.119	1.415	1.895	2.365	2.998	3.499	5.405
8	0.130	0.262	0.399	0.546	0.706	0.889	1.108	1.397	1.860	2.306	2.896	3.355	5.041
9	0.129	0.261	0.398	0.543	0.703	0.883	1.100	1.383	1.833	2.262	2.821	3.250	4.781
10	0.129	0.260	0.397	0.542	0.700	0.879	1.093	1.372	1.812	2.228	2.764	3.169	4.578
11	0.129	0.260	0.396	0.540	0.697	0.876	1.088	1.363	1.796	2.201	2.718	3.106	4.437
12	0.128	0.259	0.395	0.539	0.695	0.873	1.083	1.356	1.782	2.179	2.681	3.055	4.318
13	0.128	0.259	0.394	0.538	0.694	0.870	1.079	1.350	1.771	2.160	2.650	3.012	4.221
14	0.128	0.258	0.393	0.537	0.692	0.868	1.076	1.345	1.761	2.145	2.624	2.977	4.140
15	0.128	0.258	0.393	0.536	0.691	0.866	1.974	1.341	1.753	2.131	2.602	2.947	4.073
16	0.128	0.258	0.392	0.535	0.690	0.865	1.071	1.337	1.746	2.120	2.583	2.921	4.015
17	0.128	0.257	0.392	0.534	0.689	0.863	1.069	1.333	1.740	2.110	2.567	2.898	3.965
18	0.127	0.257	0.392	0.534	0.688	0.862	1.067	1.330	1.734	2.101	2.552	2.878	3.922
19	0.127	0.257	0.391	0.533	0.688	0.861	1.066	1.328	1.729	2.093	2.539	2.861	3.883
20	0.127	0.257	0.391	0.533	0.687	0.860	1.064	1.325	1.725	2.086	2.528	2.845	3.850
21	0.127	0.257	0.391	0.532	0.686	0.859	1.063	1.323	1.721	2.080	2.518	2.831	3.819
22	0.127	0.256	0.390	0.532	0.686	0.858	1.061	1.321	1.717	2.074	2.508	2.819	3.792
23	0.127	0.256	0.390	0.532	0.685	0.858	1.060	1.319	1.714	2.069	2.500	2.807	3.767
24	0.127	0.256	0.390	0.531	0.685	0.857	1.059	1.318	1.711	2.064	2.492	2.797	3.745
25	0.127	0.256	0.390	0.531	0.684	0.856	1.058	1.316	1.708	2.060	2.485	2.787	3.725
26	0.127	0.256	0.390	0.531	0.684	0.856	1.058	1.315	1.706	2.056	2.479	2.779	3.707
27	0.127	0.256	0.389	0.531	0.684	0.855	1.057	1.314	1.703	2.052	2.473	2.771	3.690
28	0.127	0.256	0.389	0.530	0.683	0.855	1.056	1.313	1.701	2.048	2.467	2.763	3.674
29	0.127	0.256	0.389	0.530	0.683	0.854	1.055	1.311	1.699	2.045	2.462	2.756	3.659
30	0.127	0.256	0.389	0.530	0.683	0.854	1.055	1.310	1.697	2.042	2.457	2.750	3.646
40	0.126	0.255	0.388	0.529	0.681	0.851	1.050	1.303	1.684	2.021	2.423	2.704	3.551
60	0.126	0.254	0.387	0.527	0.679	0.848	1.046	1.296	1.671	2.000	2.390	2.660	3.460
120	0.126	0.254	0.386	0.526	0.677	0.845	1.041	1.289	1.658	1.980	2.358	2.617	3.373
∞	0.129	0.253	0.385	0.524	0.674	0.842	1.036	1.282	1.645	1.960	2.326	2.576	3.291

Source: Table 3.1 is taken from Table III of Fisher and Yates: *Statistical Tables for Biological, Agricultural and Medical Research,* published by Longman Group Ltd., London (1974), 6th edition (previously published by Oliver & Boyd Ltd., Edinburgh), and by permission of the authors and publishers.

EXAMPLE 3.6 ————————————————————————————————

Two other samples are drawn from the same two batches of steel. The various values of modulus of elasticity (E) are given in the table below.

Measurement number	Batch 1 E (10^6 psi)	Batch 2 E (10^6 psi)
1	30	29
2	31	33
3	29	33
4	28	30
5	31	30
6	31	31

What inference can be drawn about the equality of strength between these two batches of steel?

$$H_0 : \mu_1 = \mu_2$$

$$\bar{X}_1 = 30 \times 10^6 \text{ psi} \quad \bar{X}_2 = 31 \times 10^6 \text{ psi}$$

$$\Sigma' X_1^2 = 8.0 \quad \Sigma' X_2^2 = 14.0$$

From Equation 3.12

$$t = \frac{|30 - 31|}{\sqrt{\dfrac{8 + 14}{6 + 6 - 2}} \sqrt{\dfrac{1}{6} + \dfrac{1}{6}}} = 1.168$$

$$DF = 6 + 6 - 2 = 10$$

$$t_{0.3,10} = 1.093$$

$$t_{0.2,10} = 1.372$$

Therefore, the hypothesis of equality of means can be rejected at the 0.3 significance level, but not at the 0.2 level. Prudence would dictate that we treat both of these steels as if they have a modulus of elasticity of 30×10^6 psi, or take more data to determine if a higher level of significance can be obtained.

Regression and Correlation

Some phenomena defy analytical description because of their complexity, or because of the lack of knowledge about all of the relevant independent variables influencing the phenomena. However, being able to estimate some approximate relationship from observed data would be useful. Relationships based solely on observed data are called empirical relationships. Other phenomena can be described by analytical mathematical models, but oftentimes these models contain parameters (coefficients and constants) that must be estimated if the model is to adequately describe a particular situation. Again, the best method for estimating these parameters is from observed data. In many cases, some form of regression analysis will accomplish the above described tasks. Correlation and other statistical tests can assist in making inferences about the accuracy of the regression equation that has been derived.

Development of Regression Equations   Regression is the process of relating a dependent variable to pertinent independent variables through the use of experimental or field data. It is basically the process of minimizing the error between the observed data and the proposed functional relationship which is supposed to describe the phenomena under study. The analyst must propose the form of the regression equation or regression function. Methods exist for estimating the parameters for linear regression functions with one or more independent variables, or for nonlinear regression functions. The analytical development in this text will be limited to linear regression with one independent variable. Multiple linear regression will be described through an example of computer-aided statistical analysis. An example of a case when nonlinear regression would be required would be when the exponential function $(1 - e^{-kX})$ is included in the regression equation. Nonlinear regression is described in the *Biomedical Computer Programs (BMDP) Manual* (Dixon, 1977).

A linear regression function would have the form

$$\hat{Y} = \sum_{k=1}^{m} a_k X_k + b \tag{3.13}$$

where
$\hat{Y}$ = estimated value of the dependent variable
m = number of independent variables
X_k = particular value of independent variable k
a_k and b = parameters of the regression function that must be estimated from the acquired data set for the dependent and independent variables

Regression analysis will make it possible to find the best estimates of all a_k and b, and to make inferences as to how good these best estimates are. It will also indicate between what limits the parameters, as well as the predicted values of $\hat{Y}$, might be expected to vary.

The simplest form of linear regression involves only one independent variable, and it is often referred to as least squares analysis. It can be used to fit the best straight line through a set of experimental or field data. Figure 3.1 shows a typical set of experimental data with Y plotted as the dependent variable and X as the independent variable.

For linear regression with one independent variable, the regression model becomes

$$\hat{Y} = aX + b \tag{3.14}$$

where a = the slope of the best fit line
b = the Y intercept when $X = 0$

Coefficient a is sometimes referred to as the regression coefficient.

The objective in this case is to minimize the cumulative error between data points and the regression line to achieve the "best fit." This is done by minimizing

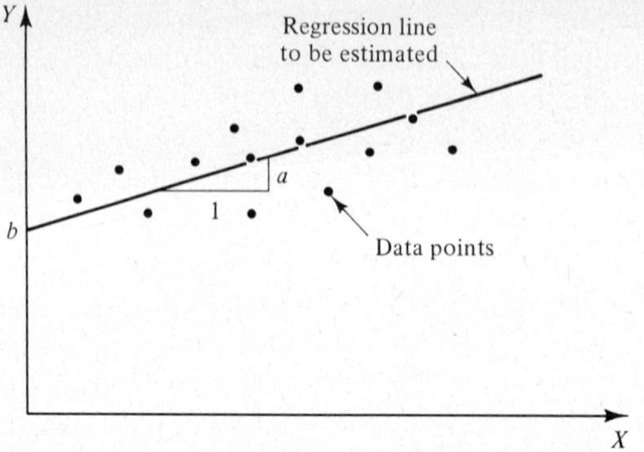

Figure 3.1   Typical experimental or field data.

the sum of squared deviations between the measured data points, Y_i, and the predicted value for $\hat{Y}_i$ from the regression line, using the corresponding measured X_i.

$$\text{objective to minimize SSD} = \sum_{i=1}^{n} (Y_i - \hat{Y}_i)^2$$

or $$\text{SSD} = \sum_{i=1}^{n} (Y_i - b - aX_i)^2 \tag{3.15}$$

It is important to note that SSD is a function of a and b, not X_i and Y_i. X_i and Y_i are now parameters.

To minimize the cumulative squared error the partial derivatives of SSD should be taken with respect to a and b. Both of these expressions should then be set equal to zero.

$$\frac{\partial \text{SSD}}{\partial a} = -2 \sum_{i=1}^{n} X_i(Y_i - b - aX_i) = 0 \tag{3.16}$$

$$\frac{\partial \text{SSD}}{\partial b} = -2 \sum_{i=1}^{n} (Y_i - b - aX_i) = 0 \tag{3.17}$$

or $$-2 \sum_{i=1}^{n} X_iY_i + 2b \sum_{i=1}^{n} X_i + 2a \sum_{i=1}^{n} X_i^2 = 0 \tag{3.18}$$

$$-2 \sum_{i=1}^{n} Y_i + 2b \sum_{i=1}^{n} (1) + 2a \sum_{i=1}^{n} X_i = 0 \tag{3.19}$$

and $$\sum_{i=1}^{n} X_iY_i - b \sum_{i=1}^{n} X_i - a \sum_{i=1}^{n} X_i^2 = 0 \tag{3.20}$$

$$\sum_{i=1}^{n} Y_i - nb - a \sum_{i=1}^{n} X_i = 0 \tag{3.21}$$

Equation 3.21 can be further transformed into:

$$\frac{\sum_{i=1}^{n} Y_i}{n} = b + \frac{a \sum_{i=1}^{n} X_i}{n} \tag{3.22}$$

and
$$\bar{Y} = b + a\bar{X} \tag{3.23}$$

or
$$b = \bar{Y} - a\bar{X} \tag{3.24}$$

If Equation 3.24 is substituted back into Equation 3.20, it becomes

$$0 = \sum_{i=1}^{n} X_i Y_i - \bar{Y} \sum_{i=1}^{n} X_i + a\bar{X} \sum_{i=1}^{n} X_i - a \sum_{i=1}^{n} X_i^2 \tag{3.25}$$

$$0 = \sum_{i=1}^{n} X_i Y_i - \bar{Y} \sum_{i=1}^{n} X_i + a \left(\bar{X} \sum_{i=1}^{n} X_i - \sum_{i=1}^{n} X_i^2 \right) \tag{3.26}$$

Solving Equation 3.26 for a, we get

$$a \left(\sum_{i=1}^{n} X_i^2 - \bar{X} \sum_{i=1}^{n} X_i \right) = \sum_{i=1}^{n} X_i Y_i - \bar{Y} \sum_{i=1}^{n} X_i \tag{3.27}$$

$$a = \frac{\sum_{i=1}^{n} X_i Y_i - \bar{Y} \sum_{i=1}^{n} X_i}{\sum_{i=1}^{n} X_i^2 - \bar{X} \sum_{i=1}^{n} X_i} \tag{3.28}$$

It can now be seen that a and b are calculable strictly from the measured data points. Equation 3.28 can now be solved for a and this value can be substituted back into Equation 3.24 to solve for b. Equation 3.24 also has other implications. It shows that a least squares regression line always passes through $\bar{X}$ and $\bar{Y}$ of the measured data. This property will be more fully explored when confidence testing is discussed.

Example 3.7 will illustrate the application of linear regression to real data.

It should also be noted that linear regression can be used to analyze certain nonlinear functions if they can be transformed into a linear form. For instance, if the proposed regression function is

$$Y = bX^a \tag{3.29}$$

then, taking natural logs of each side would yield

$$\ln Y = \ln b + a \ln X \tag{3.30}$$

To utilize Equation 3.30, all data values would be converted to natural logs, and the regression analysis would proceed as before. When the regression is finished, the equation could be reconverted to its original form by taking the antilog of $\ln b$, and writing the equation again as $Y = bX^a$.

EXAMPLE 3.7 ―――――――――――――――――――――――――――――――

The following dissolved oxygen (DO), temperature, and flowrate data are given for a station on the Connecticut River. Since dissolved oxygen is a function of temperature in pure waters, it is believed that a model of the form

$$\widehat{DO} = a(T) + b$$

might be useful in predicting what the dissolved oxygen level in the river would be for various water temperatures. The flowrate data will not be used at this time, but they will be used in a later section.

DO (mg/l)	T (°C)	Q (ft³/s)
8.5	23.8	10.7
7.9	21.6	6.0
8.5	15.0	9.6
10.2	8.1	10.5
10.9	2.0	12.0
13.4	2.5	19.4
11.9	12.5	34.7
9.4	20.5	10.9
8.5	22.6	7.1
9.7	26.4	1.1
8.0	22.5	9.0
12.9	1.5	13.6
7.8	15.3	9.3
8.9	9.6	3.7
12.9	1.5	13.6
12.9	5.5	67.0
7.8	20.0	9.6
9.0	24.4	1.0
9.3	25.0	1.5
7.5	21.5	8.9
8.6	18.5	4.3
7.9	14.0	11.6
14.1	1.0	18.5

For this problem $n = 23$.

$$\overline{DO} = \frac{226.50}{23} = 9.848 \qquad \bar{T} = 14.58$$

$$\sum_{i=1}^{n} DO_i T_i = 2975.57$$

$$\sum_{i=1}^{23} T_i = 335.3 \qquad \sum_{i=1}^{23} T_i^2 = 6607.19$$

$$a = \frac{\Sigma \overline{DO}(T) - \overline{DO}\Sigma T}{\Sigma T^2 - \bar{T}\Sigma T} = \frac{2975.57 - 9.848(335.3)}{6607.19 - 14.58(335.3)} = -0.190$$

$$b = \overline{DO} - a\bar{T} = 9.848 + 0.19(14.58) = 12.62$$

$$\therefore \widehat{DO}_i = 12.62 - 0.19(T_i)$$

Checking the limit of the regression line or using scatterplot analysis (which will be described in the applications section) can be useful in uncovering any extreme values.

The model $\widehat{DO} = -0.19(T) + 12.62$ can be shown to be realistic. It is well established that dissolved oxygen concentration decreases with increasing water temperature. Also, since dissolved oxygen saturation at 0°C is 14.6 mg/l, the model intercept of 12.62 mg/l is reasonable. At 20°C, the saturation concentration for dissolved oxygen is 9.2 mg/l, while the regression model predicts 8.8 mg/l, again, a reasonable value.

Measures of Goodness of Fit for Regression Equations   At this point, it is important to be able to estimate how well the regression line fits the measured data points. Inferences can then be made as to how accurate predicted values may be. These will form confidence bounds on the results.

A perfect fit for a regression line would be when all data points lie right on the line. In practicality, this rarely happens. Therefore, statistics that describe the goodness of fit of the regression line must be developed. These statistics quantify the scatter of the measured data about the regression line. Several new definitions will lead to the development of these statistics.

To begin, define the sum of squares of deviation of the measured Y values, Y_i, from the average value of Y, $\bar{Y}$, as

$$\Sigma' \, Y^2 = \sum_{i=1}^{n} (Y_i - \bar{Y})^2 \tag{3.31}$$

the sum of squares of deviation of the measured X values, X_i, from the average value of X, $\bar{X}$, as

$$\Sigma' \, X^2 = \sum_{i=1}^{n} (X_i - \bar{X})^2 \tag{3.32}$$

and the sum of squares of deviation of the measured Y values, Y_i, from the estimated Y values, $\hat{Y}_i$, for the same X_i, as

$$\Sigma' \, \hat{Y}^2 = \sum_{i=1}^{n} (Y_i - \hat{Y}_i)^2 \tag{3.33}$$

If $aX_i + b = \hat{Y}_i$ is substituted into Equation 3.33, algebraic manipulation will yield

$$\Sigma' \, \hat{Y}^2 = \Sigma' \, Y^2 - a^2 \, \Sigma' \, X^2 \tag{3.34}$$

However, previously it was shown that

$$a = \frac{\Sigma XY - \bar{Y}\Sigma X}{\Sigma X^2 - \bar{X}\Sigma X}$$

which, through algebraic manipulation becomes

$$a = \frac{\Sigma' \, XY}{\Sigma' \, X^2} \tag{3.35}$$

where $\Sigma' \, XY = \Sigma_{i=1}^{n} (X_i - \bar{X})(Y_i - \bar{Y})$.

If Equation 3.35 is substituted into Equation 3.34, it becomes

$$\Sigma' \, \hat{Y}^2 = \Sigma' \, Y^2 - a \, \Sigma' \, XY \tag{3.36}$$

In this expression, $\Sigma' \, Y^2$ is the total sum of squared deviation, $a \, \Sigma' \, XY$ is the portion of the total sum of squared deviation that is explained by the regression line, and $\Sigma' \, \hat{Y}^2$ is the remainder of the sum of squared deviation which can be attributed to error. It is sometimes referred to as s_e^2. The ratio of the portion of the sum of squared deviation explained by regression to the total sum of squares of the original data is a measure of the goodness of fit of the regression function, and is called the coefficient of determination.

$$\text{coefficient of determination} = r^2 = \frac{a \, \Sigma' \, XY}{\Sigma' \, Y^2} = \frac{a^2 \, \Sigma' \, X^2}{\Sigma' \, Y^2} \tag{3.37}$$

The square root of the coefficient of determination is called the correlation coefficient.

$$\text{correlation coefficient} = r = \sqrt{\frac{a^2 \, \Sigma' \, X^2}{\Sigma' \, Y^2}} = a \, \sqrt{\frac{\Sigma' \, X^2}{\Sigma' \, Y^2}} \tag{3.38}$$

It can be seen from Equations 3.37 and 3.38 that the coefficient of determination is always positive, while the correlation coefficient will be either positive or negative, depending on the sign of a. Positive correlation indicates that Y increases when X increases, whereas, negative correlation indicates that Y decreases with increasing X. Values for r^2 will vary from zero to positive 1, while the correlation coefficient varies between minus 1 and plus 1. Perfect correlation would have an $r = \pm 1.0$. An r value of zero would indicate no relationship between the dependent variable and the independent variable.

When calculating either r or r^2, it is helpful to remember that

$$\Sigma' \, XY = \sum_{i=1}^{n} (X_i - \bar{X})(Y_i - \bar{Y}) = \sum_{i=1}^{n} X_i Y_i - n\overline{XY} \tag{3.39}$$

$$\Sigma' \, X^2 = \sum_{i=1}^{n} X_i^2 - n\bar{X}^2 \tag{3.40}$$

and
$$\Sigma' \, Y^2 = \sum_{i=1}^{n} Y_i^2 - n\bar{Y}^2 \tag{3.41}$$

EXAMPLE 3.8 ____

Calculate the correlation coefficient and coefficient of determination for the regression developed in Example 3.7.

$$\bar{T} = 14.58°C \qquad\qquad a = -0.190$$

$$\overline{DO} = 9.848 \text{ mg/l} \qquad \Sigma' \, T^2 = 1719.1$$

$$n = 23 \qquad\qquad \Sigma' \, DO^2 = 98.14$$

$$\sum_{i=1}^{n} T_i DO_i = 2975.57$$

$$r = a \sqrt{\frac{\Sigma' \, T^2}{\Sigma' \, DO^2}} = -0.190 \, \sqrt{\frac{1719.10}{98.14}} = -0.795 = -0.80$$

$$r^2 = 0.63$$

or

$$r^2 = \frac{a \, \Sigma' \, T(DO)}{\Sigma' \, DO^2} = \frac{(-0.190)[2975.57 - 23(14.58)(9.848)]}{98.14}$$

$$r^2 = 0.63$$

Both the coefficient of determination, r^2, and the correlation coefficient, r, are used to measure the goodness of fit of a regression function to the data set used to generate it. The coefficient of determination has the advantage that it is a direct measure of the proportion of the error that is explained by the regression. It is also always smaller in magnitude than the correlation coefficient, and may therefore engender more caution in putting too much confidence in predicted results. The value of r or r^2 that might be acceptable in a regression analysis is a function of several factors, including the number of data points used and the end use of the predictions to be made by the regression function. A coefficient of determination of 0.6 might be sufficient for one analysis, while a value of 0.8 might not be sufficient under different circumstances. There are several statistical tests that can assist in evaluating the quality of a regression line.

The first of these is a test of the null hypothesis, $H_0 : r = 0$. In other words, the hypothesis that there is no correlation (dependent and independent variables are completely unrelated) is tested using a variation of Student's t test. From the previous discussion, it can be seen that Student's t distribution provides a measure of statistical significance. The distribution can be used to develop tables similar to Table 3.2. Degrees of freedom would be $n - 2$ in this case, because both parameters a and b have been estimated.

Values of r given in the table are the maximum values of the correlation coefficient that could be expected by chance for the amount of data involved if there were no correlation. The probability level indicates the chance of getting a value of r as large as the tabulated value when there is no correlation. The 0.10 level means there is only a 10% chance of getting a value of r as large as those in the 0.10 column when no correlation exists.

The F test is similar to the test just described for the correlation coefficient, except that the F test utilizes the coefficient of determination. It tests the hypothesis, $H_0 : r^2 = 0$. The same inferences can be drawn from the results as those that were presented in Examples 3.9 and 3.10. A more detailed discussion of the F test can be found in engineering statistics texts such as *Applied Statistics for Engineers,* by Volk (1969).

Confidence intervals can also be developed for the estimated slope of the regression line, a, and the predicted values, $\hat{Y}_i$. A confidence interval allows the analyst to state with a certain level of probability that the actual value of the parameter or variable lies in a symmetrical band around the estimated value. Specific examples will be developed for each case.

TABLE 3.2   CORRELATION COEFFICIENT SIGNIFICANCE

DF	\multicolumn{5}{c}{Probability of larger value of r}
	0.1	0.05	0.02	0.01	0.001
1	0.988	0.997	1.0	1.0	1.0
2	0.900	0.950	0.980	0.990	1.0
3	0.805	0.878	0.934	0.959	0.991
4	0.729	0.811	0.882	0.917	0.974
5	0.669	0.754	0.833	0.874	0.951
6	0.622	0.707	0.789	0.834	0.925
7	0.582	0.666	0.750	0.780	0.898
8	0.549	0.632	0.716	0.765	0.872
9	0.521	0.602	0.685	0.735	0.847
10	0.497	0.576	0.658	0.708	0.823
12	0.458	0.532	0.612	0.661	0.780
14	0.426	0.497	0.574	0.623	0.742
16	0.400	0.468	0.542	0.590	0.708
18	0.378	0.444	0.516	0.561	0.679
20	0.360	0.423	0.492	0.537	0.652
25	0.323	0.381	0.445	0.487	0.597
30	0.296	0.349	0.409	0.449	0.554
35	0.275	0.325	0.381	0.418	0.519
40	0.257	0.304	0.358	0.393	0.490
45	0.243	0.288	0.338	0.372	0.465
50	0.231	0.273	0.322	0.354	0.443

Source: Table 3.2 is abridged from Table VII of Fisher and Yates: *Statistical Tables for Biological, Agricultural and Medical Research,* published by Longman Group Ltd., London (1974), 6th edition (previously published by Oliver & Boyd Ltd., Edinburgh), and by permission of the authors and publishers.

EXAMPLE 3.9

Test the correlation coefficient found in Example 3.8 for significance.

$$n = 23$$

$$DF = n - 2 = 21$$

$$r = -0.80$$

Interpolating in Table 3.2, it can be found that the given value of r would only be exceeded if there were more columns to the right of 0.001. Therefore, the hypothesis that there is no correlation, $H_0 : r = 0$, can be rejected with less than a 0.001 chance of being wrong. It is 99.9% certain that there is significant correlation.

EXAMPLE 3.10

What can be inferred about the significance of the following regression? Let

$$n = 12 \quad \text{and} \quad r = 0.60 \qquad DF = 10$$

From Table 3.2, 0.60 falls between the 0.05 and 0.02 probabilities for a larger value of r. Therefore, the hypothesis, $H_0 : r = 0$, can be rejected with less than a 0.05

chance of being wrong but with a greater than 0.02 chance. It could be said with 95% certainty that there is significant correlation, but not with 98% certainty.

The confidence interval is given by the

$$\text{estimated value} \pm ts \tag{3.42}$$

where t = value selected from Student's t distribution for proper degrees of freedom $(n - 2)$ and desired probability level

s = estimated standard deviation for the parameter or variable = $\sqrt{\text{variance}}$

For the slope of the regression equation, the best estimate of the variance is given by

$$s^2(a) = \frac{s^2(\hat{Y})}{\Sigma' X^2} \tag{3.43}$$

since

$$s^2(\hat{Y}) = \frac{\Sigma' \hat{Y}^2}{n - 2} = \frac{\Sigma' Y^2 - a^2 \Sigma' X^2}{n - 2} \tag{3.44}$$

then

$$s^2(a) = \frac{\Sigma' \hat{Y}^2/(n - 2)}{\Sigma' X^2} = \frac{\Sigma' Y^2}{(n - 2)\Sigma' X^2} - \frac{a^2}{(n - 2)} \tag{3.45}$$

The estimated variance for any predicted value, $\hat{Y}_i$, is given by the equation

$$s^2(\hat{Y}_i) = s^2(\hat{Y}) \left[1 + \frac{1}{n} + \frac{(\bar{X} - X_i)^2}{\Sigma' X^2} \right]$$

$$= \frac{\Sigma' Y^2 - a^2 \Sigma' X^2}{n - 2} \left[1 + \frac{1}{n} + \frac{(\bar{X} - X_i)^2}{\Sigma' X^2} \right] \tag{3.46}$$

Previously, it was shown that $\Sigma' X^2 = \Sigma X_i^2 - n\bar{X}^2$ and $\Sigma' Y^2 = \Sigma Y_i^2 - n\bar{Y}^2$. These expressions provide a computationally more efficient way of finding $\Sigma' X^2$ and $\Sigma' Y^2$.

EXAMPLE 3.11 ____

For the water quality problem described in Example 3.7, what is the 95% confidence interval for the slope (a)?

$$t_{0.05,21} = 2.080$$

$$s^2(a) = \frac{\Sigma' \widehat{DO}^2/(n - 2)}{\Sigma' T^2} = \frac{\Sigma' DO^2}{(n - 2)\Sigma' T^2} - \frac{a^2}{(n - 2)}$$

$$= \frac{98.14}{21(1719.10)} - \frac{(0.190)^2}{21} = 9.994 \times 10^{-4}$$

$$s(a) = 3.161 \times 10^{-2}$$

$$\text{confidence interval} = -0.190 \pm 2.08(3.161 \times 10^{-2}) = -0.190 \pm 0.066$$

$\therefore$ It can be said that a lies between -0.124 and -0.256, with 95% certainty. If the 98% confidence interval is calculated,

$$t_{0.02,21} = 2.518$$

$$-0.190 \pm 2.518(3.161 \times 10^{-2})$$

$$-0.190 \pm 0.080$$

In this case, it can be said with 98% certainty that a lies between -0.110 and -0.270. Notice how the confidence interval expands as the level of certainty rises. This is a logical result since it is being stated with more certainty that the actual value lies in that interval.

EXAMPLE 3.12 ____

For a given value of water temperature ($T_i = 20.58°C$), use the model developed in Example 3.7 to predict dissolved oxygen concentration for this temperature. What is the 95% confidence interval for this predicted value?

$$\widehat{DO}_i = -0.190(20.58) + 12.62 = 8.71 \text{ mg/l}$$

$$t_{0.05,21} = 2.080$$

$$s^2(\widehat{DO}_i) = s^2(\widehat{DO})\left[1 + \frac{1}{n} + \frac{(\bar{T} - T_i)^2}{\Sigma' \, T^2}\right]$$

$$s^2(\widehat{DO}) = \frac{\Sigma'(\widehat{DO})^2}{n - 2} \qquad T = \text{temperature; DO} = \text{dissolved oxygen}$$

$$= \frac{\Sigma'(DO)^2 - a^2 \, \Sigma' \, T^2}{n - 2}$$

$$= \frac{98.14 - (0.190)^2(1719.10)}{21} = 1.718$$

$$\Sigma' \, T^2 = 1719.10 \qquad \bar{T} = 14.58$$

$$s^2(\widehat{DO}_i) = (1.718)\left[1 + \frac{1}{23} + \frac{(14.58 - 20.58)^2}{1719.10}\right] = 1.829$$

$$s(\widehat{DO}_i) = 1.352$$

$$\widehat{DO}_i - t_{0.05,21}[s(\widehat{DO}_i)] < DO_i < \widehat{DO}_i + t_{0.05,21}[s(\widehat{DO}_i)]$$

$$8.71 - 2.08(1.352) < DO_i < 8.71 + 2.08(1.352)$$

$$5.90 < DO_i < 11.52$$

Thus, it is 95% certain that DO_i actually lies between 5.90 and 11.52.

What if it is necessary to know how certain it is that the actual value of DO_i will lie between 7.71 and 9.71; that is $\widehat{DO}_i \pm 1.0$?

$$1.0 = s(\widehat{DO}_i)(t)$$

$$1.0 = 1.352(t_{?,21})$$

$$0.740 = (t_{?,21})$$

From Table 3.1,

$$t_{0.5,21} = 0.686$$

$$t_{0.4,21} = 0.859$$

Therefore, it is more than 50% but less than 60% certain that DO_i lies between 7.71 and 9.71. Interpolating to get approximate value

$$0.5 - \left(\frac{0.740 - 0.686}{0.859 - 0.686}\right)(0.1) = 0.469$$

$\therefore$ it is about 53% certain that DO_i lies between 7.71 and 9.71.

A plot of the confidence interval for a given certainty level (95%, for example) would appear as is shown in Figure 3.2. The confidence interval forms a parabolic envelope about the regression line with the minimum width of that envelope occurring about the point $\bar{Y}, \bar{X}$. The confidence interval expands outward from the regression line on either side of $\bar{X}$. Also, Figure 3.2 shows that the confidence interval for $\hat{Y}$ at $X = 0$ would give the confidence interval for the computed value of b.

The standard error of estimate, often used to estimate the quality of calibration of a mathematical model, is one final measure of the goodness of fit of a regression line. It is given by

$$\text{Standard error of estimate} = S_{Y,X} = \sqrt{\frac{\sum_{i=1}^{n}(Y_i - \hat{Y}_i)^2}{n - 2}} \qquad (3.47)$$

Again, the $n - 2$ is used because it gives a better estimate for small sample sizes. The standard error of estimate has properties similar to the standard deviation. For linear regression, if lines are drawn parallel to and on each side of the regression line at one, two, and three standard errors of estimate, and if n is large, then approximately 68%,

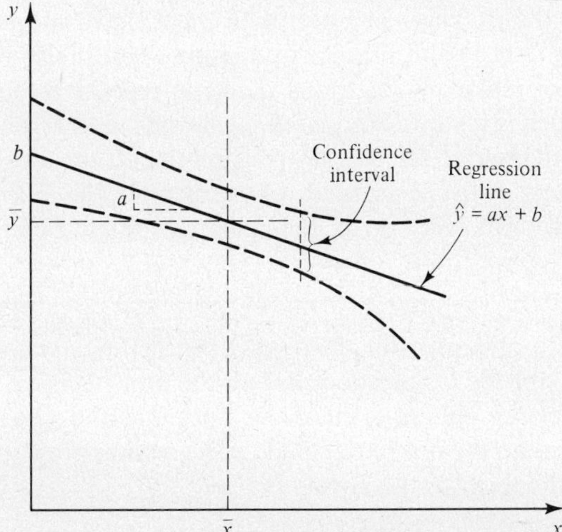

Figure 3.2 Confidence interval envelope.

95%, and 99.7% of the data points would be contained within the respective bands. Further explanation of this can be found in Appendix C.

Proper application of regression analysis takes good judgment and a basic knowledge of which independent variables might influence the dependent variable. It is possible to propose independent variables that have little or no physical relationship with the dependent variable, but which might produce an acceptable level of correlation for a given data set. Anyone familiar with the phenomena involved would have little confidence in predictions made with such a regression equation. It may also be possible that the same variables would not correlate well under slightly different conditions. Attempting to predict traffic density using air temperature, for example, would immediately create skepticism. At times, due to high recreational traffic volume, there might be both high temperature and high traffic density; low traffic density might occur simultaneously with low temperature because of adverse weather conditions. However, at other times, different factors would cause the opposite to be true. Therefore, one data set might show high correlation between temperature and traffic volume, even though temperature by itself really was not a primary causative factor. Other data sets might show low correlation. Such erroneous analyses are referred to as spurious correlations.

Computer-Aided Statistical Analysis

Theoretical methods to derive multiple linear regression equations can be developed from an extension of the theory developed in the previous section. Application of these methods to practice problems with several independent variables and large data bases is extremely tedious. Therefore, computer applications packages have been developed to accomplish the repetitive calculations. Two of the most widely used of these are the *Statistical Package for the Social Sciences* (SPSS) (Nie et al., 1975) and the *Biomedical Computer Programs* (BMDP) (Dixon, 1977). Each of these applications packages has many capabilities in addition to regression. In order to effectively use any of these packages, the user first has to understand the theory behind any computations being done by the package. The user then must study and understand the instructions for inputting data and running the program. It must also be emphasized that these packages do not make any decisions. They do repetitive calculations, and output estimates of coefficients, sums of squared deviations, and certain other easily calculable statistics. It is then up to the user to interpret the results for reasonableness and to make any inferences as to the accuracy of the results. The following example will give a brief introduction to how one of these packages might be utilized.

EXAMPLE 3.13 ──

In Example 3.7, data were presented for dissolved oxygen (DO), water temperature (TE) and rate of flow (Q) for the Connecticut River. It is now necessary to determine if a multiple linear regression equation involving both TE and Q as independent variables would be better than the one variable linear regression previously discussed. The model would be of the form

$$\widehat{DO} = A_1(TE) + A_2(Q) + B$$

Table 3.3 shows the input instructions for SPSS for this analysis. Various portions are annotated to show what is being asked for. Table 3.4 shows selected output with annotations.

An analysis of these results reveals that temperature is the primary independent variable for predicting dissolved oxygen concentration. It accounts for 63% of the total sum of squared deviation. When flowrate is also considered, 68% of the total sum is accounted for. Flowrate alone only accounts for 31%. Although it can be

TABLE 3.3   SPSS SAMPLE INPUT

RUN NAME	SPSS TEST CASE, DISSOLVED OXYGEN DATA	
VARIABLE LIST	DO, TE, Q	Telling the program what variables will be used and what they mean
VAR LABELS	DO DISSOLVED OXYGEN, MG PER LITER/	
	TE TEMPERATURE, DEGREES C/	
	Q FLOWRATE, CFS	
INPUT FORMAT	FIXED (3F10.0)	Data input format
N OF CASES	23	Number of cases
REGRESSION	VARIABLES = DO, TE, Q/	Telling the program what tasks to do. In this case it will do four regressions:
	REGRESSION = DO WITH TE, Q(2)/	$\widehat{DO} = A_1(TE) + A_2(Q) + B$
	REGRESSION = DO WITH TE(2)/	
	REGRESSION = DO WITH Q(2)/	$\widehat{DO} = A(TE) + B$
	REGRESSION = TE WITH Q(2)	$\widehat{DO} = A(Q) + B$
STATISTICS	ALL Telling the program to compute all available statistics	$TE = A(Q) + B$

```
      READ INPUT
         DATA
   8.5    23.8    10.7
   7.9    21.6     6.0
   8.5    15.0     9.6
  10.2     8.1    10.5
  10.9     2.0    12.0
  13.4     2.5    19.4
  11.9    12.5    34.7
   9.4    20.5    10.9
   8.5    22.6     7.1
   9.7    26.4     1.1
   8.0    22.5     9.0
  12.9     1.5    13.6
   7.8    15.3     9.3
   8.9     9.6     3.7
  12.9     1.5    13.6
  12.9     5.5    67.0
   7.8    20.0     9.6
   9.0    24.4     1.0
   9.3    25.0     1.5
   7.5    21.5     8.9
   8.6    18.5     4.3
   7.9    14.0    11.6
  14.1     1.0    18.5
SCATTERGRAM    DO  TO  Q
FINISH
```

(Input data)

TABLE 3.4 SPSS SAMPLE OUTPUT

STATISTICAL PACKAGE FOR THE SOCIAL SCIENCES

SPSS FOR B6700, VERSION H, RELEASE 7.2, LEVEL 72.001.028.005

DEFAULT SPACE ALLOCATION. .   ALLOWS FOR. .   50 TRANSFORMATIONS
WORKSPACE   17500 WORDS   400 RECODE VALUES + LAG VARIABLES
TRANSPACE   2500 WORDS   600 IF/COMPUTE OPERATIONS

RUN NAME	SPSS TEST CASE, DISSOLVED OXYGEN DATA
VARIABLE LIST	DO, TE, Q
VAR LABELS	DO DISSOLVED OXYGEN, MG PER LITER
	TE TEMPERATURE, DEGREES C
	Q FLOWRATE, CFS
INPUT FORMAT	FIXED (3F10.0)

ACCORDING TO YOUR INPUT FORMAT, VARIABLES ARE TO BE READ AS FOLLOWS

VARIABLE	FORMAT	RECORD	COLUMNS
DO	F10.0	1	1–10
TE	F10.0	1	11–20
Q	F10.0	1	21–30

THE INPUT FORMAT PROVIDES FOR 3 VARIABLES. 3 WILL BE READ
IT PROVIDES FOR 1 RECORDS ('CARDS') PER CASE. A MAXIMUM OF 30 'COLUMNS' ARE USED ON A
RECORD.

N OF CASES	23
REGRESSION	VARIABLES = DO, TE, Q/
	REGRESSION = DO WITH TE, Q(2)/
	REGRESSION = DO WITH TE(2)/
	REGRESSION = DO WITH Q(2)/
	REGRESSION = TE WITH Q(2)
STATISTICS	ALL

********** REGRESSION PROBLEM REQUIRES   57 WORDS WORKSPACE, NOT INCLUDING RESIDUALS *********

READ INPUT DATA. .

Note: This initial output is a recapitulation of input instructions which allows input coding errors to be found.

SPSS TEST CASE, DISSOLVED OXYGEN DATA

VARIABLE	MEAN	STANDARD DEV	CASES
DO	9.8478	2.1121	23
TE	14.5783	8.8397	23
Q	12.7652	13.8333	23

These are the mean and standard deviations for each of the variables.

CORRELATION COEFFICIENTS

	DO	TE	Q
DO	1.00000	−0.79468	0.55945
TE	−0.79468	1.00000	−0.46552
Q	0.55945	−0.46552	1.00000

These are the bivariate correlations between the variables. Notice that the correlation coefficient for DO with TE is −0.80, the same as for the bivariate linear regression.

*** MULTIPLE REGRESSION***

DEPENDENT VARIABLE. .   DO   DISSOLVED OXYGEN, MG PER LITER

VARIABLE(S) ENTERED ON STEP NUMBER 1. .   Q
                                          TE

MULTIPLE R	0.82302	—Similar to r and r^2 for simple linear regression
R SQUARE	0.67736	
STANDARD ERROR	1.25823	←—Standard error of estimate

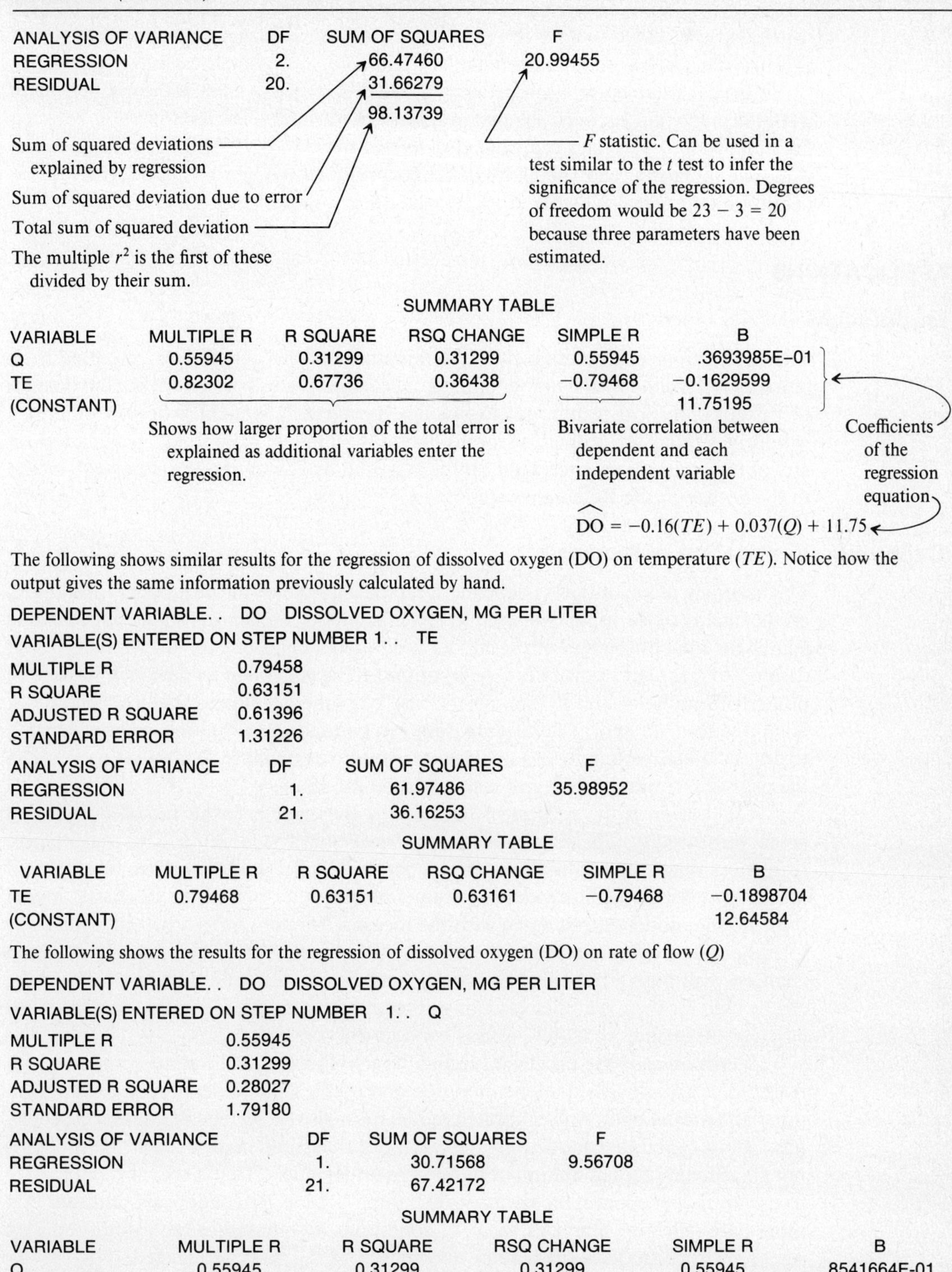

ANALYSIS OF VARIANCE	DF	SUM OF SQUARES	F
REGRESSION	2.	66.47460	20.99455
RESIDUAL	20.	31.66279	
		98.13739	

Sum of squared deviations explained by regression

Sum of squared deviation due to error

Total sum of squared deviation

The multiple r^2 is the first of these divided by their sum.

F statistic. Can be used in a test similar to the *t* test to infer the significance of the regression. Degrees of freedom would be $23 - 3 = 20$ because three parameters have been estimated.

SUMMARY TABLE

VARIABLE	MULTIPLE R	R SQUARE	RSQ CHANGE	SIMPLE R	B
Q	0.55945	0.31299	0.31299	0.55945	.3693985E–01
TE	0.82302	0.67736	0.36438	−0.79468	−0.1629599
(CONSTANT)					11.75195

Shows how larger proportion of the total error is explained as additional variables enter the regression.

Bivariate correlation between dependent and each independent variable

Coefficients of the regression equation

$$\widehat{DO} = -0.16(TE) + 0.037(Q) + 11.75$$

The following shows similar results for the regression of dissolved oxygen (DO) on temperature (*TE*). Notice how the output gives the same information previously calculated by hand.

DEPENDENT VARIABLE. .   DO   DISSOLVED OXYGEN, MG PER LITER

VARIABLE(S) ENTERED ON STEP NUMBER 1. .   TE

MULTIPLE R	0.79458
R SQUARE	0.63151
ADJUSTED R SQUARE	0.61396
STANDARD ERROR	1.31226

ANALYSIS OF VARIANCE	DF	SUM OF SQUARES	F
REGRESSION	1.	61.97486	35.98952
RESIDUAL	21.	36.16253	

SUMMARY TABLE

VARIABLE	MULTIPLE R	R SQUARE	RSQ CHANGE	SIMPLE R	B
TE	0.79468	0.63151	0.63161	−0.79468	−0.1898704
(CONSTANT)					12.64584

The following shows the results for the regression of dissolved oxygen (DO) on rate of flow (*Q*)

DEPENDENT VARIABLE. .   DO   DISSOLVED OXYGEN, MG PER LITER

VARIABLE(S) ENTERED ON STEP NUMBER   1. .   Q

MULTIPLE R	0.55945
R SQUARE	0.31299
ADJUSTED R SQUARE	0.28027
STANDARD ERROR	1.79180

ANALYSIS OF VARIANCE	DF	SUM OF SQUARES	F
REGRESSION	1.	30.71568	9.56708
RESIDUAL	21.	67.42172	

SUMMARY TABLE

VARIABLE	MULTIPLE R	R SQUARE	RSQ CHANGE	SIMPLE R	B
Q	0.55945	0.31299	0.31299	0.55945	.8541664E–01
(CONSTANT)					8.757464

concluded that flowrate does improve the regression, additional investigation would probably be called for to determine if it would contribute under different conditions.

The bivariate correlation between Q and TE is -0.47. This indicates that there is some correlation between the two independent variables. High correlation between the independent variables indicates that one of them will add little to the regression when the other is already being used. While a value of 0.47 is not high, further investigation might reveal a higher cross correlation.

APPLICATIONS

Introduction

The applications examples given in the remainder of this chapter are excerpted from research work done by the author and his colleagues. They demonstrate the application of various statistical techniques to practical problems. Thought processes described would be similar for problems from different disciplines. Additional information on any of the applications presented can be acquired by consulting the papers attributed to the author in the References section.

Calibration/Verification of Models

The accuracy of any model is dependent on the correctness of the model formulation, on the quality of the input data used to drive the model, and on the ability to determine values for the parameters of the model accurately. The need for high-quality input data is obvious; but in order for useful output to be generated by a model, it must be properly formulated, and its parameters must be properly adjusted for local conditions. High-quality field and/or laboratory data can be used in calibration and verification to adjust model parameters to achieve as accurate as possible results and to evaluate the precision (reproducibility of results) of the model.

Calibration is the process of minimizing the error between model output and actual measured system output for the same measured system input. This may involve regression analysis to estimate empirical parameters for a deterministic or empirical model or successive adjustment of parameters of a deterministic or stochastic model to make the model output agree with the measured system output within some error tolerance. Sensitivity analysis of the model will help to identify parameters that, when changed, will have little effect on model output. Average values from the literature can be used for these parameters. This will reduce the number of possible combinations of parameter values that will provide good fit with observed data.

Verification is the process of testing how well the model output compares with reality. It is accomplished after calibration and uses a separate set of measured system input and output data. Any calibrated model must also be verified, because it is usually possible to calibrate a model for a given set of measured data, even if the model is not an adequate representation of the system under study. That is why it is imperative to use an independent data set for verification. A poor verification will indicate that more data reflecting a greater range of conditions are needed in the calibration data set, or that the model is inadequate and should be modified or rejected.

The following discussion of a methodology for calibrating stormwater models will serve to illustrate the processes of calibration and verification as applied to an actual situation. Concepts and procedures that are developed in this application are equally applicable to other modeling situations.

Introduction The impetus for the author's research began with the primary goal of the Water Pollution Control Act Amendments of 1972 (PL 92-500): the achievement of water quality that would allow fishing and swimming in all of our nation's water by 1983. Section 208 of PL 92-500 requires the preparation of areawide waste management plans that define the best mix of point and nonpoint pollution control strategies to meet this goal.

Point sources of pollution are identifiable sources such as discharges from industrial plants or effluent discharges from sewage treatment plants. Nonpoint sources are more diffuse, such as urban stormwater runoff or agricultural runoff. Nonpoint sources may enter receiving water bodies from many locations, or they may be collected in a conveyance system and discharged similarly to a point source. Development of viable plans is contingent upon knowing the nature of nonpoint pollutants. In urban areas, nonpoint pollutants are carried by urban stormwater runoff; therefore, defining the characteristics of stormwater runoff is of primary importance.

Difficulties involved in gathering storm runoff data, as well as time and financial constraints, make the complete characterization of urban stormwater runoff from each small drainage basin (catchment) nearly impossible. Therefore, increased emphasis is being placed on the use of computer models for predicting quality of urban stormwater runoff. These computer models are cost effective and reasonably accurate substitutes for extensive field data-gathering programs, provided that the models are calibrated and verified for conditions existent on the particular catchment being studied.

Calibration involves minimization of deviation between measured field conditions and model output by adjusting parameters within the model. Some minimum amount of field data is required to accomplish a reasonable calibration. Verification is the process of checking the model calibration using an independent set of data. Ideally, the verification data set should be as large as the calibration data set. However, if limited data are available, it is the usual practice to use the larger portion of the data for calibration and the smaller portion of the data for verification.

Methods of Calibration Most models for predicting quality of stormwater runoff are coupled with a quantity simulation model. Calibration of these models has consisted of calibration of the quantity and quality models sequentially, using the output from the calibrated model as input to the quality model. Sometimes only the quantity portion has been calibrated and default, or estimated, values have been used for quality parameters. Also, some projects have used data from a single-storm event to calibrate models.

One of the motivations for the study described here was a belief that the state of the art of stormwater management modeling does not allow single-storm calibration. Random storm variations plus the difficulty in characterizing parameters for a particular basin make it important to use several storms for calibration and, whenever possible, several more storms for verification. Once measured quantity and quality data from several storms have been input to a model, the model could be used to predict urban stormwater pollutant loads with greater confidence. The weakness of using a single storm for calibration soon became apparent.

Rational Approach Toward Calibration   During this study, a new approach toward calibration was developed and applied to an urban basin. In developing this approach, a basic premise was that quantity and quality portions of models should be separated and calibrated independently. Following calibration, the model could be recombined and used in its normal manner.

Because each portion of the model is being calibrated separately, it is not necessary to use the same data set for both. This provides some flexibility as to what field data set is representative of the average response of the basin.

The U.S. Environmental Protection Agency Storm Water Management Model (SWMM) was chosen to illustrate the approach. SWMM is a comprehensive model used to predict quantity and quality of runoff from urban surfaces during storm events; then route the stormwater through a user-defined combination of conveyance system storage, and treatment; and estimate the effects of resulting discharges on receiving waters. It was chosen because of its availability to consultants and planners and because it is representative of the state of the art in stormwater modeling. Quantity and quality subroutines of SWMM were separated and calibrated using storm event data taken from the 1,014-acre (4.1-km^2) Maple Brook basin (separate sewers) in Greenfield, Massachusetts (see Fig. 3.3). Measured rainfall and sewer flow data were used as input to the quality subroutine and the predicted pollutant mass emission rates (mass/unit area/unit time) were calibrated against measured mass emission rates. Model parameters were adjusted to produce minimum error in total runoff and mass emissions and sums of peak flowrates across all calibration storms. Standard errors of estimate and normalized standard errors of estimates were tabulated for each run to indicate the

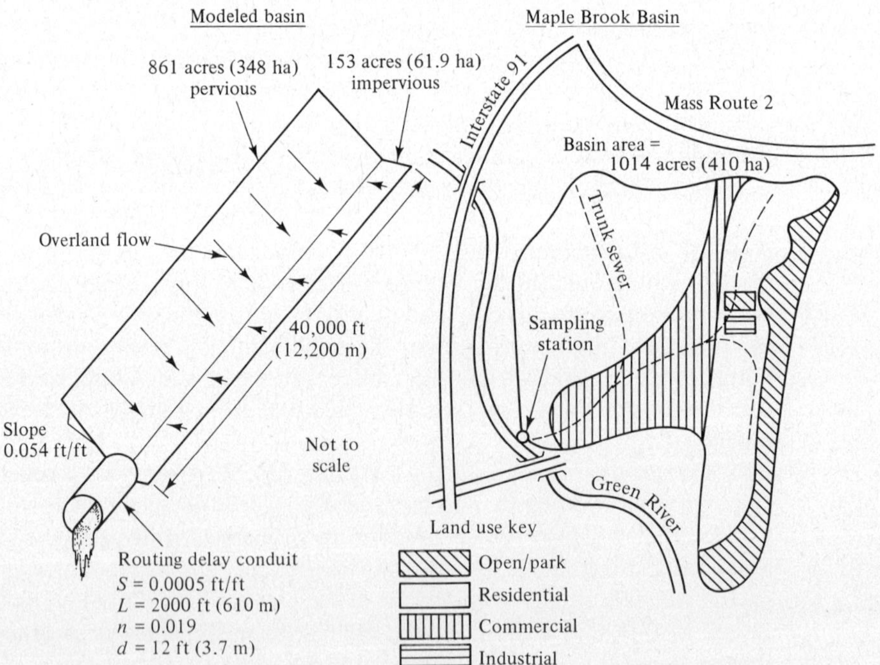

Figure 3.3   Comparison of Maple Brook Basin with modeled basin.

relative accuracy of the calibration between runs. Independently calibrated subroutines were recombined and a preliminary verification was accomplished using an independent set of data. The SWMM Runoff Block (quantity simulation) was then run in a continuous simulation mode to estimate yearly pollutant loadings from stormwater runoff in Greenfield. Pollutants considered were suspended solids, BOD_5 (five-day biochemical oxygen demand), total phosphorus, cadmium, lead, and zinc.

Greenfield Study Area   The town of Greenfield is located in the northwest central portion of Massachusetts, at the confluence of the Green, Deerfield, and Connecticut Rivers. Greenfield has a total land area of 21 mi^2 (54 km^2) and a population of 18,500. Land use varies from open, undeveloped land to the concentrated central business district typical of most New England towns. Light industry is scattered throughout several portions of the town.

Approximately 80% of the urbanized area of Greenfield is drained by the Maple Brook storm sewer system. A schematic of the trunk sewer system is shown in Figure 3.3. Average slope, percentage impervious area (percent of the basin covered by impervious surfaces such as parking lots and building roofs), and street gutter length were computed from data acquired for an earlier system study that divided the basin into 76 subcatchments. Characteristic outflow width from the basin was taken as twice the length of the main drainage channel through the basin. The length of the main drainage channel is the 40,000-ft dimension of the modeled basin in Figure 3.3. This is the width over which runoff is assumed to leave the surface and enter the sewer system. No sewer routing was included in the initial configuration.

Quantity Calibration   The objective of both the quantity and quality calibrations was to fit the model to average catchment conditions. During calibration it was not necessary to achieve close agreement of measured and predicted data for individual storm events. The emphasis was on results integrated over the entire calibration data set. If the assumption of representative data was valid, the calibrated model would give accurate predictions of average annual runoff and pollutant loadings.

Calibration criteria established were less than 1% error between measured and predicted total volume of runoff (V_m/V_p) and less than 1% error between the sums of measured and predicted peak flow rates ($\Sigma P_m/\Sigma P_p$). The first criterion was important because it measured long-term volumetric discharge accuracy, which was important in the prediction of receiving waters response. The second criterion indicated accuracy of short-term volume predictions that would be important for estimating storage/treatment requirements.

To aid in calibration, an auxiliary computer program was written to compare measured and predicted flow and pollutant data and to compute representative statistics. A flowchart of this program is presented as Figure 3.4. This program could be used to compare quantity data alone, quality data alone, or both together. Subroutine QUAL, used to predict mass emission rates, contained the same logic found in subroutine QSHED1 of SWMM.

The standard error of estimate (SEE), Equation 3.47, was chosen as a representative statistic to measure the accuracy of fit between the measured and predicted data. A small SEE would indicate good agreement between the measured and predicted data, and thus an accurate calibration.

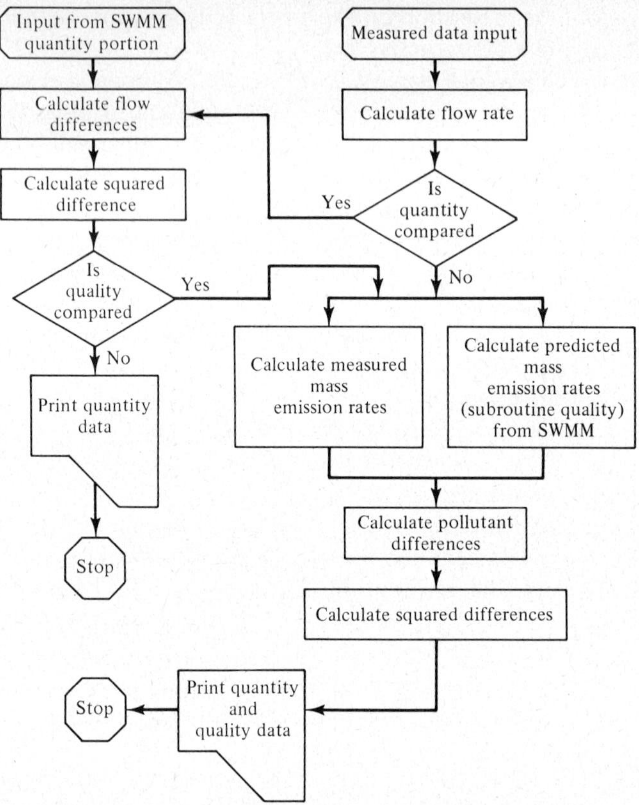

Figure 3.4   Flowchart—SWMM calibration program.

Flow data for calibration were gathered at the outfall of the Maple Brook sewer using a sharp crested, suppressed weir. Data from six storm events were used for quantity calibration. Since a wide range of rainfall intensities and durations were used, the calibration should be representative of conditions occurring on the basin. Total rainfall of 3.94 in. (100 mm) represented approximately 10% of the average annual rainfall for Greenfield. The average runoff coefficient, weighted by storm event rainfall, was 0.14.

Initial basin parameters for input to SWMM were derived from an earlier, detailed discretization of the Maple Brook basin used for sewer system analysis. Discretization is the breaking up of the basin into sub-basins of fairly uniform characteristics to aid in estimating parameters for model input. The theory is that the finer the discretization, the more accurately the model will represent the real basin.

Percentage impervious area and slope were weighted according to area to arrive at representative values for the 1014-acre (4.1-km^2) basin. The average slope remained constant throughout calibration at 0.054 ft/ft. Impervious area was initially calculated as 23.3%. Characteristic width was initially taken as twice the length of the main drainage channel through the basin, or 40,300 ft (12,290 m). Default estimates found in SWMM were used for the frictional resistance to overland flow over surfaces (overland flow resistance factors), water stored in small depressions and cracks (depression storage depths), and the rate that water soaks into the soil (infiltration rate). If necessary, these parameters could be adjusted during calibration.

Prior to beginning calibration, it was necessary to ascertain model output sensitivity to changes in input parameters. The parameter being tested was varied through a range about the accepted mean value. Total predicted runoff volume was the test model output variable. Percentage impervious area was the most sensitive parameter. Sensitivity of the other parameters was somewhat variable depending on the characteristics of the basin modeled and the range of parameter values used. With respect to total volume, the output was insensitive to pervious area depression storage (volume stored in small depressions, such as cracks and potholes) and impervious area Manning's n (frictional resistance), and was quite insensitive to characteristic width (a measure of the width of the overland flow plane). The latter two parameters may, however, have had some effect on hydrograph shape. Impervious area depression storage may have a significant effect on total volume of runoff, especially for low-rainfall volume storm events. Model output was sensitive to infiltration rates and pervious area Manning's n (overland flow resistance factor) for some basins; however, analysis has shown that pervious areas of the Maple Brook basin did not contribute a significant amount to runoff. Thus, the model output for the Greenfield basin was insensitive to any changes in pervious area parameters.

From these sensitivity analyses, it was deduced that percentage impervious area would be the primary volume calibration factor. Characteristic width, impervious area Manning's n, or possibly a routing conduit could be used to change the shapes of the hydrographs and improve agreement between sums of measured and predicted peak flows.

No program has been written that will optimize parameter values within a multiparameter model such as SWMM. Thus, model calibration remains a subjective process, aided by knowledge of model sensitivity and experience gained in previous calibrations.

Seven calibration runs were made before the criteria stated previously were met. Table 3.5 gives the parameters changed and the results for each run. Figure 3.5 shows

TABLE 3.5   QUANTITY CALIBRATION DATA

Parameter	Run 1	Run 2	Run 3	Run 4	Run 5	Run 6	Run 7
(a) Subcatchment parameters							
Impervious area, as a percentage	23.3	14.0[a]	14.6[a]	14.5[a]	14.8[a]	15.1[a]	15.1
Characteristic width (ft)	40,300	30,000[a]	30,000	30,000	40,000[a]	40,000	40,000
Impervious area, Manning's n	0.013	0.013	0.03[a]	0.03	0.03	0.013[a]	0.013
(b) Conduit parameters							
Diameter (ft)				8.0	12.0[a]	12.0	12.0
Manning's n				0.013	0.025[a]	0.025	0.019[a]
(c) Criteria							
V_m/V_p	0.626	1.044	0.922	1.023	1.017	1.007	1.000
$\Sigma P_m/\Sigma P_p$	0.477	0.774	0.852	1.066	1.165	1.080	1.007
SEE (ft^3/s)	21.00	13.50	11.43	8.05	7.56	7.88	8.40

[a] Indicates parameter has been adjusted from previous run.

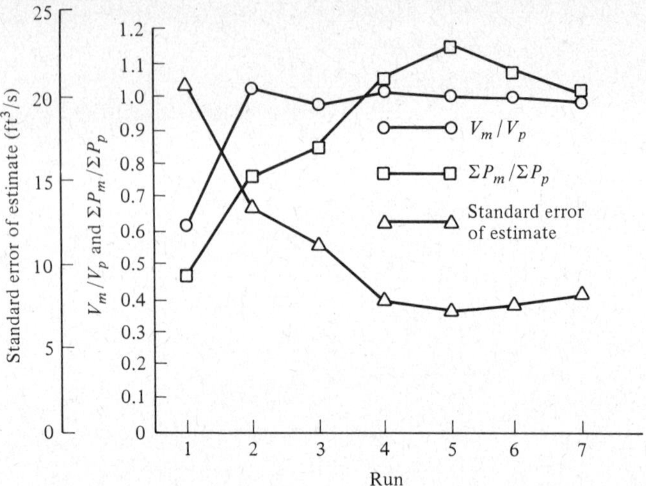

Figure 3.5　Calibration criteria and statistics.

the convergence of the volume and sum of peaks criteria and reduction in standard error of estimate.

The uncalibrated model (run 1) predicted 60% more flow than was measured. Predicted peak flowrates were occurring before measured peak flowrates and the sum of the predicted peak flowrates was 110% high. To improve total volume agreement, the percentage impervious area was reduced by a factor equal to the measured volume to predicted volume (V_m/V_p) ratio. This method of adjusting percentage impervious area was carried through subsequent calibration runs. To improve temporal and magnitude agreement of peak flowrates, characteristic width and impervious area Manning's n were adjusted during the second and third runs, respectively. It became clear during subsequent runs that reasonable adjustment of these parameters would not produce agreement between sums of measured and predicted peak flowrates. Therefore, during the fourth run, a single, 2,000-ft (610-m) conduit with a slope of 0.0005 ft/ft was added to provide routing delay. Size and Manning's n of the conduit were adjusted until the criterion of ±1% between sums of measured and predicted peak flows was met. The final ratio of predicted and measured volumes (1.000) and the final ratio of sums of measured and predicted peak flowrates (1.007) were both within the ±1% criteria. Note that the SEE had a minimum value after run 5 then increased slightly through run 7. Although the SEE is an indicator of accuracy of fit, it is not neccesarily minimized by meeting the calibration criteria set. This is consistent with the objective of calibrating for long-term agreement rather than individual storm event reproduction. Although both criteria have been met, there is still some time lag between predicted and measured peaks. If these time lags could be eliminated, it would improve the SEE considerably. To demonstrate, all predicted flowrates were lagged by 5 min and the resulting SEE was 7.87. However, artificially shifting results to achieve better looking statistics would do nothing for the long-term simulation.

Examination of measured and predicted volume of runoff for individual storms indicated that single-column calibration could result in significantly different total runoff predictions than were generated by the model calibrated across six storms. For the calibrated model, the overall ratio of measured to predicted volume was 1.00 while the ratios for individual storms were 1.43, 0.75, 1.20, 0.53, 0.71, and 1.01. The conclusion can be made that the single-storm event calibrated model could predict total volumes of runoff as much as 1.4 times or as little as 0.5 times the volume predicted by the model calibrated with six storm events.

Quality Calibration   As described earlier, the calibration program used in this investigation utilized an adaptation of SWMM subroutine QSHED1, driven by measured flow data, to simulate pollutant washoff from the watershed. Measured pollutant concentrations and simulated data were converted to mass emission rates, in pounds per minute, for comparison. The calibration program was stuctured so that comparisons could be made at time intervals corresponding to the sampling times of measured data.

Maintenance of the ratio of predicted to measured mass emissions (P/M) at 1.0 was the primary quality calibration criterion. All storms were weighted equally when predicted mass emissions were adjusted. The normalized SEE (the SEE divided by the mean value of the pollutant mass emission in question) was used for goodness of fit comparisons between different pollutants. This was a necessary modification because there was a disparity in the magnitudes of the emission rates among pollutants.

Samples for quality analysis were collected using a Manning S-4000 automatic sampler in combination with a Manning T-1000 dipper-transmitter stage recorder. Parameters determined for the quality analysis were suspended solids (SS), biochemical oxygen demand (BOD_5), total phosphorus, and three dissolved metals: zinc, cadmium, and lead. BOD_5 and SS were chosen because they are common indicators of the impact of urban runoff on receiving waters. Total phosphorus can also be an important measure of urban runoff pollution, particularly if the basin drains into a slow moving stream or lake where algal blooms or eutrophication may be problems. The choice of dissolved metals to be monitored was based on a trade-off between mass loadings in urban runoff and toxicity to mankind.

Data from five storms were used to calibrate the model pollutant accumulation rates for all pollutants except suspended solids. SS data from one storm were saved for preliminary verification and the SS data from another storm were not used for calibration because they were believed to be atypical. Through sensitivity analysis of the quality prediction parameters contained in SWMM, pollutant accumulation rates were found to be the primary calibration parameters.

An examination of the individual storms used for quality calibration again indicated the weakness of single-storm event calibration. For example, the BOD_5 measured to predicted mass emission ratios for the five storms used for calibration were 1.31, 1.63, 0.75, 2.67, and 0.52. Thus, if the quality portion of the model had been calibrated using only one storm, the annual BOD_5 loadings predicted would have been anywhere from 0.52 to 2.67 times those predicted by the model as calibrated over five storm events.

Verification   Data from two storms were used to verify flow and SS predictions. Data from one storm was also used to verify BOD_5 predictions.

Results of the verification were encouraging. The ratio of predicted to measured total runoff was 0.92, while the same ratio for the sum of peak runoff rates was 1.06. The ratios of total mass emission of SS and BOD_5 were 0.80 and 0.72, respectively.

It would have been better to accomplish the verification using data from four or five storms of varying magnitude. However, the field data-gathering program had been terminated, and further verification data would not be available.

Data from a third storm were not used for verification because the short antecedent dry period (2 days) produced pollutant mass emission predictions that were much lower than measured mass emissions. It was noted during the course of this study that accumulation rates for pollutant buildup were not linear, as the model assumed. A greater mass of pollutant was available on the basin soon after a storm event than was predicted by the linear accumulation function contained in SWMM. Therefore, even though a fairly good quality verification was obtained, the need for further study to ascertain the representativeness of the quality prediction algorithms used in SWMM was evident.

Application of Calibrated Model   The calibrated model was used to predict annual stormwater pollutant loadings from the 1014-acre (4.1-km) Maple Brook basin. Ten years of hourly rainfall records for Amherst, Massachusetts [14 mi (22.5 km) from Greenfield] were used to drive runoff calculations. No rainfall data tape was available for Greenfield but a check of annual precipitation indicated that the average rainfall for both locations was similar.

Summary and Conclusions of Study   The Maple Brook Basin Study illustrated a sensible way to calibrate coupled quantity–quality urban stormwater management models. After separating the quantity and quality portions of the model, it was possible to calibrate each by using only measured data. It showed that the calibration procedure does not have to be complicated to achieve accurate results. The calibrated model will be an accurate and economical tool for predicting alternate futures.

Good quantity calibration can be achieved by adjustment of a few basin parameters; and, in most cases, quality calibration involves only the adjustment of pollutant buildup rates. An advantage of the method used is that quantity and quality portions of the model do not have to be calibrated with the same data set.

It is difficult to quantitatively compare this calibration method with other previously used calibration methods because of the lack of discussion of multistorm calibrations in the literature, and because of the lack of year-long runoff quantity and quality data sets to check final outputs. If data for other multistorm calibrations were available, the normalized SEE could be utilized to compare the relative accuracy of methods.

The Storm Water Management Model was designed to be a deterministic model. Theoretically, it could be calibrated using any storm event with similar results. However, the variation among storms of measured to predicted flow volume and pollutant mass emission ratios (for the calibrated model) indicated that the model did not precisely portray actual basin conditions. Therefore, different storm events will result in different calibrations and different predictions. Calibrating for average conditions across several storms will reduce predictive error and increase confidence in results.

Development of Models

Introduction   When existing models are shown to be inadequate, statistical analysis of field and laboratory data may help in the development of more representative models. Judgment must be used to ensure that such models are reasonable and useful. This will involve carefully examining data sets to eliminate erroneous or outlying data. Data scatterplots can be helpful in this regard. Care must be taken in formulating models to ensure that any independent variables chosen are reasonable. Alternative models must be tested to evaluate which is best. The model chosen must then be tested for sensitivity to changes in independent variable values and for reasonableness at extreme values of the independent variables. If these steps are followed, the model developed should adequately reflect the cause-and-effect relationship for the system under study. The following example will illustrate model development for a particular application to stormwater quality modeling.

Example Introduction　　Research has shown that unless urban stormwater runoff can be treated or controlled, many receiving waters will be prevented from becoming "fishable and swimmable" as proposed in the U.S. Water Pollution Control Act Amendments of 1972. In order to determine where and when to control stormwater runoff, it is necessary to assess its impacts and estimate the effects of various control measures, tasks that would be extremely difficult using measured data alone.

Therefore, researchers have used simulation models to predict the quantity and quality of stormwater runoff and how it would affect the environment, under controlled and uncontrolled conditions. The quality portions of these models, however, have not fostered confidence in their predictive capabilities. Lack of confidence has been engendered by the use of unverified predictive formulations and by the variability of measured stormwater pollution data.

Despite the lack of confidence in model predictions, little has been done to improve pollutant washoff formulations found in the most used models. Most efforts have been directed toward developing methods to calibrate existing models, even though it has to be assumed that the formulations to be calibrated adequately portray washoff processes. One such method was discussed in the previous section.

There are several reasons for the lack of interest in developing improved formulations. Whipple (1977) was one of the first to point out that once a particular formulation becomes part of a large model, it is automatically assumed to be valid until proved otherwise. The repeated calibration of models tends to strengthen the assumption of validity. This is a false validation, however, because it has been found that faulty models can be calibrated with sufficient altering of parameters. Finally, there has been a great deal of money spent on developing large stormwater management models, and even more money spent on studies to predict the impact of stormwater on the environment through use of these models. Any questioning of the model formulations would also question the validity of the conclusions of these studies.

Existing Formulations　　Two of the most widely used stormwater management models, SWMM and the Corps of Engineers' Storage, Treatment, Overflow, and Runoff Model (STORM) use similar pollutant washoff prediction formulations. Examination of other models reveals that, with few exceptions, they use formulations identical or nearly identical to those found in SWMM and STORM.

SWMM and STORM utilize a nonlinear pollutant washoff predictive equation

$$P_{i,n} = M_{i,n}(1 - e^{-k_i R_n \, \Delta t}) \qquad\qquad (3.48)$$

where $P_{i,n}$ = mass of pollutant i washed off the subcatchment during time step n

$M_{i,n}$ = mass of pollutant i on subcatchment at start of simulation time step = $M_{i,\,n-1} - P_{i,\,n-1}$

k_i = washoff decay coefficient for pollutant i

R_n = runoff rate during time step n

Δt = simulation time step; constant throughout simulation

By applying Equation 3.48 in a stepwise manner, it can be shown that the total mass of a certain pollutant washed off during a storm event is

$$P_{i,T} = M_{i,1}(1 - e^{-k_i R_v}) \tag{3.49}$$

where $M_{i,1}$ = mass of pollutant i available on the subcatchment at the start of the storm event, and

R_v = total volume of runoff

The mass of pollutant available at the beginning of the storm event can be replaced by a generation function that predicts the buildup of pollutant during interstorm periods. Latest versions of SWMM provide the option of using a linear buildup function, which when combined with Equation 3.49 yields

$$P_{i,T} = (\text{drydays})(L_i)(1 - e^{-k_i R_v}) \tag{3.50}$$

where   drydays = length of interstorm period (days)

L_i = accumulation rate for pollutant i (lb/acre/day)

or a nonlinear buildup function which produces

$$P_{i,T} = M_{i,u}(1 - e^{-k_{b,i}(\text{drydays})})(1 - e^{-k_i R_v}) \tag{3.51}$$

where $M_{i,u}$ = ultimate pollutant accumulation

$k_{b,i}$ = accumulation decay coefficient (days^{-1})

Formulation Testing   Equations 3.50 and 3.51 could not be statistically verified as adequately portraying the urban stormwater pollutant washoff process. (Jewell et al., 1980b). Nonlinear regression analysis and other techniques were used to test the formulations' ability to predict storm event total loadings and instantaneous fluxes (amount/time step) of chemical oxygen demand, suspended solids, and ammonia nitrogen. Stormwater data from 258 storm events for 26 basins in 12 geographic areas were used in the analysis. Nonlinear regression estimates for the parameters of Equations 3.50 and 3.51 (L_i, k_i, $M_{i,u}$, and $k_{b,i}$) were found to be unrealistic. These unrealistic estimates were verified through analysis of the sum of squared deviation contours (lines of equal sums) for measured and predicted pollution data. Therefore, the conclusion was made that new formulations should be developed that could be statistically verifiable.

A number of linear and linear transform formulations were tested using stepwise linear regression. These formulations were of the general form

$$\text{multiple linear} \quad P = A + \sum_{i=1}^{n} B_i X_i \tag{3.52}$$

$$\text{semilog transform} \quad P = A + \sum_{i=1}^{n} B_i \ln(X_i) \tag{3.53}$$

$$\text{log–log transform} \quad P = A \prod_{i=1}^{n} X_i^{Bi} \tag{3.54}$$

where

$$X_i = \text{selected independent variables}$$
$$A \text{ and } B_i = \text{estimated regression coefficients}$$
$$P = \text{estimated pollutant flux or total pollutant load}$$

Regression significance was tested using the F test. The F statistic was calculated from the equation

$$F = \frac{SS_{reg}/k}{SS_{res}/(n - k - 1)} = \frac{R^2/k}{(1 - R^2)/(n - k - 1)} \tag{3.55}$$

where

$$SS_{reg} = \text{sum of squares explained by entire regression equation}$$
$$k = \text{number of independent variables in the equation}$$
$$SS_{res} = \text{residual (unexplained) sum of squares}$$
$$n = \text{number of data points}$$

It was used to test the null hypothesis that the multiple correlation coefficient for the regression was equal to zero. The value computed with Equation 3.55 was compared with tabulated values of the F distribution for the proper degrees of freedom at various levels of significance.

Quality of fit among the formulations was compared using the multiple coefficient of determination, R^2. Analyses were performed using the whole data file and nine subfiles for different basins for which there were sufficient data. R^2 tables were generated for all basin/formulation combinations for suspended solids, chemical oxygen demand, ammonia nitrogen, and total phosphorus.

Through analysis of these R^2 tables, it was found that no one formulation was consistently better than others in predicting stormwater pollutant washoff. Often, different formulations were the best predictors for different pollutants. When the same formulation fit best for several basins, the estimated parameter values varied significantly among the basins. This was true even among basins in the same geographic areas and was true for both storm event total loadings and instantaneous fluxes (mass/unit area/unit time at a particular point in time).

An important additional finding was the inconsistency of the influence of the interstorm dry period on the mass of pollutant washed off per unit of runoff, both for storm event total loadings and instantaneous fluxes. This was especially significant since existing formulations predict the mass of pollutant available at the beginning of a storm as a function of the interstorm dry period.

Thus, it was concluded that, rather than trying to calibrate a given model, an improved practice would be for stormwater modelers to gather data from each basin to be studied and develop formulations from these data to predict stormwater pollution washoff.

Recommended Modeling Strategy　　Having established that formulations contained in existing models could not be used with any degree of confidence, a recommended stormwater quality modeling methodology was developed. This methodology evolved from the procedures used for formulation testing.

Data Acquisition Data acquisition programs seem expensive, but their cost is small relative to the value of improved facilities that may be proposed based on modeling results. Output of the program will be a synchronous record of rainfall, runoff, and quality data with a minimum of sampling error. Enough data must be generated to allow derivation of regression relationships having a level of statistical significance that is acceptable to all concerned parties. Of equal importance is the requirement that the data set gathered be representative of conditions existing on the basin. Regression relationships developed from an unrepresentative data set will be unrepresentative even though they may exhibit good statistical significance. Table 3.6 lists some independent variables that may influence buildup and washoff of stormwater pollutants.

To improve representativeness of the data set gathered, it is recommended that the data acquisition program span at least a 1-yr period, with storms sampled during every part of the year when sampling is physically possible. The mean and distribution fit of the entire rainfall record during the data acquisition program can be compared with the same statistics for a long-term rainfall record to infer the representativeness of the meteorological conditions during the sampling period. The rainfall record of the actual storms sampled should also be checked for representativeness.

The type and density of data gathered will depend on the objectives of the modeling study. Flow measurements should be made with the highest degree of accuracy possible under good engineering practice. Rainfall, flow measurement, and sampling records must be time synchronized. Stormwater quality sampling should be accom-

TABLE 3.6 INDEPENDENT VARIABLES AND PARAMETERS INFLUENCING STORMWATER POLLUTION WASHOFF

Dynamic variables		Parameters
Storm event totals	Instantaneous flux	
1. Time since last storm event (days)	1. Runoff intensity (cm/hr)	1. Land use
2. Street cleaning practices	2. Cumulative volume of runoff (cm)	2. Area (hectares)
3. Total volume of runoff (cm)	3. Time from start of storm (min)	3. Percent impervious area
4. Storm duration (min)	4. Rainfall intensity (cm/hr)	4. Length of overland flow (m)
5. Total volume of rainfall (cm)	5. Cumulative volume of rainfall (cm)	5. Percent street and parking area
6. Average rainfall (cm/hr)		6. Length of streets/hectare (m/hectare)
7. Average runoff (cm/hr)		7. Population density (pop/hectare)
		8. Particulate fallout rate (kg/hectare/day)
		9. Number of catchbasins/hectare
		10. Climatological data a. Temperature b. Rainfall

plished in a manner that will ensure a representative sample. The type of sampling done will depend on the type of model to be developed. Analysis of samples should be performed according to *Standard Methods* (1980), noting any deviations in procedure. Through application of the best sampling and data-gathering program possible, systematic errors should be minimized.

Scatterplots After completion of data gathering but before detailed analysis is undertaken, scatterplots should be generated to check for outlying data. Any data that stand apart from the mass should be checked carefully. Data that are found to be of questionable validity should be excluded from the data set. Figure 3.6(a) shows a classic example of an outlier, a storm loading two orders of magnitude higher than any of the others. As a result, the rest of the data is compressed along the bottom of the scatterplot and does not show any trend. Examination of the raw data file revealed that this outlier was caused by a typing error that had not been detected in previous data screening. Rather than rerun the data-processing program to correct this one error, the storm event was removed from the data file. Figure 3.6(b) shows the results of removing this storm. There is still scatter in the data, but it is much more reasonable than before.

Scatterplots can also be used to identify possible formulations for further analysis. Figure 3.6(c) shows a strong linear association between the natural log of suspended solids mass loadings and the natural log of average runoff intensity. This plot suggests that a model of the form

$$P = AX^B \tag{3.56}$$

might adequately depict the functional relationship between suspended solids mass loading (P) and runoff intensity (X).

Correlation Matrix After any unacceptable data have been removed, and before testing any models, a correlation matrix (matrix of correlation coefficients for combinations of variables and parameters) should be generated. The Statistical Package for the Social Sciences (Nie et al., 1975) provides an option that accounts for missing data, something that stormwater modelers always have to contend with, and gives a realistic correlation matrix. This will allow examination of the bivariate correlation between any two variables and parameters that are going to be used for analysis.

Examination of the bivariate correlations will facilitate exclusion of some variable and/or parameter combinations from further analysis because of low correlation. Good bivariate correlations between pollutants will identify possible indicator–indicatee relationships that could be developed into pollutant estimators.

Regression Analysis There are several levels of regression analysis possible. At opposite ends of the spectrum are bivariate linear regression and nonlinear regression. A simple linear formulation will not be adequate. There are too many factors involved in the buildup and washoff of pollutants for a one independent variable formulation to explain a sufficient amount of variation in the data.

Nonlinear regression can be applied to formulations that cannot be made linear with respect to their parameters, no matter what type of transformations are applied.

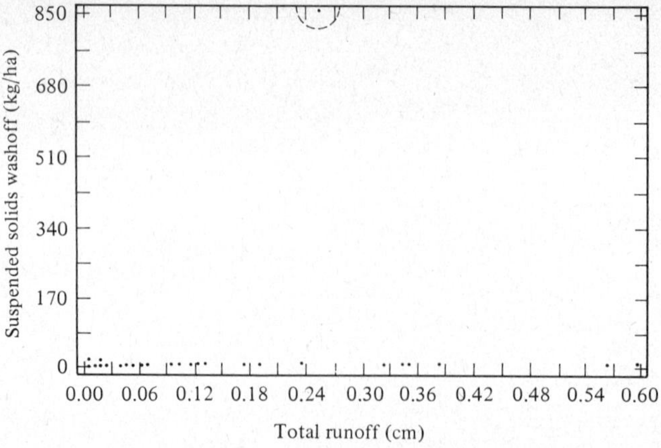

(a) Scatterplot before outlier removal.

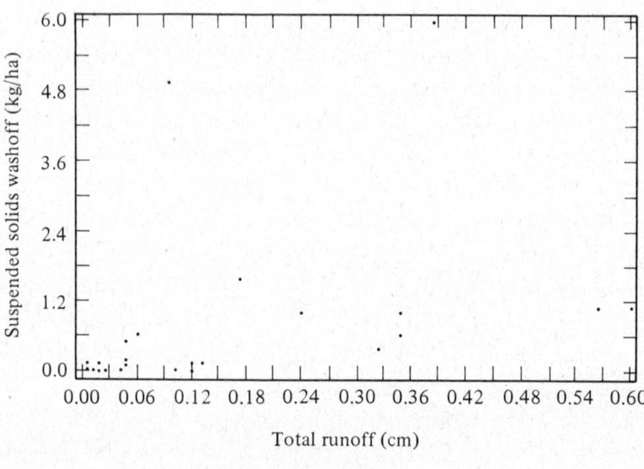

(b) Scatterplot after outlier removal.

Figure 3.6 Scatterplot analysis.

Application of nonlinear formulations, however, is significantly more difficult than the application of linear or linear transform formulations. Estimation of formulation parameters involves use of sophisticated numerical techniques that are sensitive to initial parameter values and do not always give satisfactory results. Some problems associated with unrealistic parameter estimates for nonlinear formulations have already been discussed in reference to Equation 3.51.

Between the extremes are the three types of linear and linear transform regression formulations given by Equations 3.52, 3.53, and 3.54. The R^2 analysis conducted as part of the formulation testing indicated that log–log transform formulations (Equation 3.54) predict urban stormwater pollutant washoff better than the other two types for a majority of cases. However, the variables that appear in the equation when it is fit to basin–pollutant data are not consistent. Also, multiple linear and semilog transform

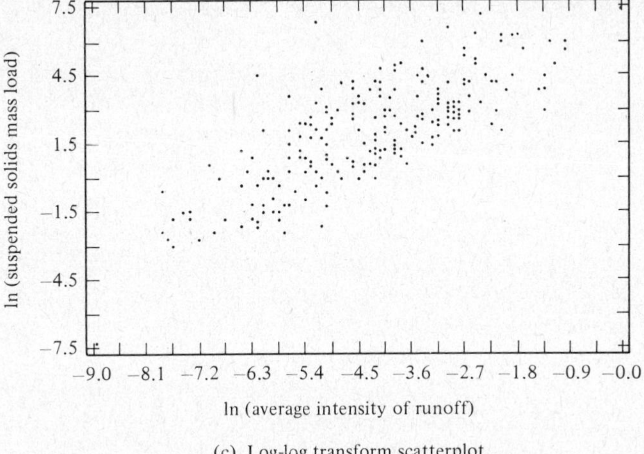

(c) Log-log transform scatterplot.

Figure 3.6 *(Continued)*

formulations fit best for some basin–pollutant combinations. Since it has been shown that linear and linear transform formulations predict stormwater quality with a reasonable degree of accuracy, and the quality of fit of these formulations can be compared to select the best one, linear and linear transform formulations should be tested for inclusion in models.

Criteria for Selecting a Formulation A suitable criterion for choosing the best formulation is the coefficient of determination (R^2), as this is a measure of the portion of the data variance explained by the formulation. It may happen, however, that two or three formulations, each of which is reasonable from a physical standpoint, may give nearly the same R^2 value for a particular pollutant. In that case, the independent variables for the formulations should be examined to see which combination is the easiest to measure accurately. If the choice was between time since last storm event and total volume of runoff, the former would be the best. Also, if two formulations have approximately equal R^2 and one contains fewer independent variables, it would be the best choice.

The sensitivity of formulations considered for selection must also be checked. The effects of independent variable values near the limits of or outside the range of measured data must be investigated. In the course of using a stormwater management model, extreme events are nearly certain to appear as input. If a formulation gives unrealistic estimates for extreme values, it should not be considered for adoption.

Before any formulation is chosen to predict the washoff of a particular pollutant, the reasonableness of the formulation should be checked. Average values should be substituted for the independent variables and the estimated value of the pollutant washoff checked to make sure that it is in the expected range of values.

Application of Models Not only must good pollutant washoff relationships be derived, but also, they must be applied in a reasonable manner if useful results are to be obtained. Modelers must also remember that the quality of output of a stormwater management model can be no better than the quality of the input data. Attention must be paid to the method of predicting runoff because any uncertainty in runoff estimation will have a multiplicative effect on uncertainty of pollutant washoff estimates.

The best storm event input data for pollutant washoff models would be measured runoff. Since runoff records are not generally available, rainfall records must be filtered through a runoff model for input to quality models. Wherever possible, an actual long-term record, be it runoff or rainfall, should be used over any statistically derived design storm. Use of a design storm is still warranted for detailed analysis to identify local trouble spots in the collection and conveyance system. For overall modeling of the effects of stormwater pollutant washoff, however, the best estimates can be made through continuous, long-term simulation.

Two types of applications are simulation and prediction. Simulation is the analysis of possible stormwater pollution effects based on an historical record of rainfall or runoff. Prediction is the estimation of future effects based on presently occurring runoff or rainfall. Simulation is useful for analyzing alternative courses of action, while prediction is most useful in estimating what the effects of present conditions will be on receiving water quality in the near future. Simulation has been the prevalent type of application in the past. Prediction is going to be utilized more as areas install stormwater treatment facilities. Prediction can be accomplished through pollutant versus storm characteristics models or through models also using indicator pollutants as independent variables.

Conclusions The preceding study demonstrated some techniques that can be used when working with an established mathematical model. It showed the importance of not accepting an established model at face value simply because it has been used frequently and gives impressive-looking results. No established model should be used until it can be shown to be compatible with the present situation that is being studied by the engineer. Manipulation of an existing model to make it fit the situation at hand, or even the development of a new model, may be necessary.

A methodology has been outlined that facilitates development of formulations that can be used with confidence to predict urban stormwater pollutant washoff. The methodology consists of testing several linear and linear transform multiple regression formulations for estimating pollutant washoff loadings as a function of storm and basin characteristics, using measured data. The best formulation can be chosen for each pollutant using R^2 and other criteria. This methodology eliminates the need to estimate both buildup and washoff, processes whose linkages are not well defined.

In the study in question, it was shown that until basic research can uncover more of the underlying processes involved in stormwater pollution generation, the methodology developed becomes a practical alternative that can be used now to provide more confidence in results than before. Confidence limits can be calculated and null hypotheses tested, something that could not be done with deterministic formulations. The modeler must be careful, however, not to ask more of regression analysis than it can deliver. It is only a tool, and as such has to be understood to be used effectively. The only disadvantage of the method is that storm event data have to be gathered for each basin that is to be modeled. However, this is the only way that modeling can be undertaken with any degree of confidence.

Experiences Encountered in Data Acquisition Programs

Shelley (1975) wrote two manuals describing some of the pitfalls that can be encountered during data acquisition for flow measurement and water quality sampling programs. Similar problems could be encountered in almost any type of data acquisition program, whether the program is carried out in the laboratory or in the field. Shelley described programs where, due to either sludge blockage or inadequate turbulence, tracer dyes did not mix properly, and therefore, did not yield accurate results. Devices used for measuring flow velocity in storm drains were prone to being fouled by debris. Blockages of input ports for liquid samplers, despite the fact that they had purge cycles, were encountered. Numerous cases of samples being rendered useless due to unclean sample bottles or inadequate sample fixing or preservation prior to analysis were cited. The failure of automatic sampling equipment due to dampness, severe weather, or tampering was another major problem. Electrical contacts and solenoids deteriorated, fuses were inadequate, and switches and relays failed. Weirs installed in streams caused surcharges during high-flow storm events, invalidating critical data and damaging other equipment through submergence. Weirs also tended to catch debris and sludge and had to be cleaned frequently.

These examples, as well as many similar to them, point out the necessity for strict adherence to two basic rules of data gathering. Established procedures and good principles for data gathering must be utilized at all times. Also, all measuring devices and equipment should be calibrated (in place, if possible) and checked frequently for any unforeseen problems. If these two basic steps are followed, the data acquisition program should yield reliable results.

SUMMARY

The acquisition and analysis of data are complex processes that must be accorded adequate attention. Thorough preparation and planning for any data acquisition program are essential if any useful data are to result. There is no substitute for complete attention to detail concerning equipment, acquisition sites, personnel, and processing of data. The data gatherer must be constantly alert for any circumstances that might contaminate data.

Elementary statistics, such as the mean and standard deviation of a data set, may provide all of the information needed for recommendations to be made. Comparisons can also be made to estimate whether or not two data sets have equal mean values. If data are to be used in a modeling effort, they may be used to calibrate and verify a mathematical model, or they may be used to estimate model parameter values or to develop new models through regression analysis. Additional statistics can be used to make inferences as to the quality and accuracy of a regression model. In all cases, careful analysis of data is critically important. All of the statistical techniques available do nothing but manipulate the data and compute certain characteristic parameters. It is the analyst who must interpret and make appropriate inferences from these results.

APPLICATIONS EXERCISES

3-1. Find the best fit straight line through the following data. What is the predicted value of Y for $X = 16$ and what is the 90% confidence interval for this predicted value? What is the standard error of estimate for the regression line?

X	Y	X	Y
4	10	11	11
5	7	11	14
7	10	13	12
7	12	12	15
8	14	14	14
9	10	15	17
10	15	17	15

3-2. The following data summations were calculated from a set of gathered data. Y is the dependent variable.

$$n = 10 \qquad \bar{X} = 3.45 \qquad \bar{Y} = 546$$

$$\Sigma X_i = 34.5 \qquad \Sigma Y_i = 5460 \qquad \Sigma X_i Y_i = 20{,}329.0$$

$$\Sigma(X_i - \bar{X})^2 = 27.99 \qquad \Sigma(Y_i - \bar{Y})^2 = 82{,}490.0$$

$$\Sigma(X_i - \bar{X})(Y_i - \bar{Y}) = 1492.0$$

$$\Sigma X_i^2 = 147.0 \qquad \Sigma Y_i^2 = 3{,}063{,}550.0$$

(a) What is the regression equation for this data?
(b) What is the correlation coefficient?
(c) What is the 90% confidence interval for the slope of the regression line?
(d) What is the 95% confidence interval for the Y intercept (b)?

3-3. For the data given below, what is the line of best fit ($\widehat{DO} = aQ + b$)?

DO (mg/l)	Q (ft³/s)
7.9	6.0
10.2	10.5
13.4	19.4
9.4	10.9
9.7	1.1
12.9	13.6
8.9	3.7
12.9	67.0
9.0	1.0
7.5	8.9

(a) What is the coefficient of determination?
(b) How sure are you that there is significant correlation between the dependent and independent variable?

3-4. You have prepared several specimens of SAE 1005 steel ($E = 30 \times 10^6$ psi), and measured strain for various applied stresses. Your experimental results are shown below:

Stress (σ) (psi)	Strain (ϵ) (in./in.)	Stress (σ)	Strain (ϵ)
20×10^3	0.0002	44×10^3	0.0010
16×10^3	0.0003	38×10^3	0.0012
22×10^3	0.0005	44×10^3	0.0013
26×10^3	0.0005	42×10^3	0.0014
30×10^3	0.0005	48×10^3	0.0015
28×10^3	0.0007	50×10^3	0.0017
36×10^3	0.0007	48×10^3	0.0018
32×10^3	0.0008	52×10^3	0.0018
38×10^3	0.0008	50×10^3	0.0020
38×10^3	0.0010	52×10^3	0.0022

(a) Use linear regression to estimate the modulus of elasticity (E) for this steel.
(b) What is the 95% confidence interval for the value of E estimated?
(c) What is the 90% confidence interval for the stress when the strain is 0.002 in./in.?
(d) Are you at least 90% sure that the correlation coefficient is not equal to zero?
(e) Carefully plot the data given to you. Superimpose on this the estimated regression line. Use this graph to interpret your results. What conclusions can you infer about the metal tested or about the experimental method used?

3-5. You have taken two sets of wastewater samples and have analyzed them for BOD_5, with the following results:

Set 1 BOD_5 (mg/l)	Set 2 BOD_5 (mg/l)
200	205
210	210
180	235
170	203
195	201
208	220
193	225
188	196
	205
	200

(a) Estimate the standard deviation and coefficient of variation for each data set.
(b) How confident are you that there is a significant difference in the mean values for these two data sets?
(c) Estimate the population standard deviation for each data set. Comment on the differences in these estimates from those found in part (a).

3-6. The data given below show yearly motor fuel consumption in the United States between the years 1945 and 1975.

Data

Motor fuel demand

$(\times 10^6$ barrels/yr)	969	994	1330	1512	1750	2162	2385
Year	1945	1950	1955	1960	1965	1970	1975

(a) Plot the data and place an estimated best-fit line through the plotted points. From this line, estimate the coefficients of the equation $\hat{Y} = aX + b$, and predict what the consumption will be in the years 1985 and 2000.

(b) Use linear regression to estimate the same coefficients and to predict the consumption for the same two years. Also, estimate the correlation coefficient and the confidence level in the correlation.

(c) What is the 90% confidence interval for the slope of the regression line?

(d) Determine the 90% confidence interval for the estimates of consumption in 1985 and 2000. Explain any differences in the estimates.

(e) Describe any differences in the estimates found in parts (a) and (b). What could be the reasons for the differences? How could either, or both, estimates be improved?

3-7. The Metropolitan Planning Organization (MPO) has been conducting speed versus density studies on the main interstate highway through the metropolitan area. Their data are given below:

Speed, V (mph)	Density, K (vehicles/mile)
20	300
30	224
30	240
36	220
40	204
44	164
48	164
48	240
62	120
66	88
68	80
70	80
74	74
80	60

(a) Plot the data, then use linear regression to find the best fit line through the data. Estimate the correlation coefficient and the confidence level for correlation.

(b) Estimate the speed for a vehicle density of 200 vehicles/mi, and find the 90 and 95% confidence intervals for this predicted value.

3-8. Describe the types of data that would have to be gathered to quantify the quality of life factors developed for Exercise 2-13. Describe the key elements that would have to be included in a data acquisition program for each of these types of data. Are there any of the factors that could not be quantified realistically? If so, how might they be included in the analysis?

3-9. Population growth is a good analog for predicting future water demand. Population growth data over the past 30 yr for the metropolitan area described in Exercise 2-14 are given below. Use this data to predict what the population of the area might be in the years 1990, 2000, 2010, and 2020. What factors might make these estimates unrealistic?

Year	Population (×1000)
1950	200
1955	210
1960	320
1965	700
1970	800
1975	1500
1980	1900

3-10. The basin size and average annual rainfall data for the drainage basins for the three reservoirs of Exercise 2-15 are given below. Also included is the fraction of the total rainfall (runoff coefficient) that can be expected to reach each basin's particular reservoir.

Drainage basin	Size (mi^2)	Runoff coefficient
A	10	0.40
B	12	0.45
C	6	0.35

	Average annual rainfall (in.) for drainage basin		
Year	A	B	C
1970	36.2	37.0	36.0
1971	40.1	38.0	40.5
1972	28.5	33.5	27.2
1973	40.4	39.0	38.3
1974	38.0	37.1	40.4
1975	32.1	34.3	32.4
1976	34.6	36.0	33.1
1977	35.0	36.5	33.3
1978	48.9	43.9	45.6
1979	40.1	38.2	41.0
1980	29.6	33.4	31.0

Determine what the average annual yield for each reservoir would be, in units of gallons/day. Which estimates would you have the most, and least, confidence in, and why? What are some possible weaknesses in the given data, and how might they be overcome?

3-11. Federal regulations have mandated that pollution in stormwater runoff be controlled. The problem lies in estimating what pollutant load is carried by stormwater runoff from urban and suburban surfaces. Federal grants were utilized to develop large computer models designed to estimate both the quantity and quality of stormwater runoff. Dynamic input to this model is a measured or synthetic rainfall record for the basin being studied. Hydrologic and hydraulic routing theory transforms the input record into an outflow versus time record for the basin. Buildup of pollutants is modeled as a linear function of time between storm events. Washoff of pollutants is governed by an exponential function of runoff rate. Numerous parameters, such as percent impervious area, and surface resistance factors, have to be estimated before the model can be used. Alternatively, default values for many of these parameters were included in the computer package as options. When the model was developed, a limited amount of field data was available for use in testing out the model. A portion of this data was utilized to estimate various default parameter values, including pollutant accumulation rates. When the model was tested against an independent set of input and output data, results did not match well, so changes were made in the model to make output agree with this second set of data. That ended the initial model testing, and it was released to the general public for use, accompanied by a four-volume set of documentation.

The model was utilized for numerous stormwater characterization studies, with little further testing, and default values were often used for parameters with no attempt at calibration.

Discuss what was wrong in this computer program development and its subsequent usage. Would you be confident in predictions made by such a model? How could you increase your confidence in its output?

3-12. Since 1963, the American Association of State Highway Officials has suggested using a panel of raters to evaluate road conditions. The rating scheme suggests using at least 50 raters, who rate serviceability on a zero to 5 scale. After adjustment for rater bias, a passenger rating index (PRI) is assigned. A new rear axle bump measuring device has been proposed to replace the panel of raters in determining the serviceability of road surfaces. This measuring device produces an index, called the E value, for each pass over the surface. To calibrate the new device, five test sections have been designated, and a panel of raters used to estimate PRI values for each section.

You are assigned the task of determining if the new measuring device can take the place of the panel of raters. If it can, you are to develop a calibration function that will covert the E value to the more familiar PRI. The following data are given:

Test section	E value ($\times 10^{-3}$)	PRI
62	410.8	4.101
	404.4	4.101
	393.2	4.101
41	360.5	4.777
	360.7	4.777
	358.6	4.777
34	1536.3	0.667
	1555.0	0.667
	1584.7	0.667
65	997.8	2.420
	1000.9	2.420
	894.4	2.420
36	1133.0	1.561
	1142.8	1.561
	1134.2	1.561

How certain are you of your results? Also, discuss any weaknesses in the data or methodology, and how they might be remedied.

3-13. You have been given the task of modeling a certain physical process, as part of a design project. You are aware that there are two commercially available computer models that are supposed to model similar physical processes. Describe how you would go about evaluating these models to decide if either of them is adequate, and if both are, which is best. If one is chosen, what steps would you take to make sure the selected model is applied in the best possible manner?

3-14. Two sets of data are given below. How sure are you that the mean values of the sets are different? What is your overall conclusion about the sets?

Set 1	Set 2
31	33
30	31
32	35
29	37
33	29
28	
34	

3-15. You have been given the following data:

X	Y
1	2
2	2
2.2	3
3	2.5
3.5	4
4	2
4	5
5	3
5.1	7

(a) Find the best fit line through this data, and make a statement regarding the significance of the correlation.

(b) Find the standard error of estimate.

(c) How confident are you that the actual slope of the regression line (a) lies between the calculated value of $a \pm 0.5$?

(d) Find the 90% confidence interval for the estimated value of Y when $X = 4.0$.

(e) What general comments can you make about this data set?

3-16. You are a county engineer and have three year's worth of PRI data relating to a 2-mi-long highway lane with a traffic volume of 500 vehicles/hr. The data are broken down into the number of lane miles that fall into a given PRI interval.

| | Number of lane miles in year | | |
PRI interval	1980	1979	1978
5.00	0.000	0.100	0.000
4.50	0.000	0.000	0.000
4.00	0.000	0.000	0.000
3.50	0.099	0.109	0.219
3.00	0.161	0.055	0.250
2.50	0.135	0.517	0.348
2.00	0.552	0.387	0.625
1.50	0.315	0.215	0.000
1.00	0.225	0.323	0.208
0.50	0.284	0.167	0.108
0.00	0.323	0.221	0.336

What inferences can be drawn from this data? Are you confident in your inference? What weaknesses are there in the data, and how might they be remedied?

3-17. Find the mean, median, and mode of the data given below. Which do you feel is the most appropriate measure of central tendency in this case, and why?

18	9	12	10	13	14	17	12
15	14	11	13	12	33	15	14
11	10	13	12	14	11	13	17
17	14	11	12	11	15	16	10
12	13	18	11	40	15	12	13
16	19	12	10	16	25		

PART TWO

THE PLANNING/DESIGN PROCESS: EVALUATION

Chapter 4

Development of System Models

INTRODUCTION

In Chapter 1, models were defined as simplifications of reality that are utilized to approximate actual systems. Properly formulated models can be used to generate and evaluate alternative courses of action in order to produce a desired system output. It is generally much easier and less expensive to manipulate a model than to manipulate the prototype system.

The primary responsibility of the modeler is to develop a model that is sufficiently realistic to provide useful information. Beyond that, the modeler must also recognize weaknesses in theory or gaps in background information, and not try to mask these problems with an elegant model which contains basic inconsistencies. In fact, it may be impossible to develop an adequate model under existing conditions. Recognition of this inability by the modeler is in itself valuable information, and may indicate the need for further research.

The total model environment consists of the evaluative criteria plus the model that generates the alternatives to be judged under the evaluative criteria. This chapter will provide a general overview of the development of evaluative criteria and development of graphical and/or mathematical models to generate alternative courses of action. Chapter 5 will develop several specific system models and associated evaluative criteria.

EVALUATIVE CRITERIA

Evaluative criteria are the tools by which the degree of attainment of objectives can be measured. While these criteria are essentially measures of effectiveness, they also provide a means of rank-ordering alternatives, or of determining if particular alternatives measure up to a predefined standard. Whenever possible, evaluative criteria should be quantitative.

There may be several possible evaluative criteria for a particular problem. The most appropriate ones must be chosen from among this group. The choice of evaluative criteria may have a significant influence on the development of problem constraints and the formulation of alternative feasible solutions.

Problems that have more than one identifiable objective may have multiple criteria as well. Multiple criteria can be explicitly incorporated into the model and evaluated through one of the multiobjective analysis techniques that will be described in Chapter 13. In other instances, it may be possible to incorporate some of the objectives into the model as constraints; or it may be possible to identify one objective that overrides others in importance, to develop criteria for that objective, and to neglect the objectives of lesser importance.

In some cases, it may not be possible to develop quantitative evaluative criteria, and the engineer will have to fall back on subjective evaluation. In all cases, evaluative criteria will be subject to some degree of uncertainty as to their representativeness and accuracy.

Appropriate criteria can include financial considerations, protection of public interests, and accrual of benefits from public works projects. Financial considerations can include costs to achieve some goal, or rewards that would accrue by reaching some goal. A required level of pollutant removal in a waste treatment plant is a criterion that reflects protection of public interest. Benefits such as the increase in recreational potential generated by a new reservoir, or the reduction in accident losses as a result of a highway rehabilitation project are appropriate criteria for judging public works projects. Disbenefits caused by the loss of previously available resources should also be included in the criteria.

OBJECTIVE FUNCTION

The objective function is the form of the evaluative criteria used in optimization problems. It provides a measure of the effectiveness of the system being depicted by the optimization model. The objective function can be either a linear or nonlinear mathematical function.

The decision variables are the independent variables that operate through the objective function coefficients (parameters) to establish a value for the objective function. This value of the objective function is the dependent variable for the optimization model. For example, in the objective function

$$Z = 3X_1 + 4X_2 \tag{4.1}$$

Z is the dependent variable, X_1 and X_2 are the decision variables, and 3 and 4 are parameters, commonly referred to as cost or gain coefficients. Some decision variables may not appear explicitly in the objective function because they do not add or subtract from the value of the objective function.

Cost or gain per unit of independent variable has to be determined by the modeler through reliance on past experience, published unit data, or statistical analysis of measured data. Any decision variables and parameters that the engineer has some degree of control over can be considered to be design variables, in that they can be varied to improve, or optimize, the design.

ALTERNATIVES

During development of evaluative criteria, it is important to conduct a preliminary screening of possible alternatives to determine, with as much certainty as possible, if each component is technologically and economically feasible. Answers to questions such as:

Do geologic conditions allow the construction of a high rise?

Is there sufficient demand for the power from a hydroelectric plant?

Will financing be available for the project?

may remove some alternatives from further consideration, or may indicate that the whole project is not worthwhile. Considerable savings in time and effort can be realized in this manner.

Alternatively, it is important to generate and evaluate as many feasible alternatives as possible, since the only alternatives that will be considered are those that are explicitly formulated. Feasible alternatives are those alternatives that meet at least the minimum established criteria and lie within the problem boundaries established during problem definition. Generally, some type of system model is used to generate these feasible alternatives for consideration.

MODELS

A model gives form to a problem and facilitates quantitative analysis by allowing quick, systematic generation and evaluation of alternatives. Information generated by a model can be fed forward to aid in decision making for planning and design. However, models can be equally valuable in generating information that may indicate feedback is necessary, and can aid in redefining the problems or indicating where additional data are needed.

Models can be used in systems analysis to study the response of the system to varying inputs abstractly, without having to manipulate the system itself. This is assuming that the model is derived from an existing system. For systems design, the model can be used to create a new system that will achieve the stated goals. The model can serve as a blueprint for building the new system.

A model is a conceived image of reality, and is generally a simplification. Graphical models represent systems in a two-dimensional visual display; as, for example, in schematic diagrams and construction drawings. Network diagrams of highway systems or of the activities involved in a construction project are examples of graphical models that assist in planning and design decision making. The latter type of graphical models are generally associated with a related mathematical model to assist in evaluation of alternatives.

Mathematical models represent systems through the language of mathematical symbols and relationships. Mathematical models which adequately describe a system can be used to predict future states of the system, and may assist in modification of the system to control its future behavior. Many systems can be adequately described by linear mathematical models. Although few natural systems are completely linear, many are close enough to warrant the linear assumption. This is a desirable situation, because linear systems are the easiest to analyze mathematically. Some linear systems can be modeled by a set of equations with no more variables than equations. Such systems have a unique solution which can be arrived at through the application of linear algebra principles. Other linear systems are modeled by inequalities that form the constraint boundaries on the system. These systems do not have unique solutions, but their evaluative criteria (objective function) can be optimized through application of linear programming theory. More complex systems can only be modeled by nonlinear mathematical functions, a system of nonlinear functions, or a system of stochastic functions. These models are much more complex mathematically, and often do not fit into standardized solution procedures.

Mathematical models are generally of the descriptive or prescriptive type. Descriptive models analyze and present information on a certain state of a system. All interpretation of results and recommendations for action are left up to the user. Prescriptive models, on the other hand, select the most favorable alternative or alternatives based on criteria incorporated into the model by the modeler. Interpretation of results still needs to be undertaken by the user before final recommendations are made.

Development of models may require the acquisition of data by observation and measurement. At other times, the type and amount of data available may limit the type and complexity of model that can be developed. One must be wary of models developed on insufficient data. Although they may look elegant, the lack of sufficient data leaves the user powerless to confirm or reject their validity.

The term validity, as used here, refers to the ability of the model to predict relative effects of alternative courses of action with sufficient accuracy to permit decision making with an acceptable level of confidence in the results.

Development of realistic mathematical models requires a high level of technical and professional knowledge on the part of the modeler. The forces, constraints, and values at work in the problem environment must be understood and properly expressed. Modelers must be able to visualize problems and use their creative ability to synthesize past experience in developing the proper type of model for the problem at hand. The modeler must possess sufficient knowledge of mathematical theory to adequately depict the physical phenomena and processes associated with the system. Many of the mathematical models used in system analysis require skill with using matrix notation and matrix operations. Skill at modeling requires familiarity with mathematical notation so that the wording and form of a model can be recognized and understood by other engineers. Skills required to develop any type of model are acquired through repeated practice on numerous problems.

The modeler always seeks the simplest model that will adequately describe the phenomena being modeled. Often, no single model is adequate, and combinations must be used. Any model, or combination of models, like a problem statement, should be continuously tested throughout its formulation and use. Modelers should maintain a good degree of mutual skepticism for others' models, and be able to accept constructive criticism, or even to make self-criticism, of their own models. Such a healthy attitude will result in higher quality models being developed, and will help to avoid rigidity, or the stereotyping of modeling problems, to the detriment of originality and innovation.

Figure 4.1 depicts the modeling process as a graphical model. This model was adapted from a similar illustration of the modeling process presented by Meredith

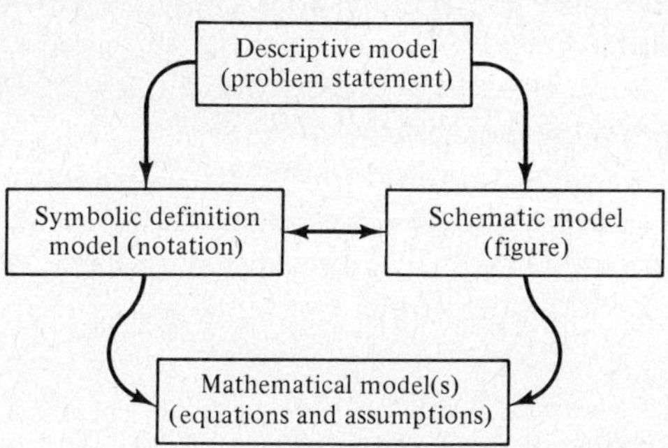

Figure 4.1   Modeling process.

(1973). This figure presents four submodels as making up the modeling process. The descriptive model is essentially the problem definition stage in which the problem is properly posed, and constraints are set on the boundaries of the problem environment. The symbolic definition and schematic models go hand in hand. The symbolic model defines the notation that will be used to identify the relevant variables from the descriptive submodel as they are used in later modeling and analysis. The schematic model represents the system graphically, using elements of the symbolic model to identify pertinent parts. The free body diagram in solid mechanics and the control volume in fluid mechanics are two examples of schematic models. Flow network models and project activity sequencing diagrams are two new types that will be developed for later usage in this text. The actual mathematical modeling stage involves development of mathematical expressions to represent the physical processes going on within the system, within the framework of the three other submodels. Assumptions may have to be made about the state of certain aspects of the system to facilitate development of a mathematical model. Such assumptions should be clearly stated so evaluation of their influence on results can be estimated.

Development of any of the submodels may require additional research into the underlying theory of the physical principles involved. Also, additional research into mathematical solution techniques may have to be undertaken if any meaningful results can be obtained from the mathematical model produced.

Once the model is developed, it must be calibrated and verified. Then it can be used to predict alternative outcomes for the system based on alternative inputs or system transformations. Outcomes are analyzed and fed forward or fed back as required.

A good model must be functional and succinct, in that it facilitates treatment of important elements of the system, free from unnecessary complications. Models must be of functional size and must be time saving. They must also be economically practical, in that the benefits to be gained by using the model outweigh the costs of model development and application. Frequently, model development costs are greater than the cost of running the model and using it for analysis. Any expenses involved in acquiring otherwise unneeded data for model calibration and verification should be included in the cost of model development.

Some common model deficiencies are that the model includes irrelevant or excludes relevant variables. Technical expertise in the physical principles involved is the best way to avoid these problems. A model can also have one or more relevant variables inaccurately evaluated. Careful research and attention to detail in gathering calibration/verification data can minimize these inaccuracies. A model can also contain incorrect functions or functional forms. Again, technical as well as mathematical expertise are required to minimize errors of this last type. The modeler must be able to recognize when outside consultation is necessary to close gaps in the available expertise.

System Constraints

All models are controlled by the physical, institutional, and economic constraints imposed on the system being modeled. The physical relationships among the components and subsystems of the system form several types of constraints. Component behavior constraints describe input–output relationships of the system. These are sometimes called the equations of state. Acceleration of a vehicle when the accelerator is depressed involves interaction among several subsystems, all of which are related by behavioral constraints. Distortion of a structural member by applied loads is also an example of a behavioral constraint. System compatibility constraints involve geometric relationships among components and subsystems. An example of this is the connection of truss members at gusset plates. Compatibility is maintained provided that the members remain connected. System equilibrium constraints indicate that the system conforms to the physical laws of conservation of mass and energy. Two examples from solid mechanics include the algebraic sum of forces at a joint equaling zero, and the sum of moments about any point in a structure equaling zero. Examples from fluid mechanics include the zero summation of head loss around a loop in a pipe distribution system and the continuity of flow at any junction point in the system.

Institutional and economic restrictions, as well as limits on the amounts of resources available can also place constraints on the system being modeled. Standards and regulations can limit system flexibility. Standards are limitations on systems properties established by some authority or agreement. They can be minimum acceptable levels of performance, such as minimum strength requirements for a certain type of steel, or maximum allowable levels such as maximum permissible discharge of pollutants into a receiving water. Standards can also place tolerance limits on physical properties. The American National Standards Institute is an example of a standard setting agency. Regulations are rules, ordinances, or laws which control conduct or performance. They may have nothing to do with the physical properties of the systems involved, but may control interaction between the system and other systems. Import–export laws and transportation carrier regulations are examples of this type.

Economic constraints can be determined either by the upper limit on the monetary resources available or by an economic feasibility study to show if the investment is warranted. Economic feasibility studies will be discussed in depth in Chapter 7. Basically, they involve estimation of all the costs and benefits associated with a system over its expected lifetime. This is often referred to as life cycle costing. Capital and operation/maintenance costs must be considered, and then contrasted with benefits and salvage value at the end of the system's useful life. Resource constraints may include material availability limitations, restrictions on the availability and quality of labor resources, and maintainability and reliability of equipment.

Graphical Models

Schematic diagrams are simple graphical models frequently used by engineers. Any two-dimensional representation of a system or component is a graphical model. Free body diagrams, isometric drawings, construction drawings, and road maps are all examples. Under certain conditions the graphical model can itself be used to generate and/or evaluate alternatives. A much more common occurrence is that the graphical model will be used to develop a mathematical model for quantitative analysis.

Graphical models should be laid out in a logical manner and should agree with nomenclature and symbolism that is generally accepted for the particular system being depicted. Enough information should be given to completely convey the existing conditions; however, extraneous material that could be confusing should be deleted.

Example 4.1 shows three different graphical models of a truss, each with a different purpose. Figure 4.2(a) merely shows the overall layout of the bridge and its surrounding environment. Figure 4.2(b) is a schematic model that can be used to develop a mathematical model to determine the reactions acting on the truss. Note that certain assumptions have been made in the development of the schematic model. The principal assumption is that the bridge is pin connected at the left end (location A) and is free to slide on the abutment at the right (location E). This is a good assumption, since bridge supports have to allow for thermal expansion and contraction. Analytically it is also a good assumption, for two pinned connections would produce a statically indeterminant structure that would be more difficult to analyze. The third graphical model, Figure 4.2(c), is an exploded free body diagram that could be used in conjunction with the method of joints to determine the force in each member of the truss. The symbolic model includes the force identifiers (F_{AB}, etc.) and the arrows indicating assumed compression or tension in each member. Note that the reactions at A and B have to be known before the method of joints model can be applied. The primary assumption made in this case is that all members are pin connected at the gusset plate, with no transfer of moment between members. This is an assumption that has been shown to be acceptable in practice. Note also that different assumptions have been made as to whether certain members are in tension or compression. When the mathematical model is developed for this system, it will have to take these differing assumptions into account.

Example 4.2 illustrates another type of graphical model that can be used to analyze resource distribution problems, more commonly referred to as transportation problems. The descriptive model includes data on needed resources, available resources, and delivered costs to each site. It also explicitly states the system goal of minimizing acquisition costs. The schematic (graphical) model depicted in Figure 4.3(b) simplifies the actual road network of Figure 4.3(a) into single arrows indicating transfer of resource from supplier to site. The cost associated with this transfer is found by determining the most economical route through the road network from any supplier to any site. Models for determining the most economical route between two points in a network such as this also involve the use of graphical models.

Example 4.3 illustrates one such problem. Fire vehicles want to use the quickest route to arrive at the location of a call. In this case the measure of economy is time of travel. In other problems the measure of economy may be cost, distance, or some combination of cost, time, and distance. Another measure of economy might be the maximization of flow through a network, subject to certain capacity constraints for the various links. The flow involved could be vehicles, a fluid, or any other medium that behaves in a similar manner. A different class of problems involved with management of nonrepetitive engineering projects uses the longest time path through a network of work activities to identify the critical activities that may delay the project. Each of these problems would use a graphical model similar to the one developed for Example 4.3.

Given the street layout shown in Figure 4.4(a), a logical model would be to reduce the street network into a network of links and nodes, as has been done in Figure 4.4(b). The links represent pertinent street segments, and the nodes, intersections where alternate paths either converge or diverge. Streets that would not appear in the solution have been left out of the model. Also, note that some links are unidirectional, indicating that the quickest route would not logically double back on itself. Other links are bidirectional, indicating possible crossover points between directed paths.

There exist graphical techniques for analyzing certain types of graphical models. However, most realistically sized problems are handled more efficiently by mathematical models and/or computerized solution techniques which are derived from the graphical model.

EXAMPLE 4.1 ⎯⎯⎯⎯⎯⎯⎯⎯⎯⎯⎯⎯⎯⎯⎯⎯⎯⎯⎯⎯⎯⎯⎯⎯⎯⎯⎯⎯⎯⎯⎯⎯

The truss bridge shown in Figure 4.2(a) is to span a small brook. Construct additional graphical models that could assist in determining the forces developed in each member of this truss, if a 10,000-lb load is applied to the joint at midspan.

EXAMPLE 4.2 ⎯⎯⎯⎯⎯⎯⎯⎯⎯⎯⎯⎯⎯⎯⎯⎯⎯⎯⎯⎯⎯⎯⎯⎯⎯⎯⎯⎯⎯⎯⎯⎯

Three construction sites each require a certain amount of resource. Four suppliers are available for the resource, but the delivered price depends on the supplier and the distance between the supplier and the construction site. The goal of the engineer in charge is to minimize the total cost of acquiring the needed resource. The following information has been gathered:

Resources available		Resources needed	
Supplier	Amount	Site	Amount
1	6	1	10
2	8	2	8
3	5	3	6
4	10		

The actual spatial relationship between the suppliers and sites is shown in Figure 4.3(a). The dashed lines indicate major highways that may be used for transport of the resource. Table 4.1 shows the delivered cost per unit of resource from each supplier to each site shown in the schematic model of Figure 4.3(b).

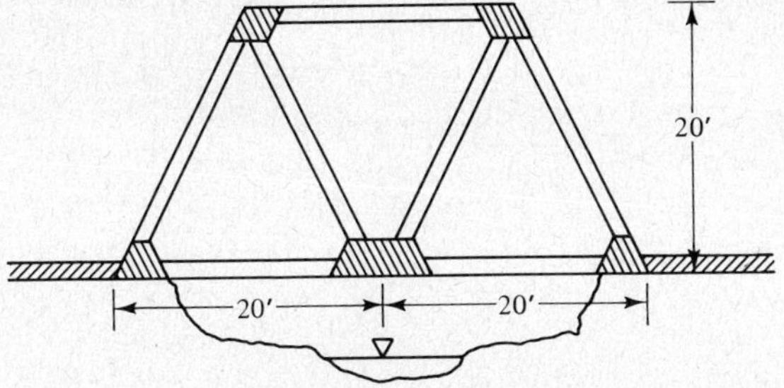

(a) Pictorial model of truss.

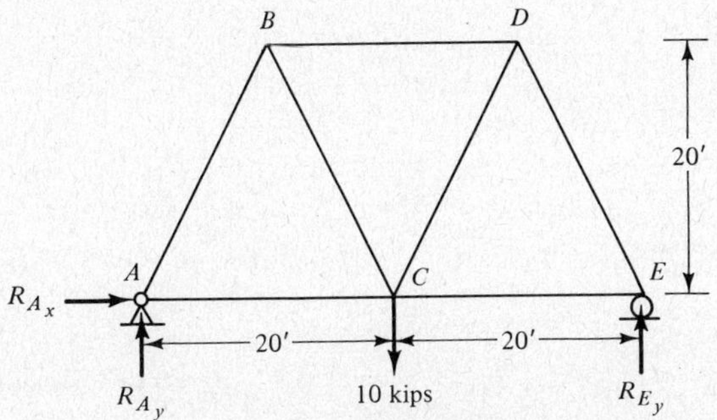

(b) Overall free body.

(c) Exploded free body

Figure 4.2 Truss analysis.

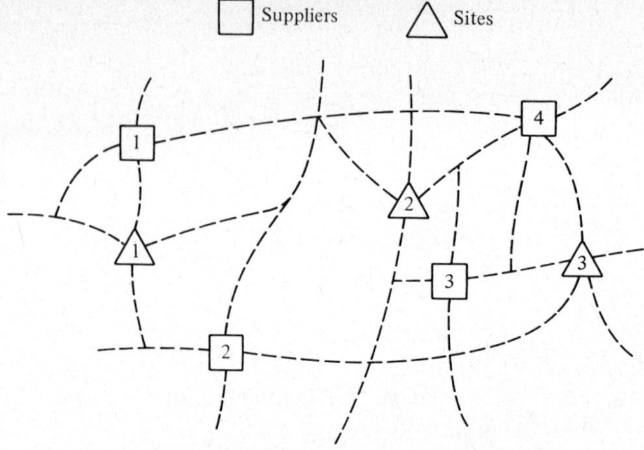

(a) Spatial relationship between suppliers and sites.

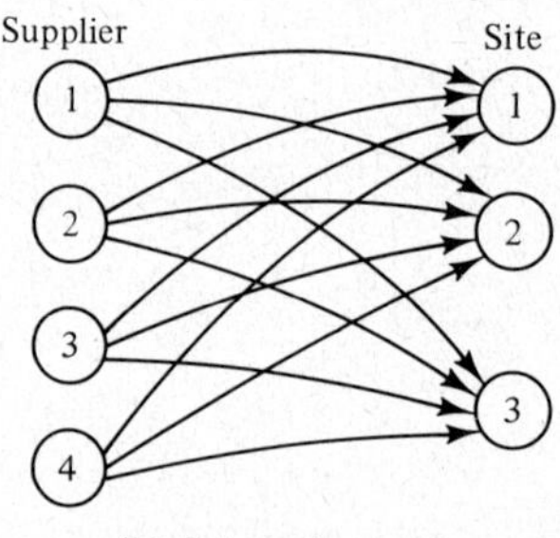

(b) Schematic model.

Figure 4.3 Transportation problem.

TABLE 4.1   DELIVERED COST/UNIT OF RESOURCE

		To site		
		1	2	3
From supplier	1	12	13	18
	2	13	15	17
	3	14	11	12
	4	15	12	13

EXAMPLE 4.3

A firehouse is located on the corner of Union Street and Scotia Boulevard. In order to speed up response time, the chief is trying to determine the quickest route from the firehouse to various destinations within the fire district. Top Notch Circle is the next destination to be routed. Using the street map given in Figure 4.4(a), develop a graphical model that could be used to find the quickest path between these two points.

Mathematical Models

The development of suitable mathematical models depends on properly posing the problem through the descriptive model, developing suitable graphical and symbolic models, and making valid assumptions that allow writing of mathematical relationships as an abstraction of the system.

Example 4.4 illustrates development of the mathematical model for the graphical, descriptive, and symbolic models for the bridge truss developed in Example 4.1. A system of equations can be written which represents the static balance of forces at each joint. This system can be solved sequentially, starting at joint A and moving to successive joints where there are no more than two unknown forces. Note that the coefficients in Equations 4.2 represent the directional cosines (or the unit vector components) of each force.

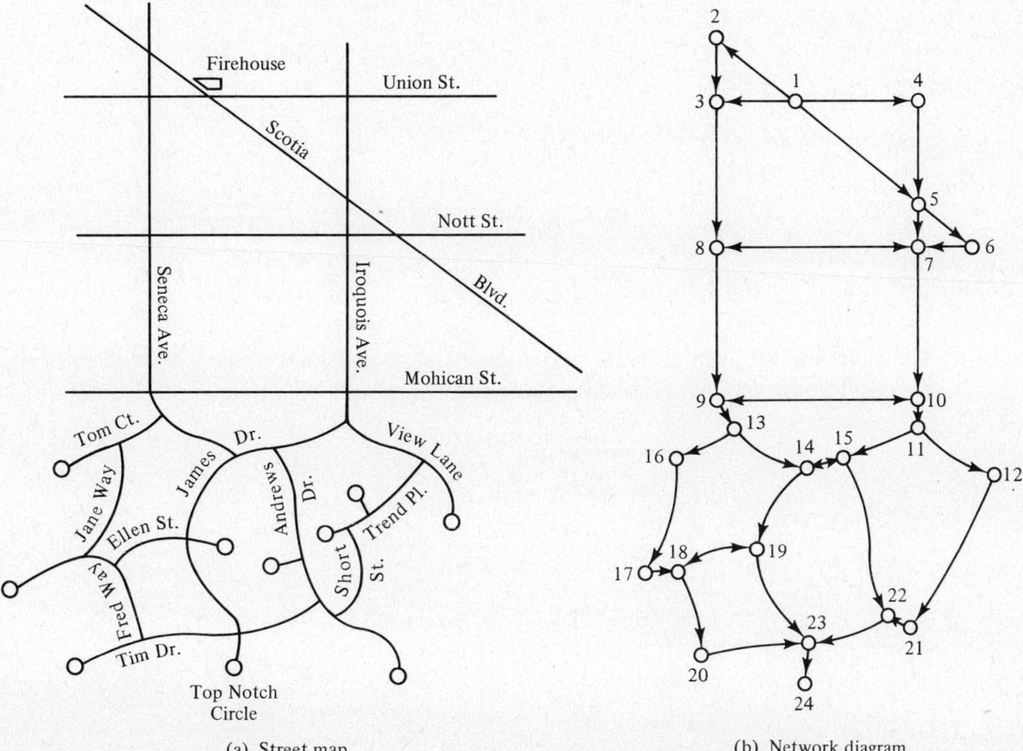

(a) Street map.   (b) Network diagram.

Figure 4.4 Network model.

EXAMPLE 4.4 ____

In Example 4.1, a graphical model for a seven-member truss was developed from descriptive and symbolic models. Assumptions made in developing the graphical model [Figure 4.2(c)] were also discussed. Now it is time to develop the mathematical model that can be used to analyze the forces developed in each member of the truss.

From the overall free body, the reactions at the abutments can be found through application of the theory of statics. As indicated in Figure 4.5, $R_{Ax} = 0$, $R_{Ay} = 5k$, and $R_{Ey} = 5k$. The reactions are now known force inputs for the mathematical model used to analyze the forces in the members of the truss.

Successive application of the method of joints would provide the following equations that could be used to find the forces in each member.

$$\text{At joint } A: \quad \sum F_x = 0 = F_{AC} - \frac{1}{\sqrt{5}} F_{AB}$$

$$\sum F_y = 0 = -\frac{2}{\sqrt{5}} F_{AB} + 5$$

$$\text{at joint } B: \quad \sum F_x = 0 = \frac{1}{\sqrt{5}} F_{AB} + \frac{1}{\sqrt{5}} F_{BC} - F_{BD}$$

$$\sum F_y = 0 = \frac{2}{\sqrt{5}} F_{AB} - \frac{2}{\sqrt{5}} F_{BC}$$

$$\text{at joint } C: \quad \sum F_x = 0 = -F_{AC} - \frac{1}{\sqrt{5}} F_{BC} + \frac{1}{\sqrt{5}} F_{CD} + F_{CE}$$

$$\sum F_y = 0 = -10 + \frac{2}{\sqrt{5}} F_{BC} + \frac{2}{\sqrt{5}} F_{CD}$$

$$\text{at joint } D: \quad \sum F_x = 0 = F_{BD} - \frac{1}{\sqrt{5}} F_{CD} - \frac{1}{\sqrt{5}} F_{DE}$$

$$\sum F_y = 0 = -\frac{2}{\sqrt{5}} F_{CD} + \frac{2}{\sqrt{5}} F_{DE}$$

$$(4.2)$$

The equations for joint E would be redundant, so they are not needed. A more compact and utilitarian way of expressing this problem would be through matrix notation.

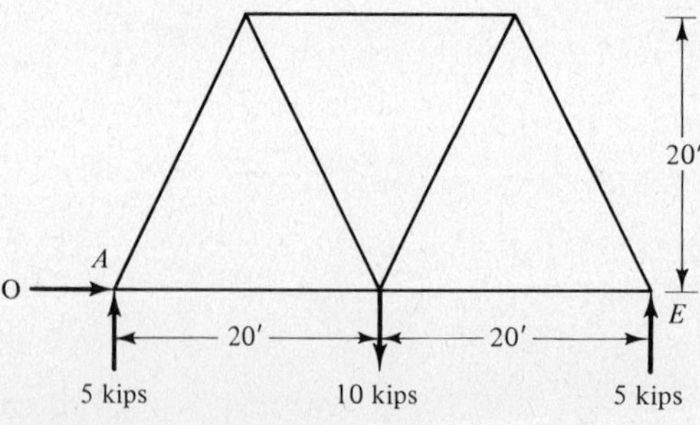

Figure 4.5 Overall free body with reactions.

Examination of the joint equations A through D reveals that one of the eight equations is also redundant since there are only seven unknown forces. Thus, the following set of seven equations completely specifies the forces generated in the truss system for the applied load given.

$$
\left.
\begin{aligned}
F_{AC} - \frac{1}{\sqrt{5}} F_{AB} &= 0 \\[4pt]
\frac{2}{\sqrt{5}} F_{AB} &= 5 \\[4pt]
\frac{1}{\sqrt{5}} F_{AB} + \frac{1}{\sqrt{5}} F_{BC} - F_{BD} &= 0 \\[4pt]
\frac{2}{\sqrt{5}} F_{AB} - \frac{2}{\sqrt{5}} F_{BC} &= 0 \\[4pt]
-F_{AC} - \frac{1}{\sqrt{5}} F_{BC} + \frac{1}{\sqrt{5}} F_{CD} + F_{CE} &= 0 \\[4pt]
\frac{2}{\sqrt{5}} F_{BC} + \frac{2}{\sqrt{5}} F_{CD} &= 10 \\[4pt]
F_{BD} - \frac{1}{\sqrt{5}} F_{CD} - \frac{1}{\sqrt{5}} F_{DE} &= 0
\end{aligned}
\right\} \quad (4.3)
$$

In matrix notation, this set of equations becomes

$$
\begin{bmatrix}
1 & -\frac{1}{\sqrt{5}} & 0 & 0 & 0 & 0 & 0 \\[6pt]
0 & \frac{2}{\sqrt{5}} & 0 & 0 & 0 & 0 & 0 \\[6pt]
0 & \frac{1}{\sqrt{5}} & \frac{1}{\sqrt{5}} & -1 & 0 & 0 & 0 \\[6pt]
0 & \frac{2}{\sqrt{5}} & -\frac{2}{\sqrt{5}} & 0 & 0 & 0 & 0 \\[6pt]
-1 & 0 & -\frac{1}{\sqrt{5}} & 0 & \frac{1}{\sqrt{5}} & 1 & 0 \\[6pt]
0 & 0 & \frac{2}{\sqrt{5}} & 0 & \frac{2}{\sqrt{5}} & 0 & 0 \\[6pt]
0 & 0 & 0 & 1 & -\frac{1}{\sqrt{5}} & 0 & -\frac{1}{\sqrt{5}}
\end{bmatrix}
\begin{bmatrix}
F_{AC} \\[6pt] F_{AB} \\[6pt] F_{BC} \\[6pt] F_{BD} \\[6pt] F_{CD} \\[6pt] F_{CE} \\[6pt] F_{DE}
\end{bmatrix}
=
\begin{bmatrix}
0 \\[6pt] 5 \\[6pt] 0 \\[6pt] 0 \\[6pt] 0 \\[6pt] 10 \\[6pt] 0
\end{bmatrix}
\quad (4.4)
$$

or
$$
\mathbf{AF} = \mathbf{L} \tag{4.5}
$$

If the problem has been properly posed and a solution exists, the inverse of the **A** matrix can be found and used to solve for the forces in the members. Utilizing the techniques described in Appendix A,

$$\mathbf{A}^{-1} = \begin{bmatrix} 1 & \dfrac{1}{2} & 0 & 0 & 0 & 0 & 0 \\[2ex] 0 & \dfrac{\sqrt{5}}{2} & 0 & 0 & 0 & 0 & 0 \\[2ex] 0 & \dfrac{\sqrt{5}}{2} & 0 & -\dfrac{\sqrt{5}}{2} & 0 & 0 & 0 \\[2ex] 0 & 1 & -1 & -\dfrac{1}{2} & 0 & 0 & 0 \\[2ex] 0 & -\dfrac{\sqrt{5}}{2} & 0 & \dfrac{\sqrt{5}}{2} & 0 & \dfrac{\sqrt{5}}{2} & 0 \\[2ex] 1 & \dfrac{3}{2} & 0 & -1 & 1 & -\dfrac{1}{2} & 0 \\[2ex] 0 & \dfrac{3}{2(\sqrt{5})} & -\sqrt{5} & -\sqrt{5} & 0 & -\dfrac{\sqrt{5}}{2} & -\sqrt{5} \end{bmatrix} \tag{4.6}$$

and
$$\mathbf{A}^{-1}\mathbf{L} = \mathbf{F} \tag{4.7}$$

provides the solution

$$\begin{bmatrix} F_{AC} \\ F_{AB} \\ F_{BC} \\ F_{BD} \\ F_{CD} \\ F_{CE} \\ F_{DE} \end{bmatrix} = \begin{bmatrix} 2.50 \\ 5.59 \\ 5.59 \\ 5.00 \\ 5.59 \\ 2.50 \\ 5.59 \end{bmatrix} \tag{4.8}$$

The most useful model is to write the system in matrix notation, find the solution by first determining the inverse of the coefficient matrix, and then, using matrix multiplication, to determine the solution through Equation 4.7. This method has a distinct advantage over the sequential solution technique. The coefficient matrix, and thus the matrix inverse, is only a function of the truss geometry. Once the inverse matrix is found, it can be used to determine member forces for any combination of loadings on the structure. The load vector, $\mathbf{L}$, is composed of the following:

$$\mathbf{L} = \begin{bmatrix} R_{Ax} \\ R_{Ay} \\ L_{Bx} \\ L_{By} \\ L_{Cx} \\ L_{Cy} \\ L_{Dx} \end{bmatrix} \tag{4.9}$$

Example 4.2 developed a graphical model for shipment of a particular resource for a number of suppliers to a number of locations where the resource is needed. Data on the availability of the resource from the four suppliers and on the needs of the three sites were given in the descriptive model for Example 4.2. Figure 4.3(b) showed the schematic diagram of possible shipping paths between suppliers and sites. Table 4.1 gave the delivered costs over each of these shipping paths. A symbolic model can be easily developed to complement the descriptive and graphical models. Let X_{ij} equal the number of units of resource that are acquired from source location i for site j. The source locations can be considered origins and the sites can be considered destinations. If c_{ij} is the delivered cost per unit of resource from origin i to destination j, and the system objective is to minimize the total delivered cost, then an objective function of the form

$$\min Z = \sum_{i=1}^{n} \sum_{j=1}^{m} c_{ij} X_{ij} \tag{4.10}$$

where n = number of suppliers
    m = number of sites

can be written with Z representing the value of the objective function (the aggregate cost of all the required resource to the proper site) for any combination of X_{ij} that satisfy the constraints of the problem. For the particular problem of Example 4.2, the objective function is

$$\min Z = c_{11}X_{11} + c_{12}X_{12} + c_{13}X_{13} + c_{21}X_{21} + c_{22}X_{22} + c_{23}X_{23}$$
$$+ c_{31}X_{31} + c_{32}X_{32} + c_{33}X_{33} + c_{41}X_{41} + c_{42}X_{42} + c_{43}X_{43} \tag{4.11}$$

or $\min Z = 12X_{11} + 13X_{12} + 18X_{13} + 13X_{21} + 15X_{22} + 17X_{23}$
$$+ 14X_{31} + 11X_{32} + 12X_{33} + 15X_{41} + 12X_{42} + 13X_{43} \tag{4.12}$$

The total amount that is shipped from any supplier is bounded by the amount available at that supply location. However, not all units of resource have to be shipped from each supplier. For this problem the total supply available is 29 units, while the total required units at the sites is 24 units. Five units will be left over at one of the suppliers after all requirements are met. On the other hand, excess resource is not going to be shipped to any site, but the required amount can be met in all cases. Thus, the objective function (and the problem) is subject to the following mathematical constraints:

$$
\begin{aligned}
\text{amount shipped } & X_{11} + X_{12} + X_{13} & & & & & \leq 6 \\
\text{from supplier 1} \\[4pt]
\text{amount shipped } & & X_{21} + X_{22} + X_{23} & & & & \leq 8 \\
\text{from supplier 2} \\[4pt]
\text{amount shipped } & & & X_{31} + X_{32} + X_{33} & & & \leq 5 \\
\text{from supplier 3} \\[4pt]
\text{amount shipped } & & & & X_{41} + X_{42} + X_{43} & & \leq 10 \\
\text{from supplier 4} \\[4pt]
\text{amount shipped } & X_{11} & + X_{21} & + X_{31} & + X_{41} & & = 10 \\
\text{to site 1 from} \\
\text{all sources} \\[4pt]
\text{amount shipped } & X_{12} & + X_{22} & + X_{32} & + X_{42} & & = 8 \\
\text{to site 2 from} \\
\text{all sources} \\[4pt]
\text{amount shipped } & X_{13} & + X_{23} & + X_{33} & + X_{43} & & = 6 \\
\text{to site 3 from} \\
\text{all sources}
\end{aligned} \tag{4.13}
$$

An additional constraint placed on this type of problem is that all $X_{ij} \geq 0$. This is imposed because negative amounts shipped have no physical significance. The transportation problem is a particular type of linear programming optimization problem. A corollary problem to the above is one in which there are intermediate transfer points between suppliers and the locations where the resource is needed, or there is transfer of resource between suppliers and possibly between destination sites. Problems of this type are referred to as transshipment problems. Other types of linear programming problems will be developed in Chapter 5. Solution techniques will be developed in Chapter 6.

Modeling of many practical systems requires the representation of graphical networks by mathematical models. Figure 4.6 shows a smaller network similar to the road network of Figure 4.4(b). The numbers on each link represent mileages. Mathematical modeling of this graphical model must identify the connectivity among links and nodes of the network, and would have to account for the undirected links between ②–③ and ③–④. A mathematical abstraction referred to as the link–node incidence matrix, sometimes referred to as the branch–node incidence matrix, can be used to

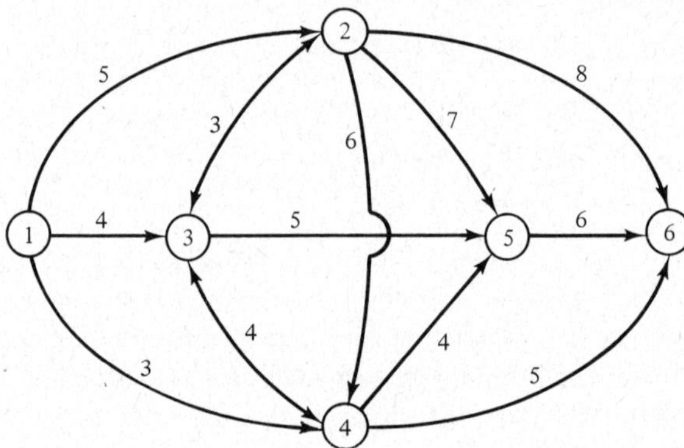

(a) Graphical network model.

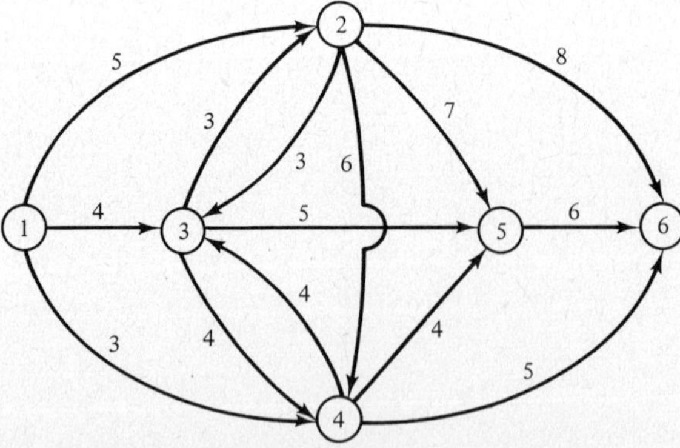

(b) Network as modeled by link-node incidence matrix.

Figure 4.6 Road network model.

represent a directed link network. This is a type of topological matrix. If the undirected paths are represented by a pair of directed paths, then a link–node incidence matrix can be developed. If a directed link leaves a node, it is assigned a value of -1. Links that enter nodes are given a value of $+1$. If a link is not incident to a particular node, it has a value of zero. Table 4.2 shows the link–node incidence matrix that can be developed for Figure 4.6(b). Inspection of this matrix will reveal that it is 1 degree column dependent. To demonstrate, if columns 1 through 5 are added together and multiplied by -1, they will equal column 6. Therefore, one of the node columns could be eliminated with no loss of information. If column 6 is eliminated, the link–node incidence matrix becomes

$$\mathbf{A} = \begin{bmatrix}
-1 & 1 & 0 & 0 & 0 \\
-1 & 0 & 1 & 0 & 0 \\
-1 & 0 & 0 & 1 & 0 \\
0 & -1 & 1 & 0 & 0 \\
0 & 1 & -1 & 0 & 0 \\
0 & -1 & 0 & 1 & 0 \\
0 & -1 & 0 & 0 & 1 \\
0 & -1 & 0 & 0 & 0 \\
0 & 0 & -1 & 1 & 0 \\
0 & 0 & 1 & -1 & 0 \\
0 & 0 & -1 & 0 & 1 \\
0 & 0 & 0 & -1 & 1 \\
0 & 0 & 0 & -1 & 0 \\
0 & 0 & 0 & 0 & -1
\end{bmatrix} \tag{4.14}$$

A mathematical model can be developed from the transpose of the incidence matrix that will be used in a subsequent optimization model. The model will be in the form

TABLE 4.2   LINK–NODE INCIDENCE MATRIX

Link	Node 1	2	3	4	5	6
1–2	−1	1				
1–3	−1		1			
1–4	−1			1		
2–3		−1	1			
3–2		1	−1			
2–4		−1		1		
2–5		−1			1	
2–6		−1				1
3–4			−1	1		
4–3			1	−1		
3–5			−1		1	
4–5				−1	1	
4–6				−1		1
5–6					−1	1

$$\mathbf{A}^{\mathrm{T}}_{(5\times14)}\mathbf{X}_{(14\times1)} = \mathbf{S}_{(5\times1)} \tag{4.15}$$

where

$$\mathbf{X} = \begin{bmatrix} X_{12} \\ X_{13} \\ X_{14} \\ X_{23} \\ X_{32} \\ X_{24} \\ X_{25} \\ X_{26} \\ X_{34} \\ X_{43} \\ X_{35} \\ X_{45} \\ X_{46} \\ X_{56} \end{bmatrix} \qquad \mathbf{S} = \begin{bmatrix} 1 \\ 0 \\ 0 \\ 0 \\ 0 \end{bmatrix} \tag{4.16}$$

and all X_{ij} are either $= 1$ or 0. If link ij is used, then $X_{ij} = 1$. Otherwise $X_{ij} = 0$. Vector $\mathbf{S}$ indicates that no links enter node 1, that is, it is the starting point. Nodes 2 through 5 are intermediate nodes; therefore, the algebraic sum of entering and leaving paths must be zero. Note that the model of Equation 4.15 does not contain any information about the length of each link. That will be added when this model is modified to become an optimization model.

Another example of mathematical model development is illustrated in Example 4.5. After definition of system goals and objectives, models are developed to assess the degree of objective attainment. All four types of submodels are included. Alternative mathematical models are evaluated to determine which would be the most appropriate to analyze the state of the system. If different goals and objectives had been defined, different models would have been appropriate. This example also illustrates the importance of acquiring sufficient system response data to confidently evaluate alternative models.

The preferred model developed in Example 4.5 predicts changes in estuary dissolved oxygen caused by entry of pollutants into the estuary. This model would be part of a larger model which would be used to evaluate how or even if the objective of maintaining dissolved oxygen levels capable of supporting aquatic life could best be met. In the larger system model, minimum dissolved oxygen level would be a constraint, as would be the maximum amounts of pollutants that could be removed by particular treatment processes. Realistic criteria would include the total costs of meeting system objectives. Alternatives would be ranked in reverse order of their costs.

EXAMPLE 4.5 ———————————————————————————————

A river estuary system is depicted in Figure 4.7. (An estuary is a portion of a river system that is under the influence of ocean tides.) Pollutants are discharged into the estuary from the tributary and primary inflows as well as from the cities. Phytoplankton and zooplankton use pollutants in their metabolic processes, while at the same time

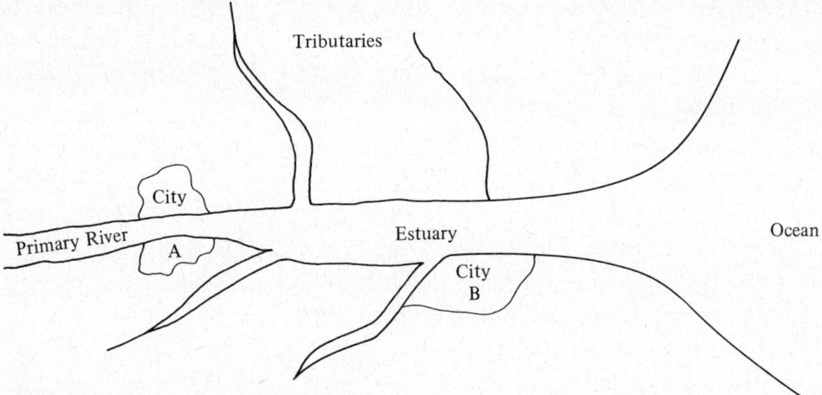

Figure 4.7 Estuary sketch map.

they are consuming dissolved oxygen from the estuary. Dissolved oxygen is replaced through diffusion through the air–water interface.

A goal of maintaining water quality at a level that will permit fishing and recreational use of the estuary has been established. Possible objectives associated with this goal are:

1. Maintaining dissolved oxygen levels that will support aquatic life.
2. Keeping levels of toxic substances within acceptable levels.
3. Maintaining an aesthetically pleasing environment.

The objectives listed are not necessarily mutually exclusive: all three could be accommodated, but each might not necessarily be optimized.

An objective function and mathematical model could be developed to assist in meeting each of these objectives. The objective function for the first objective might logically be to maintain a specified dissolved oxygen level in the estuary with minimum investment by all discharging parties. The maintenance of a level of dissolved oxygen that would provide for the maintenance of aquatic life would be a primary constraint. Other constraints would involve the dynamics of pollutant removal at each of the discharge points. Finally, a model would have to be developed to relate the levels of pollutant discharge to the concentration of dissolved oxygen expected in the estuary. The present development concentrates on this model.

The assumption is made that the river can be broken into discrete segments for analysis. Each segment is assumed to have a uniform geometry, and all lateral inflow is assumed to enter each segment only at the upstream end of that segment. Logical junction points for the segments are where there are tributaries entering the estuary or where there are significant changes in geometry. A final assumption made is that photosynthesis is not significant within the estuary.

Two competing processes are at work in the estuary. The pollutants exert a biochemical oxygen demand (BOD) that depletes the dissolved oxygen (DO) level in the estuary. This is accomplished through metabolism of the pollutants by microorganisms. At the same time, dissolved oxygen is reentering the estuary through the process of diffusion through the air–water interface, and from the inflows to the estuary.

The symbolic definition model for this problem would contain the following variables:

L_i = BOD concentration in segment i. This is also the concentration that will be transported to the downstream segment. Concentration is the demand per unit volume.

Q_i = flowrate between segments i and $i + 1$, $= Q_{i-1} + Q'_i$

P_i = pollutant concentration in tributary inflow to segment i

Q'_i = tributary flowrate into segment i

D'_i = tributary DO deficit

V_i = volume of segment i

D_i = DO deficit in segment i, $= C_{si} - C_i$

C_{si} = saturation concentration of DO in segment i

C_i = DO concentration in segment i

Schematic models for this problem are shown in Figure 4.8. Figure 4.8(a) shows a schematic of the segmented estuary. Figures 4.8(b), (c), and (d) show schematic models for three possible mathematical models to depict the interaction between the BOD and the DO concentration in each segment of the estuary. The model depicted by Figure 4.8(b) treats each segment as a completely mixed reactor. All contents within the reactor are assumed to be mixed together with an even concentration throughout. Reactions are taking place that reduce the concentration of BOD, which in turn increases the DO deficit. This is the simplest of the mathematical formulations, with the solution having the form

$$D_i = \frac{t_d D_i \text{ in}}{t_d + k_2} + \frac{k_1 t_d L_i \text{ in}}{(t_d + k_2)(t_d + k_1)} \tag{4.17}$$

where k_1 = BOD reaction rate (time^{-1}),

k_2 = DO replacement rate (time^{-1}), and

t_d = 1/(travel time through segment)

A more complicated model is the series of plug flow reactors depicted in Figure 4.8(c). For this formulation, it is assumed that each slug entering the estuary retains its identity, and reactions take place within the slug. The mathematical model would have the form

$$D_i = \frac{k_1}{k_2 - k_1} (L_i \text{ in})(e^{-k_1 t} - e^{-k_2 t}) + (D_i \text{ in})e^{-k_2 t} \tag{4.18}$$

where t is the time of travel through the segment

The most complicated of the three formulations is the series of plug flow reactors with diffusion depicted by Figure 4.8(d). In this model, the reactions are the same as in the previous model; however, there is diffusion of both BOD and DO between slugs. Solution of the simultaneous differential equations associated with the plug flow with diffusion model results in the following equation:

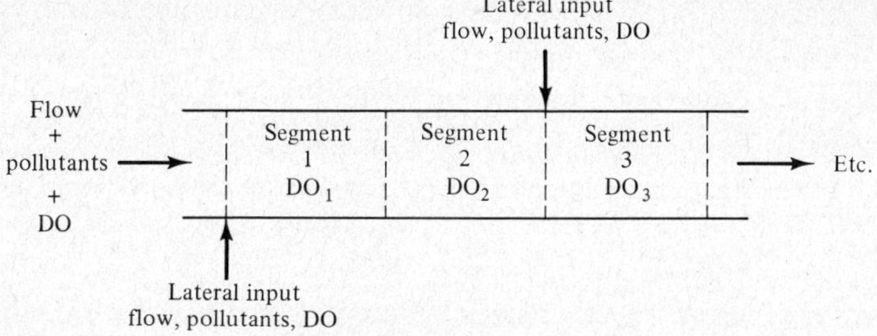

(a) Segmentation of estuary.

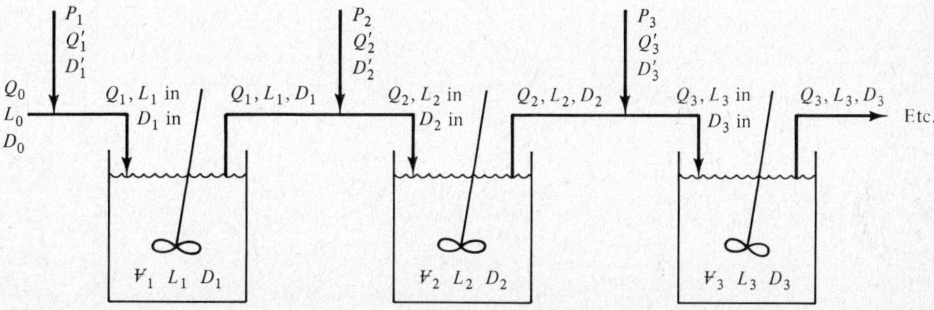

(b) Modeled as series of completely mixed reactors.

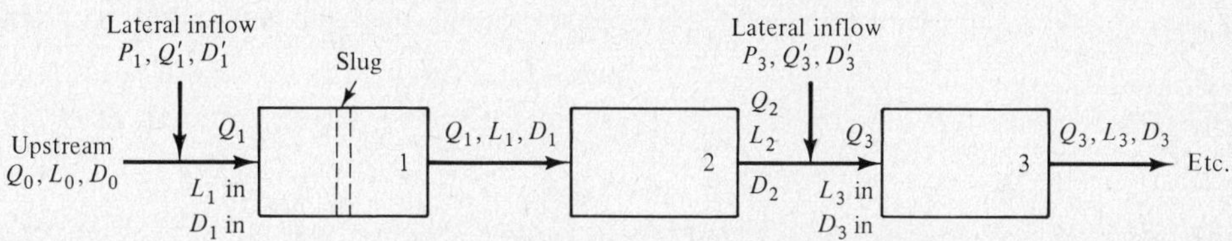

(c) Modeled as series of plug flow reactors.

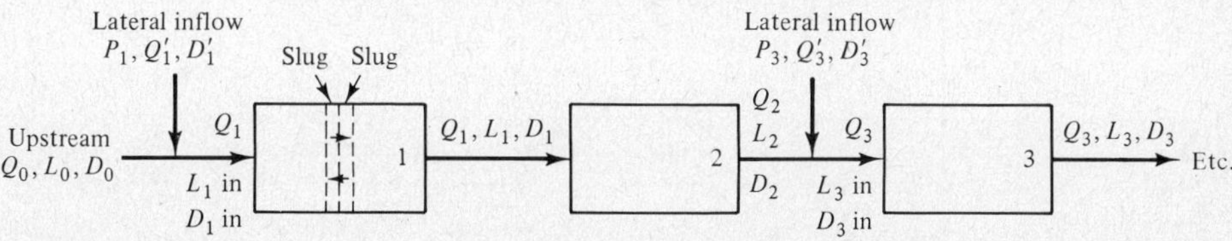

(d) Modeled as series of plug flow reactors with diffusion.

Figure 4.8 Schematic model.

$$D_i = \left\{ D_i \text{ in } - \frac{k_1 L_i \text{ in}}{(k_2 - k_1)(1 + 4k_1 E/u_i^2)^{1/2}} \right\} \exp\left\{ \frac{u^2}{2E} \left[1 - (1 + 4k_2 E/u_i^2)^{1/2} \right] t \right\}$$

$$+ \left\{ \frac{k_1 L_i \text{ in}}{(k_2 - k_1)(1 + 4k_1 E/u_i^2)^{1/2}} \right\} \exp\left\{ \frac{u^2}{2E} \left[1 - (1 + 4k_1 E/u_i^2)^{1/2} \right] t \right\} \qquad (4.19)$$

where $\quad u_i$ = net velocity of flow through segment = Q_i/A_i
$\quad\quad\quad\quad E$ = diffusion constant (time)
$\quad\quad\exp\{n\} = e^n$

Figure 4.9 shows a comparison of the three models with three known data points. Data used to generate these plots were taken from the Delaware River estuary. Two different values of the BOD reaction rate were used to generate plots (*a*) and (*b*). It appears that the plug flow with diffusion model is superior; however, there are not enough data points to make a quantitative comparison between the models. Also, no observed data values cover the critical low DO section of the estuary.

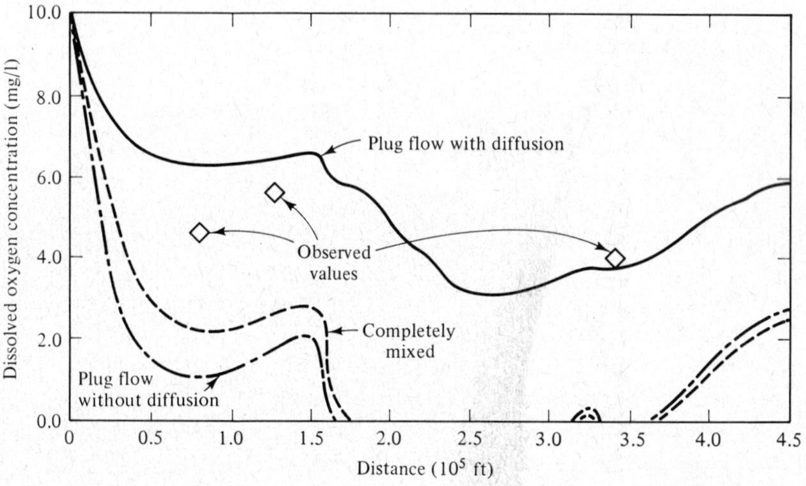

(a) Dissolved oxygen sag curve. k_1 = 0.23/day.

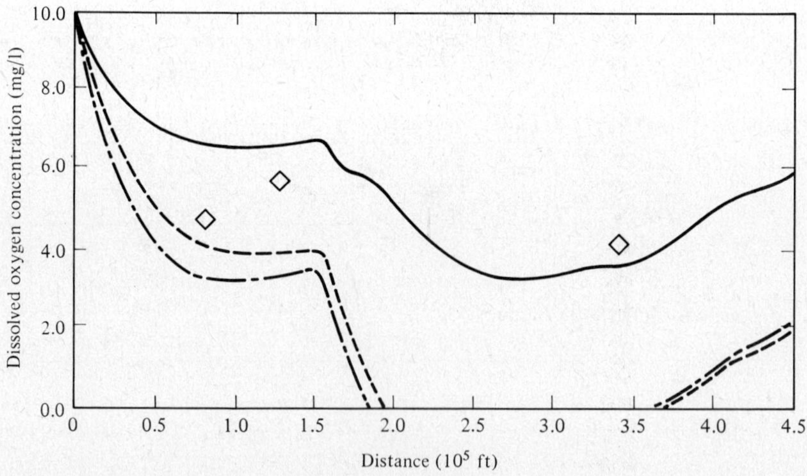

(b) Dissolved oxygen sag curve. k_1 = 0.1/day.

Figure 4.9   Comparison of models.

The two values of k_1 used are both values that appear in literature covering the Delaware River estuary. A preliminary calibration comparison indicates that a k_1 value of 0.23 seems to be slightly better; however, again more data are needed to make a quantitative judgment.

FRAMEWORK FOR ANALYSIS OF ALTERNATIVES

System models themselves form much of the framework for the analysis of alternatives. However, the engineer must also apply judgment in the use of models to avoid imprudent, or even infeasible, system recommendations.

Models should always be tested for reasonableness before doing any analysis of alternatives. This testing can be done through the formal calibration and verification processes described in Chapter 3, or may be checked against theory. The question to be answered is how closely does the model reflect the real world situation?

The best theoretical model of a system is sometimes difficult or infeasible to use from an operational standpoint. In that case, the engineer must evaluate how the model might be modified to facilitate analysis. It may be possible to approximate nonlinear functions by linear functions, either continuous, or piecewise continuous. An iterative solution may be required to correct for the approximation. Variables having ranges of values to which the solution is not sensitive may be treated as constants by assigning average values to them. Alternatively, variables to which the solution is believed to be insensitive could be left out of preliminary models. Later, it is necessary to explore the sensitivity of the solution to these variables.

Different evaluative procedures are useful for different types of problems. Therefore, the engineer must take into account the evaluative criteria developed and the mathematical form and complexity of the system model generating the alternatives when choosing an evaluative procedure. Several evaluative procedures will be discussed in subsequent chapters. Once the best procedure has been established, care must be taken to follow that procedure. The end product will be a rank ordering of alternatives; and where possible, a mathematically optimum solution subject to the given constraints.

SUMMARY

System models can be invaluable in application of the systems approach, and are virtually essential in application of systems analysis techniques. Any model is an abstraction and simplification of the prototype system. To be useful, the model must realistically portray the system, while at the same time remain simple enough to be used as a practical tool. The engineer can use the model to generate and compare numerous alternative courses of action without having to manipulate the prototype system. While recognizing the value of a model, the engineer must nevertheless demand verification of its applicability and validity, and remain cognizant of its assumptions and inherent weaknesses when evaluating and interpreting results.

Development of realistic models requires a thorough understanding of the applicable physical principles, and mastery of the tools of mathematics. Engineering experience and practice at developing models are also important prerequisites for an engineer to become an effective modeler.

APPLICATIONS EXERCISES

Develop Descriptive, Symbolic, and Schematic Models for Exercises 4–1 through 4–9

4-1. Four cities in a metropolitan area produce refuse daily. This refuse is picked up by the local refuse collection department and delivered to one of two transfer stations. At the transfer stations, the refuse is processed and shipped to one of three landfills. The objective will be to minimize total cost of disposing of the refuse.

4-2. A refuse truck has to travel over city streets to pick up refuse at various points. The truck has to stop at each pickup point, then proceed back to the starting point. There are paths of varying length between each of the pickup points. The objective will be to minimize the total distance the refuse truck has to travel while stopping at all of the designated pickup points. (Assume that there are eight pickup points for this problem.)

4-3. A water company is drilling six new wells in a well field. They want to use the minimum length of pipe required to transport the water from the six wells to the water treatment plant.

4-4. Reformulate Exercise 4-3 if the diameter of each pipe is a function of the flowrate through it, and the objective is to minimize the total cost of pipe to transport the water.

4-5. Reformulate Exercise 4-4, adding the condition that pumping costs are a function of the flowrate through the pipe and the inverse of the diameter of the pipe. The objective is to minimize the total delivery cost of the water.

4-6. Cities A and B are serviced by the road network shown in the figure for Exercise 4-6. Each link of the system has a given flow capacity and time of travel. The initial objective is to determine the shortest time path between Cities A and B.

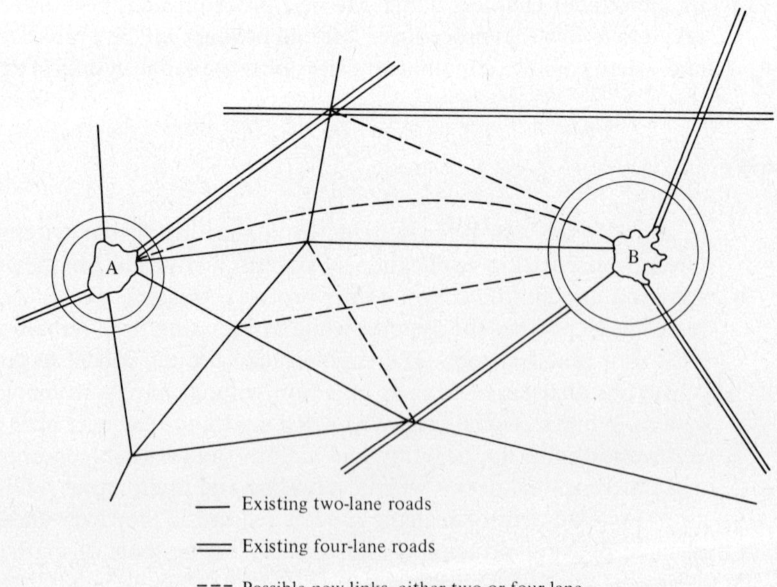

4-7. New two or four lane links have different costs per mile. In addition, any of the existing two lane roads could be converted to four lanes for a certain amount per mile. The objective is now to provide sufficient capacity for a specified total volume of traffic between Cities A and B at minimum cost.

4-8. The network shown in the figure for Exercise 4-8 is a water distribution system. The numbers shown on each link of the network represent the maximum flowrate through that pipe, in either direction. The objective is to find the maximum flow through the system from point A to point B.

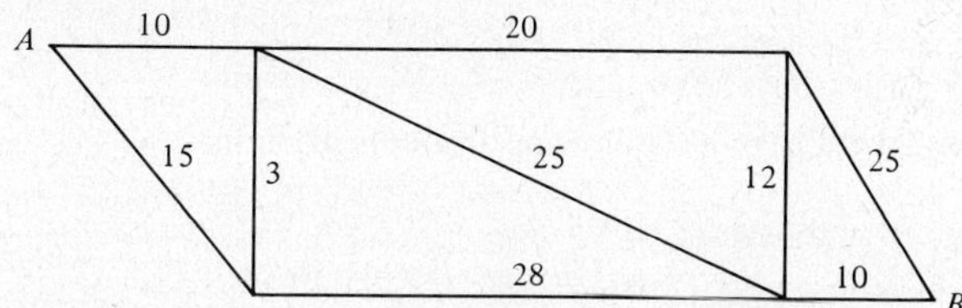

4-9. Activities A–O described below make up the different activities associated with a construction project. These activities have to be scheduled according to the precedence given to avoid conflicts among the activities. Each activity also has an associated expected completion time.

A, B, and C can be accomplished simultaneously and are the first activities in the project.

D and E can be accomplished simultaneously and follow A.

H and I can be accomplished simultaneously and follow E.

G follows C, F, and H.

F follows B and C.

J and K can be accomplished simultaneously and follow B, D, G, and I.

L and M can be accomplished simultaneously and follow J.

N follows L.

O follows K and M.

N and O can be accomplished simultaneously and are the last activities in the project.

The objective will be to find the critical (longest) time path through the activities.

Develop Descriptive, Symbolic, Schematic, and Mathematical Models for Exercises 4-10 through 4-15

4-10. The objective is to find the force in each member in the figure for Exercise 4-10.

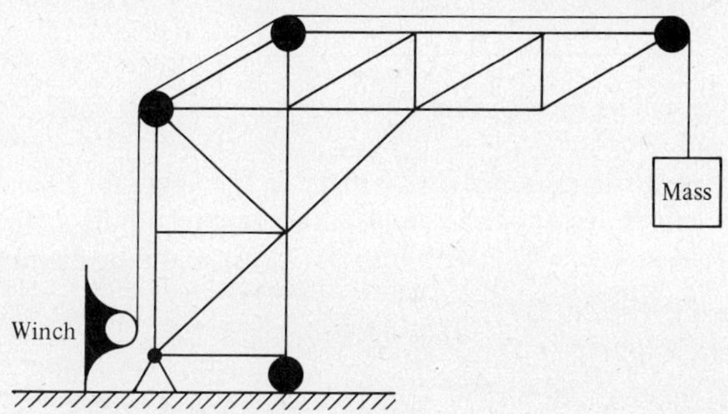

4-11. The objective is to find the forces that have to be withstood by the pins at A, B, C, and D, in the figure for Exercise 4-11.

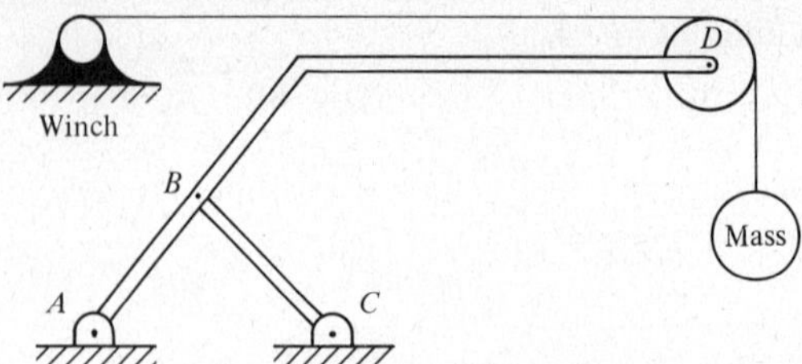

4-12. Three suppliers have the following amounts of reinforcing steel available:

Supplier	Tons available
1	20
2	10
3	30

A contractor needs the following amounts of reinforcing steel at four different job sites:

Site	Tons required
1	5
2	20
3	30
4	15

The transportation costs from each supplier to each site are given in the following table:

	Site 1	Site 2	Site 3	Site 4
Supplier 1	8	23	21	40
Supplier 2	19	10	28	21
Supplier 3	35	24	28	25

The objective will be to minimize the total transportation costs.

4-13. A small water distribution network is given in the figure for Exercise 4-13. All pipes are 10 in. in diameter. An approximate equation that relates flow (Q_i, ft³/s) in pipe i to the frictional head loss (h_{L_i}, ft), length of pipe (L_i, ft), and pipe diameter (D_i, ft) is

$$Q_i = 45\left(\frac{D_i^{2.5}}{L_i^{0.5}}\right)(h_{L_i})^{1/2}$$

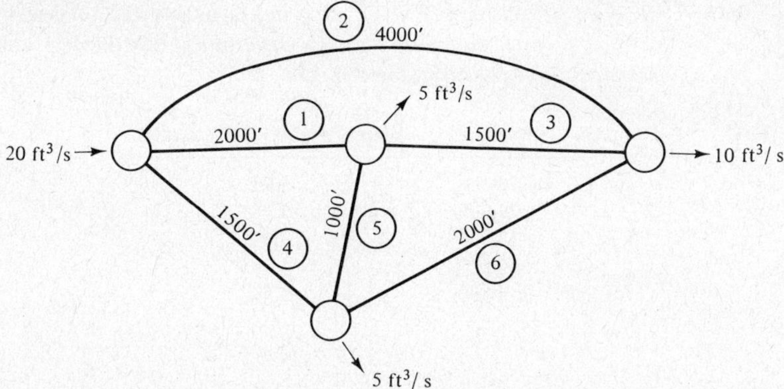

(a) Develop a mathematical model that could be utilized to solve for the flowrate in each pipe segment of the network while satisfying all of the physical constraints of the system.

(b) Suggest any simplifications or modifications that could be made in this model to assist in a solution, and suggest a solution technique.

4-14. A set of survey points is shown below. The elevation of point ① is known from a previous survey. The elevations of the other points relative to point ① are desired. Since survey instruments are not completely accurate, the elevation of each unknown point is estimated from at least two sets of elevation differences. Therefore, the following elevation differences have been measured:

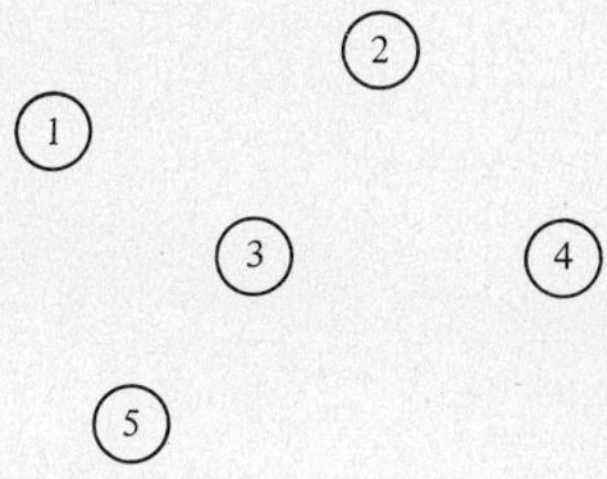

From point	To point	Measured elevation difference (ft)
1	2	+19.85
1	3	+29.8
1	5	+40.1
2	3	+10.1
2	4	+29.7
4	3	−19.8
4	5	−10.2

Assuming that all of the elevation differences have been measured with equal accuracy, and that the best estimate of the elevation of any unknown point is the average of all possible computed elevation values (elevation of adjacent point ± elevation difference) for that point, develop a general mathematical model that could be utilized to estimate point elevations. Note that recognition of a link–node incidence matrix will be helpful in generalizing this model.

4-15. Develop a mathematical model that could be utilized to determine possible ways of meeting the water demands for each community described in Exercises 3-10 and 2-15. The demands are listed in the following table:

Community	Population	Water demand (10^6 gal/day)
1	28,000	5.03
2	21,000	3.78
3	13,000	2.34
4	6,000	1.08
5	14,000	2.52

Chapter 5

Optimization Models

INTRODUCTION

Optimization models are prescriptive models, in that they suggest the best method of system operation. They do this through the objective function, which is a measure of the degree of attainment of system objectives. This objective function (equation of state) can be a linear or nonlinear function of the problem decision variables. Allowable combinations of values for the decision variables are controllable by the mathematical constraints imposed on the system. Engineers will often identify multiple objective functions that are appropriate for a particular problem. In these cases, either the most important objective function can be identified, and the others incorporated as constraints, or multiobjective optimization techniques (to be discussed in Chapter 13) can be used to try to find the optimal solution for all of the objective functions simultaneously. This overall optimal solution may not be the same as the optimal solution for any of the individual objective functions.

Although optimization models are prescriptive, proper interpretation, use, and/or implementation of their results is still dependent on the good engineering judgment, management expertise, and communications skills of the modeler. Also, the optimization model itself is only as good as the skill of the modeler who developed it.

This chapter will present several examples of the development of optimization models from different disciplines within the civil engineering profession. Methods for finding the best solutions to these models will be discussed in Chapter 6.

STRUCTURAL OPTIMIZATION MODELS

Structural optimization is a rapidly growing field. New analytical and numerical techniques are allowing designers and modelers to find optimal or near optimal solutions to more complex structural problems. Optimality is defined as determining the most economical combination of members which meets all codes and specifications pertaining to strength and layout.

Generally the descriptive model for a structural optimization model will include a formulation of the functional requirements and a conceptual design. Functional requirements are determined by the type and level of service that the structure will be subjected to. Examples are the required number of lanes for a bridge or the required floor space, by type, for an office building. Conceptual design involves the overall planning to ensure that the structural system serves its functional purposes. The designer must choose the type of structure, the overall topology, and the materials to be considered. This is the stage at which designers can exhibit their creativity and ingenuity. Also, good engineering judgment is required to avoid proposing technically infeasible conceptual designs.

Structural optimization involves developing procedures that will find the optimal, or near optimal, values of certain design variables, while considering the desired criteria and constraints. Large, high-speed computers have done much toward the automation of these procedures, and toward the applicability of these procedures to realistic structural problems. Analytical optimization methods can be used for some structural problems, but others require numerical techniques. Analytical techniques are generally most applicable to optimization of single structural components or simple structural systems, such as optimization of beam support locations or maximization of load that can be applied to a given structural system. However, modelers must be careful to apply meaningful constraints to the geometric form of the structure and to carefully interpret results if impractical solutions are to be avoided.

Numerical optimization most often produces near optimal solutions through an iterative, automated procedure. An initial best estimate (or guess) of the optimal solution is used as a starting point, and the numerical technique provides for a systematic search for better solutions. Solutions to such problems are sensitive to the form of the mathematical model developed, to the accuracy of the initial estimate, and to the numerical search technique utilized. It is quite possible for solutions to diverge if conditions are not proper. Also, when the procedure is terminated at a near optimal solution (according to specified criteria), it is not always possible to determine whether or not this is a global optimal solution (no other local optimal solutions will provide a better solution). Local optimum solutions occur where all solutions in the immediate vicinity are not as desirable as the solution point in question. However, this is not to say that a better solution will not occur at some distance. Consider all of the mountains in a state. Each mountain represents a local high point (local optimum), but there is only one tallest mountain in the state (global optimum). Numerical search routines are discussed more thoroughly in Appendix E.

Design variables associated with structural optimization problems can be of several types. They can involve mechanical or physical properties of the material. However, the state of the art of optimization techniques makes their direct application to materials selection impractical. They can, however, assist in comparing materials by finding the optimal design for alternative discrete materials. Other design variables could possibly involve the topology of the structure, either the pattern of connection of members or the number of elements in a structure. Topological design variables are normally integer variables. Examples include the existence or absence of a member, the number of spans in a bridge, or the number of columns supporting a roof system. Configuration or geometric layout of the structure can be another design variable. The coordinates of the joints in a truss or frame, the location of supports in a bridge, or the length of spans in a continuous beam are all examples of this type. Cross-sectional design variables are a fourth type, and include cross-sectional areas, member dimensions, moment of inertia of flexural members, or thickness of plates.

It is not practical to include all of these quantities as design variables in each problem. The multivariability of the resulting model would be unworkable. Therefore, most of them are presented as parameters in the model, with the appropriate entities retained as design variables. Parameters might be the material properties such as modulus of elasticity or yield stress, the topology of the structure (number of members), and the coordinates of the nodes. Design variables could be the cross-sectional areas of the members. The behavior of the structure will now be dependent on the values assigned to the design variables; however, the behavior is also subject to a set of constraints, or restrictions, that must be satisfied in order to produce a workable (feasible) design. Design constraints, or side constraints, are specified upper or lower bounds on a design variable or relationship which includes several interdependent design variables. Examples are minimum width of roadway, maximum height of a building, or minimum level of waste removal from sewage. Behavioral constraints, on the other hand, are derived from the behavioral requirements of the system. Examples are limits on maximum stresses in flexural members, maximum displacement of truss joints, and maximum axial compressive load to prevent buckling in columns.

As a first example, consider the optimum placement of supports for a simply supported flexural member. For flexural members,

$$\sigma_{max} = \frac{Mc}{I} \tag{5.1}$$

where M = bending moment in the beam
   c = maximum distance from neutral axis
   I = moment of inertia of the cross section about its centroidal axis
   σ_{max} = maximum bending stress for any particular cross section

For a given cross section, the maximum bending stress will occur when the bending moment is a maximum, either positive or negative. In design, σ_{max} is replaced by σ_{allow}, and Equation 5.1 becomes

$$\sigma_{allow} \geq \frac{Mc}{I} \tag{5.2}$$

or
$$\frac{M}{\sigma_{allow}} \leq \frac{I}{c} \tag{5.3}$$

The quantity I/c is a property of the cross section of the member, and is usually called S, the section modulus. The usual procedure for choosing a beam is as follows:

1. Compute and graph moment function.

2. Find the maximum moment, either positive or negative.

3. Using the allowable stress ($\sim$20,000 psi for A 36 steel), calculate the required section modulus

$$S_{req} = \frac{M}{\sigma_{allow}} \tag{5.4}$$

4. Choose the most economical (least weight per foot) beam from the steel tables that has $S \geq S_{req}$ about the appropriate centroidal axis, and meets any other dimensional restrictions.

Since the σ_{allow} is a material property and is constant in this analysis, the section modulus (which governs the size of beam required) is directly proportional to the maximum bending moment applied to the beam.

The usual procedure in analyzing the applied load is to develop shear and bending moment diagrams and/or functions. From these, the designer can ascertain the maximum applied bending moment, be it positive or negative.

The form and magnitude of the shear and bending moment diagram are dependent on the type and magnitude of the applied load and the placement of the support structure. Considering a given loading condition, the placement of the supports can be varied to try to minimize the maximum applied bending moment.

Example 5.1 will demonstrate the method of finding the optimum placement of supports for a simply supported, uniformly loaded beam. For the given loading condition, end supports required a 14 W 34 beam, while optimum placement of the supports required only an 8 W 13, a 62% saving in weight per foot, and a 43% saving in the required depth of the beam. This is one of the reasons many modern buildings are supported by interior columns or walls, with nonload-bearing walls on the exterior. For nonsymmetrical loadings, computer simulation can be useful in determining best placement of the supports to produce the smallest possible maximum bending moment, and thus the most economical section.

A second type of structural optimization problem involves finding the maximum load that a given structure can carry when it is loaded at several possible points. Example 5.2 shows a bar and cable structure that could represent a gymnasium apparatus. The structure can be loaded at six possible points. It is desirable to determine what combinations of loadings would maximize the total load that the structure would carry. Behavioral constraints can be developed based on the type of loading conditions encountered and the properties of the materials used. Equilibrium constraints can be utilized to rewrite the behavioral constraints in terms of the decision variables included in the objective function. The result is an objective function (Equation 5.33) which is subject to a consistent set of constraints (Equation 5.44). Both the objective function and constraints are linear in this case, so the maximum load could be found through application of linear programming. It should be noted that inclusion of the redundant constraints would not influence the optimum solution. They would only make the computational process somewhat less efficient.

The type of analysis described in Example 5.2 could also be applied to other types of loadings; however, determination of the point of maximum moment may be more difficult.

EXAMPLE 5.1 ____

Find the optimum placement of supports for the following conditions: A 20 × 20 ft deck structure (Figure 5.1) is supported by three wide flange beams. The design load for the deck is 150 lb/ft^2. The allowable stress for the steel to be used is 20,000 psi. Each beam is assumed to support a proportional amount of the load; thus, the center beam would support the load applied to 200 ft^2 of the deck, while each side beam would support 100 ft^2 of the load.

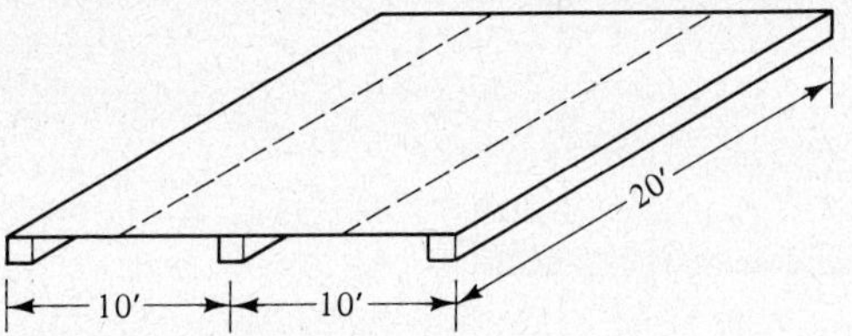

Figure 5.1 Deck structure.

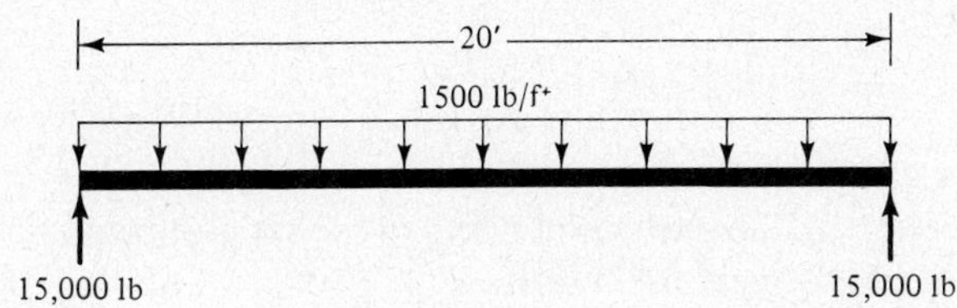

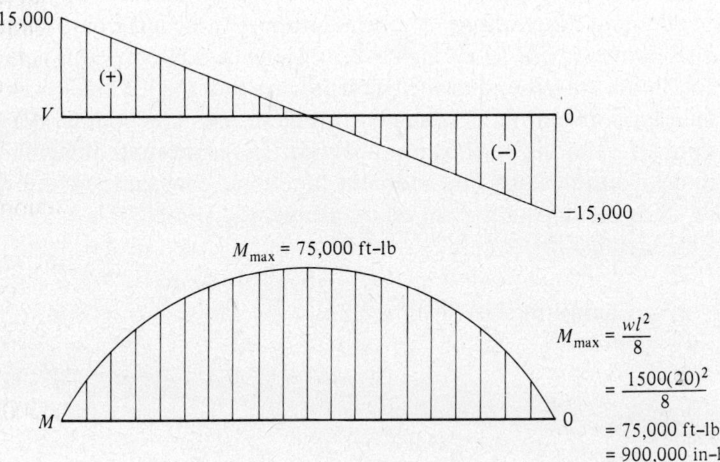

Figure 5.2 Shear and bending moment diagrams for end supports.

The load on the center beam could be reduced to a uniform load of 1500 lb/ft of beam. If the center beam is supported at each end, then the shear and bending moment diagrams shown in Figure 5.2 can be generated. For this support scheme,

$$S_{req} = \frac{900,000 \text{ in.-lb}}{20,000 \text{ lb/in.}^2} = 45 \text{ in}^3.$$

The following wide flange beams would support this load:

	S (in.³)
10 W 45	49.1
10 W 49	54.6
12 W 36	45.9
14 W 34	48.5

Depth ——— lb/ft

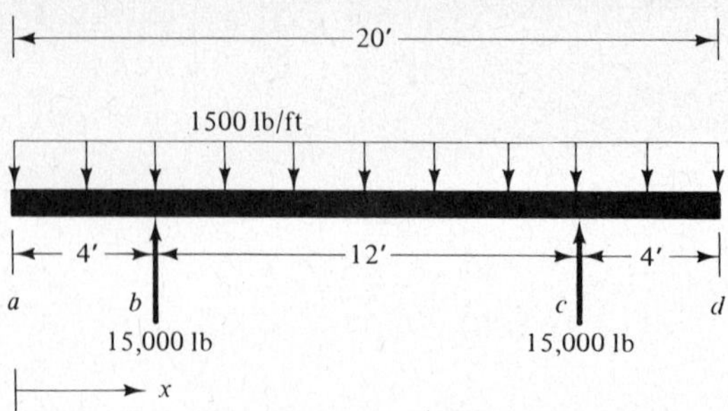

Figure 5.3 Supports moved 4 ft inboard from ends.

If there is no limitation on the maximum depth of the support beam, then the 14 W 34 would be the most economical at 34 lb/ft.

Suppose the supports are moved in 4 ft from each end, with all other aspects of the problem remaining the same. Figure 5.3 would result. To develop the shear and bending moment functions for this configuration, three imaginary cuts must be made; one in section a–b, one in section b–c, and one in section c–d. All functions will be developed with the left end of the beam as zero, and the left free body will be utilized with the sign convention as is shown in Figure 5.4. The counterclockwise moment (M) is positive and the vertical upward shear (V) is positive. Integral analysis will be used to develop the shear and moment functions. The same functions could be developed using static equilibrium relationships.

For section a–b

$$\text{Change in shear over section} = \Delta V(X)_{a-b} = \int_0^X W(X)\, dx$$

$$= \int_0^X -1{,}500\, dx = -1{,}500X \tag{5.5}$$

also,
$$V(X)_{a-b} = V_a + \Delta V(X)_{a-b} \tag{5.6}$$

Since $V_a = 0$,

$$V(X)_{a-b} = -1{,}500X \tag{5.7}$$

$$\text{Change in bending moment over section} = \Delta M(X)_{a-b} = \int_0^X V(X)\, dx$$

$$= \int_0^X -1{,}500X = -750X^2 \tag{5.8}$$

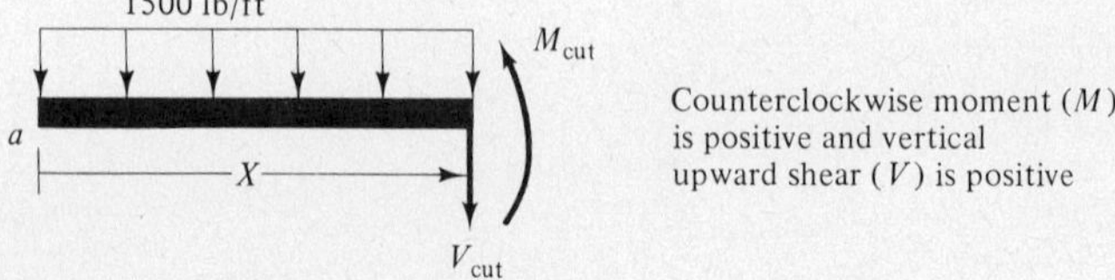

Figure 5.4 Free body for section a–b.

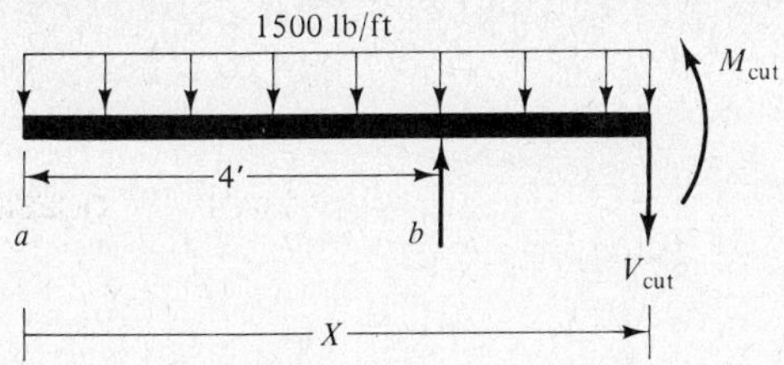

Figure 5.5 Free body for section b–c.

As before
$$M(X)_{a-b} = M_a + \Delta M(X)_{a-b} \tag{5.9}$$

Since $M_a = 0$, then

$$M(X)_{a-b} = -750X^2 \tag{5.10}$$

For section b–c, the free body in Figure 5.5 is used.

$$\Delta V(X)_{b-c} = \int_4^X W(X)\,dx = \int_4^X -1{,}500\,dx = 6{,}000 - 1{,}500X \tag{5.11}$$

At b,
$$V_b = -6{,}000 + 15{,}000 = 9{,}000 \tag{5.12}$$

$$V(X)_{b-c} = V_b + \Delta V(X)_{b-c} = 9{,}000 + 6{,}000 - 1{,}500X \tag{5.13}$$

$$V(X)_{b-c} = 15{,}000 - 1{,}500X \tag{5.14}$$

$$\Delta M(X)_{b-c} = \int_4^X V(X)\,dx = \int_4^X (15{,}000 - 1{,}500X)\,dx$$

$$= (15{,}000X - 750X^2)\big|_4^X$$

$$= -48{,}000 + 15{,}000X - 750X^2 \tag{5.15}$$

At b,
$$M_b = -12{,}000 \tag{5.16}$$

$$\therefore M(X)_{b-c} = M_b + \Delta M(X)_{b-c}$$

$$= -12{,}000 - 48{,}000 + 15{,}000X - 750X^2 \tag{5.17}$$

$$M(X)_{b-c} = -60{,}000 + 15{,}000X - 750X^2 \tag{5.18}$$

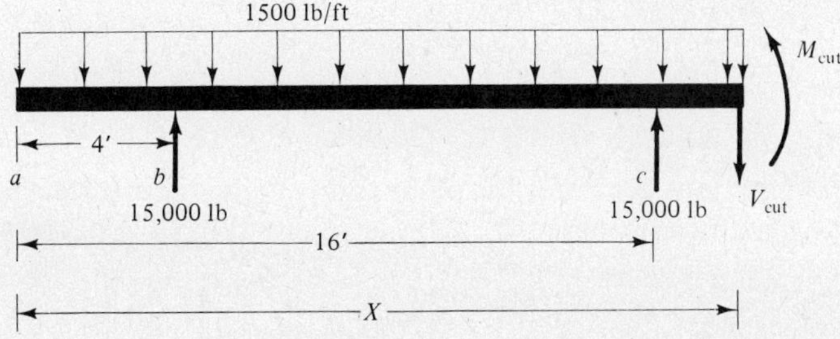

Figure 5.6 Free body for section c–d.

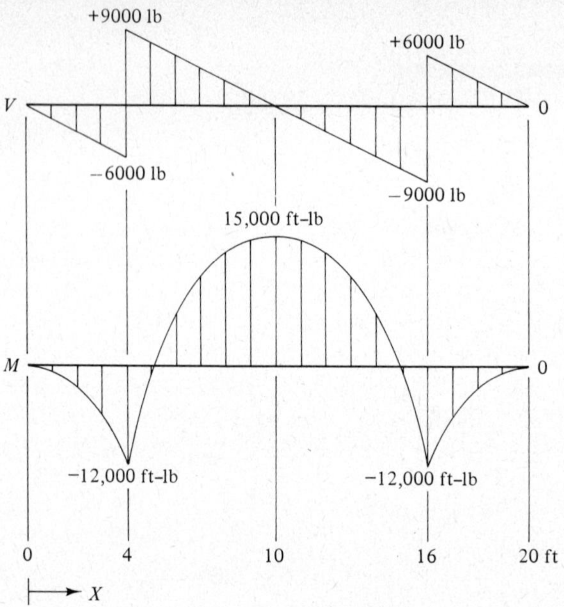

Figure 5.7 Shear and moment diagrams for interior supports.

For section c–d, the free body in Figure 5.6 is used. At c,

$$V = -9{,}000 + 15{,}000 = 6{,}000 \text{ lb}$$

$$M = -12{,}000 \text{ ft-lb} \tag{5.19}$$

$$\Delta V(X)_{c\text{-}d} = \int_{16}^{X} W(X)\, dx = \int_{16}^{X} -1{,}500\, dx = 24{,}000 - 1{,}500X \tag{5.20}$$

$$V(X)_{c\text{-}d} = V_c + \Delta V(X)_{c\text{-}d}$$

$$= 6{,}000 + 24{,}000 - 1{,}500X \tag{5.21}$$

$$V(X)_{c\text{-}d} = 30{,}000 - 1{,}500X \tag{5.22}$$

$$\Delta M(X)_{c\text{-}d} = \int_{16}^{X} V(X)\, dx = \int_{16}^{X} (30{,}000 - 1{,}500X)\, dx$$

$$= (30{,}000X - 750X^2)\big|_{16}^{X} = -288{,}000 + 30{,}000X - 750X^2 \tag{5.23}$$

$$M(X)_{c\text{-}d} = M_c + \Delta M(X)_{c\text{-}d}$$

$$= -12{,}000 - 288{,}000 + 30{,}000X - 750X^2 \tag{5.24}$$

$$M(X)_{c\text{-}d} = -300{,}000 + 30{,}000X - 750X^2 \tag{5.25}$$

Plotting these functions generates the shear and bending moment diagrams in Figure 5.7. Now the maximum bending moment is 15,000 ft-lb, or 180,000 in.-lb, and the required section modulus is

$$S_{\text{req}} = \frac{180{,}000}{20{,}000} = 9$$

This provides a choice of the following sections:

	S
6 W 15.5	10.1
8 W 17	14.1
8 W 24	20.8

A 6 W 15.5 is the most economical. Mathematically optimum placement of the supports can be determined by working with the functions for moment change over the beam sections. The objective is to minimize the maximum bending moment (either positive or negative). Proper placement of the supports should result in the maximum positive and negative bending moments being equal. Any other placement will result in one maximum being larger. Figure 5.8 shows that in order for the maximum negative moment to equal the maximum positive moment, the change in moment between b and the center of the span has to be twice as large as the change in moment between a and b. Therefore,

$$|\Delta M_{a-b}| = \tfrac{1}{2}|\Delta M_{b-10}| \tag{5.26}$$

From Equations 5.8 and 5.15,

$$|\Delta M_{a-b}| = |-750l^2| \tag{5.27}$$

$$|\Delta M_{b-10}| = |15{,}000X - 750X^2|_l^{10}| = |75{,}000 - 15{,}000l + 750l^2| \tag{5.28}$$

$$\therefore 750l^2 = 37{,}500 - 7{,}500l + 375l^2 \tag{5.29}$$

$$0 = -375l^2 - 7{,}500l + 37{,}500 \tag{5.30}$$

Solving for l, the optimum placement of the support is 4.14 ft in from the end, or 0.207 times the length of the beam. For this placement, the maximum moment is 12,870 ft-lb, and the required section modulus is 7.72. An 8 W 13 beam with a section modulus of 9.91 is adequate.

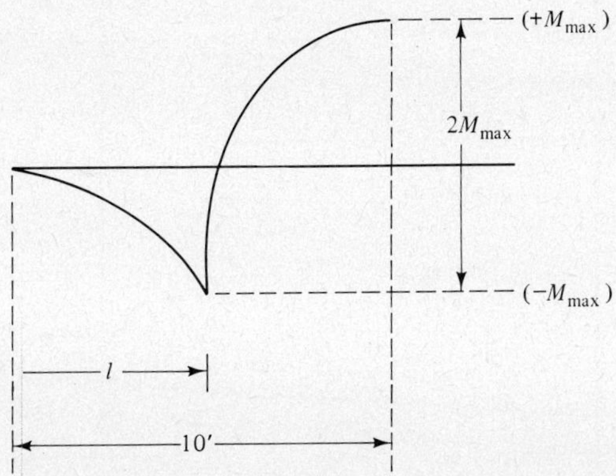

Figure 5.8 Optimum placement of supports.

EXAMPLE 5.2 ____

The gymnasium structure shown in Figure 5.9 is subjected to six possible loads, L_1 to L_6. What is the maximum load the structure will safely carry?

All of the joints, except the bases of the columns (E, F, and G), are pinned connections. The two horizontal bars have a rectangular cross section 2 in. high by 1 in. wide. Cables a, b, c, and d have a cross-sectional area of 1 in.^{2}. The horizontal beams, C and D, and the vertical columns, E, F, and G, have a rectangular cross section 4 in. high by 2 in. wide. The allowable stress in the cables is 10,000 psi, while the allowable stress for all flexural members is 20,000 psi. The modulus of elasticity for all members is 30×10^6 psi. Member and joint deflections are not to be considered in the analysis. The following supplemental criteria can be applied.

For column design,

$$P \le P_{\text{cr}} = \frac{\pi^2 EI}{L_e^2} \tag{5.31}$$

where P = axial load applied to the column

  I = moment of inertia of cross section about the weak centroidal axis = $bh^3/12$

  L_e = twice the length of the column

  E = modulus of elasticity

For beam design,

$$\sigma_{\text{max}} = \frac{Mc}{I} \le \sigma_{\text{allow}} \tag{5.32}$$

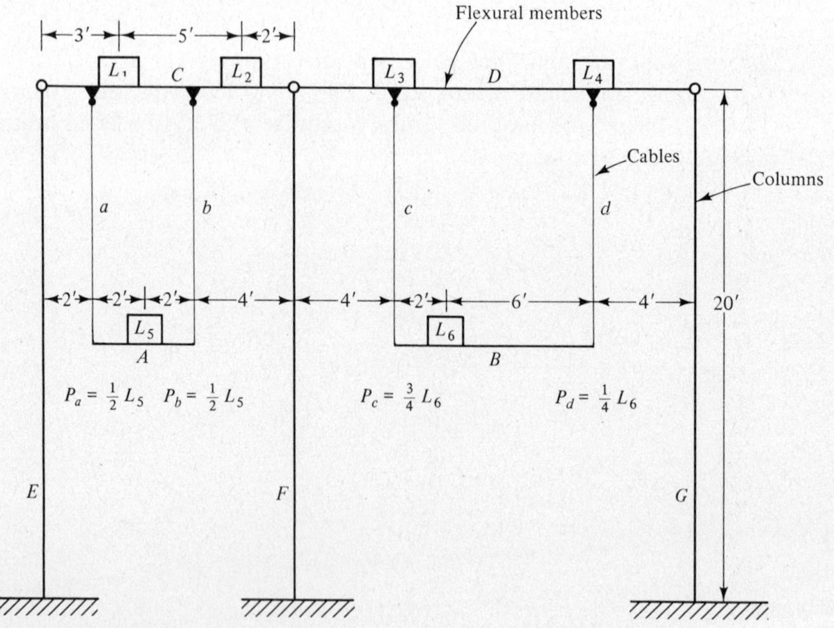

Figure 5.9 Gymnasium structure.

where σ_{max} = maximum stress generated in any cross section

M = maximum bending moment generated in the beam

c = maximum distance from neutral axis

A model will be formulated that can be used to maximize the total load that can be applied to this structural system. The objective function can be written as

$$\max Z = L_1 + L_2 + L_3 + L_4 + L_5 + L_6 \tag{5.33}$$

which is subject to the following behavioral constraints.

For the columns, if $E = 30 \times 10^6$ psi,

$$P_E \leq \frac{\pi^2 EI}{L_e^2} = \frac{\pi^2(30 \times 10^6)[4(2)^3/12]}{[40(12)]^2} \tag{5.34}$$

$$\therefore P_E \leq 3{,}427 \text{ lb}$$

and similarly,
$$P_F \leq 3{,}427 \text{ lb} \tag{5.35}$$

$$F_G \leq 3{,}427 \text{ lb}$$

Since the area of each cable is 1.0 in.^{2}, the allowable load in each cable is

$$P_a \leq 10{,}000 \text{ lb}$$

$$P_b \leq 10{,}000 \text{ lb}$$

$$P_c \leq 10{,}000 \text{ lb} \tag{5.36}$$

$$P_d \leq 10{,}000 \text{ lb}$$

Each of the horizontal members is also subjected to flexural stress and is subject to the limitation set by Equation 5.32.

For member A,

$$M_A \leq \frac{\sigma_{allow}(I)}{c} \tag{5.37}$$

$$M_A \leq \frac{20{,}000[1(2)^3/12]}{1}$$

Therefore,
$$M_A \leq 13{,}333 \text{ in.-lb}$$

and similarly,
$$M_B \leq 13{,}333 \text{ in.-lb} \tag{5.38}$$

For members C and D,

$$I = \frac{2(4)^3}{12} = 10.67 \qquad \text{and} \qquad c = 2$$

Therefore;
$$M_c \leq 106{,}657 \text{ in.-lb}$$

$$M_D \leq 106{,}657 \text{ in.-lb} \tag{5.39}$$

In order to determine the critical conditions, the point of maximum bending moment has to be determined. For bars A and B, the maximum bending moment has to occur at the point where the load is applied. Utilizing the theory of statics, these maximum moments are

$$M_A = 2P_a \le 13{,}333 \text{ in.-lb}$$

$$M_B = 2P_c \le 13{,}333 \text{ in.-lb}$$

(5.40)

Each of the upper beams is subjected to loads at more than one point. Therefore, while it is impossible to say at which loading point the maximum bending moment will occur, it is certain that the maximum will occur at one of the loading points. Figure 5.10 demonstrates why this is true. Although the magnitudes of the loads (and thus the magnitudes of the discontinuities in the shear diagram) are not known, one of these discontinuities will take the shear through zero, which is the point where the maximum moment will occur.

Separate moment equations can be written between each change in loading, and each moment equation can be subjected to the maximum bending moment constraint. For beam C,

$$M_a = 2P_E \le 106{,}667$$

$$M_{L_1} = 2P_E + 1(P_E - P_a) \le 106{,}667$$

$$M_b = 2P_E + 1(P_E - P_a) + 3(P_E - P_a - L_1) \le 106{,}667$$

$$M_{L_2} = 2P_E + 1(P_E - P_a) + 3(P_E - P_a - L_1) + 2(P_E - P_a - L_1 - P_b)$$
$$\le 106{,}667$$

(5.41)

Similarly, for beam D,

$$M_d = 4(P_G) \le 106{,}667$$

$$M_c = 4(P_G) + 8(P_G - L_4 - P_d) \le 106{,}667$$

(5.42)

Equilibrium constraints can be utilized to write equations that relate the forces in vertical members to the applied loads.

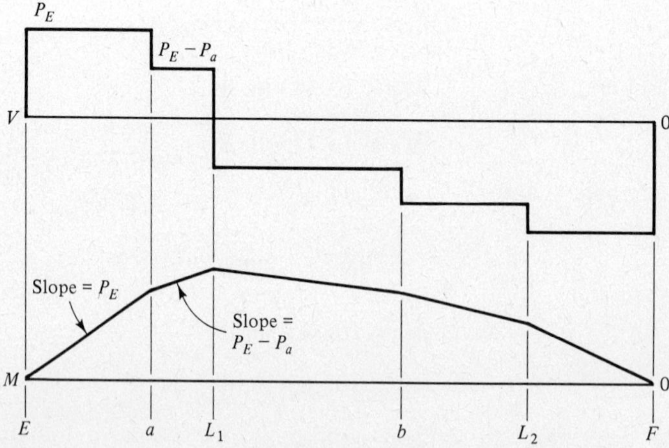

Figure 5.10 Shear and bending moment diagrams for beam C.

$$P_E = 0.7L_1 + 0.2L_2 + 0.6L_5$$

$$P_F = 0.3L_1 + 0.8L_2 + 0.4L_5 + 0.75L_3 + 0.625L_6 + 0.25L_4$$

$$P_G = 0.25L_3 + 0.375L_6 + 0.75L_4$$

$$P_a = 0.5L_5$$

$$P_b = 0.5L_5 \qquad\qquad\qquad\qquad\qquad\qquad\qquad (5.43)$$

$$P_c = 0.75L_6$$

$$P_d = 0.25L_6$$

When Equations 5.43 are combined with the previously developed constraints, the following set is produced:

For Columns

$$E \quad [1] \quad \tfrac{7}{10}L_1 + \tfrac{6}{10}L_5 + \tfrac{2}{10}L_2 \le 3{,}427 \text{ lb}$$

$$F \quad [2] \quad \tfrac{3}{10}L_1 + \tfrac{4}{10}L_5 + \tfrac{8}{10}L_2 + \tfrac{3}{4}L_3 + \tfrac{5}{8}L_6 + \tfrac{1}{4}L_4 \le 3{,}427 \text{ lb}$$

$$G \quad [3] \quad \tfrac{1}{4}L_3 + \tfrac{3}{8}L_6 + \tfrac{3}{4}L_4 \le 3{,}427 \text{ lb}$$

For Cables

$$a \quad [4] \quad \tfrac{1}{2}L_5 \le 10{,}000 \text{ lb}$$

$$b \quad [5] \quad \tfrac{1}{2}L_5 \le 10{,}000 \text{ lb}$$

$$c \quad [6] \quad \tfrac{3}{4}L_6 \le 10{,}000 \text{ lb}$$

$$d \quad [7] \quad \tfrac{1}{4}L_6 \le 10{,}000 \text{ lb}$$

For Beams A and B

$$A \quad [8] \quad L_5 \le 1{,}111 \text{ ft-lb} \qquad\qquad\qquad (5.44)$$

$$B \quad [9] \quad 1.5L_6 \le 1{,}111 \text{ ft-lb}$$

For Beam C

$$a \quad [10] \quad 1.4L_1 + 1.2L_5 + 0.4L_2 \le 8{,}888 \text{ ft-lb}$$

$$L_1 \quad [11] \quad 2.1L_1 + 1.3L_5 + 0.6L_2 \le 8{,}888 \text{ ft-lb}$$

$$b \quad [12] \quad 1.8L_1 + 1.4L_5 + 0.8L_2 \le 8{,}888 \text{ ft-lb}$$

$$L_2 \quad [13] \quad 1.2L_1 + 0.6L_5 + 1.2L_2 \le 8{,}888 \text{ ft-lb}$$

For Beam D

$$d \quad [14] \quad L_3 + 1.5L_6 + 3L_4 \le 8{,}888 \text{ ft-lb}$$

$$c \quad [15] \quad -L_3 + 0.5L_3 + 5L_4 \le 8{,}888 \text{ ft-lb}$$

Examination of the cable constraints shows that they are all redundant with constraints 8 and 9, so they can be eliminated. Constraint 10 can also be eliminated because the geometry of this particular problem prevents the maximum moment from occurring at point *a*. Thus, the objective function, Equation 5.33, could be maximized subject to the remaining 10 constraints.

The next example will add displacement constraints to the structural optimization problem. Figure 5.11 depicts a two-tiered walkway with the loads shown applied to the floor beams. The beams are to remain level during loading, and are limited in the amount they can move vertically under load. Cost for the support structure for the beams is a function of the size of the hanging members and the type of connection made between the hanging members and the floor beams. Good engineering practice dictates that the minimum cost hanging structure which meets the system behavior constraints should be adopted. Behavioral, geometric compatibility, and system equilibrium constraints are developed that relate system limitations to the objective function decision variables. Both the objective function and the system constraints contain nonlinearities in this case.

EXAMPLE 5.3 ⎯⎯⎯⎯⎯⎯⎯⎯⎯⎯⎯⎯⎯⎯⎯⎯⎯⎯⎯⎯⎯⎯⎯⎯⎯⎯⎯⎯⎯⎯

Figure 5.11 depicts a two-tiered hanging walkway support structure. Each floor beam supports a section of walkway 40 ft long by 8 ft wide. Design loading for the walkway is 150 lb/ft^2. Thus, each floor beam will have to support an equivalent 48-kip point load at midspan.

The walkways are to remain level during loading. Maximum depression of the upper walkway support under load is 0.1 in., while the lower walkway is allowed to

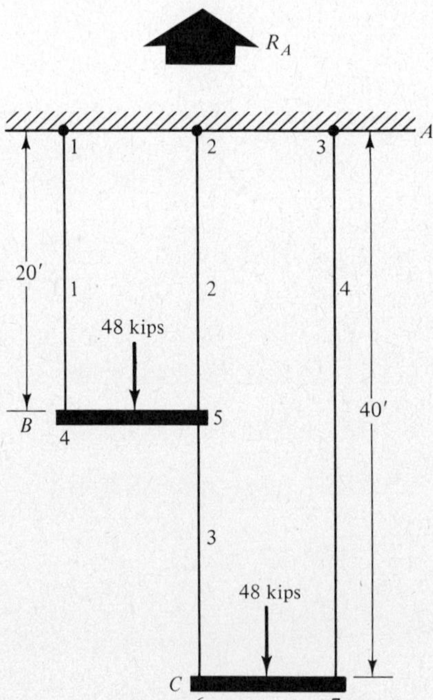

Figure 5.11   Hanging walkway support structure.

depress 0.2 in. under design load. Steel with a modulus of elasticity of 30×10^6 psi is to be used for the hangers.

The cost of the hanging members is a function of the volume of steel used in each member. Since the initial length is specified, the cost is related to the cross-sectional area of the member, A_i. Costs for end connections can be related to the circumference of the member being connected. This circumference can in turn be related to the cross-sectional area design variable by the equation

$$\text{circ}_i = 2\sqrt{\pi A_i} \tag{5.45}$$

Thus, the objective function can be written as

$$\min Z = \sum c_i A_i + \sum C c_i 2\sqrt{\pi A_i} \tag{5.46}$$

where c_i = cost of hangers per unit cross-sectional area
Cc_i = cost of connections per unit cross-sectional area of hanger

Note that each hanger has a connection at both ends, so Cc_i would have to actually reflect two connections.

Behavioral constraints can be written as follows:

$$\left.\begin{array}{c} \sigma_{\max} = 20 \text{ ksi for all members} \\[1em] d_1, d_2 \le 0.1 \text{ in.} \\[1em] d_4 \le 0.2 \text{ in.} \\[1em] d_1 = d_2 \\[1em] d_2 + d_3 = d_4 \end{array}\right\} \tag{5.47}$$

where d_i is the elongation of member i.

A system of equations must now be developed that presents the behavioral constraints in terms of the design variables. The tension in each member (T_i) is related to the elongation of that member (d_i) by the equation

$$T_i = \left(\frac{E_i A_i}{L_i}\right) d_i \tag{5.48}$$

Therefore
$$\sigma_i = \frac{E_i}{L_i} d_i \tag{5.49}$$

In matrix form,

$$\begin{bmatrix} T_1 \\ T_2 \\ T_3 \\ T_4 \end{bmatrix} = \begin{bmatrix} \dfrac{E_1 A_1}{L_1} & 0 & 0 & 0 \\ 0 & \dfrac{E_2 A_2}{L_2} & 0 & 0 \\ 0 & 0 & \dfrac{E_3 A_3}{L_3} & 0 \\ 0 & 0 & 0 & \dfrac{E_4 A_4}{L_4} \end{bmatrix} \begin{bmatrix} d_1 \\ d_2 \\ d_3 \\ d_4 \end{bmatrix} \tag{5.50}$$

and

$$
\begin{bmatrix} T_1 \\ T_2 \\ T_3 \\ T_4 \end{bmatrix} = 30 \times 10^6 \begin{bmatrix} \dfrac{A_1}{L_1} & 0 & 0 & 0 \\ 0 & \dfrac{A_2}{L_2} & 0 & 0 \\ 0 & 0 & \dfrac{A_3}{L_3} & 0 \\ 0 & 0 & 0 & \dfrac{A_4}{L_4} \end{bmatrix} \begin{bmatrix} d_1 \\ d_2 \\ d_3 \\ d_4 \end{bmatrix}
$$

$$
= \frac{30 \times 10^6}{20} \begin{bmatrix} A_1 & 0 & 0 & 0 \\ 0 & A_2 & 0 & 0 \\ 0 & 0 & A_3 & 0 \\ 0 & 0 & 0 & \dfrac{A_4}{2} \end{bmatrix} \begin{bmatrix} d_1 \\ d_2 \\ d_3 \\ d_4 \end{bmatrix}
\tag{5.51}
$$

Also,

$$
\begin{bmatrix} \sigma_1 \\ \sigma_2 \\ \sigma_3 \\ \sigma_4 \end{bmatrix} = E \begin{bmatrix} \dfrac{1}{L_1} & 0 & 0 & 0 \\ 0 & \dfrac{1}{L_2} & 0 & 0 \\ 0 & 0 & \dfrac{1}{L_3} & 0 \\ 0 & 0 & 0 & \dfrac{1}{L_4} \end{bmatrix} \begin{bmatrix} d_1 \\ d_2 \\ d_3 \\ d_4 \end{bmatrix} \leq \begin{bmatrix} 20 \text{ kips} \\ 20 \text{ kips} \\ 20 \text{ kips} \\ 20 \text{ kips} \end{bmatrix}
\tag{5.52}
$$

and
$$
\boldsymbol{\sigma} = E\mathbf{LD} \leq 20 \text{ kips}
\tag{5.53}
$$

For system equilibrium, the following equations must hold:

$$
\left.\begin{aligned}
48 \text{ kips} &= T_1 + T_2 - T_3 \\
48 \text{ kips} &= T_3 + T_4
\end{aligned}\right\}
\tag{5.54}
$$

or
$$
\begin{bmatrix} 48 \text{ kips} \\ 48 \text{ kips} \end{bmatrix} = \begin{bmatrix} 1 & 1 & -1 & 0 \\ 0 & 0 & 1 & 1 \end{bmatrix} \begin{bmatrix} T_1 \\ T_2 \\ T_3 \\ T_4 \end{bmatrix}
$$

A third equation specifying the reaction at A would be redundant with respect to the other two, and therefore is omitted.

Substituting Equation 5.51 in for $\mathbf{T}$ gives

$$\begin{bmatrix} 48 \text{ kips} \\ 48 \text{ kips} \end{bmatrix} = \begin{bmatrix} 1 & 1 & -1 & 0 \\ 0 & 0 & 1 & 1 \end{bmatrix} \frac{30 \times 10^6}{20} \begin{bmatrix} A_1 & 0 & 0 & 0 \\ 0 & A_2 & 0 & 0 \\ 0 & 0 & A_3 & 0 \\ 0 & 0 & 0 & \dfrac{A_4}{2} \end{bmatrix} \begin{bmatrix} d_1 \\ d_2 \\ d_3 \\ d_4 \end{bmatrix}$$

$$= \frac{30 \times 10^6}{20} \begin{bmatrix} A_1 & A_2 & -A_3 & 0 \\ 0 & 0 & A_3 & \dfrac{A_4}{2} \end{bmatrix} \begin{bmatrix} d_1 \\ d_2 \\ d_3 \\ d_4 \end{bmatrix} \tag{5.56}$$

Referring back to Figure 5.11, the floor beam displacements can be related to the individual member elongations by the incidence matrix.

$$\begin{bmatrix} d_1 \\ d_2 \\ d_3 \\ d_4 \end{bmatrix} = \begin{bmatrix} 1 & 0 \\ 1 & 0 \\ -1 & 1 \\ 0 & 1 \end{bmatrix} \begin{bmatrix} d_B \\ d_C \end{bmatrix} \tag{5.57}$$

where $\quad d_B =$ displacement of floorbeam B
$\qquad d_C =$ displacement of floorbeam C

If Equation 5.57 is substituted into Equation 5.56,

$$\begin{bmatrix} 48 \text{ kips} \\ 48 \text{ kips} \end{bmatrix} = \frac{30 \times 10^6}{20} \begin{bmatrix} A_1 & A_2 & -A_3 & 0 \\ 0 & 0 & A_3 & \dfrac{A_4}{2} \end{bmatrix} \begin{bmatrix} 1 & 0 \\ 1 & 0 \\ -1 & 1 \\ 0 & 1 \end{bmatrix} \begin{bmatrix} d_B \\ d_C \end{bmatrix}$$

and
$$\begin{bmatrix} 48 \text{ kips} \\ 48 \text{ kips} \end{bmatrix} = \frac{30 \times 10^6}{20} \begin{bmatrix} (A_1 + A_2 + A_3) & (-A_3) \\ (-A_3) & (A_3 + A_4/2) \end{bmatrix} \begin{bmatrix} d_B \\ d_C \end{bmatrix} \tag{5.58}$$

Therefore

$$\frac{20}{30 \times 10^6} \begin{bmatrix} (A_1 + A_2 + A_3) & (-A_3) \\ (-A_3) & (A_3 + A_4/2) \end{bmatrix}^{-1} \begin{bmatrix} 48 \text{ kips} \\ 48 \text{ kips} \end{bmatrix} = \begin{bmatrix} d_B \\ d_C \end{bmatrix} \leq \begin{bmatrix} \dfrac{0.1}{12} \\ \dfrac{0.2}{12} \end{bmatrix} \tag{5.59}$$

which provides the constraint on beam displacement. Rearranging Equation 5.53 gives

$$\mathbf{D} = (1/E)(\mathbf{L}^{-1})(\sigma) \tag{5.60}$$

If Equation 5.60 is substituted into Equation 5.56, then

$$\begin{bmatrix} 48 \text{ kips} \\ 48 \text{ kips} \end{bmatrix} = \begin{bmatrix} A_1 & A_2 & -A_3 & 0 \\ 0 & 0 & A_3 & A_4/2 \end{bmatrix} \begin{bmatrix} \sigma_1 \\ \sigma_2 \\ \sigma_3 \\ \sigma_4 \end{bmatrix} \tag{5.61}$$

also
$$\begin{bmatrix} \sigma_1 \\ \sigma_2 \\ \sigma_3 \\ \sigma_4 \end{bmatrix} \leq \begin{bmatrix} \sigma_{1\text{allow}} \\ \sigma_{2\text{allow}} \\ \sigma_{3\text{allow}} \\ \sigma_{4\text{allow}} \end{bmatrix} = \begin{bmatrix} 20 \text{ ksi} \\ 20 \text{ ksi} \\ 20 \text{ ksi} \\ 20 \text{ ksi} \end{bmatrix} \tag{5.62}$$

Equations 5.61 and 5.62 form constraints on the stress within each hanger member, and coupled with Equation 5.59 provide the constrained space for the problem.

If the geometry and loading of a truss are specified, the total weight of the truss members required to carry the specified load can be minimized. Cross-sectional areas of the truss members are used as the design variables.

A three-bar truss problem (Example 5.4) will be utilized to illustrate the model development strategy. This example was first presented by Schmit (1960) and has been used by several authors of structural optimization texts, including Kirsch (1981). The three-bar truss problem is an excellent introductory problem. With certain simplifications the model developed only contains two design variables; and it can therefore be graphed in two dimensions to illustrate the optimization procedure. This will be accomplished in Chapter 6.

Note that $\mathbf{K}^{-1}$ is a nonlinear function of the design variables. For more complicated problems, it may not be possible to express $\mathbf{K}^{-1}$ explicitly. Numerical methods may be employed to find approximate solutions in those cases. If there had been displacement constraints imposed on Example 5.4, then Equation 5.72 could have been used to establish these constraints.

The foregoing has been a brief introduction to structural optimization modeling. Complexity of models and difficulty of solution techniques increase rapidly with larger, realistic problems. Large problems can be made more manageable by decomposing the problem into smaller, interacting subproblems. These subproblems can be optimized individually, then coordinated to give an optimal, or near optimal solution to the whole problem. Engineers who contemplate using structural optimization techniques are directed to the excellent reference texts by Carmichael (1981), Galligher and Zenkiewicz (1973), and Kirsch (1981) for further study.

EXAMPLE 5.4 ———————————————————————————————

Given the three-bar truss with a single load in Figure 5.12; let A_1, A_2, and A_3 represent the cross-sectional areas of members 1, 2, and 3, respectively. Because of the symmetry involved in the truss and possible alternative loading angles, it is specified that the cross-sectional areas of member 1 and member 3 should be equal. Thus, A_1 and A_2 are the selected design variables. It will be further assumed that there are no displacement constraints associated with this problem. Behavioral constraints will involve the tensile and compressional stress limitations of the members. Design constraints would

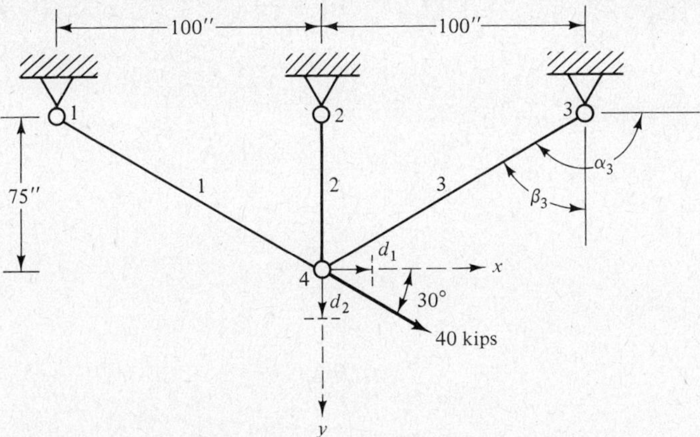

Figure 5.12 Three-bar truss.

require that the cross-sectional areas, A_1 and A_2, are greater than or equal to zero. If the allowable stress in tension is 20 ksi, the allowable stress in compression is 15 ksi, and the tensile stress is considered positive; the following design and behavioral constraints can be written

Design

$$\left.\begin{array}{c} A_1 \geq 0 \\[1em] A_2 \geq 0 \end{array}\right\} \tag{5.63}$$

Behavioral

$$\left.\begin{array}{c} \sigma_1 \leq 20 \\[0.7em] -\sigma_1 \leq 15 \\[0.7em] \sigma_2 \leq 20 \\[0.7em] -\sigma_2 \leq 15 \\[0.7em] \sigma_3 \leq 20 \\[0.7em] -\sigma_3 \leq 15 \end{array}\right\} \tag{5.64}$$

If d_1 and d_2 indicate the magnitude of the displacement of joint 4 along the X and Y axes, respectively, then equations can be written that relate the stress in the member to the displacement of the joint according to the relationship

$$\sigma_i = \frac{E}{L_i}(d_1 \cos \alpha_i + d_2 \cos \beta_i) \tag{5.65}$$

where E = modulus of elasticity (ksi)

        L_i = length of member i (in.)

       α_i, β_i = directional angles for member i defined as shown in Figure 5.12

Written in matrix form for all three members, Equation 5.65 becomes

$$
\begin{bmatrix} \sigma_1 \\ \sigma_2 \\ \sigma_3 \end{bmatrix} = E \begin{bmatrix} \dfrac{\cos \alpha_1}{L_1} & \dfrac{\cos \beta_1}{L_1} \\[2ex] \dfrac{\cos \alpha_2}{L_2} & \dfrac{\cos \beta_2}{L_2} \\[2ex] \dfrac{\cos \alpha_3}{L_3} & \dfrac{\cos \beta_3}{L_3} \end{bmatrix} \begin{bmatrix} d_1 \\ d_2 \end{bmatrix}
$$

$$
\begin{bmatrix} \sigma_1 \\ \sigma_2 \\ \sigma_3 \end{bmatrix} = E \begin{bmatrix} \dfrac{4/5}{125} & \dfrac{3/5}{125} \\[2ex] \dfrac{0}{75} & \dfrac{1}{75} \\[2ex] \dfrac{-4/5}{125} & \dfrac{3/5}{125} \end{bmatrix} \begin{bmatrix} d_1 \\ d_2 \end{bmatrix}
$$

$$
\begin{bmatrix} \sigma_1 \\ \sigma_2 \\ \sigma_3 \end{bmatrix} = E \begin{bmatrix} \dfrac{4}{625} & \dfrac{3}{625} \\[2ex] 0 & \dfrac{1}{75} \\[2ex] -\dfrac{4}{625} & \dfrac{3}{625} \end{bmatrix} \begin{bmatrix} d_1 \\ d_2 \end{bmatrix} \tag{5.66}
$$

According to the displacement method of structural analysis, the nodal displacement vector, **D**, can be related to the applied load vector, **R**, through the stiffness matrix, that is,

$$
\mathbf{KD} = \mathbf{R} \tag{5.67}
$$

Since this truss has 2 degrees of freedom (DF), the stiffness matrix will be 2×2. Elements of the stiffness matrix are computed as follows:

$$
\left.
\begin{aligned}
K_{11} &= E \sum_{i=1}^{3} \left(\frac{A_i}{L_i} \cos^2 \alpha_i \right) \\
K_{12} = K_{21} &= E \sum_{i=1}^{3} \left(\frac{A_i}{L_i} \sin \alpha_i \cos \alpha_i \right) \\
K_{22} &= E \sum_{i=1}^{3} \left(\frac{A_i}{L_i} \sin^2 \alpha_i \right)
\end{aligned}
\right\} \tag{5.68}
$$

For the given truss configuration,

$$
K_{11} = E \left[\frac{A_1}{125} \left(\frac{4}{5} \right)^2 + \frac{A_2}{75} (0)^2 + \frac{A_1}{125} \left(\frac{-4}{5} \right)^2 \right] = EA_1 \left(\frac{1.28}{125} \right)
$$

$$
K_{12} = K_{21} = E \left[\frac{A_1}{125} \left(\frac{3}{5} \right) \left(\frac{4}{5} \right) + \frac{A_2}{75} (1)(0) + \frac{A_1}{125} \left(\frac{3}{5} \right) \left(\frac{-4}{5} \right) \right] = 0
$$

$$K_{22} = E\left[\frac{A_1}{125}\left(\frac{3}{5}\right)^2 + \frac{A_2}{75}(1)^2 + \frac{A_1}{125}\left(\frac{3}{5}\right)^2\right]$$

$$= E\left[A_1\left(\frac{0.72}{125}\right) + A_2\left(\frac{1}{75}\right)\right]$$

and the stiffness matrix becomes

$$\mathbf{K} = E\begin{bmatrix} \left(\dfrac{1.28}{125}\right)A_1 & 0 \\ 0 & \left(\dfrac{0.72}{125}\right)A_1 + \left(\dfrac{1}{75}\right)A_2 \end{bmatrix} \tag{5.69}$$

the inverse of the stiffness matrix can be easily found for this example.

$$\mathbf{K}^{-1} = \frac{1}{E}\begin{bmatrix} \dfrac{1}{(1.28/125)A_1} & 0 \\ 0 & \dfrac{1}{(0.72/125)A_1 + (1/75)A_2} \end{bmatrix} \tag{5.70}$$

If Equation 5.67 is premultiplied by $\mathbf{K}^{-1}$, that is,

$$\mathbf{K}^{-1}\mathbf{K}\mathbf{D} = \mathbf{K}^{-1}\mathbf{R} \tag{5.71}$$

then
$$\mathbf{D} = \begin{bmatrix} d_1 \\ d_2 \end{bmatrix} = \frac{1}{E}\begin{bmatrix} \dfrac{1}{(1.28/125)A_1} & 0 \\ 0 & \dfrac{1}{(0.72/125)A_1 + (1/75)A_2} \end{bmatrix}\begin{bmatrix} 0.866 \\ 0.5 \end{bmatrix}\cdot(40)$$

$$\begin{bmatrix} d_1 \\ d_2 \end{bmatrix} = \frac{40}{E}\begin{bmatrix} \dfrac{0.866(125)}{1.28A_1} \\ \dfrac{0.5(125)}{0.72A_1 + 1.67A_2} \end{bmatrix}$$

$$\begin{bmatrix} d_1 \\ d_2 \end{bmatrix} = \frac{40}{E}\begin{bmatrix} \dfrac{108}{1.28A_1} \\ \dfrac{62.5}{0.72A_1 + 1.67A_2} \end{bmatrix} \tag{5.72}$$

If Equation 5.72 is substituted into Equation 5.66,

$$\begin{bmatrix} \sigma_1 \\ \sigma_2 \\ \sigma_3 \end{bmatrix} = 40\begin{bmatrix} \dfrac{4}{625} & \dfrac{3}{625} \\ 0 & \dfrac{1}{75} \\ -\dfrac{4}{625} & \dfrac{3}{625} \end{bmatrix}\begin{bmatrix} \dfrac{108}{1.28A_1} \\ \dfrac{62.5}{0.72A_1 + 1.67A_2} \end{bmatrix} \tag{5.73}$$

$$\sigma_1 = 40\left[\frac{108(4)}{(625)1.28A_1} + \frac{62.5(3)}{625(0.72A_1 + 1.67A_2)}\right]$$

$$= \frac{27.65}{1.28A_1} + \frac{12.0}{0.72A_1 + 1.67A_2}$$

$$\sigma_2 = 40\left[\frac{62.5}{75(0.72A_1 + 1.67A_2)}\right] = \frac{33.33}{0.72A_1 + 1.67A_2} \tag{5.74}$$

$$\sigma_3 = 40\left[-\frac{108(4)}{(625)1.28A_1} + \frac{62.5(3)}{625(0.72A_1 + 1.67A_2)}\right]$$

$$= -\frac{27.65}{1.28A_1} + \frac{12.0}{0.72A_1 + 1.67A_2}$$

Equations 5.74 can be substituted back into Equations 5.64 to form the system constraints governing the objective function

$$\text{min } Z = 125A_1 + 75A_2 + 125A_3 \tag{5.75}$$

$$= 250A_1 + 75A_2$$

which will minimize the total volume of material used in truss members.

WATER RESOURCES AND WATER QUALITY MODELS

The methods of systems analysis and the systems approach have been successfully applied to numerous water resources and water quality problems. Three representative types of problems will be presented here. More in-depth discussion can be found in the texts by Thomann (1972), Biswas (1976), Major and Lenton (1979), Hall and Dracup (1970), or Loucks, Stedinger, and Haith (1981). Each of these texts is devoted exclusively to application of the systems approach and systems analysis to water resources systems.

Example 5.5 illustrates a general type of problem called a blending problem in which a number of resources are mixed together to produce one or more products while remaining within resource, product, or resource combinatorial constraints. The objective is to either minimize total cost or to maximize total profit. For the particular problem of Example 5.5, the optimal solution will show how much waste treatment should be provided at each of six different locations to minimize cost while remaining within pollutant loading limitations. Example 5.6 will demonstrate another type of blending problem, mixing of two water sources to produce a supply of drinking water that meets quality criteria. This is also a two-stage problem, in that present demand is being met at minimum cost; however, a projected increase in demand will present a new set of conditions that must be met at overall minimum cost.

EXAMPLE 5.5 ____

A regional water supply and waste treatment authority has been organized for the three cities of Example 1.2. Figure 5.13 shows the schematic model of Example 1.2, with elements of the symbolic model (described in Table 5.1) and flowrates added.

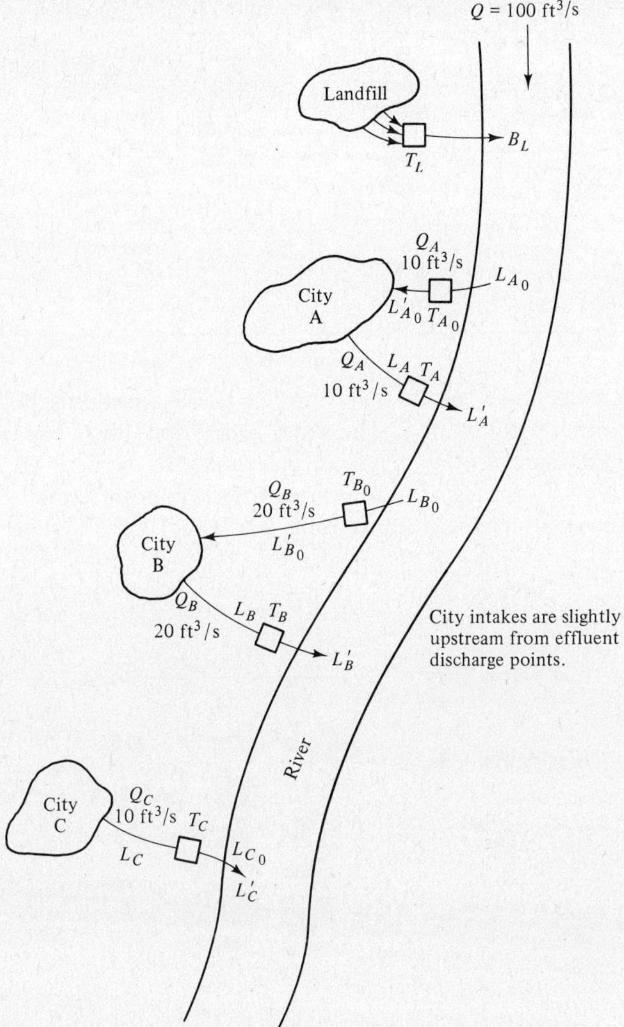

Figure 5.13   Areawide water supply, waste treatment plan.

The authority has the responsibility of providing adequate volume and quality of water supply to the three cities, and of providing sufficient waste treatment to meet water quality standards in the river. Planners wish to accomplish these services at minimum total cost. Water quality standards require that the concentration of pollutant L in the river be no more than L_L lb/ft^3 at all points within the authority's jurisdiction. The landfill is close enough to City A that almost no change in river pollutant concentration takes place between where the landfill effluent enters the river and City A draws off its water supply. However, the pollutants do become completely mixed into the river flow. Approximately 20% of the pollutant load in the river after effluent mixing from City A is removed by natural means before water enters the intake for City B's water supply. Similarly, 30% of the river loading after City B is removed prior to the outfall from City C.

TABLE 5.1 PERTINENT SYSTEM VARIABLES FOR TREATMENT OPTIONS

Location	Flowrate (cfs)	Pollutant load before treatment	Design variable (lb removed/s)	Pollutant load after treatment	Treatment costs ($/lb removed)
Landfill	~ 0	B_G (lb/s)	T_L (lb/s)	B_L (lb/s)	C_L ($/lb removed)
Water supply A	Q_A	L_{A0} (lb/ft^3)	T_{A0}	L'_{A0} (lb/ft^3)	C_{A0}
Waste treatment A	Q_A	$L'_{A0} + L_A$	T_A	L'_A	C_A
Water supply B	Q_B	L_{B0}	T_{B0}	L'_{B0}	C_{B0}
Waste treatment B	Q_B	$L'_{B0} + L_B$	T_B	L'_B	C_B
Waste treatment C	Q_C	L_C	T_C	L'_C	C_C

Drinking water codes specify that water for public consumption can contain no more than L_D lb/ft^3 of pollutant L. The water supply for City C can be assumed not to contain any pollutant L. Also, the river is clean above the City A landfill.

The following constraints can be written for the concentration of L in the river. Pertinent system variables are given in Table 5.1.

$$
\left.
\begin{aligned}
\text{Above City A} \qquad & L_{A0} = \frac{B_L}{Q} = \frac{(B_G - T_L)}{Q} \le L_L \\[2ex]
\text{Just below City A} \qquad & \frac{L'_A Q_A + L_{A0}(Q - Q_A)}{Q} \le L_L \\[2ex]
\text{Just below City B} \qquad & \frac{L'_B Q_B + L_{B0}(Q - Q_B)}{Q} \le L_L \\[2ex]
\text{Just below City C} \qquad & \frac{L'_C Q_C + L_{C0} Q}{Q_C + Q} \le L_L
\end{aligned}
\right\} \qquad (5.76)
$$

Drinking water quality for Cities A and B is subject to the constraints

$$
\left.
\begin{aligned}
L'_{A0} = L_{A0} - \frac{T_{A0}}{Q_A} \le L_{DA} \\[2ex]
L'_{B0} = L_{B0} - \frac{T_{B0}}{Q_B} \le L_{DB}
\end{aligned}
\right\} \qquad (5.77)
$$

Wastes generated by the cities and removed by the treatment process can be related to the load discharged by

$$
\left.
\begin{aligned}
(L'_{A0} + L_A) - \frac{T_A}{Q_A} = L'_A \\[2ex]
(L'_{B0} + L_B) - \frac{T_B}{Q_B} = L'_B \\[2ex]
L_C - \frac{T_C}{Q_C} = L'_C
\end{aligned}
\right\} \qquad (5.78)
$$

Removal of pollutants between cities is depicted by the equations

$$L_{B_0} = 0.80 \left[\frac{L'_A Q_A + L_{A_0}(Q - Q_A)}{Q} \right]$$
$$L_{C_0} = 0.70 \left[\frac{L'_B Q_B + L_{B_0}(Q - Q_A)}{Q} \right] \tag{5.79}$$

Equations 5.76 through 5.79 form the constrained space for the problem. The objective function can be written as

$$\min Z = C_L T_L + C_{A_0} T_{A_0} + C_A T_A + C_{B_0} T_{B_0} + C_B T_B + C_C T_C \tag{5.80}$$

EXAMPLE 5.6

A town requires the daily consumption of 4 million gal of water of a quality such that the concentration of a certain mineral pollutant must be kept below 100 mg/gal. The water can be supplied from two sources, either purchased from a local company at a cost of \$100/million gal, or pumped from a nearby stream at the cost of \$50/million gal. The concentration of the pollutant from the first source is 50 mg/gal, and that from the second source is 200 mg/gal. The water from the two sources is completely mixed before it is used. A model can be developed to find the mix of waters which meets the quality criterion at minimum cost.

Let $\quad X_1$ = millions of gallons presently being taken from the local company, and
$\quad\quad X_2$ = millions of gallons presently being taken from the stream

Then
$$X_1 + X_2 \geq 4 \tag{5.81}$$

The requirement on the concentration of pollutant is

$$\frac{(50 \text{ mg/gal})(X_1 \text{ million gal}) + (200 \text{ mg/gal})(X_2 \text{ million gal})}{(X_1 \text{ million gal}) + (X_2 \text{ million gal})} \leq 100 \text{ mg/gal} \tag{5.82}$$

thus,
$$50X_1 + 200X_2 \leq 100(X_1 + X_2)$$

or
$$50X_1 + 200X_2 - 100(X_1 + X_2) \leq 0$$

and
$$-50X_1 + 100X_2 \leq 0 \tag{5.83}$$

The model for present consumption is thus given as

$$\min Z = 100X_1 + 50X_2$$
$$\text{subject to} \quad \text{(ST):} \quad X_1 + X_2 \geq 4$$
$$-50X_1 + 100X_2 \leq 0 \tag{5.84}$$

Solution of this model will provide a minimum value of \$333.33/day for $X_1 = 2.67$ million gal/day and $X_2 = 1.33$ million gal/day.

This is the course that the town has followed, and it has signed a long-term contract to obtain 2.67 million gal/day from the local company. Now, the town foresees a need to double water supply capacity while maintaining the same finished product quality. The local water company has stated that it can supply up to an additional 2 million gal/day, but the cost will be \$120/million gal. Quality of the additional water will also be slightly degraded to 60 mg/gal. Environmental considerations limit the additional water that can be removed from the river to 3 million gal/day. The quality of the river water will remain the same; however, the cost of pumping will be \$70/million gal. If the consumption is doubled, and letting

X_3 = additional millions of gallons from the local company, and

X_4 = additional millions of gallons from the stream

then the overall model can be written as

$$
\begin{aligned}
\min Z = 100X_1 + {} & 50X_2 + 120X_3 + 70X_4 \\
\text{ST:} \quad X_1 + {} & X_2 && = 4 \\
-50X_1 + {} & 100X_2 && \leq 0 \\
X_1 + {} & X_2 + X_3 + X_4 && \geq 8 \\
& X_3 && \leq 2 \\
& X_4 && \leq 3 \\
-50X_1 + {} & 100X_2 - 40X_3 + 100X_4 && \leq 0 \\
X_1 & && = 2.67 \\
& X_2 && = 1.33 \\
& X_1, X_2, X_3, X_4 && \geq 0
\end{aligned}
\tag{5.85}
$$

The last two constraints are required to keep X_1 and X_2 at their present values. This model will be further investigated in Exercise 6-29.

Another water resources problem is the prevention of receiving water pollution by stormwater runoff. In order to meet receiving water quality standards, many urban areas may have to capture and treat stormwater runoff. Example 5.7 shows a methodology for allocating combinations of storage and treatment to accomplish this task at minimal cost. Diseconomies of scale (storage costs increase at a faster rate than the volume of storage) are included in the estimation of storage costs. Also, the constraints are presented in a tabular form that eliminates the need to write down the variable names with each constraint. Note that each constraint is identified, and the constraint coefficients are displayed in a form that facilitates their transcription to a computer data file. The first three constraints are continuity constraints for each time increment. Constraints 4 through 6 require that the storage volume accommodate the largest volume that is to be stored during any time increment. Constraints 7 and 8 bracket the allowable treatment rate, while the last five constraints relate the storage volume required to the storage diseconomies of scale.

EXAMPLE 5.7 ————————————————————————————————

A stormwater management model has been used to generate an outflow hydrograph from an urban drainage basin. Figure 5.14 shows a sketch of the system and the predicted hydrograph. If all of the outflow from the storm sewer system is to be captured

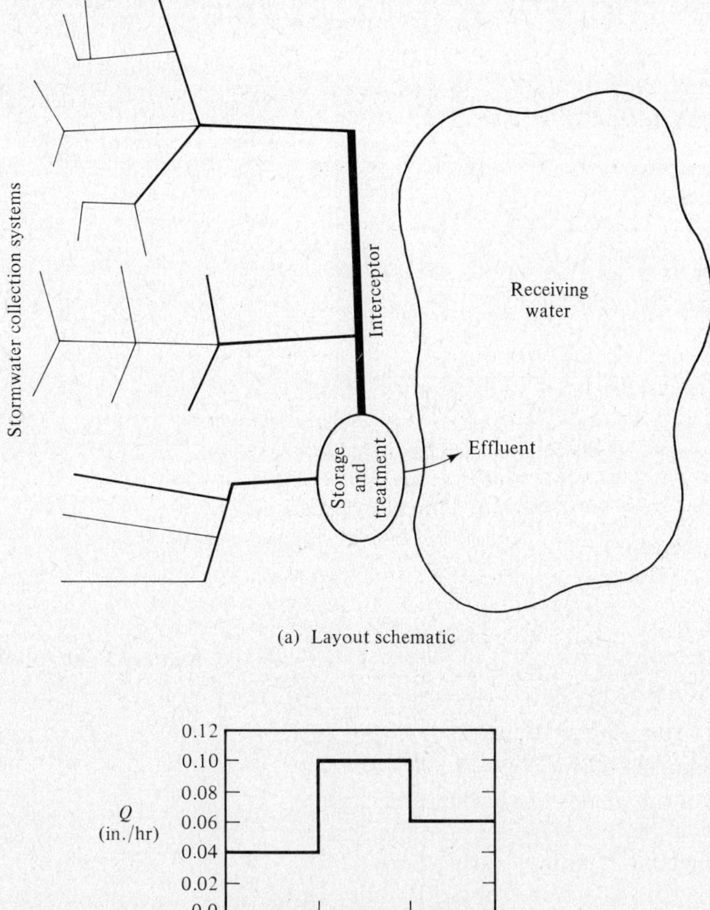

(a) Layout schematic

(b) Design storm hydrograph

Figure 5.14 Stormwater management.

and treated prior to release, develop a model that could be used to determine the optimum combination of storage and treatment that would accomplish this at minimum cost.

Because of the high cost of acquiring and preparing additional land, there are diseconomies of scale associated with increasing storage capacity. Figure 5.15 shows these diseconomies. Treatment costs are estimated by a linear function with a slope of $\$60 \times 10^6$/in. of treatment. A minimum rate of 0.01 in./hr and a maximum rate of 0.02 in./hr are placed on treatment. Economies of scale in treatment could be included in the model. The objective is to allocate the inflow during each hour to

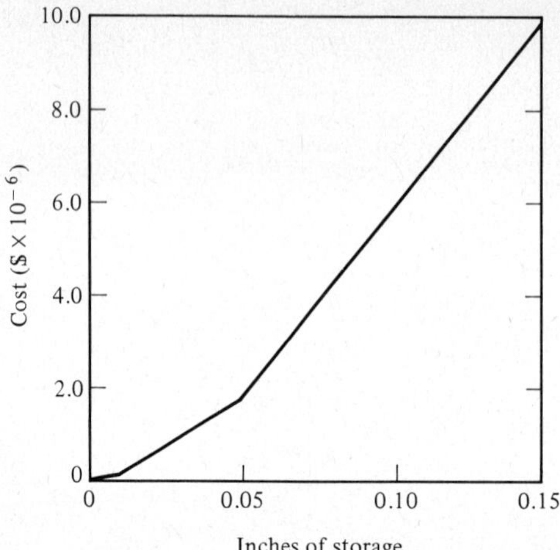

Range	Slope ($ \times 10^{-6}$/in.)
0–0.01	12
0.01–0.05	40
0.05–	80

Figure 5.15 Diseconomies of scale in storage costs.

either storage or treatment in the most cost-effective manner. The following variables are used in the model:

T = treatment rate (in./hr)
S = storage volume (in.)
IS_1 = volume in storage during time step 1
IS_2 = volume in storage during time step 2
IS_3 = volume in storage during time step 3
E_1 = portion of storage subject to unit cost of 12
E_2 = dummy variable used in calculation of E_3; E_2 is the total volume in storage, including E_1
E_3 = portion of storage subject to unit cost of 40
E_4 = portion of storage subject to unit cost of 80

The objective function is

$$\min Z = 60T + 12E_1 + 40E_3 + 80E_4 \tag{5.86}$$

with all cost coefficients scaled by a factor of 10^{-6}; and the constraints, presented in tabular form, are shown in Table 5.2.

TRANSPORTATION MODELS

Transportation optimization models involve movement of some commodity from two or more origins to two or more destinations. Equations 4.12 and 4.13 presented a model of the classic transportation problem, depicted by Figure 4.3, in which some

TABLE 5.2 TABULAR FORM OF CONSTRAINTS

Constraint		T	S	IS_1	IS_2	IS_3	E_1	E_2	E_3	E_4	Sign	RHS[a]
1.	Time step 1	1		1							$\geq$	0.04
2.	Time step 2	1		−1	1						$\geq$	0.10
3.	Time step 3	1			−1	1					$\geq$	0.06
4.			1	−1							$\geq$	0
5.	Storage size		1		−1						$\geq$	0
6.			1			−1					$\geq$	0
7.	Minimum treatment rate	1									$\geq$	0.01
8.	Maximum treatment rate	1									$\leq$	0.02
9.							1				$\leq$	0.01
10.								1			$\leq$	0.05
11.	Storage diseconomies						1	−1	1		$=$	0
12.			−1					1		1	$=$	0
13.			−1				1		1	1	$=$	0

[a] RHS = right-hand side.

resource or product is shipped from several origins to several destinations, the objective being to minimize total transportation costs while meeting the requirements of the destinations. This is a linear programming model that can be solved by general linear programming procedures. However, because of their special structure, transportation problems can also be solved by other algorithms which are more efficient computationally, but more difficult to program for computer solution. (An algorithm is a specialized method of solving a particular kind of problem. Algorithms usually require cycling and repetitions of computations.)

Another type of transportation problem involves shipping of goods or resources between origins and destinations, but can also involve transfer of the goods or resources through intermediate transfer points. Generally, this transfer is accompanied by a handling cost. In this type of problem, the definition of origins and destinations is somewhat blurred, as any of the nodes that are subject to transshipment can act as both origins and destinations. For transshipment problems, it is better to think in terms of locations, rather than origins and destinations. Example 5.8 illustrates a model to find the least cost alternative for disposing of refuse from four cities when it is possible to transfer refuse from one city through a second to a third; or, a central transfer point could be utilized. This model has the same final form as the transportation model, and can be solved by the same methods.

Example 5.9 illustrates how an engineer could minimize costs of providing cut and fill for a highway project. This model could be expanded to fit any length of highway as long as the required cut and fill are known for each section or station. Equal signs are used in the source constraints to ensure that the proper size cut is made. The assumption is made that excess material can be disposed of in the fills, so these constraints use greater than or equal constraints.

Transportation problems can also involve finding the shortest, or longest, path between two points in a road network. Figure 5.18 shows the road network of Figure 4.6, but with an additional dashed link from node 6 to node 1. This dashed, or pseudo-, link indicates that something is going to be determined in regards to a path from node 1 to node 6 through the network. Whether the longest or shortest path is to be found depends on the form of the model. The pseudolink ensures that the path starts at node 1 and finishes at node 6. Example 5.10 develops the model for finding the shortest path between these nodes.

All X_{ij} in this model will be either zero or 1. This is actually a specialized case of the assignment problem, in which a vehicle is sequentially assigned to links to travel between two locations in the network by either the shortest or longest path. The assignment problem will be discussed in the next section and specialized solution techniques will be developed in Chapter 6. It should also be noted that the project management techniques, such as the Critical Path Method, are special cases of the longest path problem in which all branches are unidirectional. These management techniques will be developed thoroughly in Chapter 9.

EXAMPLE 5.8 ───

You are an engineer with the planning department for a large metropolitan area. There are four cities within the metropolitan area that generate the following amounts of refuse each day:

City	Refuse (tons/day)
1	300
2	500
3	200
4	100

City 1 has a landfill with a 500 ton/day capacity. City 3 has a landfill with a 600 ton/day capacity. Refuse can be transferred through one city to another. In addition, there is one transfer station serving all four cities. Transportation and transfer costs are shown in Table 5.3. The costs on the diagonal are the transfer costs.

Cities 2 and 4 have net supplies of 500 and 100 tons/day, respectively, while Cities 1 and 3 have net capacities of 200 and 400 tons/day, respectively. Figure 5.16 shows a schematic model of the problem. The symbolic definition model for the transshipment problem will be as follows:

X_{ij} = quantity shipped from location i to location j,
  t_i = quantity transshipped through location i,
  c_{ij} = cost of shipping from location i to location j (note: c_{ij} may or may not equal c_{ji}),
  c_i = cost of transshipping through location i,
  a_i = net supply at location i, and
  b_i = net demand at location i

TABLE 5.3 TRANSPORTATION AND TRANSFER COSTS

Transportation costs

To location

		1	2	T	3	4
	1	4	6	4	10	15
	2	6	3	5	14	11
From location	T	4	5	2	3	6
	3	10	14	3	4	7
	4	15	11	6	7	5

T = transfer station

Diagonal shows transshipment (transfer) costs through location i

Since each node can act as either an origin or a destination, there will be two constraints for each node. The source constraint for location 1 can be given by

$$X_{12} + X_{13} + X_{14} + X_{1T} - t_1 = a_1 \tag{5.87}$$

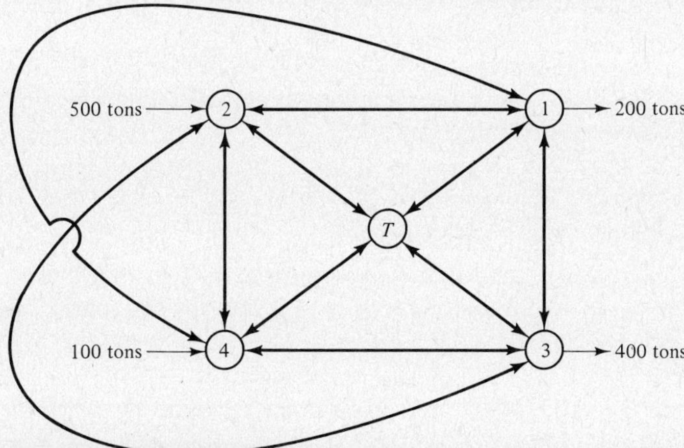

Figure 5.16 Refuse disposal optimization.

and the destination constraint can be given by

$$X_{21} + X_{31} + X_{41} + X_{T1} - t_1 = b_1 \tag{5.88}$$

It is convenient to insert a new variable, X_{ii}, into Equations 5.87 and 5.88 to represent the additional number of units that could have been transshipped through node i. If t is the upper bound on the number of units that could be transshipped through a node (t will equal the total production or demand for a balanced problem), then,

$$t_i + X_{ii} = t$$

and

$$t_i = t - X_{ii} \tag{5.89}$$

Substituting Equation 5.89 into each source and destination constraint results in the following set of constraints:

Source Constraints

$$X_{12} + X_{13} + X_{14} + X_{1T} + X_{11} = t + a_1$$

$$X_{21} + X_{23} + X_{24} + X_{2T} + X_{22} = t + a_2$$

$$X_{31} + X_{32} + X_{34} + X_{3T} + X_{33} = t + a_3$$

$$X_{41} + X_{42} + X_{43} + X_{4T} + X_{44} = t + a_4$$

$$X_{T1} + X_{T2} + X_{T3} + X_{T4} + X_{TT} = t + a_T$$

Destination Constraints

$$X_{21} + X_{31} + X_{41} + X_{T1} + X_{11} = t + b_1$$

$$X_{12} + X_{32} + X_{42} + X_{T2} + X_{22} = t + b_2$$

$$X_{13} + X_{23} + X_{43} + X_{T3} + X_{33} = t + b_3$$

$$X_{14} + X_{24} + X_{34} + X_{T4} + X_{44} = t + b_4$$

$$X_{1T} + X_{2T} + X_{3T} + X_{4T} + X_{TT} = t + b_T$$

$$\tag{5.90}$$

Equation 5.89 actually forms a companion set of constraints which relate Equations 5.90 to the objective function

$$\min Z = \sum_{i=1}^{n} \sum_{j=1}^{n} c_{ij} X_{ij} + \sum_{i=1}^{n} c_i t_i \tag{5.91}$$

For the problem above, $a_1 = 0$, $a_2 = 500$ tons, $a_3 = 0$, $a_4 = 100$ tons, $a_T = 0$, $b_1 = 200$ tons, $b_2 = 0$, $b_3 = 400$ tons, $b_4 = 0$, $b_T = 0$, and $t = 600$ tons; c_{ij} and c_i are given in Table 5.3.

EXAMPLE 5.9 ____

A highway engineer is trying to decide how best to accomplish a cut and fill operation. It costs $1.00/yd^3 to excavate fill from the borrow pit and $.50/yd^3 to dump excess material in a spoils area. Transportation costs are approximately $5.00/yd^3/mi. The

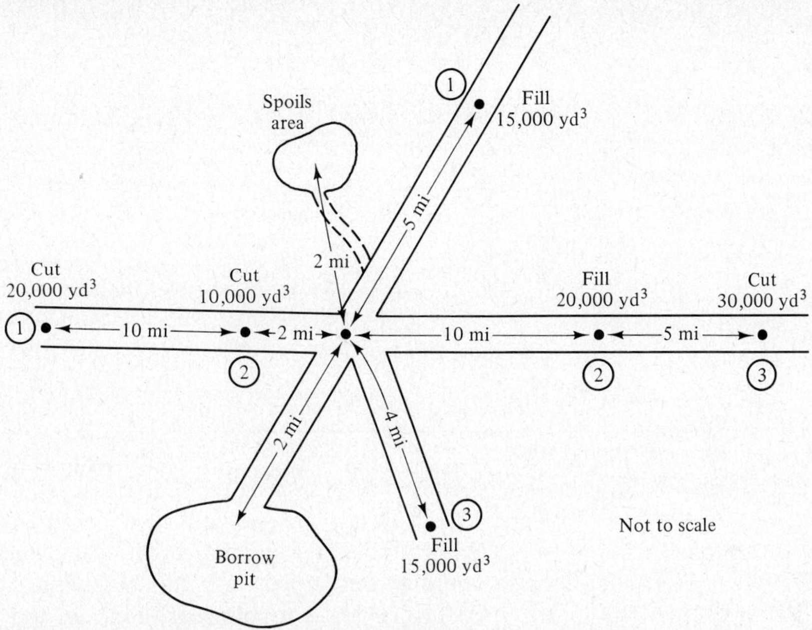

Figure 5.17 Highway cut and fill operation.

engineer's objective is to minimize the total cost while making all required cuts and fills. The costs of excavating the cuts and dumping in the fills will not be included in the model because the excavating and dumping would have to be done anyway. No more than 50,000 yd³ can be excavated from the borrow pit. The layout of the project is shown in Figure 5.17. Table 5.4 gives the pertinent costs for hauling, including excavation costs in the borrow pit and dumping costs in the spoils area.

An objective function for this problem is

TABLE 5.4 CUT AND FILL COSTS

		Destinations (sinks)			
		1	2	3	S
	1	85	110	80	70.5
	2	35	60	30	20.5
Origins (sources)	3	100	25	95	85.5
	B	36	61	31	—

$$\min Z = \sum_{i=1}^{n} \sum_{j=1}^{m} c_{ij} X_{ij} \qquad (5.92)$$

which is subject to the following constraints:

$$
\left.
\begin{array}{l}
X_{11} + X_{12} + X_{13} + X_{1S} \hspace{5.5cm} = 20{,}000 \\[4pt]
\hspace{1.2cm} X_{21} + X_{22} + X_{23} + X_{2S} \hspace{3.7cm} = 10{,}000 \\[4pt]
\hspace{3.6cm} X_{31} + X_{32} + X_{33} + X_{3S} \hspace{1.9cm} = 30{,}000 \\[4pt]
X_{11} \hspace{1.7cm} + X_{21} \hspace{1.5cm} + X_{31} \hspace{1.5cm} + X_{B_1} \hspace{0.8cm} \ge 15{,}000 \\[4pt]
\hspace{0.7cm} X_{12} \hspace{1.6cm} + X_{22} \hspace{1.5cm} + X_{32} \hspace{1.3cm} + X_{B2} \hspace{0.4cm} \ge 20{,}000 \\[4pt]
\hspace{1.5cm} X_{13} \hspace{1.6cm} + X_{23} \hspace{1.5cm} + X_{33} \hspace{1.2cm} + X_{B3} \hspace{0.2cm} \ge 15{,}000 \\[4pt]
\hspace{7.0cm} X_{B1} + X_{B2} + X_{B3} + X_{BS} \le 50{,}000 \\[4pt]
\hspace{1.2cm} X_{1S} \hspace{2.6cm} + X_{2S} \hspace{2.3cm} + X_{3S} \hspace{1.9cm} + X_{BS} \ge 0
\end{array}
\right\} (5.93)
$$

EXAMPLE 5.10

Develop a model that can be utilized to determine the shortest path between locations
(nodes) 1 and 6 of Figure 5.18. Mileages are shown for each path (link). The link–
node incidence matrix of Table 4.2 would be augmented with data for the pseudopath
6–1. An incidence matrix could be developed from this augmented table, and the
transpose of the matrix used to develop a topological model of the road network
system. Normally, the link–node incidence matrix omits one of the node columns
because of the column redundancy. However, from a modeling standpoint, it is not
incorrect to include this extra column. Equations 5.94 show the product of the transpose
of the link–node table and the solution vector, **X**.

$$-X_{12} - X_{13} - X_{14} + X_{61} = 0$$

$$X_{12} - X_{23} + X_{32} - X_{24} - X_{25} - X_{26} = 0$$

$$X_{13} + X_{23} - X_{32} - X_{34} + X_{43} - X_{35} = 0$$

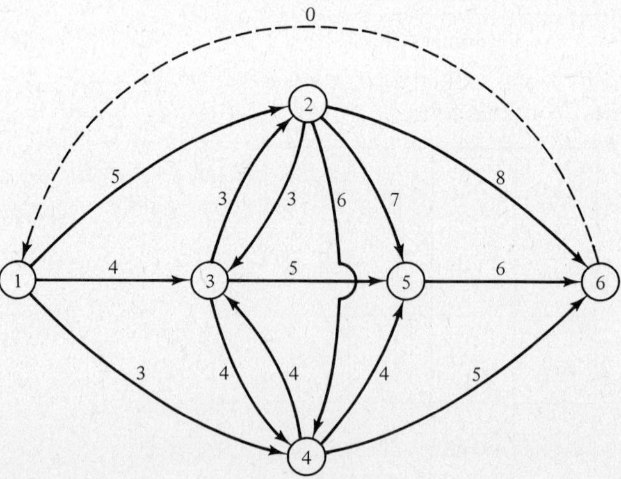

Figure 5.18 Network with pseudolink.

$$X_{14} + X_{24} + X_{34} - X_{43} - X_{45} - X_{46} = 0$$

$$X_{25} + X_{35} + X_{45} - X_{56} = 0 \tag{5.94}$$

$$X_{26} + X_{46} + X_{56} - X_{61} = 0$$

In actuality, any one of the equations in 5.94 could be eliminated with no loss in generality. One additional constraint is required. Equation 5.95,

$$X_{61} = 1 \tag{5.95}$$

specifies that link 6–1 has to be in the solution. The objective function of Equation 5.96 completes this model.

$$\min X = 5X_{12} + 4X_{13} + 3X_{14} + 3X_{23} + 3X_{32} + 6X_{24} + 7X_{25} + 8X_{26}$$
$$+ 4X_{34} + 4X_{43} + 5X_{35} + 4X_{45} + 5X_{46} + 6X_{56} + 0X_{61} \tag{5.96}$$

PRODUCTION MODELS

Production models help in making decisions about the best way to produce some end product or mix of products. The objective function is either to minimize costs or to maximize profit. A first type of production model is the product mix problem. Several products can be generated from given resources, some of which are limited in quantity. Example 5.11 illustrates a typical production model that can be used to maximize profit from a product mix at a steel fabrication plant.

Another type of production problem involves scheduling of production activities to meet demand for a certain product while optimizing costs or profits. Constraints may be placed on manpower or material resources. Certain assumptions have to be made about inventory levels at the beginning and end of the production period. The production period is divided up into accounting periods, generally 1 month long. Storage costs may be associated with carrying over inventory from one accounting period to another. Example 5.12 illustrates the scheduling of concrete sewer pipe production. Note that the model developed considers only one product out of many that the plant might be producing. The manager would also have to account for the interactions among these various products.

EXAMPLE 5.11 ————————————————————————————————————

A steel fabrication plant produces four different specialized steel assemblies. Each unit of specialized assembly is made up of the amounts of material specified in Table 5.5. The plant can realize a \$10 profit per unit of A, a \$12 profit per unit of B, an \$8 profit per unit of C, and a \$6 profit per unit of D. Assuming that whatever can be fabricated

TABLE 5.5 MATERIAL REQUIREMENTS

Specialized assembly	Angle iron (lb/unit)	Steel plate (lb/unit)	Fastening brackets (lb/unit)
A	150	200	75
B	200	250	60
C	100	125	80
D	250	175	65

can be sold, formulate a linear programming model that can be used to maximize profit, subject to the following additional constraints:

1. There are 100,000 lb of angle iron, 200,000 lb of steel plate, and 50,000 lb of strap steel available to be used in this fabrication.
2. Due to inventory problems, at least 100,000 lb of steel plate should be used.
3. Usage of the assemblies requires that two units of assembly A be produced for each unit of B.

The objective function is

$$\max Z = 10A + 12B + 8C + 6D \tag{5.97}$$

subject to the following constraints:

amount of angle iron available $\qquad 150A + 200B + 100C + 250D \le 100,000$

amount of steel plate available $\qquad 200A + 250B + 125C + 175D \le 200,000$

amount of fastening brackets available $\qquad 75A + 60B + 80C + 65D \le 50,000 \qquad (5.98)$

minimum amount of steel plate to be used $\qquad 200A + 250B + 125C + 175D \ge 100,000$

usage limitation $\qquad 0.5A - B = 0$

EXAMPLE 5.12

The manager of a concrete products plant is scheduling production of a certain size concrete sewer pipe over the next 5 months. Any pipe on hand at the beginning of the first month has been subtracted from the first month's requirement. The manager wants to meet the projected demand over the 5-month period at minimum cost. Constraints are imposed on the monthly regular rate processing time and the amount of raw material available each month. Table 5.6 gives cost and availability data for the problem.

TABLE 5.6 SEWER PIPE PRODUCTION DATA

	Month				
	1	2	3	4	5
Pipe required (units)	20	30	40	20	40
Cost of regular processing time ($/unit)	150	150	150	160	160
Cost of overtime processing ($/unit)	170	170	170	185	185
Regular processing time available (hr)	2000	2000	2000	2000	2000
Material availability (possible units)	40	50	50	10	30

It takes 75 hr of processing time to produce one unit of sewer pipe, and costs approximately \$20 to store one unit of pipe for 1 month. If it is assumed that no pipe will be carried over after the fifth month, the following model can be developed:

Let XR_i = units of pipe produced during month i on regular time
XO_i = units of pipe produced during month i on overtime
XS_i = units of pipe stored from month i to month $i + 1$

The objective function reflects the total cost of regular time and overtime processing for the 5 months, plus storage costs.

$$\min Z = 150XR_1 + 150XR_2 + 150XR_3 + 160XR_4 + 160XR_5 + 170XO_1$$
$$+ 170XO_2 + 170XO_3 + 185XO_4 + 185XO_5 + 20XS_1$$
$$+ 20XS_2 + 20XS_3 + 20XS_4 \tag{5.99}$$

During any particular month, the number of units available at the beginning of the month plus the number of units produced during the month has to equal the demand during that month plus the number of units carried over to the next month as storage. Thus, the following compatibility constraints can be developed:

$$\left.\begin{aligned}
XR_1 + XO_1 - XS_1 &= 20 \\
XR_2 + XO_2 + XS_1 - XS_2 &= 30 \\
XR_3 + XO_3 + XS_2 - XS_3 &= 40 \\
XR_4 + XO_4 + XS_3 - XS_4 &= 20 \\
XR_5 + XO_5 + XS_4 &= 40
\end{aligned}\right\} \tag{5.100}$$

The regular processing time and material availability constraints cannot be violated. Note that it is assumed that material available during 1 month is not carried over to the next month. Thus,

$$\left.\begin{aligned}
75XR_1 &\leq 2000 \\
XR_1 + XO_1 &\leq 40 \\
75XR_2 &\leq 2000 \\
XR_2 + XO_2 &\leq 50 \\
75XR_3 &\leq 2000 \\
XR_3 + XO_3 &\leq 50 \\
75XR_4 &\leq 2000 \\
XR_4 + XO_4 &\leq 10 \\
75XR_5 &\leq 2000 \\
XR_5 + XO_5 &\leq 30
\end{aligned}\right\} \tag{5.101}$$

A final constraint is that all variables have to be greater than or equal to zero, since a negative amount produced or stored would have no physical meaning.

Managers can also be concerned with allocating certain machines to different tasks, when any machine can be allocated to any task, but certain combinations of man and machine are more efficient. Example 5.13 develops a model for assignment of excavators to tasks in an open pit coal mine. These assignment problems are special forms of transportation problems, which in turn are special forms of the general linear programming problem. Special solution techniques for assignment models will be developed in Chapter 6.

EXAMPLE 5.13 ───────────────────────────────

The site manager of an open pit coal site has five different types of excavator available for allocation to five tasks, each of which is different in nature by virtue of its position or the nature of the material to be moved. The cost per cubic yard output for the machines at each of the five tasks (A, B, C, D, and E) is shown in Table 5.7. Costs are in pennies per cubic yard. To what jobs should each excavator be allocated in order to minimize production costs?

The objective function will be

$$\min Z = \sum_{i=1}^{n} \sum_{j=1}^{n} c_{ij} X_{ij} \tag{5.102}$$

in which c_{ij} = cost per cubic yard for the ith job to be done by the jth excavator, and

X_{ij} = indicator variable showing that the ith job will be accomplished by the jth machine. All X_{ij} will equal either zero or 1

TABLE 5.7   ASSIGNMENT COSTS

		Excavators				
		1	2	3	4	5
	A	30	17	25	18	28
	B	23	28	16	28	29
Tasks	C	17	15	21	14	33
	D	27	41	35	40	37
	E	16	32	16	32	25

Constraints preventing any task from being done twice are

$$
\left.
\begin{aligned}
X_{11} + X_{12} + X_{13} + X_{14} + X_{15} &= 1 \\
X_{21} + X_{22} + X_{23} + X_{24} + X_{25} &= 1 \\
X_{31} + X_{32} + X_{33} + X_{34} + X_{35} &= 1 \\
X_{41} + X_{42} + X_{43} + X_{44} + X_{45} &= 1 \\
X_{51} + X_{52} + X_{53} + X_{54} + X_{55} &= 1
\end{aligned}
\right\}
\qquad (5.103)
$$

and constraints preventing any excavator from being used more than once are

$$
\left.
\begin{aligned}
X_{11} + X_{21} + X_{31} + X_{41} + X_{51} &= 1 \\
X_{12} + X_{22} + X_{32} + X_{42} + X_{52} &= 1 \\
X_{13} + X_{23} + X_{33} + X_{43} + X_{53} &= 1 \\
X_{14} + X_{24} + X_{34} + X_{44} + X_{54} &= 1 \\
X_{15} + X_{25} + X_{35} + X_{45} + X_{55} &= 1
\end{aligned}
\right\}
\qquad (5.104)
$$

SUMMARY

Optimization models can be developed for a wide variety of civil engineering applications, including structural, water resource and quality, transportation, and production problems. They can vary from fairly simple to extremely complex mathematical formulations. Successful development of optimization models will depend upon the ability to establish reasonable objective functions and constraints that are both comprehensive and realistic. The engineer developing the model must ensure that the information to be gained from applying a model will be worth the expense of developing and using it. With practice, the modeler will learn to recognize problems that, when expressed in mathematical terms, may have the same form as other problems for which solution techniques have been developed, even though the system being modeled may be quite different. For example, a mixing model could be applied to a stream water quality problem or to an industrial process control problem.

APPLICATIONS EXERCISES

5-1. You are in charge of a large construction project that involves two sites. Currently, you need 50 tons of aggregate at site 1 and 22 tons of aggregate at site 2. Two suppliers are available. Supplier 1 can deliver 29 tons and supplier 2 can deliver 43 tons. Restrictions on the type of work to be done require that at least 20 tons used at site 1 have to come from source 2 and no more than 10 tons of aggregate from source 1 can be used at site 2. Delivered aggregate costs ($/ton) are as follows:

Site Source	1	2
1	1	2
2	1.5	1.25

Formulate this as a linear programming problem to minimize the cost of aggregate.

5-2. You are a planning engineer for a metropolitan area. You want to minimize the cost of landfilling the refuse produced by Cities A, B, C, and D under your jurisdiction. Three of the cities (A, B, and D) also have landfills, but their capacities do not match up with the amount of waste produced in each city. Each city produces the following amount of refuse:

$$
\begin{array}{ll}
\text{A:} & 100 \text{ tons/day} \\
\text{B:} & 50 \text{ tons/day} \\
\text{C:} & 60 \text{ tons/day} \\
\text{D:} & 40 \text{ tons/day}
\end{array}
$$

The landfills have the following capacities:

$$
\begin{array}{ll}
\text{A:} & 50 \text{ tons/day} \\
\text{B:} & 75 \text{ tons/day} \\
\text{D:} & 150 \text{ tons/day}
\end{array}
$$

The cost per ton for hauling and landfilling are given in the following matrix:

		Landfills	
	A	B	D
A	10	20	30
B	18	12	16
C	22	18	26
D	28	20	8

Cities

Formulate this as a linear programming problem to minimize the total cost of disposing of all trash.

***5-3.** A contractor is planning a job that will require a large amount of gravel and sand. The estimates for the necessary gravel and sand are:

Coarse gravel	20,000 yd³
Fine gravel	29,000 yd³
Sand	20,000 yd³

There are two pits from which material can be obtained, and the plan is to haul from these two pits and to separate (screen) on the job whatever material is needed. Analysis shows that the material at each pit has the following composition:

	Pit A	Pit B
Coarse gravel	20%	30%
Fine gravel	14%	50%
Sand	25%	20%
Waste	41%	0%

It costs \$8/yd³ for material and hauling from pit A and \$16/yd³ from pit B. Set this up as a linear programming problem to minimize delivered cost.

5-4. The figure for Exercise 5-4 shows a portion of a highway construction project involving cut and fill operations. Develop a mathematical model that could be used to determine the minimum cost method of meeting all requirements. Borrow pit 1 has 200,000 yd³ of fill available, and could accept up to 100,000 yd³ additional fill, while pit 2 has 150,000 yd³ available, and could accept up to 200,000 yd³ additional.

* From Meredith, Wong, Woodhead, and Wortman, *Design and Planning of Engineering Systems,* © 1973, p. 160. Reprinted by permission of Prentice-Hall, Inc., Englewood Cliffs, NJ.

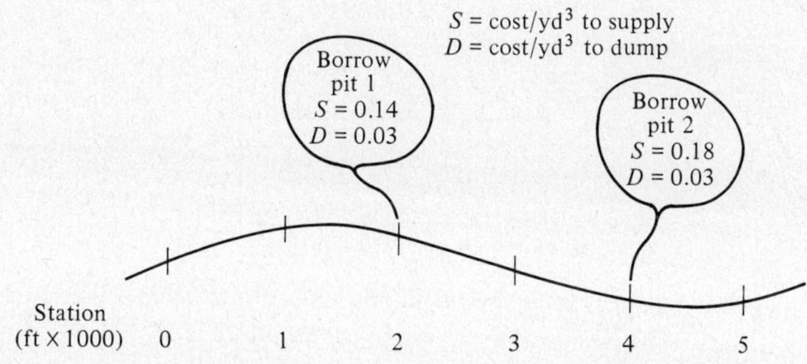

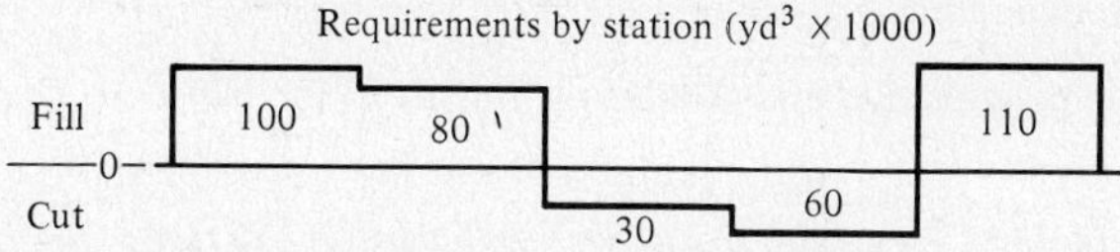

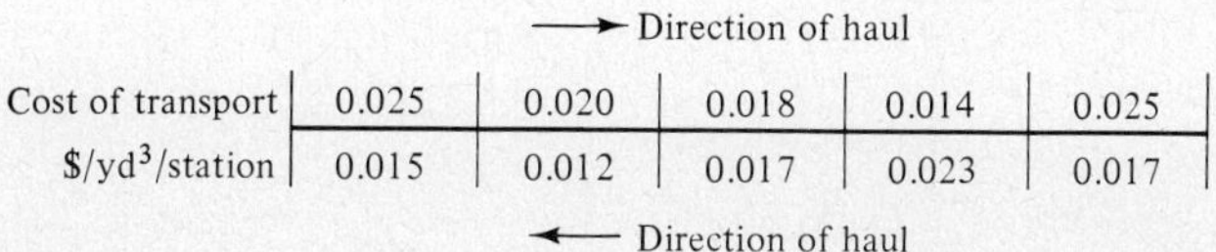

Cost of transport	0.025	0.020	0.018	0.014	0.025
\$/yd³/station	0.015	0.012	0.017	0.023	0.017

5-5. You are an engineer working for a construction firm that plans to build an apartment complex. You have a choice of building some combination of one-, two-, and three-bedroom apartments. A one-bedroom apartment requires 1000 board ft of lumber, 10 yd^3 of concrete, and 20 rolls of insulation. A two-bedroom apartment requires 1400 board ft of lumber, 12 yd^3 of concrete, and 24 rolls of insulation. A three-bedroom apartment requires 1600 board ft of lumber, 13 yd^3 of concrete, and 26 rolls of insulation. A one-bedroom apartment rents for $250 per month, a two-bedroom apartment for $300 per month, and a three bedroom for $325 per month. Local ordinances require at least one three-bedroom apartment for every three one-and-two bedroom apartments. Your firm has set aside 100,000 board ft of lumber, 1200 yd^3 of concrete, and 2500 rolls of insulation for this project. Formulate this as a linear programming problem to maximize rental income. Give objective function and constraints.

5-6. You are in charge of a large construction project involving three separate sites some miles apart. Site 1 requires 1000 yd^3 of concrete; site 2 requires 3000 yd^3; and site 3 requires 2000 yd^3. You have three concrete batching plants to draw from: plant 1 can supply 2000 yd^3, plant 2 can provide 1000 yd^3, and plant 3 can provide 2000 yd^3. If the three batching plants cannot supply sufficient concrete, you can acquire concrete from an outside source at a cost of $20/$yd^3$ for site 1, $22/$yd^3$ for site 2, and $24/$yd^3$ for site 3. Normal concrete costs (delivered) are as follows:

		Site 1	Site 2	Site 3	
	1	10	15	20	
Batching plant	2	12	10	16	$/$yd^3$ delivered
	3	15	7	22	

Formulate this problem so that you can minimize your concrete costs.

5-7. A steel supplier has two plants used to supply three warehouses. The steel is to be shipped to each warehouse via the least expensive route. Transshipment is allowed only between the two plants or between the three warehouses.

The diagram for this problem is shown in the figure for Exercise 5-7.

Shipping Costs from Location i *to Location* j *($/ton)*

$$c_{12} = c_{21} = 10 \qquad c_{23} = 30 \qquad c_{34} = c_{43} = 20$$

$$c_{13} = 15 \qquad c_{24} = 40 \qquad c_{45} = c_{54} = 10$$

$$c_{14} = 30 \qquad c_{25} = 25 \qquad c_{35} = c_{53} = 25$$

$$c_{15} = 40$$

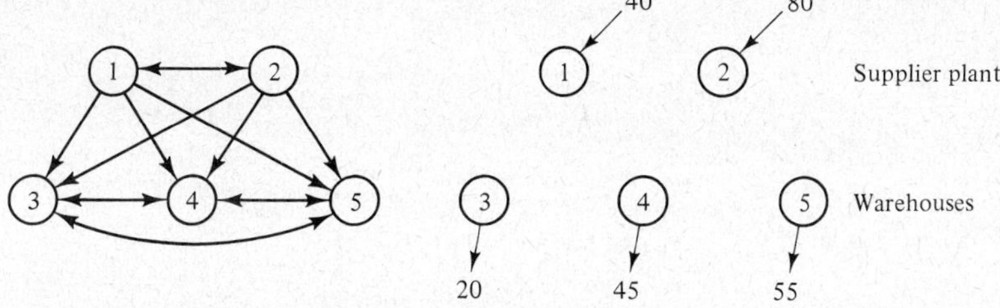

Transshipping Costs Through Location i ($/ton)

$c_1 = c_2 = 2$ net supplies: $a_1 = 40$, $a_2 = 80$

$c_3 = 3$ net demands: $b_3 = 20$, $b_4 = 45$, $b_5 = 55$

$c_4 = 2$

$c_5 = 4$

Develop a model that could be utilized to determine the minimum total shipping cost.

5-8. A construction firm has supply centers located in Atlanta, Chicago, and New York City. These centers have available 40, 20, and 30 units of a particular resource, respectively. The firm's job sites require the following number of units: Cleveland, 25; Louisville, 10; Memphis, 20; Pittsburgh, 30; and Richmond, 15. The shipping cost per unit in dollars between each center and job site is given in the following table:

	Cleveland	Louisville	Memphis	Pittsburgh	Richmond
Atlanta	55	30	40	50	40
Chicago	35	30	100	45	60
New York	40	60	95	35	30

Formulate a model that could be used to minimize the total shipping cost.

5-9. You are given the road network in the figure for Exercise 5-9, with the distances for each link shown. Develop a model that could be utilized to find the minimum length path from node 1 (location A) to node 7 (location B).

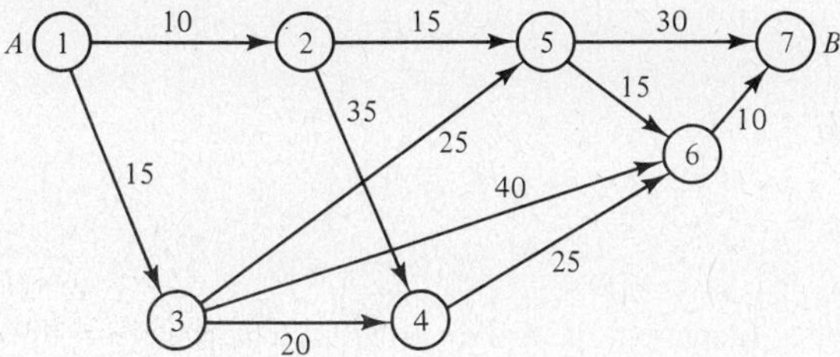

5-10. Given the following information, develop a model that could be used to find the minimum transportation costs for supplying each project.

Project	Requirement (truckloads/week)	Plant	Production (truckloads/week)
A	45	W	35
B	50	X	40
C	20	Y	40

Transportation costs

From	To: Project A	Project B	Project C
Plant W	5	10	10
Plant X	20	30	20
Plant Y	5	8	12

5-11. Three cities produce the excess refuse shown in the figure for Exercise 5-11. This excess refuse can be shipped through either (or both) of two transfer stations enroute to any one of three landfills. Landfill capacities are also shown. Refuse quantities are in units of tons per day. Shipment costs in dollars per ton are given alongside each route arrow. Transshipment costs are given above each applicable node. These are the costs per unit of refuse transferred through a node to another node. Note that there is no transshipment between the sink nodes, and some of the arrows are unidirectional.

     Develop a model that could be utilized to find the minimum cost of disposing of the excess refuse.

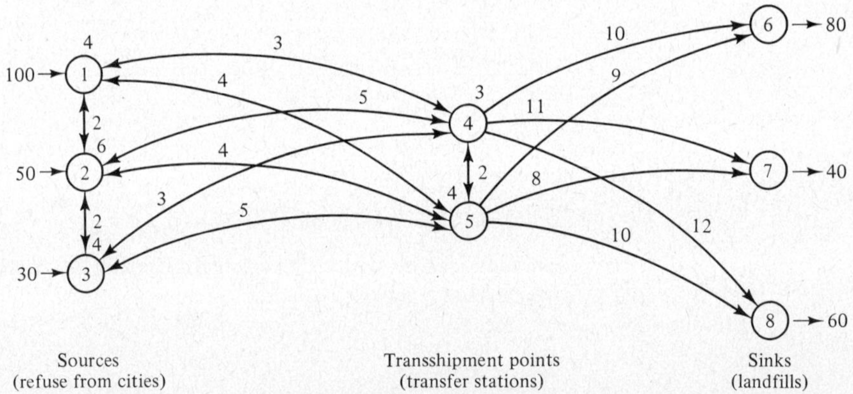

5-12. A company produces two products and the net profits from producing X units of each are given by

$$\text{profit } P_1 = 20X_1 - 10^{-6}X_1^2$$

$$P_2 = 5X_2 - 10^{-6}X_2^2$$

(These equations represent the diseconomies of scale which occur when the plant becomes overloaded.)

(a) Develop a mathematical model that could be used to find the optimum production quantities.

(b) If product 2 is to be produced in quantities always exactly 10 times product 1, show what difference this makes in the mathematical model.

5-13. A design calls for 520 yd^3 of type X granular material for use as a subbase for a roadway, and 70 yd^3 of type E for use as a base underlayment for sidewalks along the roadway. Three gravel beds are available to draw from. Two of these are very low due to a busy summer and the third has limited draw because it is already being tapped by an ongoing project. The availability and cost for each of the sources is as follows:

> pit A has available: 250 yd^3 of type X @ $\$$ 6.80/yd^3
>
> 25 yd^3 of type E @ \$15.75/yd^3
>
> pit B has available: 250 yd^3 of type X @ $\$$ 7.00/yd^3
>
> 35 yd^3 of type E @ \$16.50/yd^3
>
> pit C has available: 400 yd^3 of type X @ $\$$ 7.50/yd^3
>
> 25 yd^3 of type E @ \$16.00/yd^3

Pits A, B, and C also have a minimum purchase order of \$1500 each.

Develop a model for determining the minimum cost way of meeting the requirements.

5-14. Preliminary investigations have shown that the hazardous waste treatment facilities discussed in Exercise 2-15 should be centralized at several regional facilities. Factors that have to be considered when picking sites include impact on the environment, seismic risk, impact on municipalities, proximity of site to incompatible land uses, population density near the site, and distances from points of waste generation.

Develop a mathematical model that could be used to provide a first estimate of the optimal location of treatment facilities, based on distance between waste generation locations and proposed facility sites. Assume that the center of mass of waste generation is near the largest city in each county, and that possible sites for treatment facilities exist in the vicinity of the population centers indicated. The model should be able to determine where facilities should be located, how large each facility should be, and the distribution of wastes from sources to treatment facilities. Additional information pertinent to the problem is given below and in the figure for Exercise 5-14 (on p. 182).

After developing the model, study it to determine if there are any aspects of the model that seem unrealistic. If there are, describe them, and modify the model to accommodate these additional factors. Would any additional information be required?

Additional Information

(a) Approximate hauling costs for industrial chemical hazardous waste haulers are:

> average size chemical transport vehicle = 5,000 gal
>
> transportation cost = \$50/hr for the vehicle, plus
>
> \$11/hr for the operator

Assume vehicles travel at 55 mph without rest stops. Neglect unloading time.

(b) The smallest facility capacity that is deemed to be cost effective is 3×10^6 gal/yr. Approximate annual treatment costs for different size plants are:

Capacity (10^6 gal/yr)	Annual treatment costs (10^6 \$/yr)
3	1.5
10	3.0
60	7.0

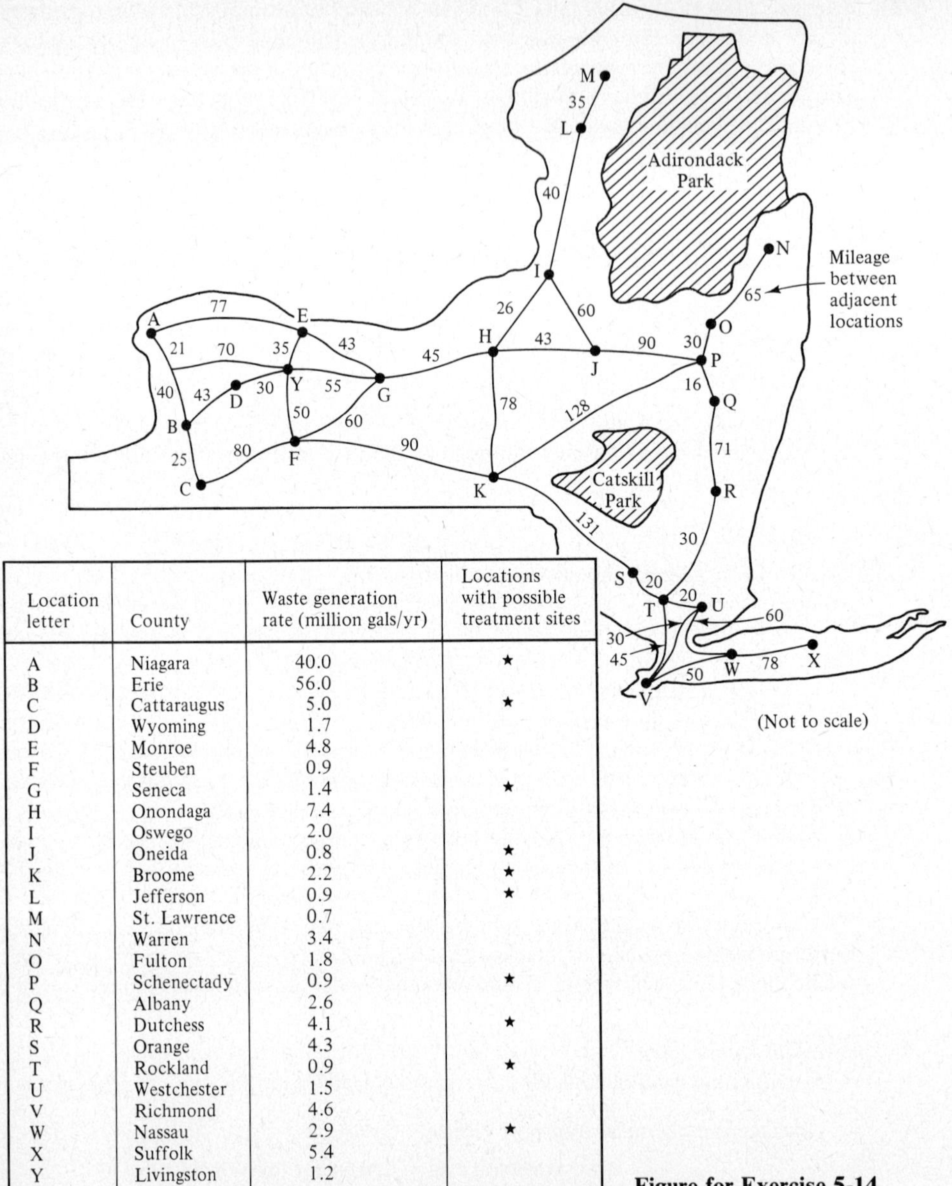

Location letter	County	Waste generation rate (million gals/yr)	Locations with possible treatment sites
A	Niagara	40.0	★
B	Erie	56.0	
C	Cattaraugus	5.0	★
D	Wyoming	1.7	
E	Monroe	4.8	
F	Steuben	0.9	
G	Seneca	1.4	★
H	Onondaga	7.4	
I	Oswego	2.0	
J	Oneida	0.8	★
K	Broome	2.2	★
L	Jefferson	0.9	★
M	St. Lawrence	0.7	
N	Warren	3.4	
O	Fulton	1.8	
P	Schenectady	0.9	★
Q	Albany	2.6	
R	Dutchess	4.1	★
S	Orange	4.3	
T	Rockland	0.9	★
U	Westchester	1.5	
V	Richmond	4.6	
W	Nassau	2.9	★
X	Suffolk	5.4	
Y	Livingston	1.2	

Figure for Exercise 5-14

5-15. Develop a model or models that could be used to minimize the total delivered cost of water to the five communities of Exercise 4-15. Use realistic evaluative criteria. Indicate what additional data would have to be acquired before the models could be utilized, and describe how that data could be gathered. Outline a solution procedure that could be utilized to produce useful information from the model(s).

5-16. You are an engineer with the planning department for a county that contains two major metropolitan areas. The two cities are located in a flood plain of a broad, meandering river with mountains overlooking the valley. The cities are under the jurisdiction of a single sewage authority that provides service for the entire area. There are five wastewater treatment plants that treat sewage for both of the metropolitan areas. The treatment plants may ship sludge to any of five disposal sites. The disposal sites include two landfills (LF), two land disposal (LD) sites, and one incinerator (INC). A diagram of the problem and information of waste production and waste disposal have been provided in the figure for Exercise 5-16. Develop schematic, symbolic, and mathematical models that could be used to minimize the sludge disposal costs.

What information might be useful in further refining the definition of this problem?

Wastewater treatment plants	Sludge (lb/day)
1	2500
2	6000
3	4000
4	6800
5	3000

Disposal sites		Sludge (lb/day)
LF	1(1)	6700
LF	2(2)	7300
LD	1(3)	3000
LD	2(4)	5000
INC	(5)	2000

Transportation costs
$(\$ \times 10^{-2}/\text{lb})$

Sinks

Sources	1	2	3	4	5
1	20	40	35	15	18
2	20	38	34	16	20
3	35	17	19	40	22
4	18	33	27	17	15
5	25	23	20	26	16

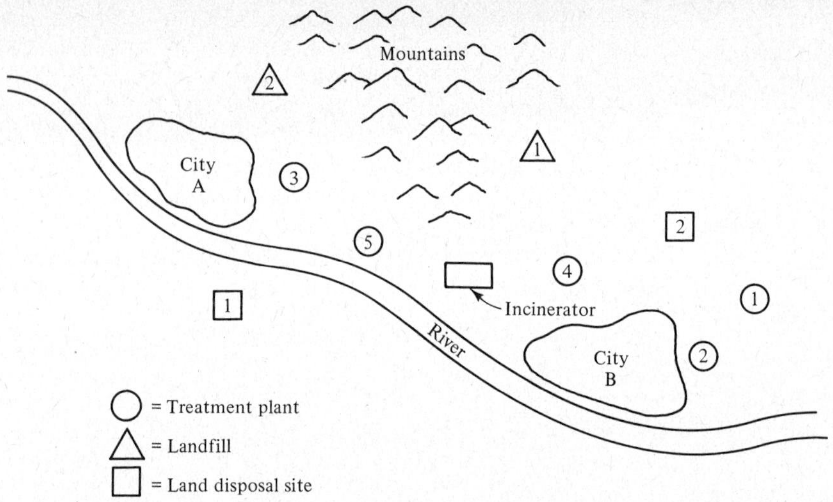

5-17. In order to keep to a construction schedule, you must place 1000 tons of pavement a day. Because of other projects in the area, you will be unable to purchase all of your pavement from one batch plant. In fact, you will have to purchase from three plants in the area because of the high demand. Plant A can supply you with 275 tons/day, plant B with 350 tons/day, and plant C with 595 tons/day. You have two pavers, one laying top (1) and one applying binder (2). Paver 1 will do 575 tons/day and paver 2 425 tons/day. The cost from A to 1 is \$32/ton; A to 2, \$36/ton; B to 1, \$30/ton; B to 2, \$33/ton; C to 1, \$31/ton; and C to 2, \$35/ton. Develop schematic, symbolic, and mathematical models that could be utilized to find the minimum cost way of acquiring the necessary material.

5-18. The manager of a paper mill receives an order for low, medium, and high grade paper. The mill consists of two factories which are each capable of producing each style of paper, but in differing quantities per day. Pertinent data are given below.

(a) Formulate a model that could help the manager determine how to fill the order at minimum time.

(b) Formulate a model that could help the manager determine how to fill the order at minimum cost.

Data
Present order

Low grade	16 tons
Medium grade	5 tons
High grade	20 tons

Factory outputs

Factory 1 (F_1)		Factory 2 (F_2)	
Low grade	8 tons/day	Low grade	2 tons/day
Medium grade	1 ton/day	Medium grade	1 ton/day
High grade	2 tons/day	High grade	7 tons/day

	Daily overhead
Factory 1	$1000/day
Factory 2	$2000/day

5-19. A landowner has decided to develop a piece of property and has hired a developer to manage the project. Land clearing and grubbing operations are completed and one of two bridges on the property is completed. The second bridge will be completed in 6 weeks. An engineering firm has been engaged by the developer to do all design work as well as perform all necessary field services and quality control of all construction operations. The engineering design requires fill to be placed, and spread by bulldozer, in the three areas shown on the plot plan for Exercise 5-19. One bulldozer is presently assigned to each fill area. Fill is available from suppliers 1, 2, and 3 at a cost of $1.25/yd^3 at a distance up to 5 mi. Any hauling distance beyond 5 mi will incur an additional cost of $.05/yd^3/mi. Supplier 4 would have to lease trucks to supply the site (up to six trucks maximum) and charges a flat rate of $2.00/yd^3. Each bulldozer can handle a maximum of 8 trucks/hr, dumping fill. Each truck has a capacity of 10 yd^3.

(a) Develop a model that could be used to find the minimum cost way of meeting the fill requirements if the bridge under construction was not finished.

(b) Develop a model that could be used to find the minimum cost way of meeting the fill requirements if the bridge under construction was in place.

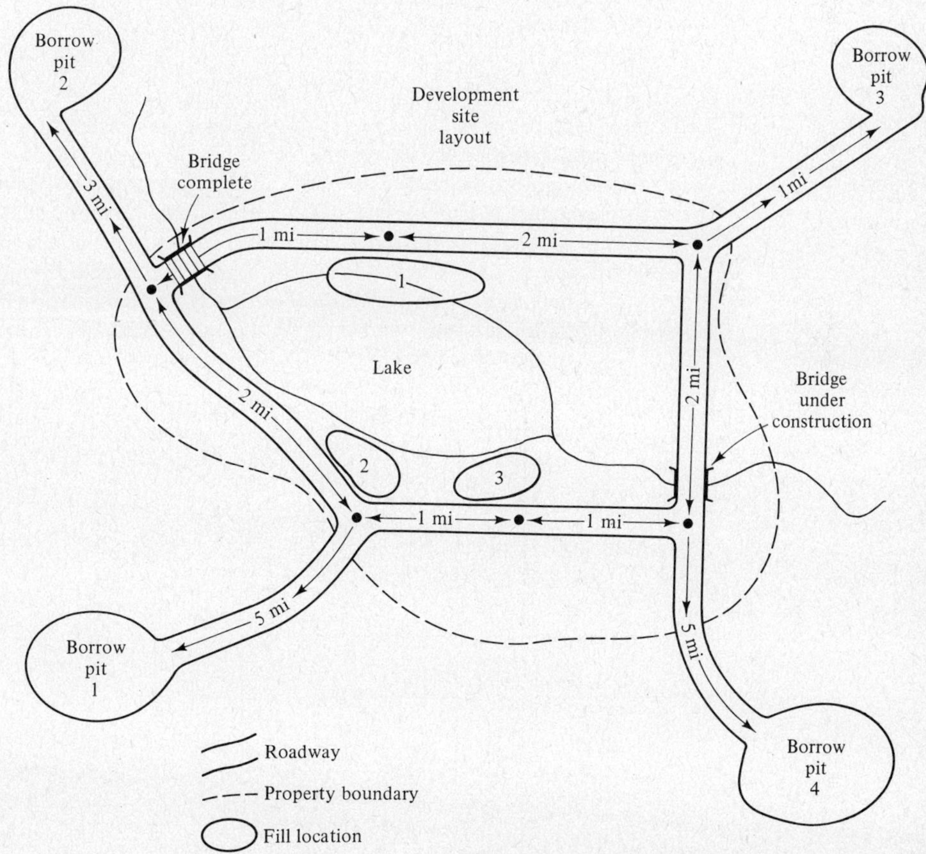

(c) Each round trip for a truck involves about 30 min of loading and unloading time. Average speed over the access roads is 20 mph. Six hours are available for transport each day, and 5 work days per week are scheduled. Develop a third model that could be used to find the overall minimum cost, taking into account the completion time of the second bridge.

Fill requirements

Site	yd^3
1	120,000
2	100,000
3	80,000

Fill supply

Borrow pit	yd^3 available	Trucks available
1	60,000	4
2	40,000	4
3	180,000	10
4	100,000	6

5-20. Streets A, B, and C of the figure for Exercise 5-20 all dump traffic onto Boulevard D. There are three types of vehicles that use the roadways, and each type requires a different amount of time traversing the intersection. The objective is to time the light cycle so that the maximum number of vehicles per hour can enter Boulevard D. Assume that Avenue D can accommodate as many vehicles as can be routed onto it, but that only one of the streets A, B, or C can be flowing traffic at one time. Excess traffic on the converging streets will seek alternate routes. A traffic light cycle should not exceed 3 min, and each street should have a green light for at least 40 s/cycle. Develop a mathematical model that could be used to determine the optimum timing for the light. The following data will be useful.

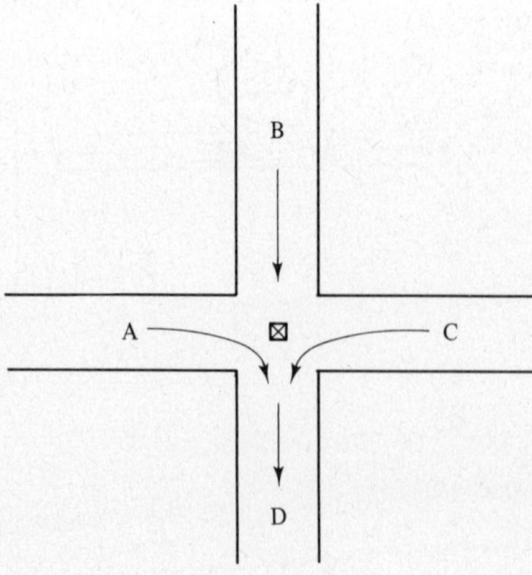

Route	Maximum capacity (vehicles/hr)	Fractional traffic mix by type		
		Type 1	Type 2	Type 3
A	1000	0.4	0.4	0.2
B	1400	0.7	0.2	0.1
C	800	0.8	0.1	0.1

Type of traffic	Time allocated to traverse intersection (s)
1	2
2	3
3	5

5-21. A rapid transit system serving a metropolitan area has five stations in the suburbs. Each morning, cars are dispatched to these terminals from three carbarns near the center of the area. Data for car availability at each carbarn, and car requirements for each station, along with the costs of dispatching a car from any carbarn to any station are given below. Develop a model that could be utilized to find the least cost method of dispatching the required cars. How would the model change if more cars were available than were needed at the stations? What if more were needed at the stations than were available in the carbarns?

Station	Number of cars required
1	30
2	20
3	10
4	10
5	15

Carbarn	Number of cars available
1	40
2	20
3	25

Dispatching costs ($) to station

From carbarn	1	2	3	4	5
1	40	30	20	30	60
2	20	20	40	50	70
3	60	40	50	30	30

5-22. Mt. Abstract has just been purchased by a development company that wishes to maximize its profit. However, strict zoning laws and restrictions have to be obeyed. Also, proposed environmental impact laws may put additional requirements on the developer. The company wants to develop a range of alternatives to estimate what their return on investment might be for varying environmental impact limitations. Nine hundred acres were purchased, and have been initially subdivided into nine, 100-acre tracts. Zoning restrictions require that building lots be at least 10 acres. The development company has priced the lots according to their position on the mountain, with the choicest lots being closest to the summit. The general layout and expected price per lot are shown in the figure for Exercise 5-22. Also included is an environmental index for each lot, which is an indicator of the amount of environmental impact each developed lot will have. The impact increases with the distance away from Route 66 and as the terrain steepens. The total amount of impact from the development is what will be regulated. The developer wishes to estimate the maximum profit to be expected under different levels of allowed total environmental impact.

Other practical restrictions have also been identified. Due to the roughness of the terrain, it is important that parcel 1 be developed before, or coincidentally with, parcels 4 and 5; similarly, 5 before 9 and 4 before 7. Eagle Brook flows from parcel 8, through parcel 5, to parcel 3. Effluent from septic systems may degrade the stream, so upstream development is limited to 70% of the development level in the next downstream parcel. Develop a model that could be used to determine the expected return for total environmental impacts of 5000, 4000, 2000, 500, and 200 units.

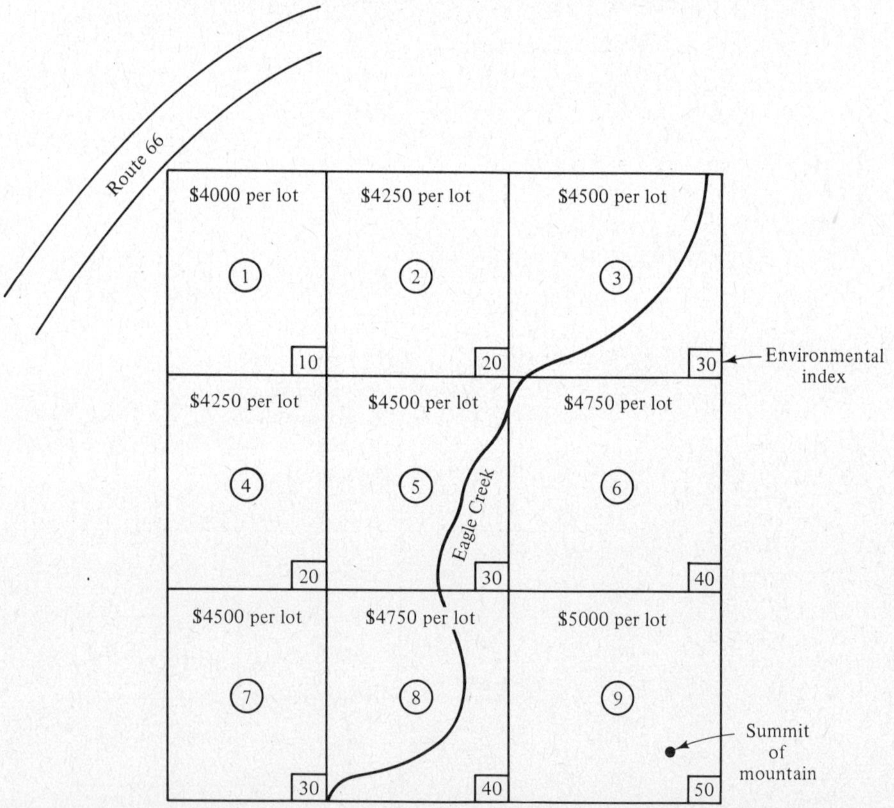

5-23. Develop a model that could be utilized to determine the maximum weight that could be carried by the overhead crane system shown in the figure for Exercise 5-23. The crane rails at A and B can each support 30 kips, while the cables have a working strength of 10 kips.

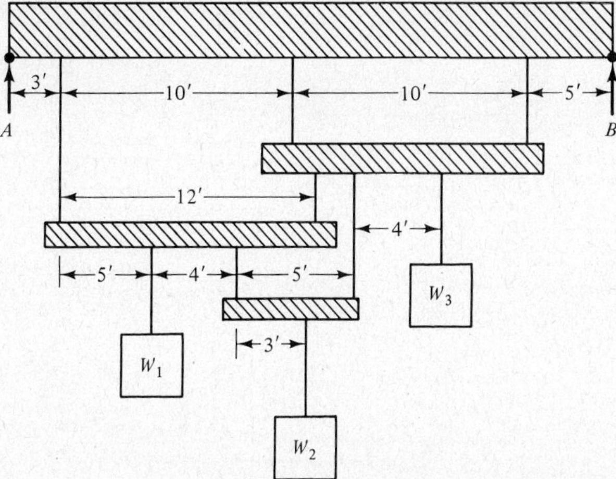

5-24. Refine the model developed in Exercise 5-23 so that placement of the loads along each lifting yoke is a variable, with the objective again being to maximize the total load that can be carried. Would it be possible also to use this model to find the placement combination that would yield the smallest total load? Describe how this could be done.

5-25. Through 5-30. Develop a model that could be utilized to find the optimum placement of the supports for each of the loading conditions shown in the figures for Exercises 5-25 through 5-30. The objective is to minimize maximum bending moment.

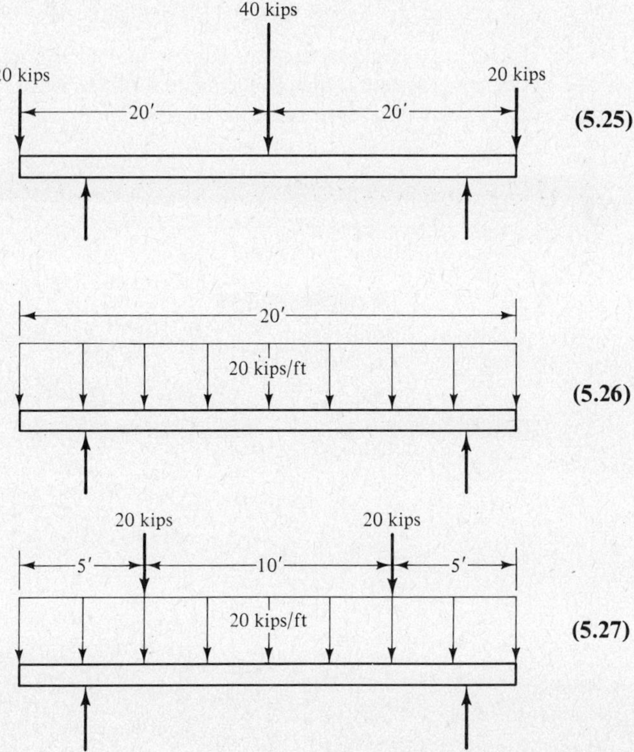

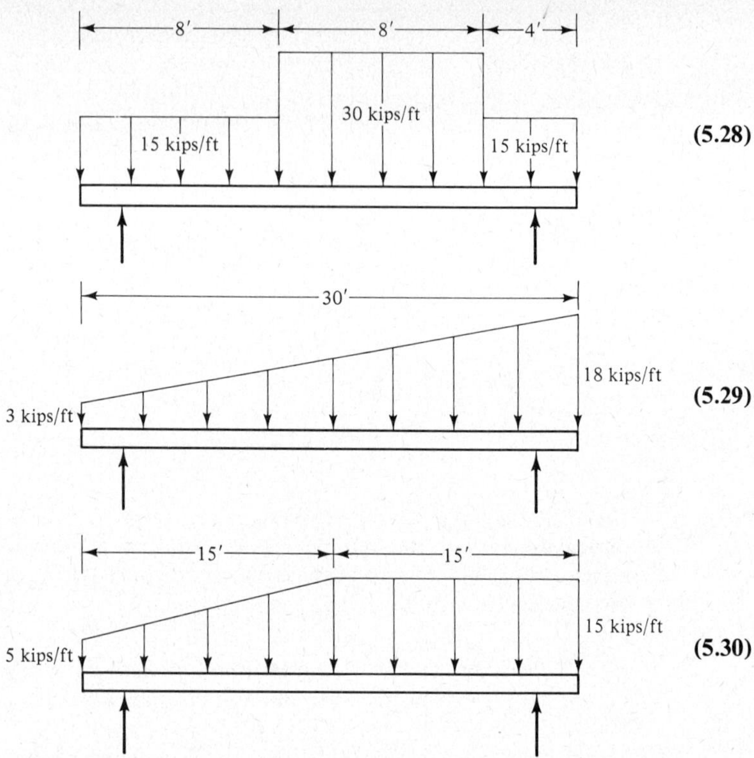

5-31. Develop a model that could be used to estimate the maximum total load that could be carried by the crane system shown in the figure for Exercise 5-31. Each rail can support a maximum of 100 kips. Design strengths for the various cables are given on the figure.

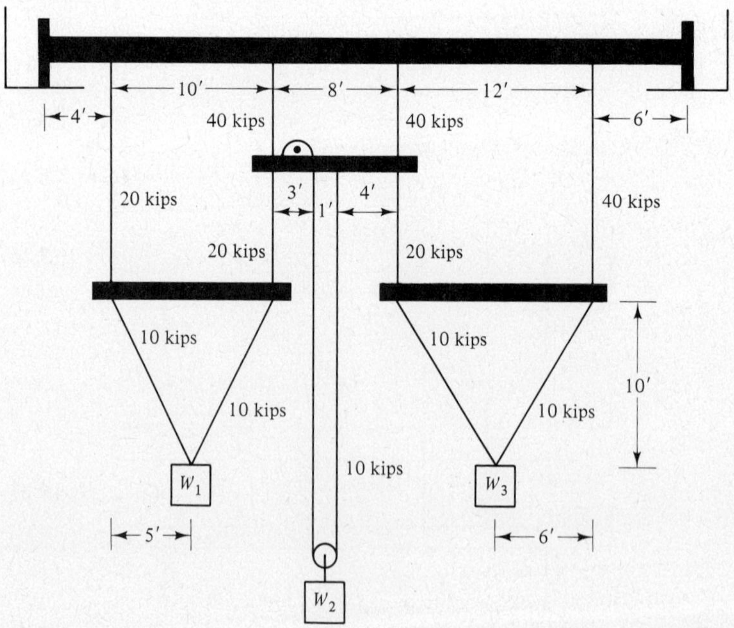

5-32. Four cities discharge wastes containing biochemical oxygen demand (BOD) into a river. The cities are far enough apart that combined treatment is not practical. However, a regional authority wishes to minimize the overall costs of treating the wastes so water quality requirements in the river are met. Different limitations are put on the allowable BOD load in the river because of different expected uses of the water. The BOD load in the river above City 1 is 0.0003 lb/ft^3. Removal of BOD by natural processes in the river follows an exponential decay function

$$R_{i,\,i+1} = \exp(-kt_{i,\,i+1})$$

in which $R_{i,\,i+1}$ = the fraction of BOD remaining after passage from City i to City $i + 1$, k = BOD exertion rate, and $t_{i,\,i+1}$ = time of river travel from City i to City $i + 1$. Streamflow is to be considered to be constant between cities. The maximum efficiency of BOD removal at any of the cities is 85%. Formulate a model that could be used to determine the minimum cost method of achieving the required river quality. Pertinent data are given below.

City	Miles upstream from City 4	Velocity in river (mi/day)	BOD load (lb/day)	BOD exertion rate (day^{-1})	Treatment cost ($/lb removed)	Streamflow (ft^3/s)
1	50		2,000	0.24	0.26	40
2	35	20	10,000	0.26	0.15	50
3	20	15	6,000	0.22	0.22	55
4	0	10	8,000	0.25	0.20	60

Allowable BOD load in river	
Above City 1	0.0004 lb/ft^3 of streamflow
Between Cities 1 and 2	0.0012 lb/ft^3 of streamflow
Between Cities 2 and 3	0.0008 lb/ft^3 of streamflow
Between Cities 3 and 4	0.0012 lb/ft^3 of streamflow
Below City 4	0.0012 lb/ft^3 of streamflow

5-33. A company has four machines that can accomplish three jobs. Each job can be assigned to one, and only one, machine. The objective is to assign the jobs to the machines in the most cost-efficient manner. Costs for each job–machine combination are given in the following table. Develop a model that can be used to find the optimum assignment.

		Machine		
Job	1	2	3	4
1	19	23	28	31
2	7	14	16	19
3	10	15	20	22

5-34. You have bought an excavation company which sells the sand from its sandpit to two cement companies. These companies are Atlas Cement, which is 15 mi away ($\frac{1}{2}$ hr for your trucks) and Barnaby's Concrete, 24 mi away (40 min by truck). Atlas needs six loads of sand per day, and Barnaby's needs five. You have three dump trucks which you use to transport the sand. Last year's records supply the following figures on your trucks:

Truck	Miles per gallon	Maintenance cost	Depreciation ($/mi)	Mileage in year
1	6	2050	0.09	12,000
2	6	500	0.09	10,000
3	8	500	0.10	10,000

These records will be used to help schedule the trucks for this year. You have three drivers, each of whom is paid $6.50/hr for an 8-hr day. They do not work overtime or more than 8 hr/day. It takes half an hour to load the trucks and 15 min to unload. Gas cost is $1.15/gal.

(a) Develop a model that could be used to find the best schedule for the trucks.

(b) Develop a model that could be used to find the worst schedule for the trucks.

5-35. Reformulate the model of Example 5.4 using a vertical dimension of 150 in. rather than 75 in. All other data are to be the same as in the original example.

5-36. Reformulate the model of Example 5.4 using a horizontal load of 80 kips applied at point 4. All other data are to be the same as the original example.

5-37. Reformulate the model of Example 5.4 using a vertical load of 80 kips applied at point 4. All other data are to be the same as the original example.

Chapter 6

Optimization Techniques

INTRODUCTION

Optimization is the process of finding the best solution to a problem within an established set of constraints. It will generally involve minimization of cost or material, or maximization of gain. However, optimization is only possible if there is a range of choices available. It is important, therefore, that the engineer does not apply too restrictive a set of constraints during the problem definition stage. The success of optimization depends primarily on the ability of the engineer to effectively and realistically define the problem, and to develop a reasonable model of the problem system. These processes have been described in Chapters 1–5.

Although the model developed may be unique to the problem to be analyzed, it will fall into one of several general types. Some of these types have established optimization procedures, while others may require unique procedures or may only be amenable to subjective optimization. It is preferable to use an analytical or mathematical programming optimization procedure that can be shown to give an overall optimal solution to a properly posed problem. Analytical optimization procedures include using the methods of calculus or Lagrange multipliers (which will be discussed later in this chapter) to develop mathematical relationships that represent the conditions necessary to find an optimal solution directly. Mathematical programming techniques include linear and nonlinear programming as well as numerical methods for finding approximate solutions for mathematical relationships. Mathematical programming techniques are designed to start with a solution, then work in discrete steps toward progressively better solutions.

If analytical or mathematical programming procedures are not available for the model chosen, a combinatorial approach can be undertaken. This approach involves a comparison of model predictions for various combinations of values for the design variables. The drawback of this approach is the large number of combinations that must be investigated for problems of any realistic size. If there are n design variables that each can take on m values, then there are $(m)^n$ different combinations. A problem with 5 design variables, each of which could take on 10 discrete variables, would require the examination of 100,000 combinations. Since it is generally impractical to look at this many solutions, the engineer is required to limit the analysis to a narrower, more manageable, set of alternatives. The subset of solutions can be selected using the best judgment available; however, there is no guarantee that the solution chosen from among this subset is the overall optimum solution.

Subjective optimization is the third type of optimization. It is the least quantitative of the three types described; however, often it is the only type available to the decision maker. The importance of subjective optimization should not be underestimated. A good subjective optimization will be more valuable than a poorly performed analytical optimization. Subjective optimization is optimization based on sound engineering judgment which draws upon past experience, foresight, and intuitive feel. Attention to detail is required, and reliance on the systems approach can help to ensure that adequate alternatives are examined. Subjective optimization can be used effectively to assist in day-to-day decision making both on the personal and professional level. This chapter will deal with analytical and mathematical programming methods of optimization.

OPTIMIZATION BY METHODS OF CALCULUS

Some problems that can be modeled by continuous functions can be optimized through the methods of calculus. These methods are described in detail in calculus texts, so a short example will suffice here. In Example 6.1, it is possible to solve the system of partial differential equations by substitution. This may not always be possible, and other methods for solving systems of nonlinear equations must be used. Lagrange multipliers, for example, can be used for optimizing this type of problem. The theory of Lagrange multipliers will be introduced in the nonlinear optimization section of this chapter.

EXAMPLE 6.1 ____

Minimize the cost of a rectangular box to hold 3 yd^3 of sand, with the constraint that the width of the bottom of the box must be equal to 8 ft.

> *Additional Information*
> 1. There is no top on the box.
> 2. Unit cost of sides and front end = $R/ft^2.
> 3. Unit cost of second end = $1.5R/ft^2.
> 4. Unit cost of bottom = $2R/ft^2.

$$X_1 = \text{length of box}$$
$$X_2 = \text{height of box}$$

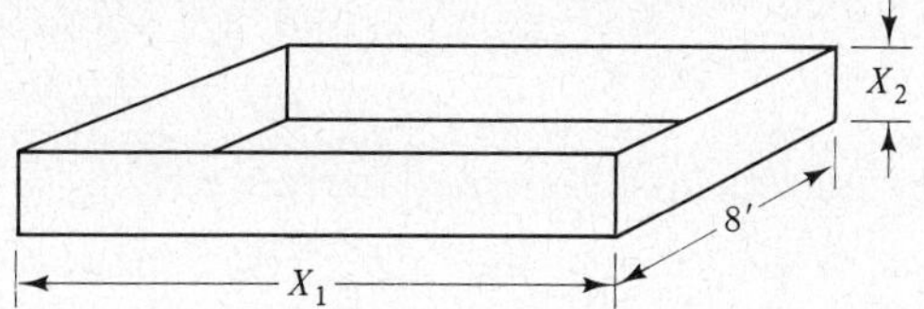

Goal   The goal is to minimize the cost of the box to hold 3 yd^3.

Objective   (Objective Function)

$$\min C = \$R(2X_1X_2 + 8X_2) + \$1.5R(8X_2) + \$2R(8X_1)$$

Criteria   (Constraints)

$$\text{Subject to (ST):} \quad 8X_1X_2 = 81 \text{ ft}^3$$

or
$$\min C = 2RX_1X_2 + 20RX_2 + 16RX_1 \tag{6.1}$$

$$\text{ST:} \quad 8X_1X_2 = 81 \tag{6.2}$$

Solving Equation 6.2 for X_1 and substituting back into Equation 6.1, gives

$$\min C = 20.25R + 20RX_2 + \frac{162R}{X_2} \tag{6.3}$$

Since there is only one independent variable in Equation 6.3, the optimum value can be found by taking the derivative with respect to X_2 and setting it equal to zero.

$$\frac{dC}{dX_2} = 0 = 20R - \frac{162R}{X_2^2} \tag{6.4}$$

$$20X_2^2 = 162$$

$$X_2 = 2.85 \text{ ft}$$

Substituting back into Equation 6.2,

$$X_1 = 3.56 \text{ ft}$$

and minimum cost,

$$C = \$134.25R$$

The second derivative could be used to confirm that this is the minimum cost.

Suppose now that the width of the box had not been specified. Let X_3 = the width of the box. Then the problem becomes

$$\min C = \$R(2X_1X_2 + X_2X_3) + \$1.5R(X_2X_3) + \$2R(X_1X_3) \tag{6.5}$$

$$\text{ST:} \quad X_1X_2X_3 = 81 \tag{6.6}$$

Solving Equation 6.6 for X_1 and substituting into Equation 6.5 yields

$$\min C = \$R\left(\frac{162}{X_3} + X_2X_3\right) + \$1.5R(X_2X_3) + 2R\left(\frac{81}{X_2}\right) \tag{6.7}$$

There are two independent variables now, so partial derivatives have to be taken and set equal to zero.

$$\frac{\partial C}{\partial X_2} = RX_3 + 1.5RX_3 - \frac{162R}{X_2^2} = 0 \tag{6.8}$$

$$\frac{\partial C}{\partial X_3} = -\frac{162R}{X_3^2} + RX_2 + 1.5RX_2 = 0 \tag{6.9}$$

Solving Equation 6.8 for X_3 yields

$$X_3 = \left(\frac{162}{2.5}\right)\frac{1}{X_2^2} \tag{6.10}$$

which when substituted into Equation 6.9 yields

$$-\frac{162R}{(162/2.5)^2(1/X_2^4)} + 2.5RX_2 = 0 \tag{6.11}$$

or
$$X_2 = 4.02 \text{ ft}$$

Substituting back into Equation 6.10 yields

$$X_3 = 4.02 \text{ ft}$$

and Equation 6.6 yields

$$X_1 = 5.01 \text{ ft}$$

and minimum cost

$$C = \$120.96R$$

Second partial derivatives will reveal that the rate of change in the gradient is positive at all points in the vicinity of the optimal solution; therefore, it is confirmed that this is a minimum solution. Note that in this case the minimum cost is less than when the box width was specified. This indicates a more general solution. However, the reality of the situation may dictate that the width be specified.

LINEAR SYSTEMS

Many real world systems can be modeled by a linear objective function subject to a set of linear constraints. The objective function is a measure of the efficiency or effectiveness of the system, and the constraints are the bounds on the system. Transportation, product mix, blending, scheduling, and network problems are a few types that can be described by linear systems. These are referred to as linear programming models. Development of these models has been described in Chapters 4 and 5. This chapter will concentrate on methods of finding the optimal solution for these models.

The simplex algorithm is one of several procedures for optimizing the general linear model. Certain specialized linear models, such as transportation, assignment, and project management models, are more efficiently optimized by specialized algorithms. Procedures for implementing the simplex algorithm and specialized algorithms will be developed in subsequent sections.

LINEAR PROGRAMMING AND THE SIMPLEX ALGORITHM

Form of Problem

A linear programming problem consists of an objective function and a set of constraints. These constraints can either be equations or inequalities. Given this set of m linear inequalities or equations in n variables, the goal is to find non-negative values of these variables which will maximize or minimize (according to the problem definition) the linear objective function of the variables. Negative values of the independent variables are excluded from the solution of linear programming problems by the simplex algorithm. Negative variables can be represented by two positive variables, with one being subtracted from the other.

The general form of the linear programming problem is

$$\text{max or min } Z = c_1 X_1 + c_2 X_2 + \cdots + c_n X_n \tag{6.12}$$

$$\text{subject to: } a_{i1} X_1 + a_{i2} X_2 + \cdots + a_{in} X_n \left(\begin{array}{c} \leq \\ \geq \\ = \end{array} \right) b_i \tag{6.13}$$

and
$$X_1, X_2, \ldots, X_n \geq 0 \tag{6.14}$$

for
$$i = 1 \text{ to } m$$

where i = constraint identifier
       m = total number of constraints
       n = total number of model-independent variables
       a_{ij} = constraint coefficients (structural coefficients)
       j = independent variable identifier
       b_i = constraint right-hand-side limitations (constraint stipulations)
       c_j = cost coefficients of objective function

Although the c_j are called cost coefficients, they actually have different meanings for minimization and maximization problems. For minimization problems, c_j represents the cost of introducing one unit of variable X_j into the solution. For maximization problems, c_j represents the gain in the objective function value to be achieved by introducing one unit of variable X_j into the solution.

Equation 6.14 is necessary because the simplex algorithm will not accommodate negative values for the independent variables. The condition of non-negativity is automatically taken care of by the simplex algorithm, so Equation 6.14 need not explicitly appear in the system of constraint equations. For models that require a variable to be unrestricted as to sign, the variable in question, X_i, can be replaced by $X_i - X'_i$, both of which will have positive values, but their combination will be unrestricted.

In order to solve this type of problem, the inequalities have to be changed into equations by adding an extra variable to each inequality constraint. For a particular constraint, this extra variable will represent the difference between the sum of the terms on the left side of the constraint and the right-hand-side value for that constraint. These extra variables are called slack and surplus variables, respectively. The process for adding them, as well as their physical meaning, will be explained in a subsequent section.

The introduction of these additional variables will cause the total number of variables to be analyzed to be greater than the number of constraints, thereby preventing an explicit solution for the values of the independent variables. If r represents the number of slack and surplus variables added to the system, then $n + r$ will be the total number of variables to be evaluated in the m constraints. The usual procedure is to specify values for $n + r - m$ variables, then solve for the other m. Linear programming theory requires that the $n + r - m$ variables be set equal to zero, and the other m solved for. Without further direction, the choice of which m variables to solve for is arbitrary, so every combination would have to be evaluated to determine which solution produced the best value for the objective function. However, to examine every combination of $n + r$ variables taken m at a time would be impractical except for small-scale problems. Statistically the number of combinations of $n + r$ things taken m at a time is given by

$$\binom{n + r}{m} = \frac{(n + r)!}{(n + r - m)!(m)!} \tag{6.15}$$

If the total number of variables to be analyzed (including original model variables and added variables to facilitate the solution process) is 10, and the number of constraints is 6, then

$$\frac{10!}{(10 - 6)!6!} = \frac{7 \cdot 8 \cdot 9 \cdot 10}{1 \cdot 2 \cdot 3 \cdot 4} = 210$$

possible combinations would result. The simplex algorithm will provide a method of finding the optimum solution to the linear model without the necessity of examining every combination.

Geometric Interpretation of the Linear Programming Problem

An understanding of the geometric interpretation of the linear programming problem can assist in comprehension of the simplex method and algorithm. Suppose that a problem could be represented by the following two-variable linear programming model:

$$\max Z = X_1 + 2X_2$$

$$
\begin{aligned}
-X_1 + 3X_2 &\leq 10 \qquad \text{constraint 1} \\
X_1 + X_2 &\leq 6 \qquad \text{constraint 2} \\
X_1 - X_2 &\leq 2 \qquad \text{constraint 3} \\
X_1 + 3X_2 &\geq 6 \qquad \text{constraint 4} \\
X_1, X_2 &\geq 0
\end{aligned}
\qquad (6.16)
$$

Each of the constraints 1–4 can be plotted as shown in Figure 6.1. The line is drawn as if the constraint is an equality, then the side of the line that satisfies the particular constraint is determined by evaluating any point off the line. For example: for constraint 1, $X_1 = 2$, and $X_2 = 2$ is a point below the constraint line, and for these values of the variables, the constraint equals 4, which is less than 10. Thus, all points on or below the line for constraint 1 satisfy the constraint. Arrows are drawn on the figure to indicate areas satisfying the other constraints. The union of these areas is the shaded portion of Figure 6.1. This region is called the feasible region, and any values of X_1 and X_2 within this region simultaneously satisfy the constraints.

The objective function can also be plotted as shown in Figure 6.1. The objective function is a sliding scale that will change position as Z varies. It will always be parallel to the line shown, but as Z increases, the line will move up and to the right. Thus, it can be seen that the present point of contact between the objective function and the feasible region provides the largest value of Z for values of X_1 and X_2 that satisfy the constraints. The optimal solution would be $X_1 = 2$, $X_2 = 4$, and $Z = 10$. Any other set of X_1 and X_2 within the feasible region would give a lower value of Z.

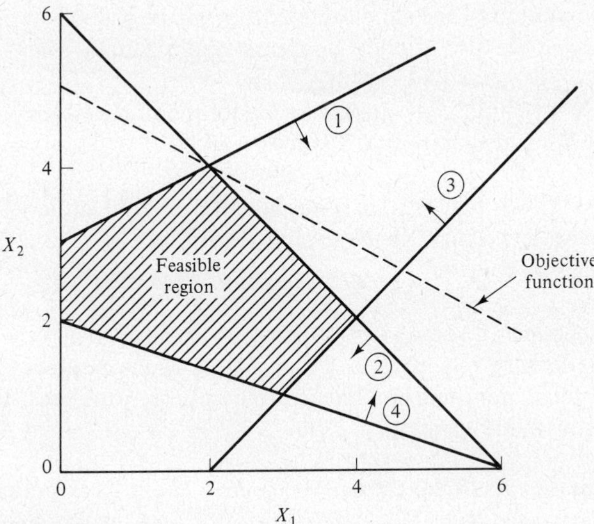

Figure 6.1 Geometric interpretation of linear programming problem.

If this had been a minimization problem, the lowest point of contact between the feasible region and the objective function would give the optimum solution. If the objective function had been parallel to one of the sides of the feasible region at the optimal point, then all X_1, X_2 values on that side would be optimum solutions. This situation is referred to as alternate optima.

The theory of linear programming can be utilized to show that the optimal solution will occur at one of the points of the feasible region (the intersections of the constraints). The simplex algorithm starts at one of these points and proceeds to adjacent points in a manner that seeks to improve the solution with each step. Higher-order problems are solved in the same manner and have an analogous geometric interpretation; however, they cannot be visualized because of the number of variables involved. For a five-variable problem, the feasible region would be five-dimensional instead of two-dimensional.

Definitions

An understanding of the definitions of the following terms is necessary for the comprehension of the simplex algorithm. The reader will want to refer back to these definitions when studying the simplex algorithm material.

Feasible Region   The feasible region consists of all of the points that satisfy the constraints. This includes the non-negativity constraints of Equation 6.14. The feasible region is bounded by the constraints, and is referred to as the feasible region hyperspace. The feasible region formed by the constraints is a convex hyperspace, which means that a line joining any two points on or within the hyperspace will be completely contained within the hyperspace. Convexity of the hyperspace guarantees that any optimum solution found by the simplex algorithm will be a global optimum. This will be illustrated during the discussion of nonlinear optimization.

Feasible Point $X = (X_1, X_2, X_3, \ldots, X_n)$ is a feasible point if all X_i in X simultaneously satisfy the constraints of Equations 6.13 and 6.14.

Global Minimum   Given that $y(X)$ is some function of X; if some $y(X^*)$ is less than all other $y(X)$ for all X except X^* in the feasible region, then $y(X^*)$ is a global minimum and X^* is the optimal solution.

Multiple Global Minima   If some $y(X^*)$ is less than or equal to all other $y(X)$ for all X except X^* in the feasible region, then X^*, and all X which produce a $y(X) = y(X^*)$, form multiple global minima. For linear programming problems, this means that alternate optimum solutions exist.

Global Maximum and Alternate Global Maximum   These have similar but opposite meaning to the previous two.

Solution   Any linear programming model will have $n + r$ variables (unknowns) and m constraint equations where $m < n + r$. In matrix form this can be written as

$\mathbf{AX} = \mathbf{B}$. A solution is any $n + r$ dimensional vector, $\mathbf{X}$, that simultaneously satisfies the constraint equations. $\mathbf{A}$ will be of dimension $m \times (n + r)$, while $\mathbf{B}$ is of dimension $m \times 1$. Therefore, $\mathbf{X}$ is of dimension $(n + r) \times 1$.

Boundary Solution   A solution that lies on the boundary of the feasible region hyperspace.

Basic Solution   A solution in which $n + r - m$ variables are set equal to zero. Such a solution can have at most m nonzero components, that is, at most m variables from $\mathbf{X}_{(n+r)}$ are nonzero.

Nondegenerate Basic Solution   A basic solution with exactly m nonzero components.

Degenerate Basic Solution   A basic solution with less than m nonzero components.

Feasible Basic Solution   A basic solution that lies within the feasible region.

Infeasible Basic Solution   A basic solution that has one or more negative components.

Simplex Algorithm

Procedures for applying the simplex algorithm to find the optimum solution to a linear programming problem will be presented through examples. Formal proof of the method will not be attempted because of space limitations. However, the reasoning behind critical steps will be explained in the context of linear algebra and linear programming theory. Formal proof of the algorithm can be found in linear programming texts by Hadley (1962) or Gass (1975).

There are several commonly used variations of the simplex algorithm, all of which accomplish the same end. The particular variation to be presented in this text is fairly straightforward, and lends itself to explanation of why certain computations and manipulations are being undertaken. The algorithm described will be for a maximization problem; however, it can be utilized for minimization problems by merely multiplying through the objective function by -1.0, and using this negative of the objective function in all computations. In other words, if some solution $(\mathbf{X}^*)$ maximizes its corresponding objective function, $Z(X)$, then $\mathbf{X}^*$ will also minimize $-Z(X)$. An example using the quadratic equation will illustrate this.

EXAMPLE 6.2 ——

Let $Z(X) = -X^2 + 4X + 6$. Show that the $\mathbf{X}^*$ that maximizes $Z(X)$ also minimizes $-Z(X)$.

$$dZ(X) = -2X + 4$$

$$d^2Z(X) = -2$$

$\therefore$ a maximum $Z(X) = 10$ occurs at $X = 2$.

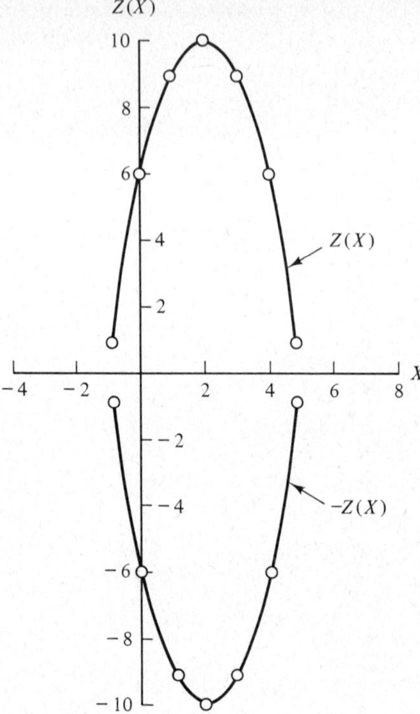

Figure 6.2   Objective function inversion.

$$\text{Taking } -Z(X) = X^2 - 4X - 6$$

$$dZ(X) = 2X - 4$$

$$d^2Z(X) = 2$$

$\therefore$ a minimum $Z(X) = -10$ occurs at $X = 2$. Figure 6.2 also illustrates this phenomenon.

Organization of Constraints   It is useful when learning the simplex algorithm to re-arrange the constraints with all of the $\leq$ constraints first, followed by all of the $\geq$ constraints, followed by all of the $=$ constraints. This practice can be discontinued when larger problems are modeled, and computer applications packages are used to find the optimal solution. Also, all of the constraint stipulations, b_i, must be positive. Any b_i that are negative must be inverted by multiplying the constraint by -1 and reversing the inequality. It is good practice to identify and record the physical signif-icance of each constraint. This will assist in understanding the meaning of the variables to be added to the system for solution and will be necessary to evaluate the final solution and undertake appropriate postoptimality analysis.

Less than or equal ($\leq$) constraints indicate that there is an upper limit on some resource or product that can be applied toward the optimum solution of a problem. Not all of the resource or product has to be used, however. Two examples of this are the total amount of sand available to a project or a limit on the amount of a particular type gravel that can be used in formulating a certain gravel mix. Other examples have been developed in previous chapters. Less than or equal constraints must be converted into equalities for analysis. This is accomplished by adding an additional variable to the constraint. The new variable is called a slack variable and represents the amount of the resource or product that is not used in the optimal solution to the problem. If the first k constraints are $\leq$, then a typical constraint will be

$$a_{i1}X_1 + a_{i2}X_2 + \cdots + a_{in}X_n \leq b_i \qquad (6.17)$$

for which i, the constraint identifier, is between 1 and k. The slack variable for this constraint will be defined as X_{n+i}; and when added, the equation becomes

$$a_{i1}X_1 + a_{i2}X_2 + \cdots + a_{in}X_n + X_{n+i} = b_i \qquad (6.18)$$

A total of k slack variables will be introduced in this manner.

Greater than or equal ($\geq$) constraints indicate that at least a minimum level of a certain resource or product has to be used in the solution of a problem. In certain cases, however, more than the minimum level will be used in arriving at the optimal solution. The requirement that at least a given percentage of a certain pollutant be removed from a waste stream is an example of a greater than or equal constraint. An additional variable is subtracted from each $\geq$ constraint to make it an equation. These new variables are called surplus variables. The surplus variable for a particular $\geq$ constraint reflects the excess amount of the resource or product described by that constraint that is applied to the problem to achieve the optimum solution. If the $k + 1$st through lth constraints are $\geq$, then a typical constraint is of the form

$$a_{i1}X_1 + a_{i2}X_2 + \cdots + a_{in}X_n \geq b_i \qquad (6.19)$$

for which i is the constraint identifier and is between $k + 1$ and l. The surplus variable for this constraint will be defined as X_{n+i}, and when subtracted, the equation becomes

$$a_{i1}X_1 + a_{i2}X_2 + \cdots + a_{in}X_n - X_{n+i} = b_i \qquad (6.20)$$

A total of $l - k$ surplus variables will be defined in this manner. Up to this point a total of $n + l$ variables have been defined for the problem.

After addition of slack and surplus variables, Equation 6.16 would become

$$
\left.
\begin{aligned}
\max Z = \quad & X_1 + 2X_2 + 0X_3 + 0X_4 + 0X_5 - 0X_6 \\
\text{ST:} \quad & -X_1 + 3X_2 + X_3 && = 10 \\
& X_1 + X_2 && + X_4 && = 6 \\
& X_1 - X_2 && && + X_5 && = 2 \\
& X_1 + 3X_2 && && && - X_6 = 6
\end{aligned}
\right\} \qquad (6.21)
$$

Equations 6.21 have six variables for the four equations. Two variables must be specified before the others can be solved for. The simplex algorithm will facilitate this process. Note that the cost/gain coefficients for the slack and surplus variables are all zero because they do not add or detract anything from the value of the objective function.

Simplex Algorithm for Maximization Problem with $\leq$ Constraints The simplex algorithm will be developed in two stages. First the maximization problem with $\leq$ constraints will be used to explain the procedures of the algorithm, then the procedures will be expanded to include the minimization problem with all three types of constraints.

As with most computational procedures, the simplex algorithm must have an initial starting point from which to proceed. In the case of the simplex algorithm, this starting point has to be a basic feasible solution of the constraint system. For a problem with all $\leq$ constraints, there will be $n + m$ variables for the m constraints. Therefore, to find a basic solution to the problem, n variables have to be specified before the remaining m can be solved for. To make this a basic feasible solution, all of the variable values have to be greater than or equal to zero. Consider the following linear programming problem:

$$\left. \begin{array}{ll} \max Z = 6X_1 + 7X_2 \\ \text{ST:} \quad 2X_1 + 3X_2 \leq 12 \\ \quad\quad\ 2X_1 + \ X_2 \leq 8 \end{array} \right\} \tag{6.22}$$

Note that the requirement that X_1 and $X_2 \geq 0$ is not explicitly stated, but is understood. If slack variables are added to this problem, it becomes

$$\left. \begin{array}{ll} \max Z = 6X_1 + 7X_2 \\ \text{ST:} \quad 2X_1 + 3X_2 + X_3 \quad\quad = 12 \\ \quad\quad\ 2X_1 + \ X_2 \quad\quad + X_4 = 8 \end{array} \right\} \tag{6.23}$$

For this case, $n = 2$, $m = 2$, $\therefore$ two variables have to be set equal to zero. If X_1 and X_2 are set equal to zero, the system can be solved for X_3 and X_4. Following the procedures outlined thus far, X_3 and X_4 would form an initial basic feasible solution with $X_3 = 12$ and $X_4 = 8$. This initial basic feasible solution is easy to find and is the one used by convention to initiate the simplex algorithm.

The information contained in the objective function and the modified constraints is organized into a tabular format for solution as is shown in Table 6.1. This tableau will provide a convenient format for performing the operations required to approach optimality. The solution variables column identifies which variables are in the solution

TABLE 6.1 INITIAL SIMPLEX TABLEAU

c_j			0	0	6	7
	Solution variables	B	X_3	X_4	X_1	X_2
0	X_3	12	1	0	2	3
0	X_4	8	0	1	2	1
	Z_j $c_j - Z_j$					

at a particular stage. In this case, X_3 and X_4 form the initial basic feasible solution. The c_j values are the cost/gain coefficients and they are placed in the horizontal row above the appropriate variable and in the vertical column next to the corresponding variable in the solution. The B column initially holds the constraint stipulations, but it also contains the values for the corresponding solution variable at any stage in the computations. Each of the remaining columns contains the constraint coefficients for the variable identified at the top. Generally the variables in the initial basic feasible solution are placed to the left next to the B column, but this is not essential. Variables are identified in general by the subscript j, as they were previously.

The value for Z_j to be placed in the B column represents the objective function value at any stage of the optimization. Call this Z_B, then

$$Z_B = \sum_{j=1}^{p} c_j X_j \qquad (6.24)$$

in which p = total number of variables in the problem. Note that this is a completely general equation because the only X_j that can have nonzero values are those that are presently in the basic feasible solution. Thus, Z_B can be calculated using the values of the solution variables in the B column and the corresponding c_j values from the left-hand column. For Table 6.1, $Z_B = 0(12) + 0(8) = 0$.

Z_j for the variable columns indicates how much the objective function value would be reduced if variable j replaced one of the variables presently in the basic feasible solution, per unit of variable j entered. These Z_j are computed according to the equation

$$Z_j = \sum_{i=1}^{m} c_i a_{ij} \qquad (6.25)$$

where m = number of constraints
$\quad\quad i$ = constraint identifier (row identifier in tableau)
$\quad\quad c_i$ = cost/gain coefficient in leftmost column corresponding to row i
$\quad\quad a_{ij}$ = constraint coefficient for variable j and constraint (row) i

For the first tableau shown in Table 6.1, all Z_j values are zero because both cost/gain coefficients for the variables in the solution are zero.

The row below the Z_j row is labeled $c_j - Z_j$. This is an important row in carrying out the simplex algorithm. Z_j represents the amount the objective function value would be reduced if X_j were inserted into the basic feasible solution; however, the corresponding c_j for that variable shows how much the objective function would be increased if variable X_j were inserted. Therefore $c_j - Z_j$ represents the marginal gain to be realized if X_j were introduced into the solution; that is, the net gain or loss per unit of X_j introduced. It can be calculated by taking the c_j at the top of column j, and subtracting from it the Z_j value for that column.

Since $c_j - Z_j$ represents the marginal gain to be achieved by introducing a new variable into the solution, any variables with a positive $c_j - Z_j$ would improve the solution, and the variable with the largest positive $c_j - Z_j$ would improve the solution at the fastest rate. This leads to Rule 1:

Rule 1. The variable with the most positive $c_j - Z_j$ should enter the basic feasible solution during the next step. If there are no variables for which the $c_j - Z_j$ is positive, then the optimum solution has been found.

The second part of Rule 1 follows directly from the fact that the space enclosed by the constraints is a convex space. If the objective function cannot be improved by moving in any direction from a particular solution point, then that point has to be the global optimum solution. Table 6.2 shows the Z_j and $c_j - Z_j$ values for the tableau shown in Table 6.1, and indicates the variable that should enter the solution, in this case X_2. X_1 could also enter the solution, but it would not improve the objective function value as much per unit as X_2.

Once the variable that will enter the solution has been chosen, one of the variables presently in the solution will have to be removed. If done properly, the exchange of variables will form a new basic feasible solution that will be at the least no worse than the previous solution. This follows from the theorem of linear algebra dealing with the substitution of vectors into and out of the set of basic vectors for the Euclidian m space. In this case, the dimension, m, is equal to the number of constraints. Rule 2 shows how the variable to leave the basic feasible solution should be chosen.

Rule 2. The variable to leave the solution should be the one with the minimum ratio of b_i, the solution variable value for the variable in the ith row, divided by a_{ik}, the constraint coefficient in the ith row in the column corresponding to the variable (X_k) to enter the solution. Only positive a_{ik} are used.

$$\Theta = \min \frac{b_i}{a_{ik}} \quad \text{for} \quad a_{ik} > 0$$

TABLE 6.2   INITIAL SIMPLEX TABLEAU WITH COMPUTATIONS

c_j				0	0	6	7
	Solution variables	B	X_3	X_4	X_1	X_2
Leaves solution → 0	X_3	12	1	0	2	3
0	X_4	8	0	1	2	1
	Z_j	0	0	0	0	0
	$c_j - Z_j$		0	0	6	7

$$\frac{b_1}{a_{12}} = \frac{12}{3} = 4$$

$$\frac{b_2}{a_{22}} = \frac{8}{1} = 8$$

Enters solution (X_2)

For the tableau of Table 6.1, a_{i2} in the X_2 column would be considered. Rule 2 is referred to as the theta rule, and in this case $\Theta = 4$ for variable X_3. Therefore, X_3 should leave the solution as X_2 enters. Computations to find Θ are shown in Table 6.2. Taking ratios for only positive a_{ik} ensures that a new basic feasible solution *can* be formed. Choosing the variable with the minimum of the calculable ratios to be the variable to leave the solution will ensure that the new solution *will* be a basic feasible solution.

If there are no a_{ik} for a variable to enter the solution which are greater than zero, it means that no new basic feasible solution can be formed. This is called an unbounded solution. It usually indicates an error in formulating the constraints. Physically it means that the constraint hyperspace is not closed, and the solution is allowed to vary endlessly along the edge of the unclosed region. Finding an unbounded solution is analogous to falling into a bottomless well: the situation is beyond recovery. The only course is to go back to the problem definition and model development stages, after checking to be sure that no numerical or transcription errors have been made.

The next step is to actually insert the new variable into the solution and remove the old variable. This involves elementary row operations similar to those used in the solution of simultaneous equations by methods such as Gauss–Jordan elimination. The columns of coefficients of the variables presently in the solution form an identity matrix. In the simplex algorithm, the rows of the tableau are operated on so that the column under the variable to enter the solution becomes the element of the identity matrix that used to be in the column under the variable that is going to leave the solution. For the example shown in Table 6.2, the column under X_3 contains the vector $[\begin{smallmatrix}1\\0\end{smallmatrix}]$. This vector will appear in the column under X_2 in the next tableau. These manipulations are utilizing the product form of the inverse to insert a new variable into the problem and form a new solution for the next tableau.

The first step in the row operations is to identify the pivot element at the intersection of the row containing the variable to leave the solution and the column of the variable to enter the solution. The pivot element is a_{12} for the initial tableau of Table 6.2. Divide each a_{ij} and b_i for the row containing the pivot element by the value of the pivot element, or 3 for the example shown. At the same time, change the solution variable identifier for the pivot row to the variable entering and insert the appropriate c_j for the entering variable in the c_j column. These actions are reflected in Table 6.3(a).

The second step is to use row operations to make all of the remaining a_{ik} for the column under the variable entering the solution equal to zero. This is accomplished by adding or subtracting some multiple of the pivot row to each of the other rows to make the remaining a_{ik} in the pivot column equal to zero. For the example in Table 6.2, subtracting the pivot row from the second row will eliminate a_{22}. Table 6.3(b) reflects this operation. After completing the above operations for any remaining rows, Z_j and $c_j - Z_j$ values can be computed and the variables to enter and leave on the next iteration can be determined.

Table 6.3(c) shows the results of the final computations. X_1 is the only variable that has a $c_j - Z_j$ value greater than zero, so it will enter the solution. The value of the objective function for this iteration is 28 and the variables in the solution have values $X_2 = 4$ and $X_4 = 4$. Note that the $c_j - Z_j$ values for variables in the solution

TABLE 6.3   SECOND SIMPLEX TABLEAU

(a) Identifying pivot element and operating on its row

c_j			0	0	6	7	
	Solution variables	B	X_3	X_4	X_1	X_2	
7	X_2	4	1/3	0	2/3	①	Pivot element
	Z_j $c_j - Z_j$						

(b) Completing row operation

c_j			0	0	6	7
	Solution variables	B	X_3	X_4	X_1	X_2
7	X_2	4	1/3	0	2/3	1
0	X_4	4	−1/3	1	4/3	0
	Z_j $c_j - Z_j$					

are always zero. Also, the $c_j - Z_j$ of $-\frac{7}{3}$ for X_3 indicates that the objective function value would go down if X_3 were forced into the solution. Theta equals 3 for this tableau, so X_4 would leave the solution.

At no time during application of the simplex algorithm should any of the elements of the B column be negative. This would indicate that one of the X_j variables would have a negative value, which is an infeasible solution. If this should happen, errors must have been made somewhere in the computational process. One or more of the variables in the solution may, however, be zero. Thus, some zeros could appear in the B column. This is a degenerate solution because less than m variables have values greater then zero. Considerable effort has been expended on theoretical problems of cycling within the simplex algorithm caused by degeneracy; however, in practical

Sensitivity analysis provides a way of identifying the most critical parameters. Errors in estimating these parameters might cause significant errors in the final solution variables and the objective function value. Therefore, additional care is advised in their estimation. At the same time, however, sensitivity analysis is only valid if the model proposed is valid. If the model leaves out essential features of the system, sensitivity analysis will not uncover the omission.

Alternate Optima The geometric interpretation of alternate optima has already been discussed. Alternate optima are desirable from the decision maker's standpoint, because they present a variety of alternatives that will all result in the same net benefit. The existence of alternate optima can be determined by examining the optimum tableau. If there are any $c_j - Z_j$ that are equal to zero for variables that are not in the solution, then that variable could enter the solution without changing the value of the objective function. Changes in the solution variables could be determined by application of the theta rule and by row operations. There are no alternate optima for the problem depicted in Table 6.4.

Simplex Algorithm for Minimization Problem with $\leq$, $\geq$, and = Constraints Several new concepts will be introduced through a minimization problem. It is important to realize that the simplex algorithm procedures are identical to those described earlier. Consider the following problem.

Exactly 150 yd^3 of gravel are needed for a project, and two mix qualities are available. Trucks bringing mix X_1 can carry 5 yd^3/load and each truck load costs two units. Trucks bringing mix X_2 can carry 10 yd^3/load and each truckload costs eight units. Mix X_1 is of lower quality; therefore, no more than 20 truckloads can be used. At least 14 truckloads of the higher quality mix, X_2, must be used. The objective is to minimize the total cost of acquiring the proper amount and mix of gravel.

Formulation of the information as a linear programming problem would result in the system

$$
\left. \begin{aligned}
\min Z = 2X_1 + \; & 8X_2 \\
\text{ST:} \qquad X_1 \qquad\quad & \leq 20 \\
X_2 & \geq 14 \\
5X_1 + 10X_2 & = 150
\end{aligned} \right\} \tag{6.28}
$$

X_1 and X_2 represent the number of truckloads of each mix that will be required. Note that the units for each constraint have to be consistent; however, there is no need that the same units be used for all constraints. In this case the units of the first two are truckloads while the units for the third are cubic yards.

The third constraint is an equality constraint. Linear programmers must be careful in the specification of equality constraints. Improper equality constraints can overconstrict the problem and prevent the formation of any basic feasible solutions, or restrict the solution possibilities so an optimal solution is found that is not the global optimum. Generally, the only time an equality constraint should be used is when the constraint represents some stipulation that has to be met exactly. For instance, the total number of hours in a week must add up to 168. For any other conditions, it is better to use $\leq$ or $\geq$ constraints.

To perform a minimization according to the simplex algorithm previously developed, the negative of the objective function must be taken. An alternate method involves reversing the decision criterion for switching variables into and out of the solution; however, it is less straightforward, and will not be covered here.

Slack and surplus variables must be added as before. However, not all the elements of the identity matrix will be available in the constraint system to define the initial basic feasible solution. The problem after objective function inversion and slack and surplus variable addition would be

$$\left.\begin{array}{llll}
\max Z = -2X_1 - & 8X_2 - 0X_3 - 0X_4 \\
\text{ST:} \quad\quad X_1 & \quad + X_3 & = 20 \\
& X_2 \quad\quad - X_4 & = 14 \\
5X_1 + & 10X_2 & = 150
\end{array}\right\} \quad (6.29)$$

The slack variable, X_3, comprises the only element of the identity matrix

$$\begin{bmatrix} 1 \\ 0 \\ 0 \end{bmatrix}$$

for the constraint system. The negative sign associated with the surplus variable, X_4, prevents it from forming part of the identity matrix for the initial basic feasible solution. Two more elements are required. To form these elements, two additional variables are added to the constraint system. Their sole purpose is to form an initial basic feasible solution, rather than having to search for one. They are called artificial variables because they have no physical meaning in relation to the problem, and will never appear in the optimum solution. To force them out of the solution, artificial variables are given a large cost coefficient. For maximization problems, this is a large negative cost coefficient. For minimization problems, it is a large positive cost coefficient. However, when the objective function is inverted, the artificial variables again become negative. The end result is that the signs for the cost coefficients of artificial variables will always be negative in the simplex tableau. (This may not be the case for alternative methods for minimizations that do not involve objective function inversion.) It is best to use a designator M to identify the large cost coefficient of artificial variables. That way there is no confusion with real problem variables, and if the M cost coefficient appears in any optimal solution, it signals that an error in application of the algorithm has occurred. Most computer applications packages automatically account for artificial variables, so the user does not have to add them to the constraints. Additionally, most computer applications packages also automatically add slack and surplus variables after the user provides the original constraint and supplies the proper inequality.

With addition of artificial variables, Equations 6.30 become

$$\left.\begin{array}{llll}
\max Z = -2X_1 - & 8X_2 - 0X_3 - 0X_4 - MX_5 - MX_6 \\
\text{ST:} \quad\quad X_1 & \quad + X_3 & = 20 \\
& X_2 \quad - X_4 + X_5 & = 14 \\
5X_1 + & 10X_2 \quad\quad\quad + X_6 & = 150
\end{array}\right\} \quad (6.30)$$

X_5 and X_6 are the artificial variables added, and they form the second and third elements of the identity matrix.

Table 6.5 shows the initial and subsequent tableaux for the present problem.

TABLE 6.5 SIMPLEX TABLEAU FOR MINIMIZATION PROBLEM

(a) Tableau 1

c_j			0	$-M$	$-M$	-2	-8	0	
	Solution mix	B	X_3	X_5	X_6	X_1	X_2	X_4	Σ
0	X_3	20	1	0	0	1	0	0	22
Out → $-M$	X_5	14	0	1	0	0	1	-1	15
$-M$	X_6	150	0	0	1	5	10	0	166
	Z_j	$-164M$	0	$-M$	$-M$	$-5M$	$-11M$	M	
	$c_j - Z_j$		0	0	0	$-2 + 5M$	$-8 + 11M$	$-M$	

In (↑ under X_2)

$$\min \frac{b_i}{a_{ij}} = 14 \quad \text{for} \quad X_5 \quad \therefore X_5 \text{ out}$$

$X_3 = 20 \qquad X_5 = 14 \qquad X_6 = 150$

(b) Tableau 2

c_j			0	$-M$	$-M$	-2	-8	0	
	Solution mix	B	X_3	X_5	X_6	X_1	X_2	X_4	Σ
0	X_3	20	1	0	0	1	0	0	22
-8	X_2	14	0	1	0	0	1	-1	15
Out → $-M$	X_6	10	0	-10	1	5	0	10	16
	Z_j	$-112 -10M$	0	$-8 + 10M$	$-M$	$-5M$	-8	$8 - 10M$	
	$c_j - Z_j$		0	$8 - 11M$	0	$-2 + 5M$	0	$-8 + 10M$	

In (↑ under X_4)

$$\min \frac{b_i}{a_{ij}} = 1 \quad \text{for} \quad X_6 \quad \therefore X_6 \text{ out}$$

$X_3 = 20 \qquad X_2 = 14 \qquad X_6 = 10$

(continued overleaf)

TABLE 6.5   *(Continued)*

(c) Tableau 3

c_j			0	$-M$	$-M$	-2	-8	0	
	Solution mix	B	X_3	X_5	X_6	X_1	X_2	X_4	Σ
0	X_3	20	1	0	0	1	0	0	22
-8	X_2	15	0	0	1/10	1/2	1	0	16.6
Out→ 0	X_4	1	0	-1	1/10	1/2	0	1	1.6
	Z_j	-120	0	0	$-8/10$	-4	-8	0	
	$c_j - Z_j$		0	$-M$	$-M + 8/10$	2↑	0	0	

In

$$\min \frac{b_i}{a_{ij}} = 2 \quad \text{for } X_4 \qquad \therefore X_4 \text{ out}$$

$$X_3 = 20 \qquad X_2 = 15 \qquad X_4 = 1$$

(d) Tableau 4

c_j			0	$-M$	$-M$	-2	-8	0	
	Solution mix	B	X_3	X_5	X_6	X_1	X_2	X_4	Σ
0	X_3	18	1	2	$-2/10$	0	0	-2	18.8
-8	X_2	14	0	1	0	0	1	-1	15
-2	X_1	2	0	-2	2/10	1	0	2	3.2
	Z_j	-116	0	-4	$-4/10$	-2	-8	4	
	$c_j - Z_j$		0	$-M + 4$	$-M + 4/10$	0	0	-4	

all $c_j - Z_j < 0$ $\quad \therefore$ optimal solution

$$X_1 = 2 \qquad X_2 = 14 \qquad X_3 = 18 \qquad X_4 = 0 \qquad X_5 = 0 \qquad X_6 = 0$$

$$Z = 116 \text{ units for } 150 \text{ yd}^3$$

Note that a new column, the Σ column, has been added. This is a summation of all the entries across that row, except the c_j value. If the same operations are performed on the Σ entries as are performed on the rest of the row, then the Σ value provides a check against numerical errors. For example, in Table 6.5(b), $(10)\cdot(15)$ subtracted from 166 gives 16 for the Σ value in the third row. This agrees with the new sum across the row. A check of this sort is important when doing the simplex algorithm by hand because it is not a self-correcting procedure. Any numerical or procedural error that is introduced will remain with the problem.

Note that the Z_B value for each iteration is negative because of the inversion, and that it is being maximized by getting less negative. However, after reinversion, the resulting Z_B value is being minimized. The final solution gives an optimal objective function value of 116 units for 150 yd^3 using two truckloads of mix X_1 and 14 truckloads of mix X_2. Eighteen of the possible 20 yd^3 of mix X_1 were not used, indicated by the slack variable X_3. There is a shadow price for the surplus variable X_4. If the second constraint, which contains X_4, could be relaxed by one unit (≥ 13), the objective function value would drop by four units to 112. There are no alternate optima for this problem and the optimum solution is nondegenerate.

EXAMPLE 6.3

Use the simplex algorithm to solve the linear programming problem illustrated by Equations 6.16. This was the problem used to illustrate the geometric interpretation of linear programming. The problem is restated below:

$$
\begin{aligned}
\max Z = \quad & X_1 + 2X_2 \\
\text{ST:} \quad & -X_1 + 3X_2 \leq 10 \\
& X_1 + X_2 \leq 6 \\
& X_1 - X_2 \leq 2 \\
& X_1 + 3X_2 \geq 6
\end{aligned}
$$

The problem with the addition of slack, surplus, and artificial variables is

$$
\begin{aligned}
\max Z = \quad & X_1 + 2X_2 + 0X_3 + 0X_4 + 0X_5 + 0X_6 - MX_7 \\
\text{ST:} \quad & -X_1 + 3X_2 + X_3 = 10 \\
& X_1 + X_2 + X_4 = 6 \\
& X_1 - X_2 + X_5 = 2 \\
& X_1 + 3X_2 - X_6 + X_7 = 6
\end{aligned} \right\} \quad (6.31)
$$

X_3, X_4, and X_5 are slack variables.

X_6 is a surplus variable.

X_7 is an artificial variable.

Table 6.6 gives the tableaux for the solution. The optimum solution is $Z = 10$ for $X_1 = 2$, $X_2 = 4$, $X_5 = 4$, and $X_6 = 8$. All other $X_j = 0$. This is the same solution found graphically. There are no alternate optima in this problem. The first two constraints have shadow prices associated with them, while the second two do not.

TABLE 6.6 SIMPLEX TABLEAUX FOR EXAMPLE 6.3

(a) Initial tableau

c_j			0	0	0	$-M$	1	2	0	
	Solution variables	B	X_3	X_4	X_5	X_7	X_1	X_2	X_6	Σ
0	X_3	10	1	0	0	0	-1	3	0	13
0	X_4	6	0	1	0	0	1	1	0	9
0	X_5	2	0	0	1	0	1	-1	0	3
Out → $-M$	X_7	6	0	0	0	1	1	3	-1	10
	Z_j	$-6M$	0	0	0	$-M$	$-M$	$-3M$	M	
	$c_j - Z_j$		0	0	0	0	$1 + M$	$2 + 3M$	$-M$	

In

$\Theta = 2$ for X_7 $\therefore X_7$ leaves the solution

(b) Tableau 2

c_j			0	0	0	$-M$	1	2	0	
	Solution variables	B	X_3	X_4	X_5	X_7	X_1	X_2	X_6	Σ
Out → 0	X_3	4	1	0	0	-1	-2	0	1	3
0	X_4	4	0	1	0	$-1/3$	2/3	0	1/3	17/3
0	X_5	4	0	0	1	1/3	4/3	0	$-1/3$	19/3
2	X_2	2	0	0	0	1/3	1/3	①	$-1/3$	10/3
	Z_j	4	0	0	0	2/3	2/3	2	$-2/3$	20/3
	$c_j - Z_j$		0	0	0	$-M - 2/3$	1/3	0	2/3	

In

$\Theta = 4$ for X_3 $\therefore X_3$ leaves the solution

TABLE 6.6 *(Continued)*

(c) Tableau 3

c_j			0	0	0	$-M$	1	2	0	
	Solution variables	B	X_3	X_4	X_5	X_7	X_1	X_2	X_6	Σ
0	X_6	4	1	0	0	-1	-2	0	①	3
0	X_4	8/3	$-1/3$	1	0	0	4/3	0	0	14/3
0	X_5	16/3	1/3	0	1	0	2/3	0	0	22/3
2	X_2	10/3	1/3	0	0	0	$-1/3$	1	0	13/3
	Z_j	20/3	2/3	0	0	0	$-2/3$	2	0	
	$c_j - Z_j$		$-2/3$	0	0	$-M$	5/3	0	0	

Out → X_4

In ↑ (under X_1)

$\Theta = 2$ for X_4 $\therefore X_4$ leaves the solution

(d) Tableau 4

c_j			0	0	0	$-M$	1	2	0	
	Solution variables	B	X_3	X_4	X_5	X_7	X_1	X_2	X_6	Σ
0	X_6	8	1/2	3/2	0	-1	0	0	1	10
1	X_1	2	$-1/4$	3/4	0	0	①	0	0	7/2
0	X_5	4	1/2	$-1/2$	1	0	0	0	0	5
2	X_2	4	1/4	1/4	0	0	0	1	0	3/2
	Z_j	10	1/4	5/4	0	0	1	2	0	
	$c_j - Z_j$		$-1/4$	$-5/4$	0	$-M$	0	0	0	

TRANSPORTATION PROBLEMS AND THE TRANSPORTATION ALGORITHM

Transportation Problems

Transportation optimization problems are specialized forms of linear programming problems that can be solved more efficiently through application of specific algorithms. Some problems that are labeled as transportation problems may not involve the movement of goods or people, but they have the generalized form of the transportation problem. These problems involve the distribution of some resource from two or more sources (suppliers or origins) over some type of pathway network to two or more sinks (customers or destinations). The objective is to meet sink requirements for the resource at minimum cost. Measures of cost may be in monetary value, time, or distance.

In the general problem, there will be m sources and n sinks. Depending on the definition of the problem, a certain entity may act as either a source or a sink. For example, warehouses may act as sinks for factories, but they may also act as sources for markets. Similarly, transfer stations can be destinations for municipal sources of refuse, while acting as sources for regional landfills.

A schematic model of the general transportation problem is shown in Figure 6.3. The symbolic model could use the following definition of variables:

X_{ij} = the number of units of resource shipped from source i to sink j,
c_{ij} = the cost of shipping one unit of resource from source i to sink j,
a_i = the number of units of resource available at source i, and
b_j = the number of units of resource needed by sink j

A problem with three sources and three sinks will be used to illustrate the general form of the mathematical model. The objective function is

$$\min Z = c_{11}X_{11} + c_{12}X_{12} + c_{13}X_{13} + c_{21}X_{21} + c_{22}X_{22}$$
$$+ c_{23}X_{23} + c_{31}X_{31} + c_{32}X_{32} + c_{33}X_{33} \qquad (6.32)$$

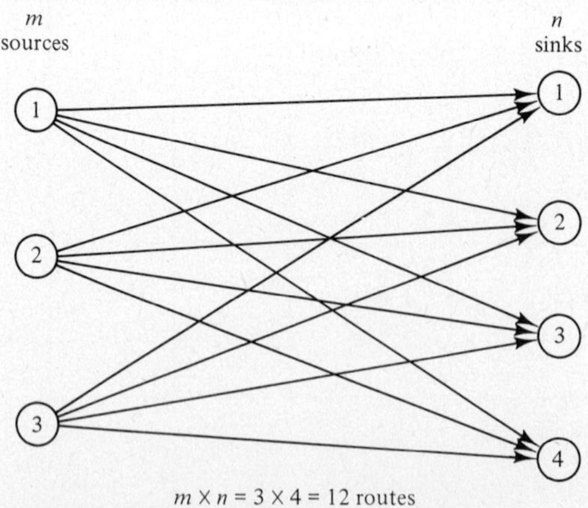

Figure 6.3   General transportation problem.

which is subject to the constraints

$$\left.\begin{array}{l} X_{11} + X_{12} + X_{13} \hspace{6cm} \leq a_1 \\ \hspace{2cm} X_{21} + X_{22} + X_{23} \hspace{3cm} \leq a_2 \\ \hspace{5cm} X_{31} + X_{32} + X_{33} \leq a_3 \\ X_{11} \hspace{2.5cm} + X_{21} \hspace{2.2cm} + X_{31} \hspace{2.5cm} = b_1 \\ \hspace{1cm} X_{12} \hspace{2.5cm} + X_{22} \hspace{2.5cm} + X_{32} \hspace{1.5cm} = b_2 \\ \hspace{2cm} X_{13} \hspace{2.5cm} + X_{23} \hspace{2.5cm} + X_{33} = b_3 \end{array}\right\} \quad (6.33)$$

Each of the first m equations in the set of constraints represents the amount shipped from a particular source to each of the sinks. The second n equations represent the amount shipped from each source to a particular sink, to meet the requirements of that sink.

Note that all coefficients of the constraint matrix are either zero or 1. Also, each variable appears only once in the first m constraints and once in the last n constraints. It is these properties that make it possible to find the optimal solution for the model using the transportation algorithm. Utilization of this specialized system is computationally more efficient than finding the optimum solution through the simplex algorithm.

Transportation Algorithm

Transportation problems must be balanced if the transportation algorithm is to be used to optimize the model. A model is balanced if the total demand (amount required at all sinks) has to be equal to the total supply (amount available at all sources). This limitation, however, does not reduce the generality of the algorithm, because unbalanced models can be slightly modified to make them balanced. Unbalanced conditions will be discussed in a later section.

Consider the following balanced problem:

Origin available units	Destination requirements
$a_1 = 10$	$b_1 = 12$
$a_2 = 12$	$b_2 = 14$
$a_3 = 10$	$b_3 = 6$
Total = 32	Total = 32

Cost matrix c_{ij}

<table>
<tr><td></td><td></td><td colspan="3" align="center">j</td></tr>
<tr><td></td><td></td><td>1</td><td>2</td><td>3</td></tr>
<tr><td rowspan="3">i</td><td>1</td><td>4</td><td>8</td><td>4</td></tr>
<tr><td>2</td><td>9</td><td>10</td><td>3</td></tr>
<tr><td>3</td><td>14</td><td>6</td><td>5</td></tr>
</table>

The linear programming model is given by

$$\min Z = 4X_{11} + 8X_{12} + 4X_{13} + 9X_{21} + 10X_{22}$$
$$+ 3X_{23} + 14X_{31} + 6X_{32} + 5X_{33} \qquad (6.34)$$

subject to the constraints

$$\left.\begin{array}{l} X_{11} + X_{12} + X_{13} \qquad\qquad\qquad\qquad\qquad = 10 \\ \qquad\qquad X_{21} + X_{22} + X_{23} \qquad\qquad\qquad = 12 \\ \qquad\qquad\qquad\qquad X_{31} + X_{32} + X_{33} = 10 \\ X_{11} \qquad\qquad + X_{21} \qquad\qquad + X_{31} \qquad\qquad = 12 \\ \qquad X_{12} \qquad\qquad + X_{22} \qquad\qquad + X_{32} \qquad = 14 \\ \qquad\qquad X_{13} \qquad\qquad + X_{23} \qquad\qquad + X_{33} = 6 \end{array}\right\} \qquad (6.35)$$

The equality signs are used because this is a balanced problem. It can be seen that any constraint in the balanced transportation model can be written as a combination of the rest of the constraints. Thus, one constraint is redundant, and at most $n + m - 1$ variables can be greater than zero in the model solution. This would be five variables for the model given above.

Initial Solution by Northwest (NW) Corner Rule   As with the simplex algorithm, transportation problems require that an initial, basic feasible solution be found. To do this, the available resources have to be allocated over the possible routes to satisfy the requirements at the sinks. The NW corner rule is the simplest way of finding an initial basic feasible solution.

To initiate the NW corner solution, the following distribution matrix is developed:

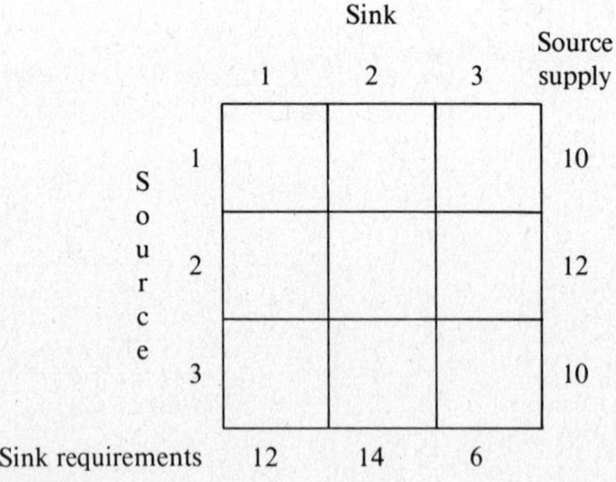

and the following steps are undertaken:

1. Start in NW corner and satisfy smaller of first row and first column requirements.

10			0
			12
			10
2	14	6	

2. Then finish the alternate requirement.

10			0
2			10
			10
0	14	6	

3. Continue to stairstep until all requirements are met.

10			0
2	10		0
			10
0	4	6	

10			0
2	10		0
	4		6
0	0	6	

10			0
2	10		0
	4	6	0
0	0	0	

The final distribution matrix shows the number of units to be shipped from each source to each sink, and what path each will take. For the above example, $X_{11} = 10$, $X_{21} = 2$, $X_{22} = 10$, $X_{32} = 4$, $X_{33} = 6$, and all other $X_{ij} = 0$.

The NW corner rule provides an initial basic feasible solution; however, it is not usually optimum.

What if the origin available units had been distributed differently?

Origin available units
$$X_{1j} = 12$$
$$X_{2j} = 12$$
$$X_{3j} = 8$$

The NW corner rule would give

12			12
	12		12
	2	6	8
12	14	6	

$$X_{11} = 12$$
$$X_{22} = 12$$
$$X_{32} = 2$$
$$X_{33} = 6$$
all others $= 0$

This is a degenerate solution because there are only four nonzero variables. Degenerate solutions do not present conceptual problems; however, slight modifications will have to be made in the transportation algorithm during the next iteration. This will be illustrated during discussion of the algorithm.

When applying the NW corner rule, it is sometimes possible to avoid an initial degenerate solution by starting at a different point than the northwest corner. If, for this particular case, the first entry had been made in the southwest corner, then, the following solution, with five nonzero variables, will result.

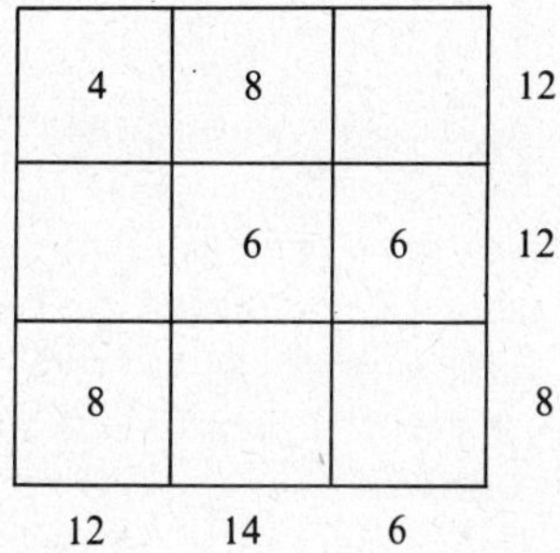

Initial Solution by Vogel Approximation Method A more rational way of developing the initial, basic, feasible solution is through application of the Vogel approximation method. The objective in this method is to avoid the most costly paths, thereby, producing an initial solution that is closer to optimum. The rules for applying the Vogel approximation method are as follows:

1. Determine the penalty for each row and column. The penalty is the absolute value of the difference between the smallest and the next smallest cost coefficients for the row or column.
2. Locate the largest penalty (either row or column). If a tie develops between two or more row or column penalties, the choice of which one to use is arbitrary.
3. Place the variable in the cell with the lowest cost for the row or column with the largest penalty. The value of this variable should be large enough to satisfy the row or column requirement associated with that cell.
4. a. If the row requirement has been satisfied, recompute column penalties.
   b. If the column requirement has been satisfied, recompute row penalties.
5. Repeat steps 2 through 4 until all resources are used up.

Example 6.4 will illustrate application of the Vogel approximation method.

If all of the entries in the cells of any row or column are added up, they must equal the original requirement for that row or column. This is true at any stage of the solution of transportation problems, and it provides a convenient check for numerical errors.

The Vogel approximation in Example 6.4 results in a degenerate solution. This is not a problem, however, as the transportation algorithm will accommodate degenerate solutions. For this particular example, the Vogel approximation also is the optimal solution. However, the analyst would have no way of knowing that at this point. The transportation algorithm would have to be applied to discover that the solution is optimal. Students should do this as a practice exercise after studying the transportation algorithm. Although it may at first seem contrary, there is nothing wrong with having a degenerate optimal solution.

EXAMPLE 6.4 ——

The following transportation problem is given:

Sources	Sinks
$X_{1j} = 16$	$X_{i1} = 12$
$X_{2j} = 6$	$X_{i2} = 14$
$X_{3j} = 10$	$X_{i3} = 6$

j

	1	2	3	
1	4	8	4	Cost matrix
2	9	10	3	
3	14	6	5	

(row labels indexed by i)

The NW corner rule would give the following initial, basic feasible solution:

12	4		16
	6		6
	4	6	10
12	14	6	

For the Vogel approximation method, the cost coefficients are added to the matrix to facilitate computation of the row and column penalties.

	Cost coefficients		Source supply	Row penalties
4	8	4	16	0
9	10	3 (6)	6	6
14	6	5	10	1

Largest penalty (row 2)

Sink requirements: 12 14 6
Column penalties: 5 2 1

The penalty for row 1 is zero because the smallest and next smallest cost coefficients are both 4. In other words, there is no penalty for shipping over either of these alternative routes. Row 2 has the largest penalty for shipping over the second least expensive route. Therefore, as much as possible should be shipped over the least expensive route of the row. The smallest cost coefficient is 3 for cell 2, 3. Row and column requirements for this cell are both 6; therefore, six units can be shipped from source 2 to sink 3. This will simultaneously satisfy both a row and column requirement, and both row and column penalties have to be recomputed.

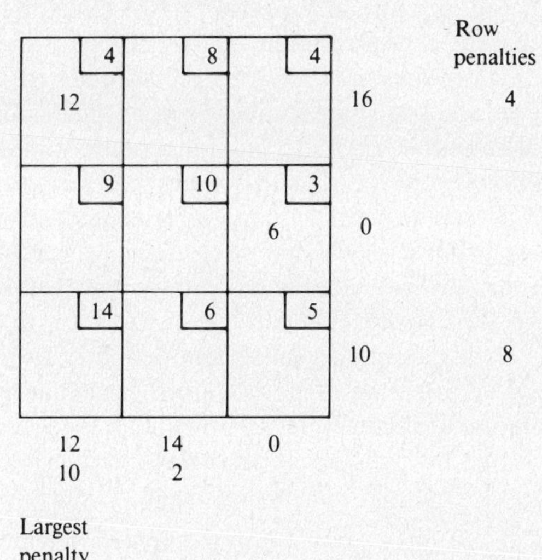

Column 1 now has the largest penalty, and as many units as possible should be shipped over route 1, 1. Twelve units will satisfy the sink 1 requirements, but there will still be four units available from source 1. Since a column requirement has been met, the row penalties have to be recomputed. However, since two of the three column requirements have been met, no new row penalties can be computed. Sources 1 and 3 can now be used to meet the requirements of sink 2 in the only way possible: 4 units from source 1, and 10 units from source 3. Note that all supply units have been shipped and all sink requirements have been met. The NW corner solution would give an objective function value ($\sum_{i=1}^{m} \sum_{j=1}^{n} c_{ij}X_{ij}$) of 194, while the Vogel approximation would result in a value of $4(12) + 8(4) + 3(6) + 6(10) = 158$.

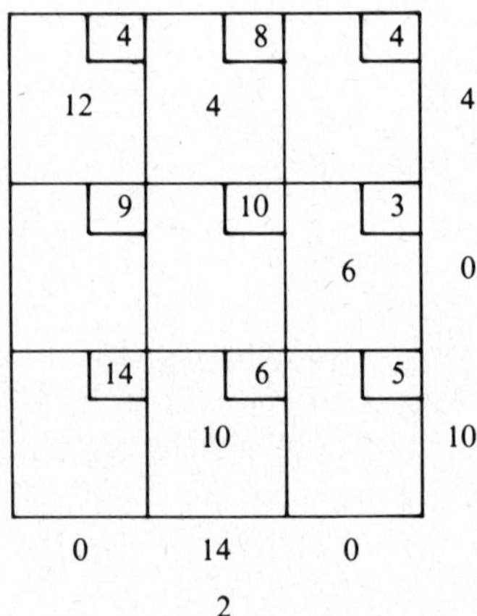

Unbalanced Conditions What if the available resource at the sources is not equal to the required amount of resource for the sinks? The inequality may go either way. There may be excess resource at the source, or there may be excess demand at the sinks. Either of these cases can be accommodated.

If the amount at the sources exceeds the demand, then another column is added to the distribution matrix to provide a pseudosink that can absorb the excess supply. Entries in this column will show where the excess supply will remain after the demand is met. Cost coefficients for this column are generally zero; however, if there is a storage cost, then the cost coefficients could reflect this cost.

If the demand exceeds the supply, then an extra row can be added as a pseudosource. Cost coefficients in this case should represent the cost of acquiring resource from alternate sources to satisfy the demand. Example 6.5 demonstrates the use of the Vogel approximation method during unbalanced conditions. Note that the value of the objective function after completion of the Vogel approximation can be computed by adding up the products of the values in the cells and the corresponding cost coefficients. For Example 6.5, this would be

$$Z = 12(4) + 4(8) + 8(10) + 6(3) + 2(6) + 8(0) = 190$$

The same method can be used during application of the transportation algorithm to find the value of the objective function at any stage of the procedure, and to determine the optimal objective function value after the optimum solution has been found. If the NW corner rule was applied to Example 6.5, the following distribution matrix would result:

<table>
<tr><td>12</td><td>4</td><td></td><td></td></tr>
<tr><td></td><td>10</td><td>4</td><td></td></tr>
<tr><td></td><td></td><td>2</td><td>8</td></tr>
</table>

which has an objective function value of 202. The Vogel approximation method is the preferable method to find an initial, basic, feasible solution because it will generally result in a solution that is closer to optimum than the NW corner rule.

EXAMPLE 6.5

The following transportation problem is given:

Sources	Sinks
$X_{1j} = 16$	$X_{i1} = 12$
$X_{2j} = 14$	$X_{i2} = 14$
$X_{3j} = 10$	$X_{i3} = 6$
Total = 40	Total = 32

Since the source supply is greater than the sink demand, an additional column is added. The following distribution matrices illustrate the application of the Vogel approximation method to this problem:

				Source supply	Row penalties
4	8	4	0		
12				16	4
9	10	3	0		
				14	3
14	6	5	0		
				10	5

Sink requirements	12	14	6	8
Column penalties	5	2	1	0

Tableau 1

				Source supply	Row penalties
4 / 12	8	4	0	4	4
9	10	3	0	14	3
14	6	5	0 / 8	10	5

Sink requirements: 0 14 6 8
Column penalties: 2 1 0

Tableau 2

				Source supply	Row penalties
4 / 12	8	4	0	4	4
9	10	3 / 6	0	14	7
14	6	5	0 / 8	2	

0 14 6 0
2 1

Tableau 3

				Source supply	Row penalties
4 / 12	8 / 4	4	0	4	0
9	10 / 8	3 / 6	0	8	0
14	6 / 2	5	0 / 8	2	0

0 14 0 0
2

$Z = 190$

Application of Transportation Algorithm   After completion of the Vogel approximation method or NW corner rule to find an initial, basic, feasible solution, the transportation algorithm is used to find the optimal solution. It is possible to apply the transportation algorithm because of the special characteristics of the constraints for the transportation problem.

Shown below is the initial solution for Example 6.5 with the first iteration of the transportation algorithm applied to it. Individual aspects of the algorithm will be explained below in the order of their application.

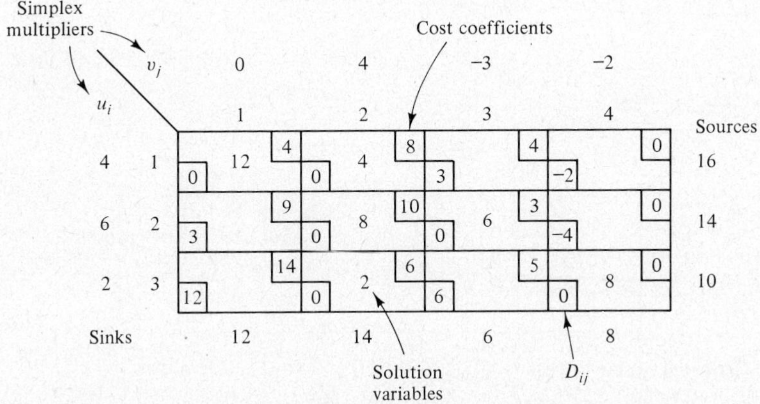

Step 1. Compute the simplex multipliers. The simplex multipliers are related to the marginal cost of bringing a particular variable into the solution. D_{ij} represents the marginal cost of bringing variable ij into the solution, and is analogous to the $c_j - Z_j$ value in the simplex algorithm. In the notation of the transportation algorithm,

$$D_{ij} = c_{ij} - Z_{ij} = c_{ij} - u_i - v_j \qquad (6.36)$$

As with the simplex algorithm, the D_{ij} value for variables in the solution has to be zero. This characteristic is used to calculate the simplex multipliers. Also, any one of the simplex multipliers has to be set equal to zero because of the redundant constraint in the transportation problem. Convention usually calls for v_1 to be set equal to zero. Thus, the following initial conditions would be given:

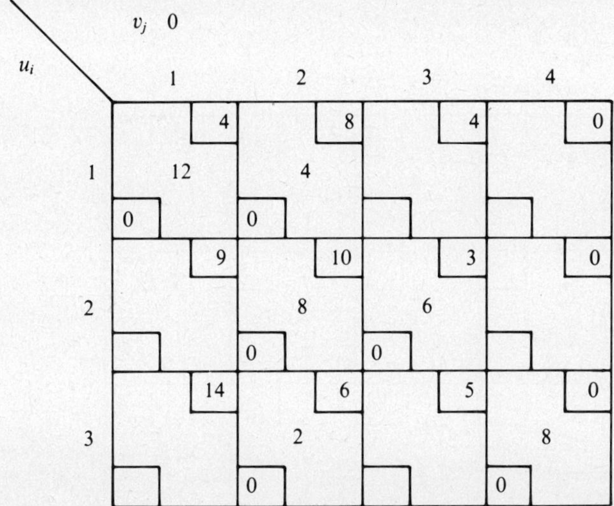

Using the equation

$$D_{ij} = 0 = c_{ij} - u_i - v_j$$

and the initial value for v_1, the rest of the u_i and v_j can be calculated. For example,

$$u_1 = c_{11} - v_1 = 4 - 0 = 4$$

Given u_1, v_2 can be calculated by using cell 12.

$$v_2 = c_{12} - u_1 = 8 - 4 = 4$$

The remaining u_i and v_j can be calculated in a similar manner to give:

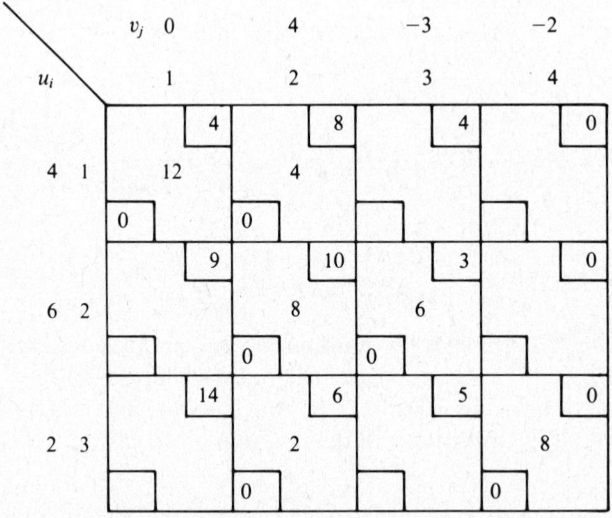

Step 2. Compute the remaining D_{ij}. Now the rest of the D_{ij} can be calculated, using

$$D_{ij} = c_{ij} - u_i - v_j$$

and the u_i and v_j previously calculated.

Step 3. Identify the variable that should enter the solution on the next iteration.

Since this is a minimization problem, negative values of D_{ij} would indicate that the objective function value would go down if the associated variable entered the solution. If there are more than one negative D_{ij}, then the most negative should be chosen, because it represents the path of steepest descent. If there are no negative D_{ij}, then the optimum solution has been found.

For this problem, there are two negative D_{ij}; however, D_{24} is the most negative, so X_{24} should enter the solution. If X_{24} is to enter the solution at some value, then other X_{ij} will have to be adjusted to maintain balance in the requirements.

Step 4. Trace a closed loop path to identify which X_{ij} must be adjusted.

	1	2	3	4	
1	12	4			16
2		8	6	+	14
3		2		8	10
	12	14	6	8	

The initial distribution matrix is given above, and a + sign has been placed in cell 24, because this is the variable that will enter the solution. If X_{24} takes on some value, then another X_{ij} in row 2 and another X_{ij} in column 4 must be reduced in value, if the requirements are to remain balanced. To identify which X_{ij} should be adjusted, form a closed loop utilizing cell 24 and other cells containing solution variables. These cells should be identified with alternating + and − signs. The path should start at 24 and end at 24. For the distribution matrix given above, the closed loop would be:

	1	2	3	4
1	12	4		
2		− 8	6	+
3		+ 2		− 8

For larger problems, the path will be more complicated, and may cross over itself, but it will always be possible to find a closed loop path.

Step 5. Determine the value at which X_{ij} will enter the solution. Retrace the closed loop path, looking only at the cells identified with a negative sign. The minimum value X_{ij} contained in these negative cells is the value at which the new variable should enter the solution. If the new variable came in at a higher level, then some X_{ij} would become negative, which would be an infeasible solution. For the given problem, both negative cells have a value of 8, so X_{24} should enter at a value of 8.

Step 6. Readjust the distribution matrix to reflect the entry of the new variable.

At this point, it is advisable to draw a new distribution matrix, and enter the new solution into it to prepare for the next iteration. All steps up to this point should have been accomplished on the original distribution matrix.

All cells which form the corners of the closed loop are adjusted by the value of the entering X_{ij}, according to the sign in each cell. Therefore, eight would be added to the value in cell 32, and subtracted from the values in cells 22 and 34. The new distribution matrix is:

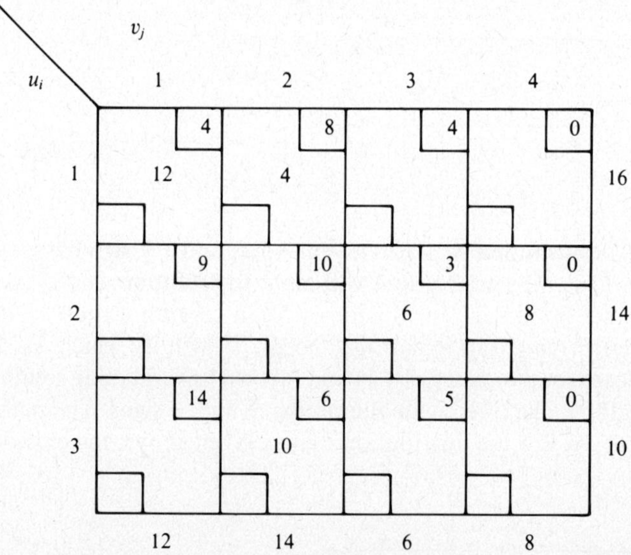

A check of row and column summations shows that the requirements are still balanced. Note that the new solution is degenerate.

Step 7. Repeat steps 1 through 6 until the optimal solution is found.

Since the new solution is degenerate, problems will be encountered in calculating the u_i and v_j values. The distribution matrix given below shows that u_1, u_3, and v_2 could be calculated as shown previously.

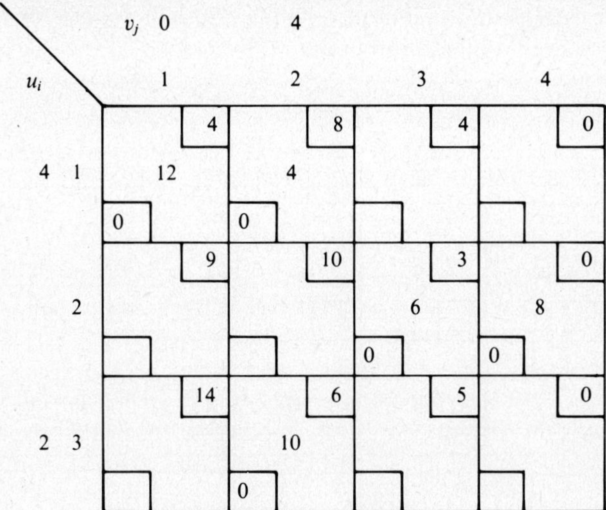

At this point, however, there is a gap that prevents calculation of the remaining u_i and v_j. Since the solution is degenerate to the first degree (five variables are greater than zero, rather than the expected 6), one additional variable can be brought into the solution at a zero value to help bridge the gap. Several X_{ij} can usually be used for this purpose, and the choice of which one to use is arbitrary. If X_{22} is brought into the solution, the remaining u_i and v_j can be calculated.

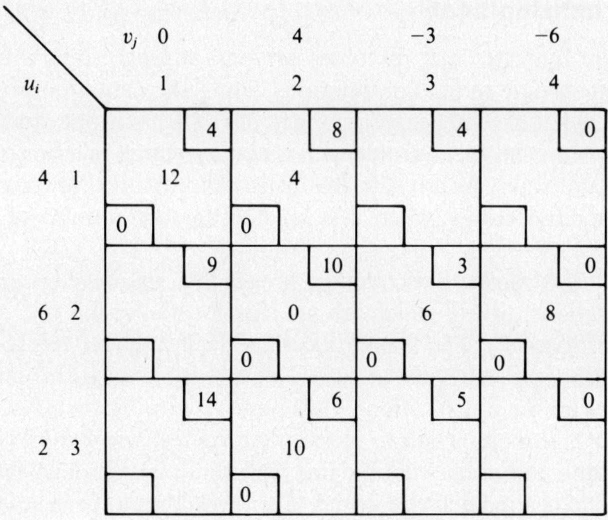

Then the remainder of the steps can be accomplished in the normal manner. If the variable in the solution at a zero level appears in the closed loop, it must be treated like any other variable in the solution. The distribution matrix for the given problem is shown below after computation of the remaining D_{ij}.

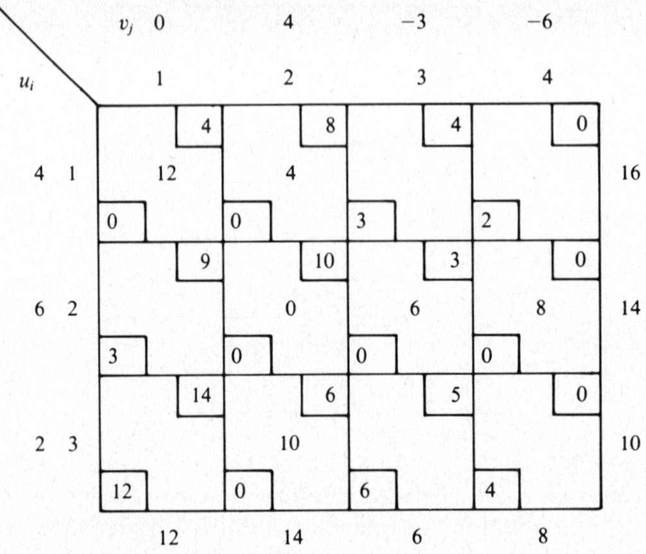

Since there are no negative D_{ij} values, this is the optimal solution, and the objective function value is 158.

Transportation with Transshipment

Transshipment indicates that resources can pass through some intermediate destinations (nodes) enroute to final destinations. Thus, the definition of sources and sinks becomes somewhat blurred because some nodes in the network can act as both sources and sinks. For transshipment problems it is better to think in terms of locations, rather than sources and sinks. As a result, the distribution matrix for transshipment will be square, and of dimension n, where n is equal to the total number of locations that are modeled.

Figure 6.4(a) shows the schematic model that was developed for Example 5.8. It has five locations, two of which are net sink locations, and one of which is a pure transshipment location. The source locations have a larger refuse production than they have landfill; thus, they act as supply locations for the distribution network. The opposite is true for the sink locations. However, it is possible for refuse to be transported from location 2, through location 4, to either the transfer station (T), or to location 3. There is some additional cost for this transshipment through the location due to vehicle transfers or time delays enroute. The model has to be balanced, and the theoretical maximum amount that could be transshipped through any particular location is the total excess supply, or the total excess demand. This upper bound on transshipment is termed t, and it is related to t_i, the amount transshiped through a node, by the equation

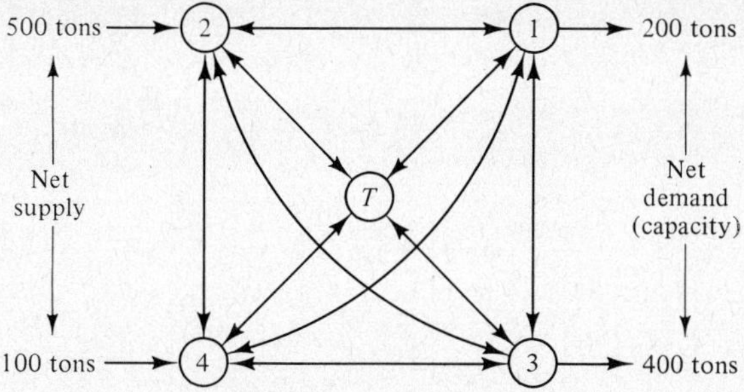

(a) Transshipment network from Example 5.8.

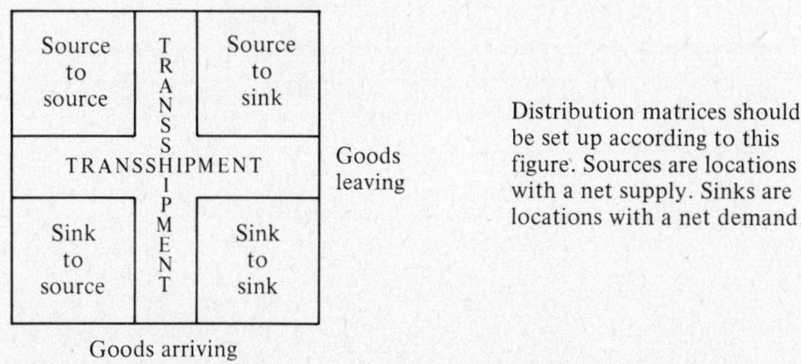

Distribution matrices should be set up according to this figure. Sources are locations with a net supply. Sinks are locations with a net demand.

(b) Distribution matrix organization.

Figure 6.4   Transportation with transshipment.

$$X_{ii} = t - t_i \qquad (6.37)$$

in which X_{ii} is a slack variable that represents the additional number of units that could have been transshipped through node i. X_{ii} will appear in the diagonal of the distribution matrix for transshipments. The corresponding cost coefficient for the X_{ii} is called c_{ii}, and it is equal to $-c_i$. This c_{ii} represents the amount that could be saved by not transshipping one unit through location i.

To assist in analysis, the distribution matrix is generally set up according to the scheme shown in Figure 6.4(b). Example 6.6 illustrates solution of a transshipment problem utilizing the transportation algorithm.

EXAMPLE 6.6 ──

The transshipment model developed in Example 5.8 can be set up in the transportation algorithm distribution matrix format, and the transportation algorithm can be utilized to find the optimum solution. Table 6.7 shows the initial distribution matrix after application of the Vogel approximation. Note the negative cost coefficients (c_{ii}), and slack variables (X_{ii}) on the diagonal. Note also that the source supplies are $t + a$, the upper bound on the transshipment plus the net supply, and the sink demands are $t + b$, the upper bound on transshipment plus net demand. Application of the Vogel approximation method proceeds as with the standard transportation algorithm; how-

TABLE 6.7   INITIAL DISTRIBUTION MATRIX, TRANSSHIPMENT PROBLEM, EXAMPLE 5.8

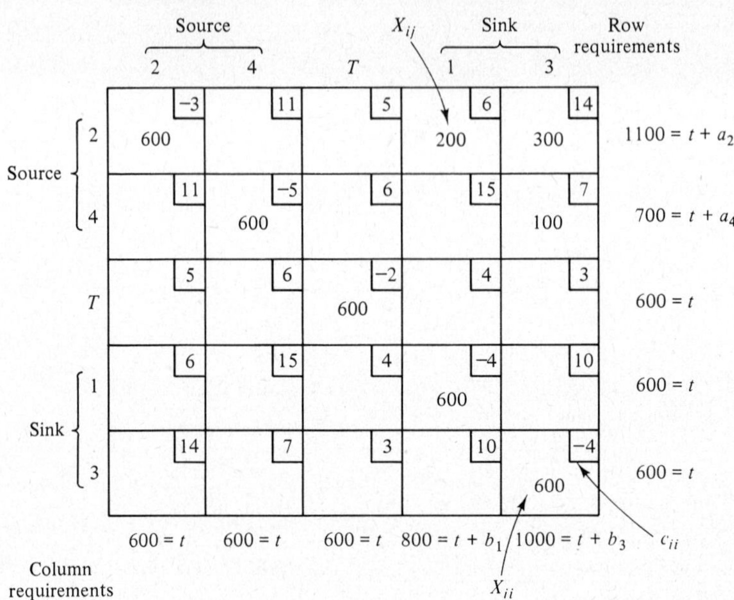

ever, the negative elements have to be taken into account when calculating the penalties. The negative value will always be the minimum cost coefficient for a row or column.

After application of the transportation algorithm, the optimal distribution matrix shown in Table 6.8 is developed. Equation 5.91, which is repeated here, can be used to determine the value of the objective function at the optimum solution.

$$Z = \sum_{i=1}^{n} \sum_{j=1}^{n} c_{ij} X_{ij} + \sum_{i=1}^{n} c_i(t - X_{ii}) \tag{5.91}$$

Substitution of appropriate values into this equation is shown at the bottom of Table 6.8.

Since $t = 600$, any cells on the diagonal that have a value of 600 in them indicate that there is no transshipment through that node. Locations 1, 2, 3, and 4 all have no transshipment. However, X_{ii} for location T (the transfer station) is 300. Therefore, $t_T = t - X_{TT}$, or $t_T = 600 - 300 = 300$.

ASSIGNMENT PROBLEMS AND THE ASSIGNMENT ALGORITHM

Problems in which there is a one-to-one correspondence between units of resource and the requirements for utilization of that resource are termed assignment problems. Examples are assigning n machines to n jobs, n individuals to n locations, n trucks to n destinations, or n methods of satisfying n requirements. The assignment of excavators to different tasks in an open pit mine, as described in Example 5.13 is a good example

TABLE 6.8 OPTIMAL DISTRIBUTION MATRIX, TRANSSHIPMENT PROBLEM, EXAMPLE 5.8

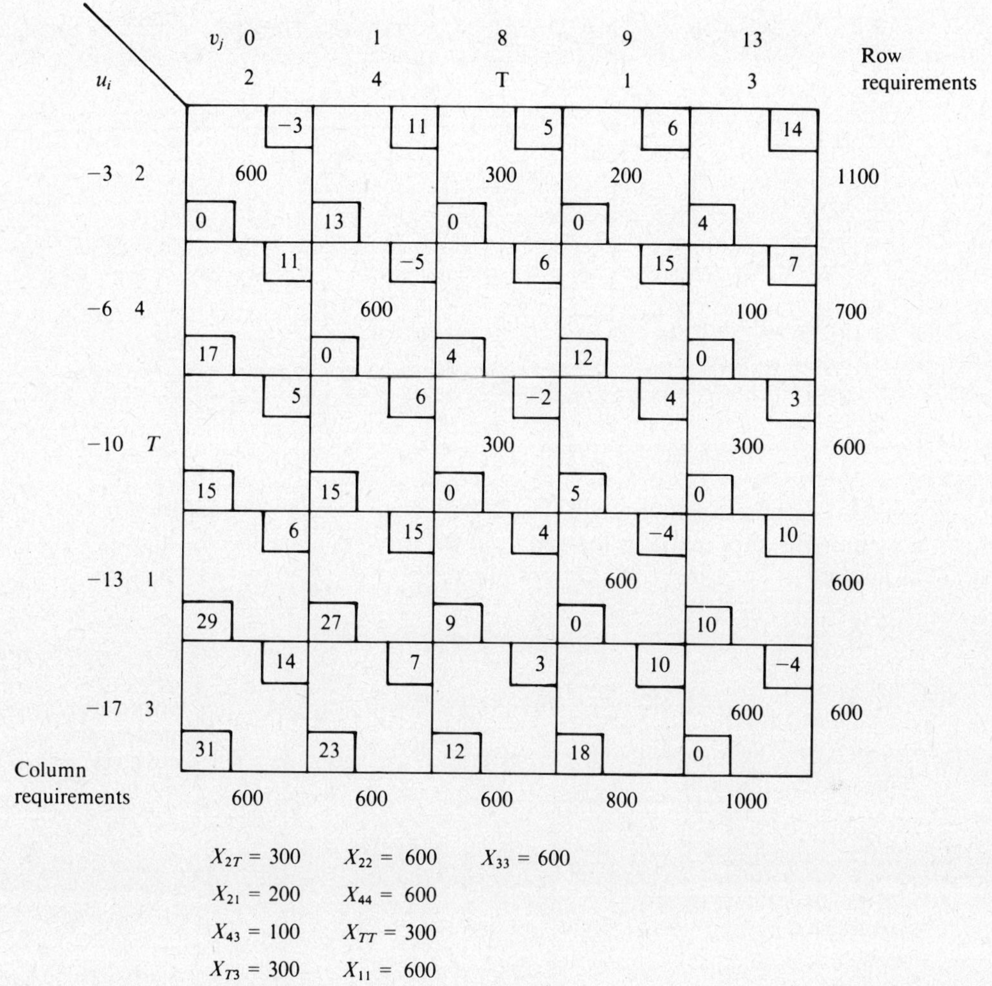

In each cell the cost is printed in the upper-right corner, the allocation (when present) in the center, and the cell evaluation in the lower-left corner.

u_i	$v_j = 0$, 2	$v_j = 1$, 4	$v_j = 8$, T	$v_j = 9$, 1	$v_j = 13$, 3	Row requirements
-3 2	-3 600 0	11 13	5 300 0	6 200 0	14 4	1100
-6 4	11 17	-5 600 0	6 4	15 12	7 100 0	700
-10 T	5 15	6 15	-2 300 0	4 5	3 300 0	600
-13 1	6 29	15 27	4 9	-4 600 0	10 10	600
-17 3	14 31	7 23	3 12	10 18	-4 600 0	600
Column requirements	600	600	600	800	1000	

$$X_{2T} = 300 \qquad X_{22} = 600 \qquad X_{33} = 600$$
$$X_{21} = 200 \qquad X_{44} = 600$$
$$X_{43} = 100 \qquad X_{TT} = 300$$
$$X_{T3} = 300 \qquad X_{11} = 600$$

$$Z = 300(5) + 200(6) + 100(7) + 300(2) + 300(3) = \$4900/\text{day}$$

of an assignment problem. Another application is the sequential assignment of a person or vehicle to segments of a network to find the shortest or longest path between two points in the network.

Solutions to assignment problems are highly degenerate, with only n variables $>$ zero. Because of the high degree of degeneracy, it is not convenient or efficient to use the transportation algorithm to find the optimal solution. The Hungarian mathematician, Konig, developed an efficient algorithm for solving this specialized type of problem, which is appropriately called the Hungarian method.

Figure 6.5 shows the transportation network of Example 5.10. The objective is

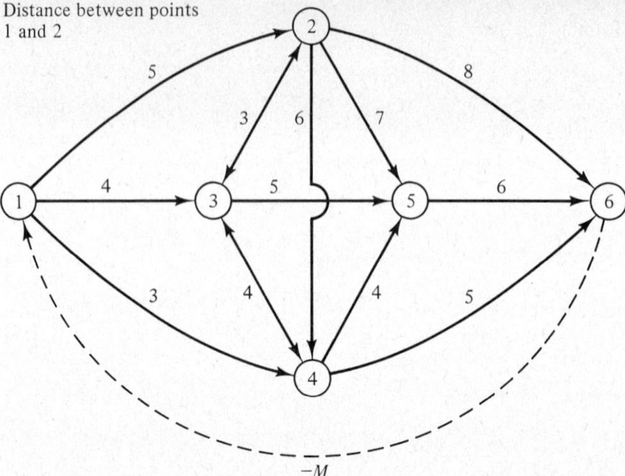

Figure 6.5 Transportation network.

to find the shortest path through the network from location 1 to location 6. An initial cost/time/distance matrix is formed from the network data, and the method proceeds as follows:

Cost/time/distance matrix

	1	2	3	4	5	6
1	0	5	4	3	M	M
2	M	0	3	6	7	8
3	M	3	0	4	5	M
4	M	M	4	0	4	5
5	M	M	M	M	0	6
6	$-M$	M	M	M	M	0

All row and column requirements are 1. $n = 6$

Indicates no path from 2 to 1

No cost to stay at node

Forces path to go from 1 to 6

1. Subtract smallest cost estimate in each row from each element in that row.
2. Check to see if optimal solution has been formed. Do n independent zeros

appear in modified cost matrix? To check, draw set of minimum number of lines through zeros, using any appropriate mix of horizontal and vertical lines. If it takes n lines, the optimum solution has been found. If not, continue. For the following matrix, # = 5, ∴ it is not optimal.

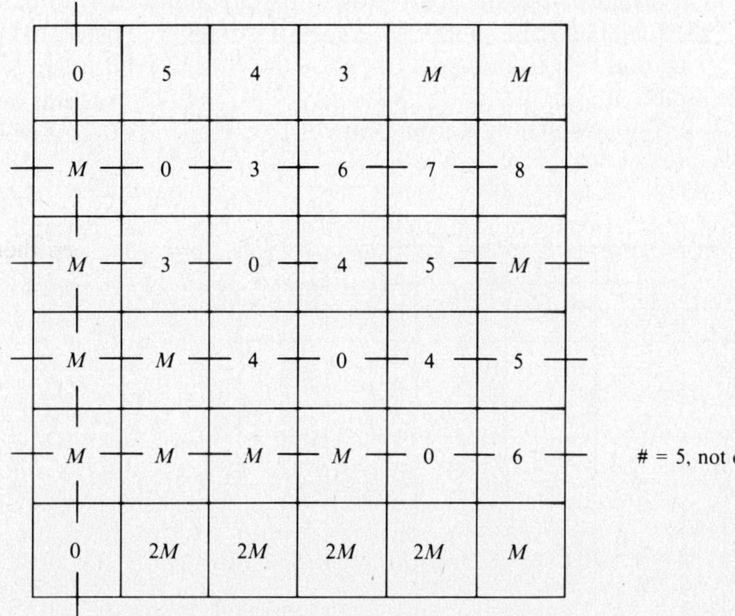

0	5	4	3	M	M
M	0	3	6	7	8
M	3	0	4	5	M
M	M	4	0	4	5
M	M	M	M	0	6
0	2M	2M	2M	2M	M

= 5, not optimal

3. Repeat steps 1 and 2 on columns

0	5	4	3	M	M−
M	0	3	6	7	3
M	3	0	4	5	M−
M	M	4	0	4	0
M	M	M	M	0	1
0	2M	2M	2M	2M	M−

= 5, not optimal

4. Call the set of elements covered by lines in step 3 matrix **C**, and the set of elements not covered matrix **N**. Find the smallest element in **N** and subtract it from all elements in **N**. Also, add this quantity to elements in **C** that lie on the intersection of two lines (if any exist). These operations will produce the next matrix. For the above matrix, the minimum entry in the **N** submatrix is 3. Note the use of $M-$ to indicate less than M. There is no need to carry the numbers in these cells, because the M value will keep them out of the solution.

The new matrix is given below:

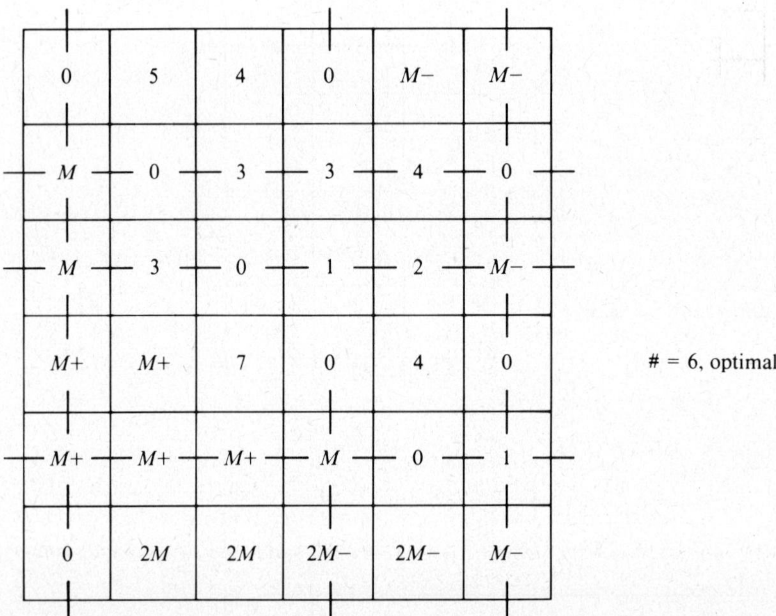

5. Construct a new minimum set of lines to check for optimality.
6. Repeat steps 4 and 5 until the optimal solution is found (until it takes n lines to cover the zeros). For this example, it takes a minimum of six lines to cover the zeros for the new matrix; therefore, the optimal distribution has been found.
7. To convert this to the actual path covered, locate a set of $6(n)$ independent zeros (if more than one set exists, there are alternate optima). Look for rows or columns with only one zero in them. These elements have to be part of the solution if the row and column requirements are to be met. Other elements can easily be determined by elimination. The set of independent zeros is identified by ones on the matrix of zeros.

	1	2	3	4	5	6	Row requirements
1	0			Ⓞ			1
2		Ⓞ				0	1
3			Ⓞ				1
4				0		Ⓞ	1
5					Ⓞ		1
6	Ⓞ						1
Column requirements	1	1	1	1	1	1	

Any ones on the diagonal are ignored. Therefore, the path is $1 - 4 - 6$, and it is 8 units long.

A second application, to find the longest path through a network, is illustrated in Example 6.7. It should be pointed out that this is the same type of network that will be encountered in project management problems. An even more specialized method, called the critical path method, can be utilized to find the longest path through these networks. It will be developed in Chapter 9.

EXAMPLE 6.7 ——————————————————————————————————

Find the longest path through the network depicted in Figure 6.6 (from node 1 to node 8) using the assignment algorithm. What is the network completion time?

This is a maximization problem, so negative time (cost) coefficients will be used. Otherwise, the procedure is identical to the shortest path (minimization) problem.

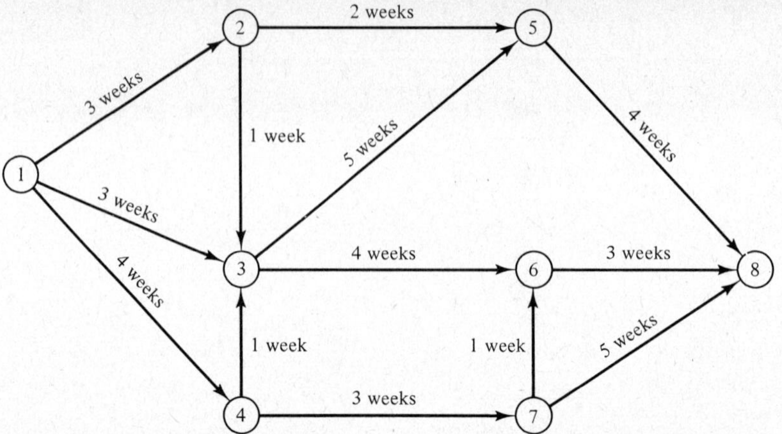

Figure 6.6 Work activity network.

	1	2	3	4	5	6	7	8
1	0	−3	−3	−4	M	M	M	M
2	M	0	−1	M	−2	M	M	M
3	M	M	0	M	−5	−4	M	M
4	M	M	−1	0	M	M	−3	M
5	M	M	M	M	0	M	M	−4
6	M	M	M	M	M	0	M	−3
7	M	M	M	M	M	−1	0	−5
8	−M	M	M	M	M	M	M	0

4	1	1	0	M	M	M	M
M	2	1	M	0	M	M	M
M	M	5	M	0	1	M	M
M	M	2	3	M	M	0	M
M	M	M	M	4	M	M	0
M	M	M	M	M	3	M	0
M	M	M	M	M	4	5	0
0	$2M$	$2M$	$2M$	$2M$	$2M$	$2M$	M

$\# = 5$

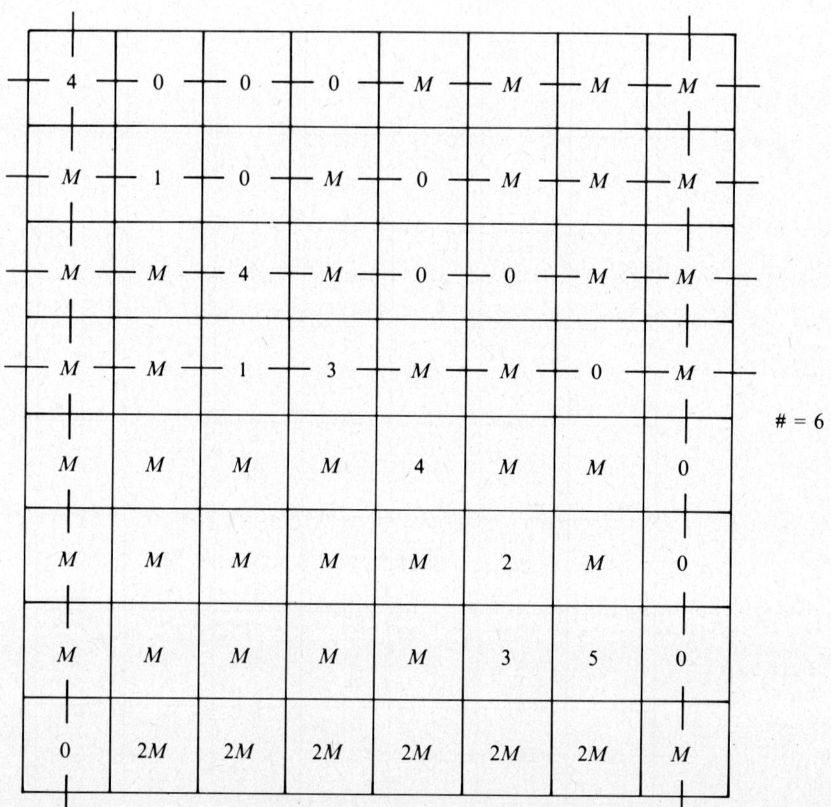
4 0 0 0 M M M M
M 1 0 M 0 M M M
M M 4 M 0 0 M M
M M 1 3 M M 0 M
M M M M 4 M M 0
M M M M M 2 M 0
M M M M M 3 5 0
0 2M 2M 2M 2M 2M 2M M
= 6

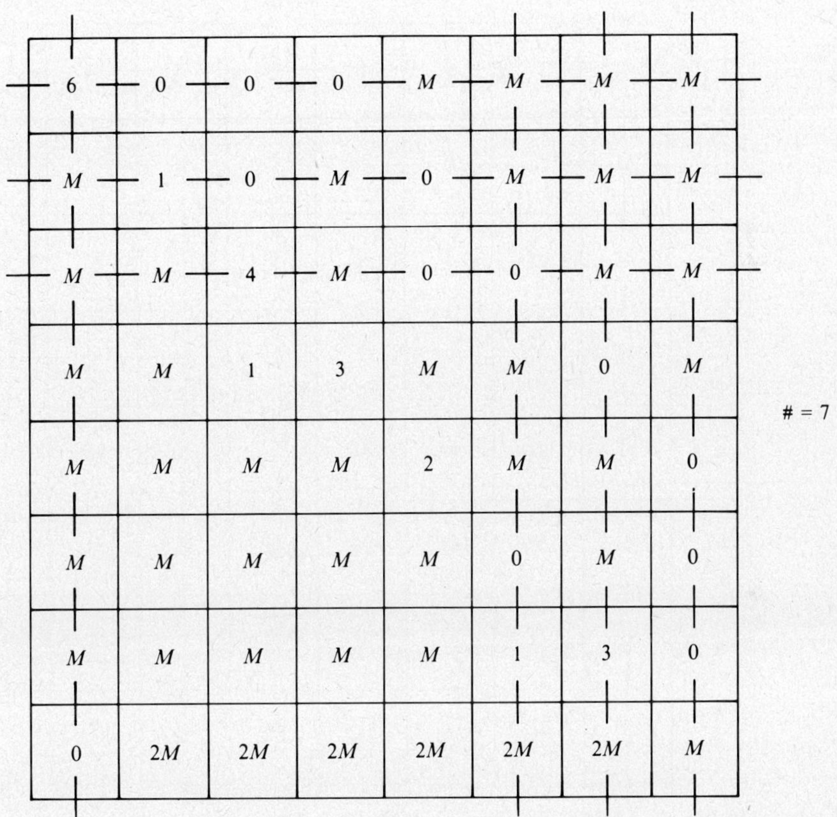
6 0 0 0 M M M M
M 1 0 M 0 M M M
M M 4 M 0 0 M M
M M 1 3 M M 0 M
M M M M 2 M M 0
M M M M M 0 M 0
M M M M M 1 3 0
0 2M 2M 2M 2M 2M 2M M
= 7

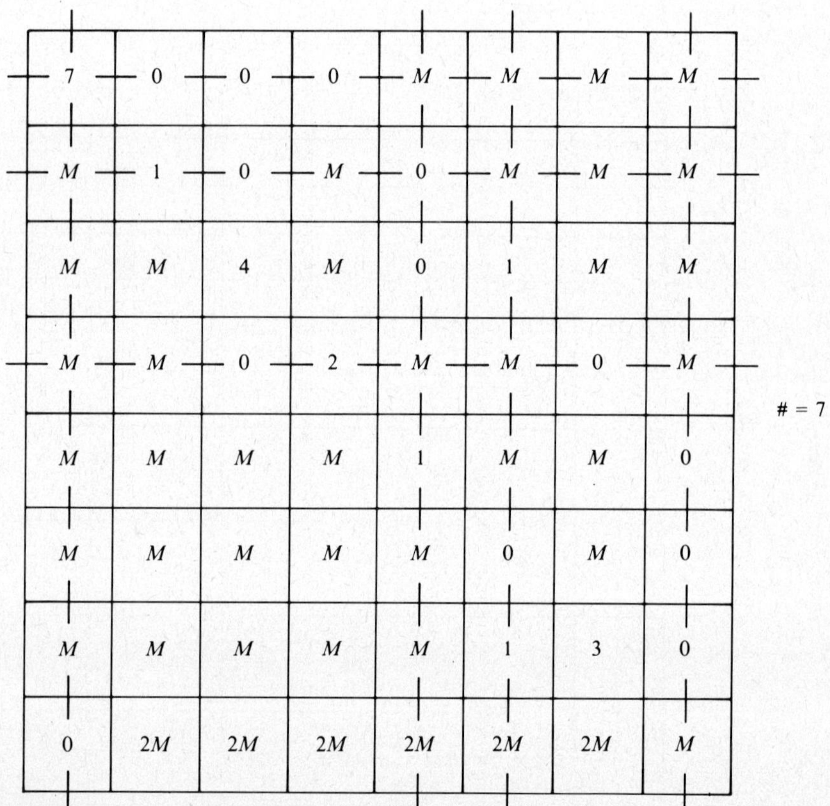

7 0 0 0 M M M M
M 1 0 M 0 M M M
M M 4 M 0 1 M M
M M 0 2 M M 0 M
M M M M 1 M M 0
M M M M M 0 M 0
M M M M M 1 3 0
0 2M 2M 2M 2M 2M 2M M
= 7

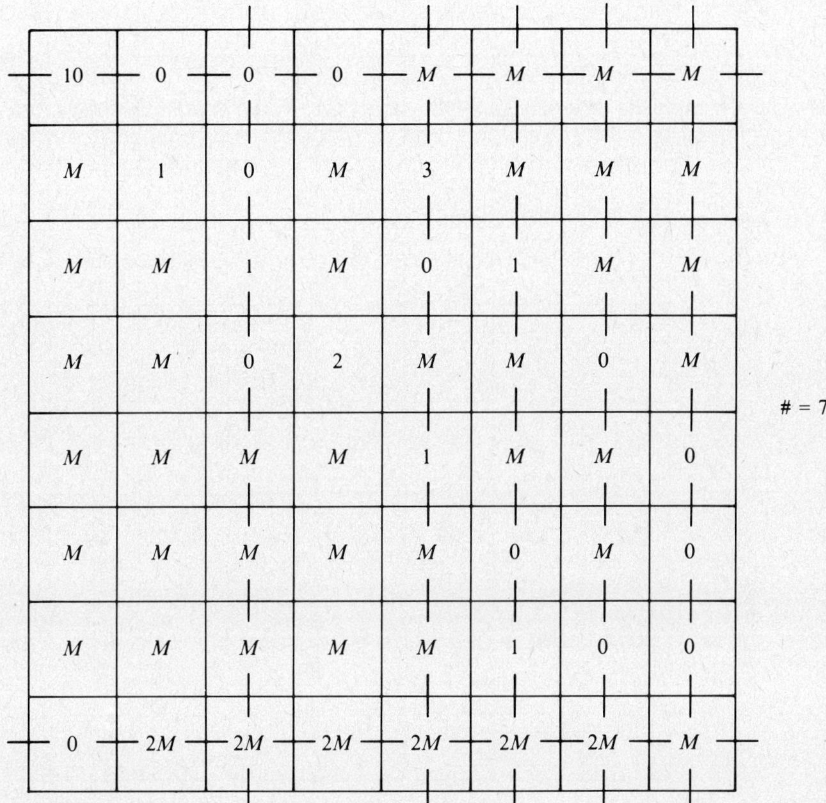
10 0 0 0 M M M M
M 1 0 M 3 M M M
M M 1 M 0 1 M M
M M 0 2 M M 0 M
M M M M 1 M M 0
M M M M M 0 M 0
M M M M M 1 0 0
0 2M 2M 2M 2M 2M 2M M
= 7

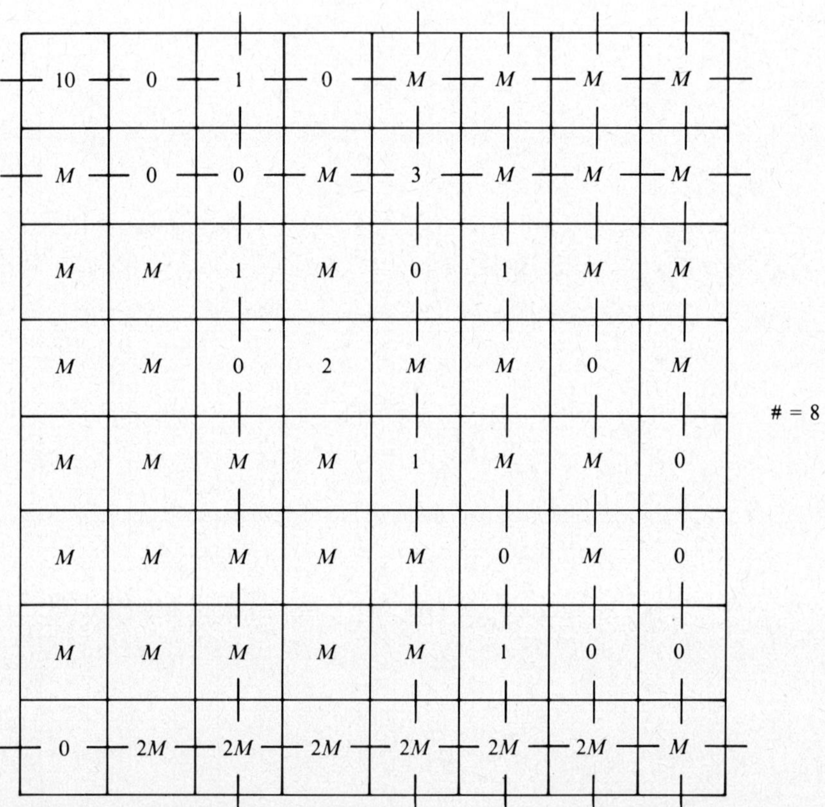
10   0   1   0   M   M   M   M
M   0   0   M   3   M   M   M
M   M   1   M   0   1   M   M
M   M   0   2   M   M   0   M
M   M   M   M   1   M   M   0
M   M   M   M   M   0   M   0
M   M   M   M   M   1   0   0
0   2M   2M   2M   2M   2M   2M   M
= 8

<table>
<tr><td></td><td>1</td><td>2</td><td>3</td><td>4</td><td>5</td><td>6</td><td>7</td><td>8</td></tr>
<tr><td>1</td><td></td><td>0</td><td></td><td>φ</td><td></td><td></td><td></td><td></td></tr>
<tr><td>2</td><td></td><td>φ</td><td>0</td><td></td><td></td><td></td><td></td><td></td></tr>
<tr><td>3</td><td></td><td></td><td></td><td></td><td>φ</td><td></td><td></td><td></td></tr>
<tr><td>4</td><td></td><td></td><td>φ</td><td></td><td></td><td></td><td>0</td><td></td></tr>
<tr><td>5</td><td></td><td></td><td></td><td></td><td></td><td></td><td></td><td>φ</td></tr>
<tr><td>6</td><td></td><td></td><td></td><td></td><td></td><td>φ</td><td></td><td>0</td></tr>
<tr><td>7</td><td></td><td></td><td></td><td></td><td></td><td></td><td>0</td><td>0</td></tr>
<tr><td>8</td><td>φ</td><td></td><td></td><td></td><td></td><td></td><td></td><td></td></tr>
</table>

Longest path: $1 - 4 - 3 - 5 - 8 = 14$ weeks.

NONLINEAR SYSTEMS

Not all systems can be modeled adequately by a linear objective function and linear constraints. Either, or both, may by necessity contain nonlinear terms. The constraints for the three-bar truss model developed in Example 5.4 contained nonlinearities, while the two-tiered walkway model (Example 5.3) contained nonlinear terms in both the objective function and constraints. A water supply model, which will be developed in Example 6.9, will have nonlinear terms in the objective function, but linear constraints. Other nonlinear models might include a waste treatment system that contains a nonlinear economy of scale with size, or a waste treatment unit operation that has a removal rate that is dependent on waste concentration.

There is no set procedure for solving all nonlinear programming problems. Problems that have two independent variables can be graphed and successfully optimized. Some problems may be broken down into linear segments, and solved by a technique called quasilinearization. Example 5.7 demonstrated a quasilinearized model.

Other algorithms may be found in the literature for solving particular types of nonlinear problems. However, successful solution to these types of problems may be sensitive to initial estimates of the variable values. Solutions using inaccurate estimates may diverge rather than converge toward an optimal solution. A thorough discussion of nonlinear programming has been presented by Wilde and Beightler (1967), McCormick (1983), and Wismer and Chattergy (1978). Kirsch (1981), Galligher and Zenkiewicz (1973), and Haug and Arora (1979) have presented nonlinear methods for optimizing structural systems.

The objective of this section is to introduce the student to nonlinear optimization, through the application of commonly used methods to two example problems. The student will be able to observe the complexities involved in the solution of even simplistic nonlinear models, and will gain an appreciation for the care that must be taken when applying nonlinear methods to larger problems.

Convexity Considerations

One of the most important considerations in any optimization technique is the convexity of the solution space. A solution space is convex if a line drawn from any boundary point of the solution space to any other boundary point lies completely within the solution space. It is desirable to have a convex solution space, because if the solution space is nonconvex, solution techniques may converge toward a local, rather than global, optimum. Properly posed constraint spaces for linear programming problems are always convex; however, this is not true for nonlinear solution (constraint) spaces. Also, convexity or concavity of nonlinear objective functions can affect convergence toward a global optimum.

The concept of convexity is best illustrated through a two-dimensional, graphical model. In terms of a function of two variables, concavity and convexity of individual constraints are defined as shown in Figure 6.7(a). Figure 6.7(b) shows a convex solution space, which can be bounded by either convex or concave constraints, depending on the direction of the inequality. The solution space shown in Figure 6.7(c) is nonconvex,

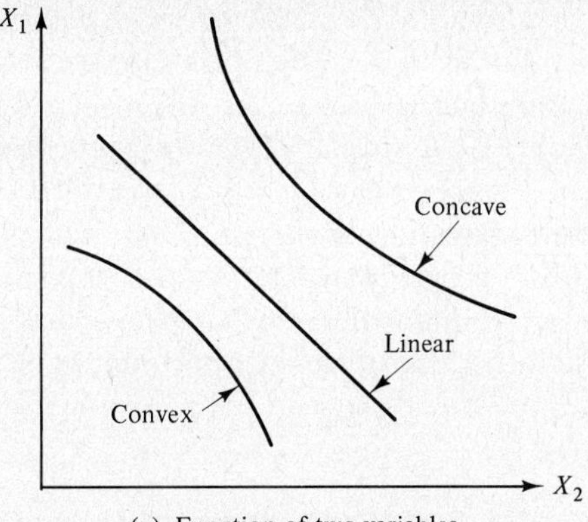

(a) Function of two variables.

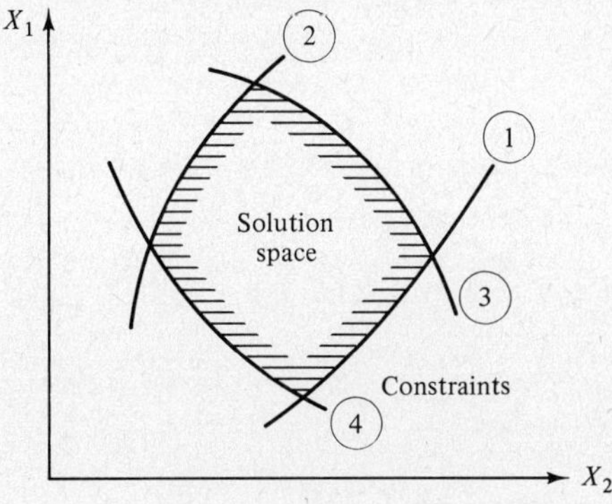

(b) Convex solution space.

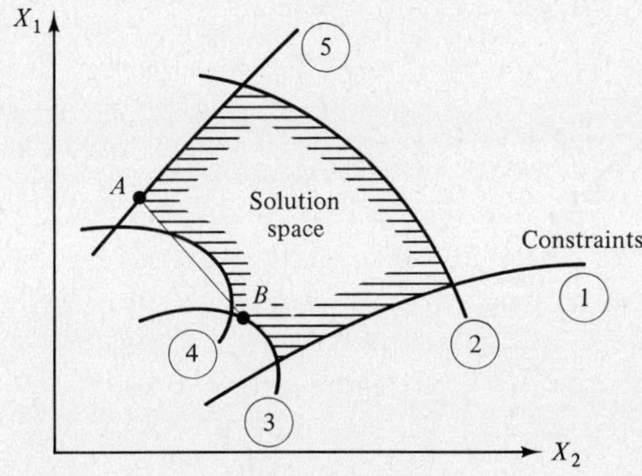

(c) Nonconvex solution space.

Figure 6.7   Convex and nonconvex solution spaces.

because the line drawn from point A to point B is not contained entirely within the solution space.

Problems can arise if the objective function is concave (nonconvex) in a minimization problem, even if the solution space is convex. This is illustrated in Figure

6.8(a). If a solution technique starts with an initial solution at point A and proceeds to point B, it will consider the solution at B to be optimum, because the objective function value increases along the boundary in either direction away from B. The actual global minimum is at point C. A similar situation is encountered with the nonconvex solution space given in Figure 6.8(b). The global optimum solution is again at point C; however, a solution progressing from point A would stop at the local minimum at point B, and a solution progressing from point E would stop at point D. Sometimes, starting from different points within the solution space can remedy the problem of converging to a local, rather than a global, optimum; however, in this case it would not.

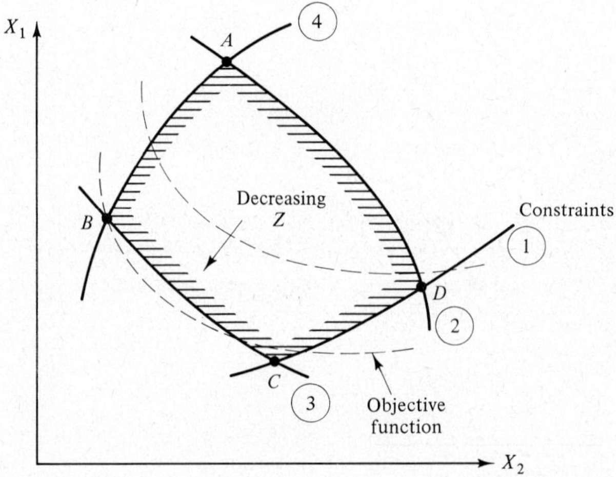

(a) Nonconvex objective function with convex solution space.

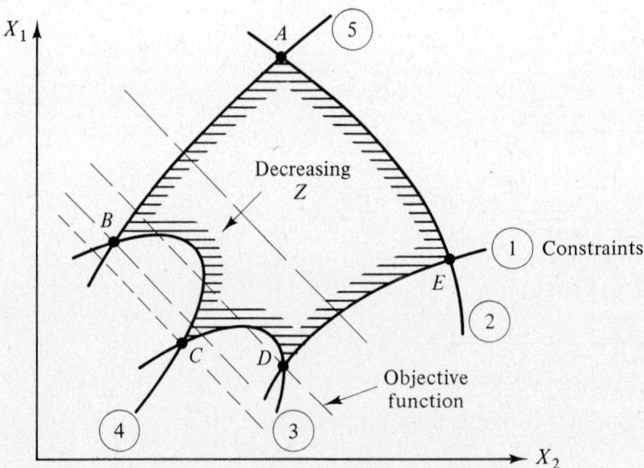

(b) Nonconvex solution space with linear objective function.

Figure 6.8 Convergence to local optimum rather than global optimum.

Convex and nonconvex considerations can be summarized as follows.

In minimization problems, a global minimum is normally obtainable if both the solution space and objective function are convex. If the objective function is concave [as defined in Figure 6.7(a)], a nonlinear programming algorithm may terminate at a local optimum. In maximization problems, a global maximum is normally obtainable if the solution space is convex, and the objective function is nonconvex. Problems may arise in maximization if the objective function is convex. Nonconvex solution spaces may produce local optimum solutions rather than global, irrespective of the shape of the objective function.

Tests for Convexity

Problems involving more than two variables cannot be solved efficiently by graphical techniques; therefore, mathematical tests for convexity must be employed. The following is an introduction to a general method for determining convexity or nonconvexity of functions and constraint spaces.

A continuously differentiable function of n variables is convex if the determinant of the Hessian matrix is ≥ 0, and all of its diagonal elements are ≥ 0. The Hessian matrix is formed by taking the partial derivatives of the function with respect to each pair of variables in the following manner. If

$$R = f(X_1, X_2, \ldots, X_n) \tag{6.38}$$

$$\text{then} \qquad \mathbf{H} = \begin{vmatrix} \dfrac{\partial^2 R}{\partial X_1^2} & \dfrac{\partial^2 R}{\partial X_1\, \partial X_2} & \cdots & \dfrac{\partial^2 R}{\partial X_1\, \partial X_n} \\[2ex] \dfrac{\partial^2 R}{\partial X_2\, \partial X_1} & \dfrac{\partial^2 R}{\partial X_2^2} & \cdots & \dfrac{\partial^2 R}{\partial X_2\, \partial X_n} \\[2ex] \vdots & \vdots & & \vdots \\[2ex] \dfrac{\partial^2 R}{\partial X_n\, \partial X_1} & \dfrac{\partial^2 R}{\partial X_n\, \partial X_2} & \cdots & \dfrac{\partial^2 R}{\partial X_n^2} \end{vmatrix} \tag{6.39}$$

If det $\mathbf{H} \geq 0$ and all $\partial^2 R / \partial X_i^2 \geq 0$ ($i = 1, 2, \ldots, n$), then R is convex. For a function of two variables,

$$\det \begin{vmatrix} \dfrac{\partial^2 R}{\partial X_1^2} & \dfrac{\partial^2 R}{\partial X_1\, \partial X_2} \\[2ex] \dfrac{\partial^2 R}{\partial X_2\, \partial X_1} & \dfrac{\partial^2 R}{\partial X_2^2} \end{vmatrix} = \dfrac{\partial^2 R}{\partial X_1^2}\dfrac{\partial^2 R}{\partial X_2^2} - \dfrac{\partial^2 R}{\partial X_1\, \partial X_2}\dfrac{\partial^2 R}{\partial X_2\, \partial X_1} \tag{6.40}$$

and the same conditions apply for convexity.

If a constraint space is defined by the following set of equations

$$\left.\begin{array}{c}
R_1(X_1, X_2, \ldots, X_n) \le b_1 \\
R_2(X_1, X_2, \ldots, X_n) \le b_2 \\
R_3(X_1, X_2, \ldots, X_n) \ge b_3 \\
R_4(X_1, X_2, \ldots, X_n) = b_4 \\
\vdots \\
R_m(X_1, X_2, \ldots, X_n) \le b_m
\end{array}\right\} \tag{6.41}$$

subject to the non-negativity constraints

$$X_j \ge 0 \qquad (j = 1, \ldots, m) \tag{6.42}$$

then the constraint space is convex if the conditions given below are met for all constraints. Less than or equal constraints must be convex functions in the region $X_j \ge 0$, greater than or equal constraints must be nonconvex functions in the region $X_j \ge 0$, and equal constraints must be linear functions.

Method of Lagrange Multipliers

The methods previously discussed can be used to determine if a global optimum solution can be found, but do not actually find any numerically optimum solutions. One method for finding optimal solutions to nonlinear models, the method of Lagrange multipliers, will be discussed here. It should be pointed out that if the method of solving the Lagrangian equations is analytic, which is possible in many cases, then the Lagrange multiplier method can be used to find all local optima, and from them the global optimum.

This is not intended to be an exhaustive presentation. There are many other methods available that can be applied to various nonlinear models. The intent is to introduce the student to what type of analytical process has to be applied to the solution of nonlinear models. It should again be emphasized that there is no one method that works for all nonlinear models. Numerical methods for finding approximately optimal solutions for nonlinear functions are presented in Appendix E.

The method of Lagrange multipliers will be presented without proof; however, the practical significance of the multipliers will be discussed. A nonlinear model can be written in the form

$$\text{max or min} \quad Z = f(X_1, X_2, \ldots, X_n)$$

$$\left.\begin{array}{l}
\text{ST:} \quad R_1(X_1, X_2, \ldots, X_n) \left(\begin{array}{c}\ge \\ \le \\ =\end{array}\right) b_1 \\[2em]
R_2(X_1, X_2, \ldots, X_n) \left(\begin{array}{c}\ge \\ \le \\ =\end{array}\right) b_2 \\[2em]
\vdots \quad \vdots \quad \vdots \qquad \vdots \quad \vdots \quad \vdots \\[1em]
R_m(X_1, X_2, \ldots, X_n) \left(\begin{array}{c}\ge \\ \le \\ =\end{array}\right) b_m
\end{array}\right\} \tag{6.43}$$

In this model, the objective function and/or some of the constraints may be linear, but they will not all be linear. The first step in the solution is to convert each of the inequalities to an equality by adding a slack or surplus variable, S_i^2. If the first l of the m constraints are inequalities, then l slack or surplus variables will be added. Each constraint is next multiplied by a non-negative constant (λ_i) called the Lagrange multiplier. Each constraint is now added to the objective function, to form what is termed the Lagrangian. The objective function will now be a function of the original variables, the S_i added, and the λ_i.

$$Z = f(X_1, X_2, \ldots, X_n, S_1, S_2, \ldots, S_l, \lambda_1, \lambda_2, \ldots, \lambda_m) \tag{6.44}$$

Partial derivatives are taken of the objective function with respect to each of its variables and set equal to zero. Simultaneous solution of the resulting $n + l + m$ equations will determine the minimum or maximum solution, depending on which type of model is being analyzed. The set of partial derivative equations will usually be nonlinear, and numerical techniques may have to be used to solve them.

Application Examples

Example 6.8 demonstrates application of Lagrange multipliers to assist in optimization of the three-bar truss model developed in Chapter 5. For this model, the objective function is linear, while the constraints are nonlinear. Examination of Figure 6.10 shows that constraint 1 is binding (it forms a critical boundary on the solution space), while constraints 2 and 3 are not. This will be confirmed when the Lagrangian multipliers are interpreted. The test for convexity indicates that constraint 1 is convex, constraint 2 is linear, and constraint 3 is nonconvex. Since not all constraints are convex, the solution space is not convex. However, in this case the nonconvexity of the solution space does not present a problem because of the slope of the objective function. The approximate solution found graphically makes a good starting point for the search technique used to find the solution to the Lagrange equations.

The Lagrange multiplier for a particular constraint (R_k) represents the rate of change of the minimum value of Z with respect to the corresponding right-hand side stipulation (b_k). Therefore, the Lagrange multipliers provide useful information about which constraint should be relaxed to gain the most change in the value of Z. They also show which constraints are binding and which are not. Note that in the solution for Example 6.8, λ_2 and λ_3 are zero, indicating that no advantage would be gained by relaxing those constraints; therefore, those constraints are not binding. λ_1 is not equal to zero, so constraint 1 is binding, which is the same interpretation that was gleaned from the graphical solution.

EXAMPLE 6.8 ————————————————————————————————

The three-bar truss used in Example 5.4 is repeated in Figure 6.9. The objective in this problem is to minimize the total volume of material used in truss members. Therefore, the objective function can be written

$$\min Z = 125A_1 + 75A_2 + 125A_3$$

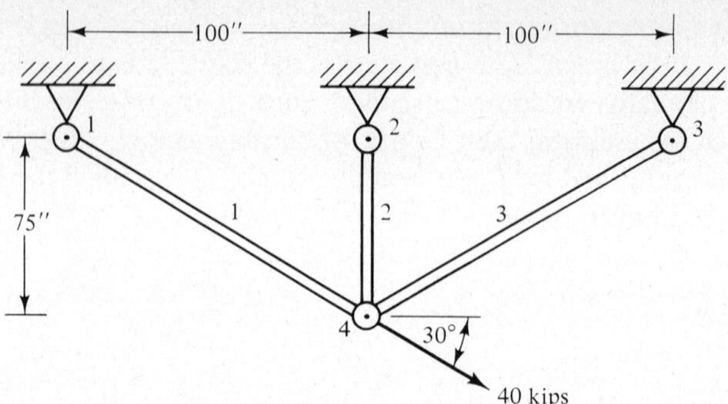

Figure 6.9 Three-bar truss.

in which A_i is the cross-sectional area of the *i*th member. Since A_1 is specified to be the same as A_3, the objective function becomes

$$\min Z = 250A_1 + 75A_2 \tag{6.45}$$

For the loading given, members 1 and 2 will be in tension, and member 3 in compression. This allows development of the following stress constraints:

$$
\left.
\begin{aligned}
\sigma_1 &= \frac{27.65}{1.28A_1} + \frac{12.0}{0.72A_1 + 1.67A_2} \le 20 \quad \binom{\text{allowable stress}}{\text{in tension}} \\[2ex]
\sigma_2 &= \frac{33.33}{0.72A_1 + 1.67A_2} \le 20 \quad \binom{\text{allowable stress}}{\text{in tension}} \\[2ex]
-\sigma_3 &= \frac{27.65}{1.28A_1} - \frac{12.0}{0.72A_1 + 1.67A_2} \le 15 \quad \binom{\text{allowable stress}}{\text{in compression}} \\[2ex]
A_1 &\ge 0; \qquad A_2 \ge 0
\end{aligned}
\right\} \tag{6.46}
$$

There are no deflection constraints imposed on this truss system; therefore, Equations 6.45 and 6.46 are sufficient to define the system model.

Since there are two independent variables in this model, an approximate solution can be found graphically. Figure 6.10 shows the three constraints and the resulting feasible region, along with a family of objective function lines. The lowest point at which the objective function touches the feasible region provides an approximate optimal solution of $A_1 = 1.43$ in.^{2}, $A_2 = 0.87$ in.^{2}, and $Z = 423$ in.^{3}.

The method of Lagrange multipliers can be used to find a mathematically optimum solution to the model. Initially, a test should be made to determine the convexity of the solution space.

For constraint 1 (σ_1; member 1),

$$\frac{\partial^2 \sigma_1}{\partial A_1^2} = \frac{43.20}{A_1^3} + \frac{12.44}{(0.72A_1 + 1.67A_2)^3} \ge 0$$

$$\frac{\partial^2 \sigma_1}{\partial A_2^2} = \frac{66.93}{(0.72A_1 + 1.67A_2)^3} \ge 0$$

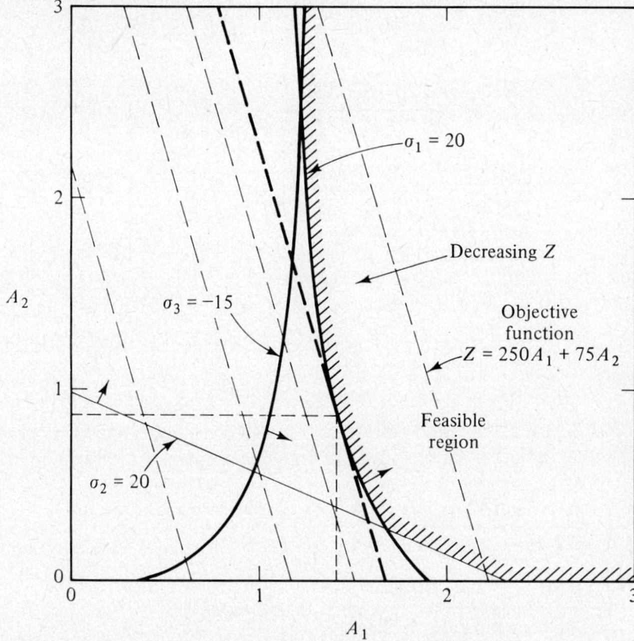

Figure 6.10 Graphical solution for three-bar truss.

$$\frac{\partial^2 \sigma_1}{\partial A_1 \, \partial A_2} = \frac{28.85}{(0.72A_1 + 1.67A_2)^3} = \frac{\partial^2 \sigma_1}{\partial A_2 \, \partial A_1}$$

$$\therefore \quad \begin{vmatrix} \dfrac{\partial^2 \sigma_1}{\partial A_1^2} & \dfrac{\partial^2 \sigma_1}{\partial A_1 \, \partial A_2} \\[2ex] \dfrac{\partial^2 \sigma_1}{\partial A_2 \, \partial A_1} & \dfrac{\partial^2 \sigma_1}{\partial A_2^2} \end{vmatrix} = \frac{43.20(66.93)}{A_1^3(0.72A_1 + 1.67A_2)^3} \geq 0 \qquad (6.47)$$

and constraint 1 is convex. The determinant of the Hessian matrix for constraint 2 (σ_2) is zero, so it also is convex. Constraint 3 (σ_3), however, has a Hessian matrix determinant of $-43.20(66.93)/A_1^3(0.72A_1 + 1.67A_2)^3$, which indicates constraint 3 is not convex, which can be verified by Figure 6.10. Therefore, the solution space is not convex, and there is no guarantee that the solution found through Lagrange multipliers will be a global minimum. In this case, however, an approximate global optimum has been found graphically, which can be used as a first approximation toward a mathematical global minimum.

Before developing the Lagrangian, all inequalities have to be converted to equalities through addition of slack or surplus variables. The constraints of Equation 6.46 now become

$$\left. \begin{aligned}
\frac{27.65}{1.28A_1} + \frac{12.0}{0.72A_1 + 1.67A_2} + S_1^2 - 20 &= 0 \\[2mm]
\frac{33.33}{0.72A_1 + 1.67A_2} + S_2^2 - 20 &= 0 \\[2mm]
\frac{27.65}{1.28A_1} - \frac{12.0}{0.72A_1 + 1.67A_2} + S_3^2 - 15 &= 0 \\[2mm]
A_1 - S_4^2 &= 0 \\[2mm]
A_2 - S_5^2 &= 0
\end{aligned} \right\} \tag{6.48}$$

Each constraint is combined with its Lagrangian multiplier, and added to the objective function to yield

$$\min Z = 250A_1 + 75A_2 + \lambda_1 \left(\frac{21.60}{A_1} + \frac{12.0}{0.72A_1 + 1.67A_2} + S_1^2 - 20 \right)$$

$$+ \lambda_2 \left(\frac{33.33}{0.72A_1 + 1.67A_2} + S_2^2 - 20 \right)$$

$$+ \lambda_3 \left(\frac{21.60}{A_1} - \frac{12.0}{0.72A_1 + 1.67A_2} + S_3^2 - 15 \right)$$

$$+ \lambda_4(A_1 - S_4^2) + \lambda_5(A_2 - S_5^2) \tag{6.49}$$

Z is a function of A_1, A_2, S_1, S_2, S_3, S_4, S_5, λ_1, λ_2, λ_3, λ_4, and λ_5. If the partial derivative of the objective function with respect to each of these variables is taken, and set equal to zero, the following set of equations is formed:

$$\frac{\partial Z}{\partial A_1} = 250 - \frac{\lambda_1(21.60)}{A_1^2} - \frac{\lambda_1(8.64)}{(0.72A_1 + 1.67A_2)^2} - \frac{\lambda_2(24.00)}{(0.72A_1 + 1.67A_2)}$$

$$- \lambda_3 \left(\frac{21.60}{A_1^2} \right) + \frac{\lambda_3(8.64)}{(0.72A_1 + 1.67A_2)^2} + \lambda_4 = 0 \tag{6.50a}$$

$$\frac{\partial Z}{\partial A_2} = 75 - \frac{\lambda_1(20.64)}{(0.72A_1 + 1.67A_2)^2} - \frac{\lambda_2(55.65)}{(0.72A_1 + 1.67A_2)^2}$$

$$+ \frac{\lambda_3(20.04)}{(0.72A_1 + 1.67A_2)^2} + \lambda_5 = 0 \tag{6.50b}$$

$$\frac{\partial Z}{\partial \lambda_1} = \frac{21.60}{A_1} + \frac{12.0}{0.72A_1 + 1.67A_2} + S_1^2 - 20 = 0 \tag{6.50c}$$

$$\frac{\partial Z}{\partial \lambda_2} = \frac{33.33}{(0.72A_1 + 1.67A_2)} + S_2^2 - 20 = 0 \tag{6.50d}$$

$$\frac{\partial Z}{\partial \lambda_3} = \frac{21.60}{A_1} - \frac{12.0}{0.72A_1 + 1.67A_2} + S_3^2 - 15 = 0 \tag{6.50e}$$

$$\frac{\partial Z}{\partial \lambda_4} = A_1 - S_4^2 \tag{6.50f}$$

$$\frac{\partial Z}{\partial \lambda_5} = A_2 - S_5^2 \tag{6.50g}$$

$$\frac{\partial Z}{\partial S_1} = 2\lambda_1 S_1 = 0 \tag{6.50h}$$

$$\frac{\partial Z}{\partial S_2} = 2\lambda_2 S_2 = 0 \tag{6.50i}$$

$$\frac{\partial Z}{\partial S_3} = 2\lambda_3 S_3 = 0 \tag{6.50j}$$

$$\frac{\partial Z}{\partial S_4} = -2\lambda_4 S_4 = 0 \tag{6.50k}$$

$$\frac{\partial Z}{\partial S_5} = -2\lambda_5 S_5 = 0 \tag{6.50l}$$

From Figure 6.8, it can be seen that both A_1 and A_2 are greater than zero at the optimal solution point; therefore, S_4 and S_5 are >0 (from Equations f and g) and λ_4 and $\lambda_5 = 0$ (from Equations 6.50k and l). Since an approximate optimal solution is known, a numerical search routine can be undertaken in that region. Values of A_1 and A_2 in this region will produce estimates of S_2 and S_3 that are >0 (from Equations 6.50d and e); thus, λ_2 and λ_3 must $= 0$ (from Equations 6.50i and j). Since $\lambda_2, \lambda_3, \lambda_4$, and λ_5 are all zero, constraints 2 through 5 of Equations 6.48 are not binding. Therefore, constraint 1 must be binding, and $S_1 = 0$, and $\lambda_1 \neq 0$. Equation 6.50c can be used to calculate an estimate of A_2, given a value of A_1. Equation 6.50b can then be used to calculate λ_1.

If Equation 6.50a equals zero when the resulting values of A_1, A_2, and λ_1 are substituted into it, then the optimal solution has been found. Successive approximations produced estimates of 1.445 in.^{2} and 0.799 in.^{2} for A_1 and A_2, respectively, and an objective function value of 421 in.^{3}.

A second example of the application of Lagrange multipliers to optimization of nonlinear systems is shown in Example 6.9. This example is adapted from one by Block and Lynn (1963). In this case, the objective function is nonlinear while the constraints are linear. Both the solution space and the objective function are convex. The concepts of Kuhn–Tucker conditions are introduced, which guarantee convergence to a global minimum when the objective function and constraints are convex. Note that this model could also have been optimized in the same manner as the model in Example 6.8.

EXAMPLE 6.9 ————————————————————————————————

A town obtains its water supply from two wells, W_1 and W_2. The cost of securing water from a particular well can be approximated by the equation

$$Z = A + BQ + CQ^2 \tag{6.51}$$

where Z = total supply costs from the well ($/day)
Q = million gal/day

A = fixed costs which are independent of flowrate such as amortization and maintenance

B = cost/million gal/day for pumping at the well and to overcome elevation differences

C = cost/(million gal/day)^2 to overcome frictional resistance in the supply pipe

For the wells in question, the costs can be approximated as follows:

$$\left.\begin{array}{l} Z_1 = 13 + 4Q_1 + Q_1^2 = 9 + (Q_1 + 2)^2 \\ Z_2 = 26 + 8Q_2 + 4Q_2^2 = 22 + 4(Q_2 + 1)^2 \end{array}\right\} \tag{6.52}$$

Therefore, the total cost of supplying water becomes

$$Z = 31 + (Q_1 + 2)^2 + 4(Q_2 + 1)^2 \tag{6.53}$$

The town needs at least 2 million gal/day, while the capacity of well 1 is 5 million gal/day, and the capacity of well 2 is 10 million gal/day. Thus the following constraints can be written:

$$\left.\begin{array}{c} Q_1 + Q_2 \geq 2 \\ Q_1 \leq 5 \\ Q_2 \leq 10 \\ Q_1 \geq 0 \\ Q_2 \geq 0 \end{array}\right\} \tag{6.54}$$

Figure 6.11(a) shows the envelope of objective function values for $Z(Q_1, Q_2)$, and Figure 6.11(b) shows the constrained solution space. If contours of equal Z are taken through the objective function envelope, they can be plotted as ellipses with center $(-2, -1)$ on the Q_1, Q_2 plane. The combined figure is shown in Figure 6.11(c). The lowest valued Z contour that touches the solution space will represent the optimum solution, which is about \$50/day at approximate flowrates of $Q_1 = 2$ and $Q_2 = 0$.

These estimates will now be refined mathematically. If it is assumed that excess supply will not be provided to the town, then the constraint

$$Q_1 + Q_2 \geq 2$$

becomes

$$Q_1 = 2 - Q_2 \tag{6.55}$$

If Equation 6.55 is substituted back into Equation 6.53, it becomes

$$\begin{aligned} Z &= 31 + (4 - Q_2)^2 + 4(Q_2 + 1)^2 \\ &= 51 + 5Q_2^2 \end{aligned} \tag{6.56}$$

The objective function is now a function of one variable (Q_2) and the minimum Z value can be found by setting

$$\frac{dZ}{dQ_1} = 0 = 10Q_2 \tag{6.57}$$

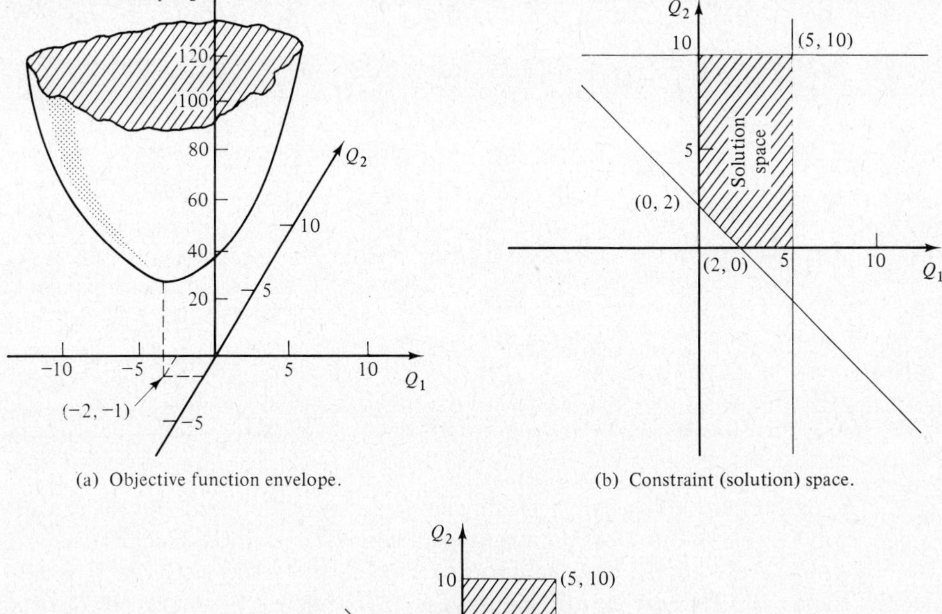

(a) Objective function envelope.

(b) Constraint (solution) space.

(c) Superposition of objective function contours on solution space.

Figure 6.11 Graphical solution of well model.

Thus, $Q_2 = 0$ at the minimum Z, $Q_1 = 2$ (from 6.55), and $Z = 51$, which verifies the graphical solution. If Q_1 and Q_2 values are substituted back into the constraints of Equation 6.54, all of them are satisfied, so this is a feasible solution.

The above solution was possible only because the objective function could be reduced to a function of one variable. A more general solution using Lagrange multipliers will now be illustrated.

The solution space in this example is bounded by linear functions, so it is convex. Convexity of the objective function can be determined by examining the determinant of the Hessian matrix and its diagonal elements.

$$\frac{\partial Z}{\partial Q_1} = 2(Q_1 + 2)$$

$$\frac{\partial Z}{\partial Q_2} = 8(Q_2 + 1)$$

$$\frac{\partial^2 Z}{\partial Q_1^2} = 2 \text{ which is } \geq 0$$

$$\frac{\partial^2 Z}{\partial Q_2^2} = 8 \text{ which is } \geq 0$$

$$\frac{\partial^2 Z}{\partial Q_1\, \partial Q_2} = \frac{\partial^2 Z}{\partial Q_2\, \partial Q_1} = 0$$

$$\therefore \mathbf{H} = \begin{vmatrix} 2 & 0 \\ 0 & 8 \end{vmatrix}$$

and det $\mathbf{H} = 16$, which is also ≥ 0. Therefore, $Z = f(Q_1, Q_2)$ is convex.

The optimum solution can be found using Lagrange multipliers; however, there are several possible solutions to the Lagrangian equations, and the global minimum must be found from among those solutions. However, since $Z(Q_1, Q_2)$ is convex, the Kuhn–Tucker theorem can be applied to verify convergence to a global minimum. The Kuhn–Tucker theorem describes a set of conditions (termed the Kuhn–Tucker conditions) which, if met, are necessary and sufficient to guarantee a global optimum has been achieved. Assuming that the objective function is differentiable, and that it is subject to a set of constraints of the form

$$\left. \begin{array}{ll} R_i(X_1, X_2, \ldots, X_n) \leq b_i & i = 1, \ldots, m \\ X_j \geq 0 & j = 1, \ldots, n \end{array} \right\} \tag{6.58}$$

the Kuhn–Tucker theorem asserts that an optimal solution $X_1^*, X_2^*, \ldots, X_n^*$ to the given model will exist only if there exist constants (Lagrange multipliers) $\lambda_1, \lambda_2, \ldots, \lambda_m$ such that all the following (Kuhn–Tucker) conditions are satisfied:

$$\left. \begin{array}{l} \text{1. If } X_j^* > 0, \quad \text{then} \quad \dfrac{\partial Z}{\partial X_j} + \sum_{i=1}^{m} \lambda_i \left.\dfrac{\partial R_i}{\partial X_j}\right|_{X_j^*} = 0; \quad (j = 1, \ldots, n) \\[2ex] \text{2. If } X_j^* = 0, \quad \text{then} \quad \dfrac{\partial Z}{\partial X_j} + \sum_{i=1}^{m} \lambda_i \left.\dfrac{\partial R_i}{\partial X_j}\right|_{X_j^*} \geq 0; \quad (j = 1, \ldots, n) \\[2ex] \text{3. If } \lambda_i > 0, \quad \text{then} \quad R_i(X_1^*, X_2^*, \ldots, X_n^*) = b_i; \quad (i = 1, 2, \ldots, m) \\[1ex] \text{4. If } \lambda_i = 0, \quad \text{then} \quad R_i(X_1^*, X_2^*, \ldots, X_n^*) \leq b_i; \quad (i = 1, 2, \ldots, m) \\[1ex] \text{5. } X_j^* \geq 0; \quad (j = 1, \ldots, n) \\[1ex] \text{6. } \lambda_i \geq 0; \quad (i = 1, \ldots, m) \end{array} \right\} \tag{6.59}$$

where $|_{X_j^*}$ means evaluated at the optimal solution.

Although the Kuhn–Tucker theorem will not be proved, some general comments about the conditions will help to attach physical meaning to them. Condition 1 expresses the condition that at the optimal solution point, the total differential of the objective function is zero. Condition 2 supplements the first condition for the case in which the optimum may be at the non-negativity boundary limit for a certain variable ($X_j^* = 0$). If the global minimum is at the boundary, then the total differential there must be positive. Condition 3 suggests that the *i*th constraint equation is a binding one, while condition 4 suggests that the *i*th constraint is not binding. Conditions 5 and 6 are the non-negativity constraints.

To apply the Kuhn–Tucker conditions, the model is written

$$\left. \begin{aligned} Z &= 31 + (Q_1 + 2)^2 + 4(Q_2 + 1)^2 \\ R_1(Q_1, Q_2) &= Q_1 \le 5 \\ R_2(Q_1, Q_2) &= Q_2 \le 10 \\ R_3(Q_1, Q_2) &= -Q_1 - Q_2 \le -2 \end{aligned} \right\} \tag{6.60}$$

The non-negativity constraints are not repeated here because they are included in the Kuhn–Tucker conditions. Partial derivatives of the constraints can be taken as

$$\left. \begin{aligned} \lambda_1 \frac{\partial R_1}{\partial Q_1} &= \lambda_1 \\[6pt] \lambda_2 \frac{\partial R_2}{\partial Q_1} &= 0 \\[6pt] \lambda_3 \frac{\partial R_3}{\partial Q_1} &= -\lambda_3 \\[6pt] \lambda_1 \frac{\partial R_1}{\partial Q_2} &= 0 \\[6pt] \lambda_2 \frac{\partial R_2}{\partial Q_2} &= \lambda_2 \\[6pt] \lambda_3 \frac{\partial R_3}{\partial Q_2} &= -\lambda_3 \end{aligned} \right\} \tag{6.61}$$

From condition 1,

$$\left. \begin{aligned} \text{if } Q_1 > 0, \quad &\text{then} \quad 2(Q_1 + 2) + \lambda_1 - \lambda_3 = 0 \\ \text{if } Q_2 > 0, \quad &\text{then} \quad 8(Q_2 + 1) + \lambda_2 - \lambda_3 = 0 \end{aligned} \right\} \tag{6.62}$$

From condition 2,

$$\left. \begin{aligned} \text{if } Q_1 = 0, \quad &\text{then} \quad \lambda_1 - \lambda_3 + 4 \ge 0 \\ \text{if } Q_2 = 0, \quad &\text{then} \quad \lambda_2 - \lambda_3 + 8 \ge 0 \end{aligned} \right\} \tag{6.63}$$

A reasonable assumption to make is that $\lambda_1 = \lambda_2 = 0$, which means that constraints 1 and 2 are not binding. Since at least one constraint has to be binding, $\lambda_3 \ne 0$ if λ_1 and λ_2 are zero. If $\lambda_3 \ne 0$, then $Q_1 + Q_2 = 2$ by condition 3. Substituting $Q_2 = 2 - Q_1$ into condition 1, results in

$$\left. \begin{aligned} \text{if } Q_1 > 0 \quad & 2Q_1 + 4 - \lambda_3 = 0 \\ \text{if } Q_2 > 0 \quad & 8Q_1 + 8 - \lambda_3 = 0 \end{aligned} \right\} \tag{6.64}$$

Solving this set of equations simultaneously results in $Q_1 = 2$ and $\lambda_3 = 8$. However, this violates the second equation of condition 1, so either Q_1 or Q_2 must be equal to zero. If $Q_1 = 0$, then $Q_2 = 2$, and $\lambda_3 = 24$; however, this violates $-\lambda_3 + 4 \geq 0$. Therefore, $Q_2 = 0$, $Q_1 = 2$, $\lambda_3 = 8$, $Z = 51$, and all Kuhn–Tucker conditions are met, which verifies this solution as a global minimum.

SUMMARY

Mathematical techniques have been described that can be helpful in optimizing systems which can be modeled by an objective function subject to constraints. Some models can be optimized by the methods of calculus. Linear systems containing equations and inequalities can be optimized by linear programming techniques. Furthermore, much additional information can be ascertained about the system through postoptimality analysis. Specialized algorithms are available for optimizing transportation, transshipment, assignment, and linear graph (network) problems. These algorithms are computationally more efficient than the simplex algorithm for particular specialized applications. Systems which contain nonlinear objective functions and/or nonlinear constraints present unique optimization problems. No general form of solution is available; however, specific techniques can be found in the literature, two of which have been illustrated through examples.

It is important to note that for complex problems, it may be exceedingly difficult or impossible to find a global optimum, or even a local optimum. In these cases, the engineer must use judgment and gather as much information as possible about the system from any analysis that can be done. Also, in applying the solution techniques, the engineer must keep in mind that the results are only as good as the data assumptions, and principles, used to develop the original model.

APPLICATIONS EXERCISES

6-1. The Ajax Sand and Gravel Company sells sand mixtures of two quality levels. The more expensive mixture has a higher proportion of fine sand, while the cheaper mixture contains more coarse sand.

The prices of sand purchased by Ajax are: fine sand, \$5/ton; coarse sand, \$2/ton. The two mixes sold by Ajax and their prices are: mixture A, \$8/ton; mixture b, \$4/ton. Ajax can sell any amount of each of these mixtures, but due to a shortage of sand, can obtain no more than 200 tons of fine sand and 400 tons of coarse sand.

Specifications state that mixture A cannot contain more than 25% coarse sand nor less than 40% fine sand. Mixture B can contain no more than 60% coarse sand and no less than 20% fine sand.

(a) Formulate this as a linear programming problem capable of offering advice to Ajax on how to mix its sand; (how many tons of mixtures A and B and their composition) so as to maximize profit.

(b) Set up the initial simplex tableau.

(c) Complete the first iteration. (Bring in new variable and take out another one.) Indicate the variables to enter and leave the basis on the second iteration.

(d) Use a computer application package to find the optimal solution to this problem.

(e) Accomplish postoptimality analysis.

6-2. Discuss fully the following tableaux to two different linear programming problems. Each is a maximization problem, and X_3 and X_4 are slack variables for each case. In your analysis, consider optimality, optimal objective value, alternate optima, shadow prices, variables in the solution and their values, degeneracy, unbounded solution, variable to enter basis (plus value) on the next iteration, and variable to leave basis.

(a)

c_j			0	0	5	10
	Solution mix	B	X_3	X_4	X_1	X_2
10	X_2	12	2	0	3	1
0	X_4	16	-2	1	2	0
	Z_j	120	20	0	30	10
	$c_j - Z_j$		-20	0	-25	0

(b)

c_j			0	0	6	10
	Solution mix	B	X_3	X_4	X_1	X_2
10	X_2	10	4	0	1/2	1
0	X_4	12	-1	1	6/10	0
	Z_j	100	40	0	5	10
	$c_j - Z_j$		-40	0	1	0

Note: for Exercise 6-2(b), also indicate what will happen to the solution on the next iteration.

6-3. Given the following linear programming problem:

$$\max Z = 5X_1 + 6X_2$$
$$\text{ST:} \quad X_1 + X_2 \leq 12$$
$$2X_1 - X_2 \geq 4$$
$$3X_1 - 4X_2 = 6$$
$$X_1, X_2 \geq 0$$

(a) Find the optimum solution graphically.

(b) Find the optimum solution using the simplex algorithm.

(c) Perform postoptimality analysis.

6-4. Tell as much as you can about each of the following tableaux. Consider the following points:

(a) Suboptimal solution; variable to enter, variable to leave.

(b) Optimal solution; alternate optima, shadow prices.

(c) Unbounded solution.

(d) Infeasible solution.

(e) Degenerate solution.

Minimization or maximization is indicated for each tableau. X_1 and X_2 are problem variables, X_3 and X_4 are slack variables, X_5 is a surplus variable, and X_6 is an artificial variable.

(a) Minimization:

c_j			0	0	−8	−6
	Solution variables	B	X_3	X_4	X_1	X_2
−8	X_1	20	−1	−3	1	0
−6	X_2	10	−2	1	0	1
	Z_j	−220	20	18	−8	−6
	$c_j - Z_j$		−20	−18	0	0

(b) Minimization:

c_j			0	0	−M	−8	−6	0
	Solution variables	B	X_3	X_4	X_6	X_1	X_2	X_5
0	X_3	−3	1	5	2	0	−3	0
−8	X_1	2	0	3	4	1	−1	0
0	X_5	5	0	1	6	0	−4	1
	Z_j	−16	0	−24	−32	−8	8	0
	$c_j - Z_j$		0	24	−M + 32	0	−14	0

(c) Minimization:

c_j			0	0	$-M$	-8	-6	0
	Solution variables	B	X_3	X_4	X_6	X_1	X_2	X_5
0	X_3	20	1	0	0	0	4	0
0	X_4	14	0	1	0	-1	5	0
$-M$	X_6	10	0	0	1	6	2	-1
	Z_j	$-10M$	0	0	$-M$	$-6M$	$-2M$	M
	$c_j - Z_j$		0	0	0	$6M - 8$	$2M - 6$	$-M$

(d) Maximization:

c_j			0	0	$-M$	$+8$	$+6$	0
	Solution variables	B	X_3	X_4	X_6	X_1	X_2	X_5
0	X_3	12	1	3	0	0	0	0
8	X_1	14	0	-2	0	1	-2	0
0	X_5	6	0	1	-1	0	-1	1
	Z_j	112	0	-16	0	8	-16	0
	$c_j - Z_j$		0	16	$-M$	0	22	0

(e) Maximization:

c_j			0	0	8	6
	Solution variables	B	X_3	X_4	X_1	X_2
0	X_3	0	1	4	0	3
8	X_1	4	0	1	1	1/2
	Z_j	32	0	8	8	4
	$c_j - Z_j$		0	-8	0	2

(f) Maximization:

c_j			0	0	$-M$	8	6	0
	Solution variables	B	X_3	X_4	X_6	X_1	X_2	X_5
8	X_1	14	3/4	0	-3	1	0	3
6	X_2	6	-1	0	-4	0	1	4
0	X_4	0	3	1	-7	0	0	7
	Z_j	148	0	0	-48	8	6	48
	$c_j - Z_j$		0	0	$-M + 48$	0	0	-48

6-5. Use the simplex algorithm to find the optimum solution:

$$\max Z = 13X_1 + 11X_2$$
$$\text{ST:} \qquad 4X_1 + 5X_2 \le 1500$$
$$5X_1 + 3X_2 \le 1575$$
$$X_1 + 2X_2 \le 420$$

6-6. Solve the following linear programming problem using the simplex method:

$$\max Z = 4X_1 + 2X_2$$
$$\text{ST:} \qquad X_1 + X_2 \le 30$$
$$X_2 \ge 10$$
$$X_1 - 2X_2 = 0$$

6-7. In a water filtration plant, the cost of filter walls can be expressed as:

$$C_T = \left[C_1 \left(\frac{N}{2}\right) \left(\frac{1}{K_l}\right)^{1/2} + C_2(N + 2)K_l^{1/2} + C_2(2N) \left(\frac{1}{K_l}\right)^{1/2} \right] \left(\frac{A}{N}\right)^{1/2}$$

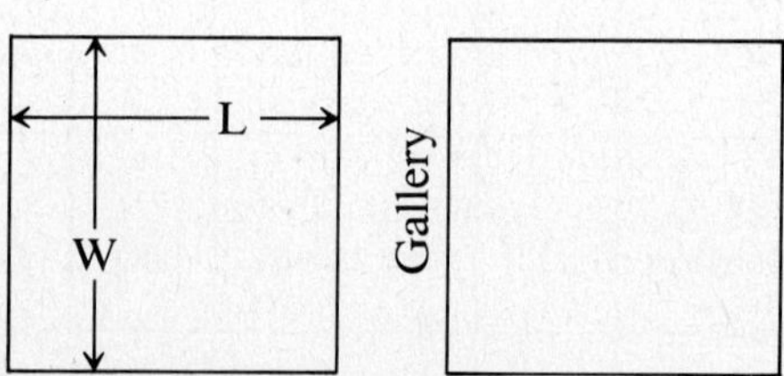

where C_1 = cost of wall parallel to gallery (\$/ft)
   C_2 = cost of wall perpendicular to gallery (\$/ft)
   N = number of filter beds (usually a multiple of 2)
   A = total surface area required
   K_l = length-to-width ratio = L/W

If
$$C_1 = \$5000/\text{ft}$$
$$C_2 = \$1500/\text{ft}$$
$$N = 4, \text{ and}$$
$$A = 3.5 \times 10^3 \text{ft}^2$$

what K_l will give minimum cost, and what is this minimum cost?

6-8. Find the minimum cost of aggregate for Exercise 5-1.

6-9. Determine the minimum total cost of disposing of all trash for Exercise 5-2.

6-10. What is the minimum delivered cost for sand and gravel in Exercise 5-3?

6-11. Determine the minimum cost way of meeting all cut and fill requirements for Exercise 5-4.

6-12. Determine the mix of units that will produce the maximum rental income for Exercise 5-5.

6-13. Find the minimum concrete costs for Exercise 5-6.

6-14. Find the optimum distribution matrix with the least total shipping cost for Exercise 5-7.

6-15. Find the minimum shipping cost for Exercise 5-8, and discuss what any value associated with a pseudosource or sink means.

6-16. Determine the longest path from A to B in Exercise 5-9. Use the Hungarian method.

6-17. Find the minimum transportation cost solution for the allocation problem given in Exercise 5-10. Use the Vogel approximation and the transportation algorithm.

6-18. Find the minimum cost alternative for disposing of the excess refuse in Exercise 5-11.

6-19. Find the optimum production quantities for both models developed in Exercise 5-12.

6-20. Find the minimum cost alternative for meeting the granular material requirements described in Exercise 5-13. Accomplish all available postoptimality analyses.

6-21. What are the basic characteristics of a problem best addressed by:
   (a) The Lagrangian multiplier?
   (b) The simplex algorithm?
   (c) The Hungarian method?
   Explain what types of problems are thus addressed and discuss the fundamental requirements and limitations which exist.

6-22. Find the minimum time and minimum cost solutions to Exercise 5-18. Describe the differences in the two solutions, and the reasons for these differences.

6-23. Determine the minimum cost way of meeting station requirements in Exercise 5-21. Determine how increased supply at any one of the carbarns might change the final solution. Explain why any changes take place. Also, determine how increased requirements at any one of the stations, without corresponding increase in the number of cars available, could be accommodated by the model. What would be the meaning of all variable values in the solution for this case?

6-24. Determine the expected level of return for each level of environmental impact restriction given in Exercise 5-22. Plot the expected return versus level of environmental impact, and discuss the implications of the plot.

6-25. Determine the maximum load that could be carried by the overhead crane system of Exercise 5-23. Examine your results to see if there are changes in the configuration that should increase the total load that the system can carry. Resolve the problem with this revised configuration to confirm your predictions.

6-26. Determine the maximum load that the crane system of Exercise 5-31 could carry. Study your solution, and suggest possible redesigns that would increase the load-carrying capacity. Verify your redesign by solving the revised model.

6-27. Find the maximum load the structure of Example 5.2 will support. Discuss the implications of your solution. What redesign of the structure might be called for?

6-28. Find the optimal waste treatment strategy for the regional authority described in Example 5.5. Accomplish postoptimality analysis, and discuss the implications and significance of your findings.

6-29. (a) Confirm the optimal solution for the first stage of the water mixing problem of Example 5.6.
  (b) Find the optimal solution to the second-stage problem with doubled capacity. If it is impossible to find the optimum solution, reevaluate the model to find out why. Suggest ways that the problem could be redefined, so that a useful model could be developed.

6-30. Determine the optimal storage–treatment combination for the storm sewer system described in Example 5.7. Evaluate your results, including postoptimality analysis.

6-31. Find the optimum solution for transporting refuse in Example 5.8. Use both the simplex algorithm (computer program) and transportation algorithm. Compare the two methods and discuss which might be more appropriate to use for computer solutions, and which might be more appropriate for hand solutions. Explain why you made your choices.

6-32. Find the optimum distribution of cut and fill for Example 5.9.

6-33. Find the shortest path between locations 1 and 6 of Example 5.10 using the simplex algorithm and the assignment algorithm. Evaluate the two algorithms for automated and hand solutions.

6-34. Find the mix of specialized assemblies that will maximize profit for the conditions described in Example 5.11.

6-35. Find the optimum assignment of excavators to tasks for the open pit coal mine described in Example 5.13.

6-36. Find the minimum cost treatment scheme to meet the water quality limitations of Exercise 5-32.

6-37. Solve the assignment problem of Exercise 5-33 by both the simplex algorithm and the Hungarian method. Comment on the comparative efficiency of the two methods, and on the meaning of the results.

6-38. Determine the best and worst truck schedules for the situation described in Exercise 5-34. How much money can be saved in 1 yr by using the best schedule rather than the worst? Discuss the implications of your solutions.

6-39. Find the minimum cost way of meeting the fill requirements using each model developed for parts (a), (b), and (c) of Exercise 5-19. Explain any differences in the solutions. If the maximum allowable time to complete the fill operations is 62.5 weeks, would the most economical solution meet this constraint? If not, how might conditions be altered so that filling could be completed in 62.5 weeks?

6-40. Find the optimum solution for Exercises 5-35, 5-36, and/or 5-37.

Chapter 7

Economic Feasibility

INTRODUCTION

Economics plays a key role in assessing any engineering project or alternative projects. If a specific task has to be accomplished, the engineer will search for the least-cost way of accomplishing it. When a project is designed to produce benefits, or return on investment, feasibility determination will be based on a positive difference between returns and project costs. Net present worth, net equivalent annual return, rate of return, or benefit/cost analysis can be used to determine economic feasibility. Each of these methods will be illustrated in subsequent sections. The basic economic formulas developed in Appendix B will have to be used in the application of these methods.

Often, two or more mutually exclusive alternatives are available for accomplishing the same output task. If the output is fixed, then return or benefits must be constant for the alternatives, and the objective will be to minimize the input costs. Present worth, equivalent uniform annual cost, rate of return, and benefit/cost analysis can be used to evaluate alternatives. Care must still be exercised to make certain that any alternative chosen is economically feasible.

For other situations, neither the level of input nor the level of output will be fixed for mutually exclusive alternatives. Mutually exclusive means that a choice has to be made among competing alternatives, not combinations of alternatives. Any alternatives that are not economically feasible can be discarded. The remaining alternatives can be rank ordered by cost, and examined incrementally to determine the advisability of increased levels of investment. Incremental rate of return or incremental benefit/cost analysis can be used to determine the economically most efficient alternative. Maximization of net present worth or equivalent uniform net annual return could also be used to accomplish the same result.

Each of the above types of projects and methods of analysis will be discussed in detail in later sections.

Frequently, it is economics that will place major limiting constraints on problem definition and the range of alternatives to be considered. For problems with fixed input, the engineer is faced with the challenge of maximizing the resulting output or return on investment. Linear programming can assist in maximizing return for constrained input.

Economic changes during the development of a project will often overshadow technical changes, possibly affecting the success or failure of the finished project. From the above discussion, it can be seen that it is extremely important for anyone who is applying the systems approach to understand the basic concepts of economic feasibility studies. Key concepts will be developed in subsequent sections of this chapter. They will rely on the basic engineering economics formulas that are presented in Appendix B.

A key component in engineering economic analysis is the quantification of the time value of money. Interest is money paid for the use of borrowed money and is therefore a good indicator of time value. The specific form of interest that is of importance to engineers is called the discount rate. It should reflect the rate of return that could have been realized if other investments had been made. Thus, the discount rate reflects the opportunity cost or attractive rate of return, and should be set at, or above, prevailing interest rates for borrowed money.

The effects of inflation on the time value of money have not been incorporated into the examples in this chapter. Inclusion of inflation would hinder understanding the concepts presented. Since the methods involved make comparisons, not absolute predictions of monetary value, the results are sufficiently accurate for most purposes. Also, over longer time horizons of 10 to 20 years, future inflation rates are extremely hard to predict. An example of the effect of inflation on the future value of money is shown in Appendix B. Essentially, the effect of inflation is to dilute the rate of return. If the inflation rate is equal to the rate of return, then the effective increase in a sum of money over time is zero. When inflation is included, all future amounts of money should be adjusted to reflect the influence of inflation before applying any of the analysis techniques that will be developed.

Engineers undertaking economic evaluation must be careful to evaluate all benefits and/or costs on an equivalent basis. Realistic costs and benefits have to be estimated, reasonable discount rates have to be adopted, and all cash flows over a period of time must be discounted to show the effects of interest on both benefits and costs in a consistent manner.

The expected life of projects for economic analysis may not be the same as the actual physical life of the components associated with the project. Several factors may make a shorter economic life preferable for planning. It may be advisable to sell components while they still have some resale value, and replace them with technologically updated components. A shorter economic life prevents being locked into a long-term commitment and leaves the planner flexibility to react to new developments. Rapid development of technology may render a component or system obsolete before it wears out. Also, varying interest rates decrease the reliability of long term projections.

This chapter will not discuss depreciation of capital equipment or income taxes, two subjects which are important to the management of engineering firms but which do not generally enter into the types of economic analysis described herein. Depreciation, the systematic allocation of the cost of a capital asset over its actual, or assumed, useful life, is important in many after tax economic analyses. Both income taxes and depreciation are covered in detail in engineering economics texts.

CASH FLOW DIAGRAMS

Cash flow diagrams can be useful in depicting the sequence of income and debits that will influence the economic feasibility of a project. They can also help in identifying the proper type of economic analysis to undertake in evaluating the project.

It is customary in engineering economic analysis to assume that initial capital costs are incurred at the beginning of the first year. Other debits (annual costs, purchase of new equipment in the future, etc.) and any income (salvage or resale) are assumed to take place at the end of the year in question. Thus, the cash flow diagram starts out at year zero, which is actually the beginning of the first year. The economic analysis formulas developed in Appendix B are also based on these assumptions. On cash flow diagrams, upward arrows indicate the receipt of money, and downward arrows represent disbursements. Time is shown on the X axis.

One year is generally the compounding period used in engineering project evaluations. However, shorter compounding periods may be necessary when considering the costs of short-term borrowing or loan payback schedules.

Examples 7.1 and 7.2 show typical cash flow diagrams, for both uniform and nonuniform series of payments. The salvage value given in Example 7.2 represents the net amount realized from resale of the system at the end of its useful life minus the cost associated with retiring or disposing of the system. If the net salvage value is negative, then the cash flow is referred to as a salvage cost.

EXAMPLE 7.1 ───

A highway bridge is expected to cost $1,000,000 to build. Annual maintenance costs will be about $20,000 and the expected life of the bridge is 20 yr. Sketch the cash flow diagram that depicts this situation.

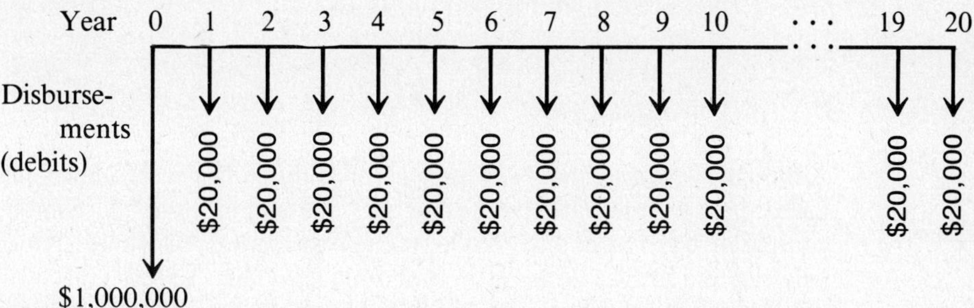

There are no receipts in this particular problem. The initial capital cost is followed by a uniform series of maintenance costs.

EXAMPLE 7.2 ____

A new minicomputer system for your engineering office will cost $300,000. The expected life of the computer system is 10 yr. Initial maintenance costs will be $30,000/yr; however, these costs will increase approximately linearly to $120,000/yr after 10 yr. The estimated salvage value of the computer system after 10 yr is $10,000. Sketch the cash flow diagram for this acquisition.

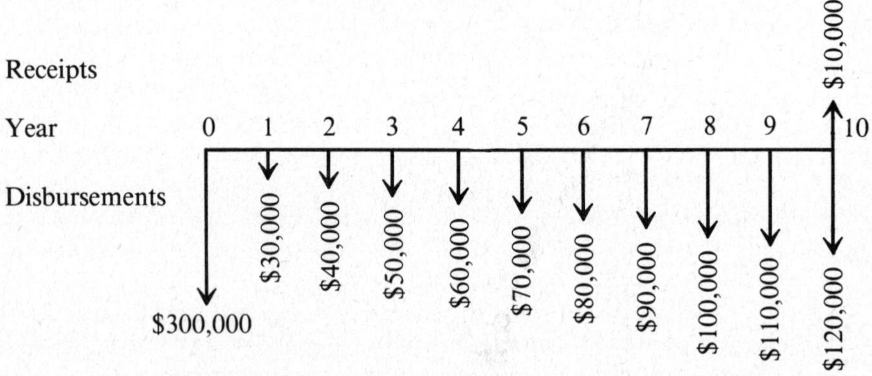

In this problem there is a small, single payment receipt after 10 yr. The initial capital cost is followed by a nonuniform series of maintenance costs.

ECONOMIC FEASIBILITY OF PROJECTS

When investigating the economic feasibility of projects, it is necessary to determine if the total benefits and return on investment are greater than the cost. In economic terms, a determination has to be made as to whether or not an alternative investment would produce a greater return. The methods for analysis of economic feasibility include:

1. Net present worth.
2. Uniform annual net return.
3. Rate of return.
4. Benefit/cost ratio.

Although these methods will appear to be dissimilar when developed, they are all based on the same underlying principles and assumptions. Each method is used for comparative purposes only, not to estimate absolute dollar values. Any of the methods will lead to a similar decision on economic feasibility. In some instances, one method may be clearly superior over another. These will be pointed out as the methods are developed. At other times, the choice of which method to use is up to the personal preference of the analyst. Also, the different forms reflect the type of information favored by decision makers from different disciplines; annual payment by budget officers, rate of return by economists, and so on.

Net Present Worth

Present worth analysis discounts future amounts of money to their present value. The minimum attractive rate of return over the period of project analysis is the discount rate that should be used in computations. The minimum attractive rate of return is the rate of return that could be expected if the funds were invested elsewhere, such as bank deposits or government bonds. Present worth represents the size of total costs and benefits over the life of the project.

The net present worth is defined as the difference between the present worth of benefits and the present worth of costs. Net present worth computations permit comparison of costs throughout the life of a project on an equivalent basis. If the net present worth of a project is positive, it is economically viable, because it exceeds the minimum attractive rate of return.

Periodic costs and benefits are treated as cash flows which are brought to equivalent present worths at the beginning of the project (time = 0). Individual future amounts of money are discounted using the present worth factor (PWF, i, n), while a series of equal amounts of money spread over n compounding periods in the future can be discounted using the uniform series present worth factor (USPWF, i, n). For projects with perpetual life, the present worth can be estimated by dividing the uniform annual amount of money by the discount rate:

$$P = \frac{A}{i} \quad \text{for} \quad n \to \infty \tag{7.1}$$

This can be confirmed by checking the USPWF for $n = \infty$. The present worth for such alternatives is termed the capitalized cost.

For a nonuniform series of payments, discounting to present worth can be accomplished by treating each payment as a single payment and summing all individual present worths. A uniform increase or decrease can be reduced to present worth using the formula:

$$P_{\text{TOT}} = A[\text{USPWF}, i\%, n] + \frac{G}{i}\left[\frac{(1 + i)^n - 1}{i} - n\right]\left[\frac{1}{(1 + i)^n}\right] \tag{7.2}$$

where G = gradient of increase or decrease in annual payments
           = \$/payment period

The first term of the above is the uniform portion of the payments and the second term accounts for the increase or decrease discounted to present worth.

Example 7.3 shows an application of Equation 7.2 to a gradient series of payments. Example 7.4 demonstrates the application of net present worth analysis to the evaluation of the economic feasibility of a typical project. Note that the net annual return of \$3500 is used to compute the present worth of the annual receipts. The cash flow diagram could also be redrawn to show the net annual return rather than separate receipts and disbursements for each year. Note also that the investment in the new equipment is advisable when the minimum attractive rate of return is 10%, but it is not advisable with a minimum attractive rate of return of 15%.

Equivalent Uniform Net Annual Return

The equivalent uniform net annual return is the amount by which uniform annual receipts (benefits) exceed the uniform annual costs. To apply the method, both benefits and costs have to be annualized using the minimum attractive rate of return. Uniform annual benefits or costs are not changed. Present-worth values are annualized by using the capital recovery factor (CRF, i, n), while future amounts of money are annualized using the sinking fund factor (SFF, i, n). Nonuniform series of future amounts of money can be treated as individual payments, while gradient series of payments can be discounted to present worth using Equation 7.2, then annualized using the CRF.

A positive difference between total annual benefits and total annual costs indicates an economically attractive project. Some analysts prefer the equivalent uniform net annual return method over the present worth method when undertaking economic feasibility studies because the former can be more closely related to annual budgeting costs. However, similar conclusions will be reached by either method.

Example 7.5 illustrates the application of equivalent uniform net annual return to determine the economic feasibility of the equipment acquisition described in Example 7.4. As before, the investment is attractive at a minimum attractive rate of return of 10%, but is unattractive at a rate of 15%.

As will be shown in later sections. equivalent uniform net annual return may be preferable over net present worth when comparing alternatives that do not have the same useful lives.

EXAMPLE 7.3 ──

Find the present worth for the gradient series of payments illustrated in Example 7.2. Use a discount rate of 12%.

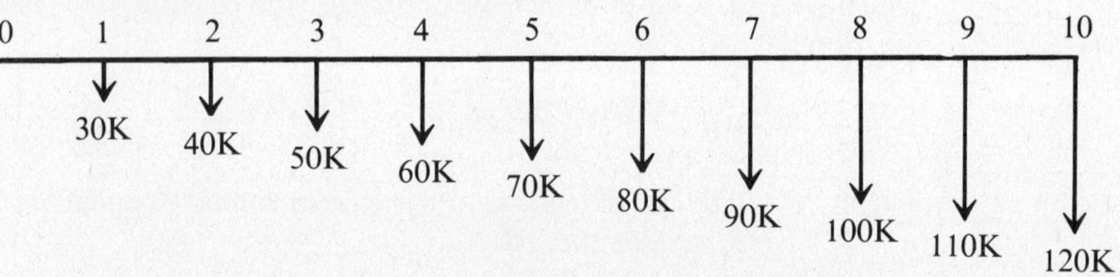

$$P = A[\text{USPWF}, 12\%, 10] + \frac{G}{i}\left[\frac{(1+i)^n - 1}{i} - n\right]\left[\frac{1}{(1+i)^n}\right]$$

$$= 30(5.650) + \frac{10}{0.12}\left(\frac{1.12^{10} - 1}{0.12} - 10\right)\left(\frac{1}{1.12^{10}}\right)$$

$$= 169.50 + 202.54$$

$$= \$372\text{K}$$

EXAMPLE 7.4 ____

You own a construction company and need to purchase some new earth-moving equipment. You have estimated the following costs:

Initial investment	$30,000
Expected life of equipment	20 yr
Salvage value after 20 yr	$ 2,500
Annual operation and maintenance costs	$ 2,500
Estimated annual income from equipment	$ 6,000

Use net present worth analysis to determine the economic feasibility of this investment, using attractive rates of return of 10 and 15%.

Cash Flow Diagram

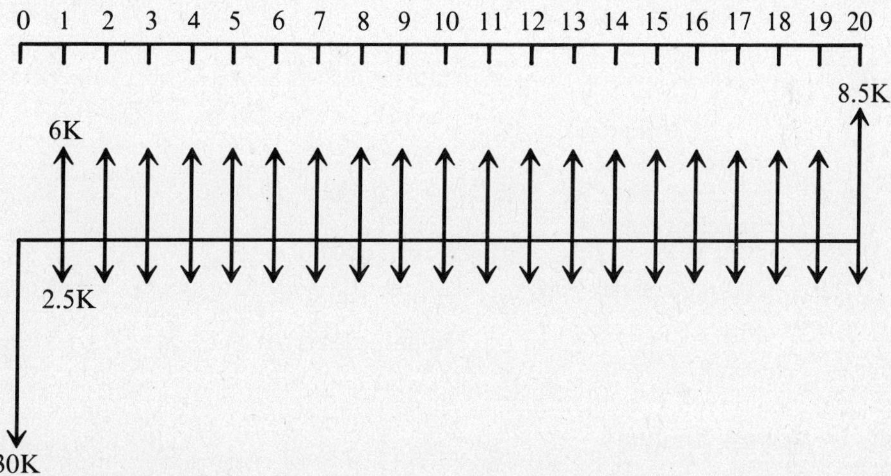

Calculations at 10%

Initial investment	−$30,000
Salvage value after 20 yr	
2500(PWF, 10%, 20) =	
2500(0.1486)	+$ 372
Net income or cost per year	
(6000 − 2500)(USPWF, 10%, 20) =	
3500(8.514)	+$29,799
Net present worth	$ 171 (attractive)

Calculations at 15%

Initial investment	−$30,000
Salvage value after 20 yr	
2500(PWF, 15%, 20) = 2500(0.0611)	+$ 153

Net income per year
  3500(USPWF, 15%, 20) =
  3500(6.259)   +$21,907
                                                              ____

                        Net present worth   −$ 7,940 (not attractive)

EXAMPLE 7.5 ____

Determine the equivalent uniform net annual return for the equipment investment proposed in Example 7.4, if the prevailing interest rates are 10 and 15%.

Calculations at 10%
Initial investment
  −30,000(CRF, 10%, 20) = −30,000(0.11746)   −$3524/yr
Salvage value after 20 yr
  2500(SFF, 10%, 20) = 2500(0.01746)   +$ 44/yr
Net income per year   +$3500/yr
                                                               ____

                        Net annual return   +$ 20/yr

Calculations at 15%
Initial investment
  −30,000(CRF, 15%, 20) = −30,000(0.15976)   −$4793/yr
Salvage value
  2500(SFF, 15%, 20) = 2500(0.00976)   +$ 24/yr
Net income per year   +$3500/yr
                                                               ____

                        Net annual return   −$1265/yr

Rate of Return Analysis

Rate of return analysis is sometimes called the internal rate of return method. It allows direct comparisons between the earning power of suggested investments and that of alternative investments. The method makes no initial assumption concerning the minimum attractive rate of return.

To accomplish the analysis, the discount rate that makes the discounted receipts (benefits) equal to the discounted costs must be computed by trial and error. Either present worths of costs and benefits or equivalent uniform annual cash flows can be utilized to determine internal rate of return, provided that all calculations are done on an equivalent basis. If the internal rate of return exceeds the minimum attractive rate of return, the project is judged to be economically feasible. Rate of return analysis is used less often than the two previously described methods for estimating economic feasibility, because it requires the previous calculation of either net present worth or equivalent uniform net annual return.

The general procedure is to find one rate of return that produces a positive net present worth or equivalent uniform net annual return, and another rate of return that produces a negative result. This brackets the internal rate of return, and interpolation can be used to estimate the true value. Once an approximate internal rate of return has been found, the estimate can be refined, if necessary, by making a bracket around the first approximation and repeating the interpolation. Using the net present worth estimates found in Example 7.5, which were $171 at 10% rate of return and negative $7940 at 15% rate of return, the approximate internal rate of return could be calculated at $10 + 5[171/(171 + 7940)]$, or approximately 10.1%. Therefore, the investment would be considered feasible for minimum attractive rates of return up to about 10.1%. Brackets of 10 and 11% could be used to refine the estimate.

Benefit/Cost Analysis

Benefit/cost analysis was first adopted by the United States Army Corps of Engineers, under congressional mandate, to show that the total benefits accruing from public works projects exceeded the costs of construction, maintenance, and operation. Since then, it has gained widespread use in evaluating the economic feasibility of government-supported public works projects, from the federal through the local level. It requires the resolution of a proposed project into its favorable and unfavorable consequences. The monetary value of each consequence has to be estimated, and this result, along with capital and operating costs, is used to evaluate the economic feasibility of a project.

Although benefit/cost analysis is often thought of as a separate form of economic analysis, it in fact relies on consistent application of one of the other methods previously described to arrive at comparable values for costs and benefits. Present worth analysis is probably the most frequently used method for finding these comparable values.

In applying benefit/cost analysis, benefits and costs resulting from the project must be identified and monetary values estimated for each. Benefits are generally taken as all positive consequences of the project, whether they accrue to the users or the initiators of the project. Costs are taken as the disbursements made by the initiators or sponsors of the project. Other costs, termed disbenefits, result from disbursements by the users of the project or losses incurred by groups directly affected by the project.

There are two equations commonly used to calculate the benefit/cost (B/C) ratio. They are

$$\frac{B - D}{C} \tag{7.3}$$

and

$$\frac{B}{C + D} \tag{7.4}$$

where B = benefits,
D = disbenefits (hardships caused by the project), and
C = costs

It is important to know which equation is being used because each can give a slightly different result. Either present worth or equivalent uniform annual costs can be used to find B, C, and D, as long as the same method is used in all calculations.

Anyone connected with benefit/cost analysis must take care in analyzing the results of the analysis. The B/C ratio can be manipulated by the analyst so that it will come out to almost any value desired. When checking B/C computations using present worth values, it must be ascertained that an equivalent lifetime is being used for all elements of the analysis. Also, all elements have to be analyzed in the same way (present worth, future worth, or whatever), and a reasonable interest rate must be used. Results are sensitive to the interest rate used; a low interest rate can make a project look economically attractive, when a more reasonable interest rate would reflect an unfavorable return on investment for the same project. This is illustrated in Example 7.6. The proposed project is economically favorable at an attractive rate of return (interest rate) of 3%, but is not favorable for rates of 6, 8, and 10%. Referring back to Examples 7.4 and 7.5, it can be seen that a B/C ratio of slightly more than 1.0 will result if an attractive rate of return of 10% is used, while a 15% rate gives a ratio considerably less than 1.0.

It is also important to ascertain exactly what costs, disbenefits, and benefits are being included in the analysis. Some disbenefits, such as loss of forest land, are easy to overlook. Indirect costs, such as required improvements in the road network leading to a project, can be left out of the analysis. Benefits can easily be inflated, or attributes that are unquantifiable can be given artificial values to increase the calculated benefits. Also, different people may assign different values to a certain attribute, or the same person may assign different values under different conditions. Some attributes may defy quantification, and they will have to be factored into the analysis in a subjective manner.

EXAMPLE 7.6 ____

A flood control dam is proposed for the Sacandaga River. It can be built for a cost of $35,000,000.00. Flood protection, recreation, and fishing are estimated to provide benefits of $3,000,000.00/yr. The loss of forest and farm crops is estimated to be a disbenefit of $1,000,000.00/yr. The life of the dam is expected to be 50 yr.

By method 1 (Equation 7.3):

$$B - D = \$2,000,000/\text{yr}$$

Interest rate (%)	PWF (50 yr)	PW	Benefit cost
3	25.730	51,460,000	1.47
6	15.762	31,524,000	0.90
8	12.233	24,466,000	0.70
10	9.915	19,830,000	0.57

By method 2 (Equation 7.4):

$$B = \$3,000,000 \qquad D = \$1,000,000$$

Interest rate (%)	Present worth			Benefit cost
	PWF	B ($)	C + D ($)	
3	25.730	77,199,000	60,730,000	1.27
6	15.762	47,286,000	50,762,000	0.93
8	12.233	36,699,000	47,233,000	0.78
10	9.915	29,745,000	44,915,000	0.66

These results show how the benefit/cost ratio can vary depending on the interest rate used; and, also, how the results using the two methods can vary.

COMPARISON OF MUTUALLY EXCLUSIVE ALTERNATIVES WITH FIXED OUTPUT

The systems or subsystems of a project may have a specific task, or output, to accomplish. Some examples of these are the heating system for a high-rise building, the powerhouse structure for a hydroelectric power project, or an interchange system between two interstate highways. Since the output or task level for these projects is constant, the benefits or return on investment would be expected to be constant. However, there may be several alternative ways to accomplish the task or achieve the level of output required. These alternatives may be different in their distribution of costs over time, so economic analysis must be undertaken to estimate equivalent costs, enroute to demonstrating the relative economic viability of alternative schemes. The objective in examining the alternatives is either to maximize the net present worth or to maximize the equivalent uniform net annual return. If only costs are considered (a reasonable assumption since return is constant) the present worth of costs or the equivalent uniform annual costs should be minimized.

For example, consider a project that requires the pumping of a certain flowrate of water against a given head for the next 5 yr. At the end of the first 5-yr period, planned expansion will require a doubling of the flowrate against the same head. Several options are available for meeting the pumping requirement. One option would be to use two large pumps, each capable of meeting the requirements. The second pump would be used as a backup whenever the primary pump was out of service. A second option would be to install a number of smaller pumps capable of collectively meeting the requirements, as well as having suitable backup capacity.

In order to compare these options, the initial cost, operation and maintenance cost, salvage value, and expected lifetime of units, would all have to be considered. A second level of analysis would have to be directed toward how best to meet the increased pumping requirements five years hence. Alternatives might include installing enough total capacity initially to meet all needs, installing part of the equipment now and the rest of it in 5 yr, or installing all of the additional equipment when it is actually needed.

Example 7.7 illustrates present worth comparison of alternatives for a freeway interchange. When comparing alternatives using present worth, it is necessary that all have the same overall life. Without this stipulation, the economic equivalency computations will be invalid. Example 7.8 will demonstrate one method of equivalencing useful lives for alternatives. Equivalencing alternative lives is not necessary when using equivalent uniform annual return.

A lack of confidence in predicting future costs and instability of interest rates will increase the uncertainty of any decisions made based on economic analysis. However, with proper attention to detail and an understanding of the limitations of economic analysis, the engineer can make more rational recommendations as to the best course of action to take.

EXAMPLE 7.7 ____

A freeway is presently being built. Along this freeway, there is an interchange that will connect with a rapid transit station that is to be built 5 yr in the future. Three alternatives have been developed.

Alternative 1 Build the interchange and station approach roads along with the freeway, for a capital cost of $3,000,000, and a yearly maintenance cost of $20,000.

Alternative 2 Build the interchange with the freeway, but wait 5 yr to build the approach roads. Building the interchange now will cost $2,000,000, and will have an annual maintenance cost of $15,000. Building the approach roads in 5 yr is estimated to cost $2,000,000 at that time.

Alternative 3 Wait 5 yr to build both the interchange and the approach roads. The estimated cost of both in 5 yr is $5,000,000.

If it is assumed that costs in years subsequent to project completion (5 yr in the future) are approximately equal for each of the three alternatives, and that all alternatives have approximately the same useful life, find the least costly of these alternatives for prevailing interest rates of 8 and 12%.

While present worth analysis will be used in these calculations, the student should verify that equivalent uniform annual costs will lead to the same conclusions.

At a minimum attractive rate of return of 8%, the following calculations can be made:

Alternative 1

Capital cost	$3,000,000
Annual maintenance cost of $20,000 for 5 yr	
20,000(USPWF, 0.08, 5) = 20,000(3.993)	$ 79,860
Total	$3,079,860

Alternative 2

Capital cost	$2,000,000
Annual maintenance cost of $15,000 for 5 yr	
15,000(3.993)	$ 59,895

Additional capital expenditure in 5 yr
  2,000,000(PWF, 0.08, 5) = 2,000,000 (0.6806) $1,361,200

                                              Total $3,421,095

Alternative 3
Capital cost $ 0
Annual maintenance cost $ 0
Capital expenditure in 5 yr
  5,000,000(0.6806) $3,403,000

                                              Total $3,403,000

Therefore, at an 8% attractive rate of return, alternative 1 is the most economical. However, at an attractive rate of return of 12%, alternative 3 becomes the most advantageous, as is shown by the following computations.

Alternative 1
Capital cost $3,000,000
Annual maintenance cost
  20,000(USPWF, 0.12, 5) = 20,000(3.605) $ 72,100

                                              Total $3,072,100

Alternative 2
Capital cost $2,000,000
Annual maintenance cost
  15,000(3.605) $ 54,075
Additional capital expenditure in 5 yr
  2,000,000(PWF, 0.12, 5) = 2,000,000(0.5674) $1,134,800

                                              Total $3,188,875

Alternative 3
Capital expenditure in 5 yr
  5,000,000(0.5674) $2,837,000

                                              Total $2,837,000

COMPARISON OF MUTUALLY EXCLUSIVE ALTERNATIVES WITH VARIABLE OUTPUT

Frequently, mutually exclusive alternatives are available which not only have different levels of input investment, but also have variable levels of accrued benefits or return on investment. Recommendations have to be made as to which alternative is the most advantageous. Any of the alternatives that are judged not economically feasible by any of the methods previously described can be eliminated from further consideration. The remaining alternatives can be compared to determine which one produces the most return for the amount invested, and thereby, to determine if increasing levels of investment are warranted. Net present worth, equivalent uniform net annual return, incremental benefit/cost analysis, or incremental rate of return analysis can be utilized to provide information upon which recommendations concerning the proper level of investment can be based.

Maximization of Net Present Worth and Equivalent Uniform Net Annual Return

In the comparison of mutually exclusive alternatives by net present worth or equivalent uniform net annual return, the best choice is the alternative with the greatest net present worth, or the greatest equivalent uniform net annual return. If net present worth comparison is to be made, any alternatives with different useful lives have to be transformed to give all alternatives the same expected life. The methodology for accomplishing this transformation will be presented in Example 7.8. The equivalent uniform net annual return method (sometimes referred to as the capital recovery method) does not require that the alternative projects have the same useful life. However, to apply either method, an assumption of continuing need for the project output must be made. In other words, it is assumed that each alternative will be replaced in kind at the end of its useful life, as many times as is necessary to extend its lifetime up to the longest lived alternative.

Example 7.8 shows the application of net present worth analysis and equivalent uniform net annual return analysis to the evaluation of two mutually exclusive projects dealing with the purchase of earth moving equipment. Alternative *A* involves the same investment that was described in Example 7.4. Again, two attractive rates of return are used to demonstrate the sensitivity of results to that rate.

EXAMPLE 7.8 ──

The construction company previously mentioned in Example 7.4 must purchase new earth moving equipment. The costs and returns for investing in one brand of equipment have been computed as:

> *Alternative A*
>
> | | |
> |---|---|
> | Initial investment | $30,000 |
> | Expected life of equipment | 20 yr |
> | Salvage value after 20 yr | $ 2,500 |
> | Annual operation and maintenance costs | $ 2,500 |
> | Estimated annual income from equipment | $ 6,000 |

However, a second alternative investment is available that has different costs and different returns.

Alternative B

Initial investment	$20,000
Expected life of equipment	10 yr
Salvage value after 10 yr	$ 1,500
Estimated cost of replacement equipment after 10 yr (different from the original because of projected product improvements)	$30,000
Annual operation and maintenance costs	$ 2,000
Estimated annual income from equipment	$ 6,000

Net present worth and equivalent uniform net annual return will be used to compare these two alternatives. The net present worth of alternative *A*, from Example 7.4, is +$171, which labels it as an economically feasible alternative. Before alternative *B* can be evaluated, a decision has to be made about the equivalency of the useful lives of the alternatives. Alternative *A* has an expected useful life of 20 yr, while alternative *B* has an expected life of 10 yr. At the end of that 10 yr, it would be the most logical choice to purchase a similar piece of equipment to provide service during years 11 through 20. The estimated cost of this replacement at 10 yr in the future is $30,000. Therefore, the net present worth of alternative *B* is found by:

Net Present Worth of Alternative B at 10%

Initial investment	−$20,000
Salvage value after 10 yr	
$\quad$ 1500(PWF, 10%, 10) = 1500(0.3855)	+$ 578
Reinvestment after 10 yr	
$\quad$ −30,000(PWF, 10%, 10) = −30,000(0.3855)	−$11,565
Salvage value after 20 yr	
$\quad$ 1500(PWF, 10%, 20) = 1500(0.1486)	+$ 223
Net income or cost per year	
$\quad$ (6000 − 2000)(USPWF, 10%, 20) = 4000(8.514)	+$34,056
Net present worth	$ 3,292

Alternative *B* is also economically feasible, and it is a better investment than alternative *A*.

At an attractive rate of return of 15%, alternative *A* was deemed economically infeasible in Example 7.4. Net present worth for alternative *B* is given by:

Net Present Worth of Alternative B at 15%

Initial investment	−$20,000
Salvage value after 10 yr	
$\quad$ 1500(PWF, 15%, 10) = 1500(0.2472)	+$ 371
Reinvestment after 10 yr	
$\quad$ −30,000(PWF, 15%, 10) = −30,000(0.2472)	−$ 7,416
Salvage value after 20 yr	
$\quad$ 1500(PWF, 15%, 20) = 1500(0.0611)	+$ 92

Net income per year
    4000(USPWF, 15%, 20) = 4000(6.259)   +$25,036

                           Net present worth   −$ 1,917

Therefore, neither alternative is feasible at a minimum attractive rate of return of 15%.

The equivalent uniform net annual return for alternative A was computed in Example 7.5 as +$20/yr at 10% and −$1265/yr at 15%. It is also possible to find the equivalent uniform net annual return for alternative B; but, since the replacement cost of the equipment is not the same as its original cost, it is necessary to base comparison on an equilized project life. At 10%, the net annual return for alternative B is given by:

Equivalent Uniform Net Annual Return for Alternative **B** *at 10%*
Initial investment
    −20,000(CRF, 10%, 20) = −20,000(0.11746)   −$2,349/yr
Salvage value after 10 years
    1500(SFF, 10%, 10) = 1500(0.06275)   +$ 94/yr
Reinvestment after 10 yr
    1. First find the present worth of $30,000 10 yr in
       the future:
       −30,000(PWF, 10%, 10) = −30,000(0.3855) =
       −11,565
    2. Then, transform this into annual payments over
       20 yr:
       −11,565(CRF, 10%, 20) = −11,565(0.11746)   −$1,358/yr
Salvage value after 20 yr
  (Accounted for by annualization over 10 years of first
    salvage value.)
Net income per year   +$4,000/yr

                           Net annual return   +$ 387/yr

If the cost of reinvestment had been $20,000, the initial investment, salvage value, and net income could have been used alone with a useful life of 10 yr.

At a minimum attractive rate of return of 15 percent, alternative B would have the following net annual return:

Equivalent Uniform Net Annual Return for Alternative **B** *at 15%*
Initial investment
    −20,000(CRF, 15%, 20) = −20,000(0.15976)   −$3,195/yr
Salvage value after 10 yr
    1500(SFF, 15%, 10) = 1500(0.04925)   +$ 74/yr
Reinvestment after 10 yr
    −30,000(PWF, 15%, 10) = −30,000(0.2472) =

$$-\$7416$$
$$-7416(CRF,\ 15\%,\ 20) = -7416(0.15976) \qquad\qquad -\$1,185/yr$$

Salvage value after 20 yr
  (Accounted for by annualization over 10 years of first
    salvage value.)

Net income per year $\qquad\qquad\qquad\qquad\qquad\qquad\qquad\qquad$ +\$4,000/yr

$\qquad\qquad\qquad\qquad\qquad\qquad\qquad$ Net annual return $\qquad$ −\$ 306/yr

The same conclusion can be drawn from these net annual return figures as were drawn from the net present worth analyses. At 10%, both alternatives are feasible, with alternative *B* being the better investment. At 15%, neither alternative is feasible.

Incremental Benefit/Cost Analysis

To evaluate the relative economic attractiveness of competing mutually exclusive alternatives using benefit/cost analysis, it is necessary to undertake incremental benefit/cost analysis. Incremental benefit/cost analysis can also be used to rank order several public works projects according to their relative merit. It must be remembered, however, that the analysis undertaken makes no assumption about the total amount of capital available for investment. Limits on available capital may eliminate some alternatives from consideration, even though they might be economically attractive.

To apply incremental benefit/cost analysis, projects are ranked and evaluated in increasing order of cost. A ratio is formed between the incremental discounted benefits of the two least expensive alternatives and the incremental discounted costs of the same two alternatives. If the incremental benefit/cost ratio is >1.0, then the additional investment in the next higher cost alternative is warranted, and it becomes the new defender alternative to be compared with higher priced alternatives. If the ratio is <1.0, then the additional investment is not warranted, and the lower cost of the two alternatives being considered remains the defender to be compared with the next higher cost alternative.

Example 7.9 illustrates incremental benefit/cost analysis of transportation improvement projects. Benefits are derived from savings in user costs that will be realized from the improvements. For highway projects, the user costs are calculated based on the types of vehicles using the highway, the physical and design characteristics of the particular type of highway, and the type of area through which the highway passes. As with any type of benefit/cost analysis, complexities involved in estimating monetary values for costs and benefits may, in practice, make application of the method difficult.

EXAMPLE 7.9 ———————————————————————————————

An engineer wishes to use incremental benefit/cost analysis to compare several mutually exclusive transportation improvement projects that involve different levels of investment, but also produce different levels of benefits. In this case, the benefits are measured as a reduction in costs to users of the transportation facilities. The costs and benefits are tabulated below. All values are in $1000 of dollars.

Alternative	Investment	Annual maintenance costs	Annual user costs
Existing conditions	—	70	2500
1	2000	30	2000
2	4000	40	1700
3	6000	30	1600
4	7000	50	1400

If the estimated lifetimes of the alternatives are all 25 yr, and the minimum attractive rate of return is assumed to be 10%, the benefit/cost ratio of each alternative relative to the existing case can be determined by the equation:

$$B/C_{i \text{ to ex}} = \frac{\text{user cost}_{ex} - \text{user cost}_i}{\text{annualized investment}_i + \text{maintenance}_i - \text{maintenance}_{ex}}$$

$$= \frac{\Delta \text{user cost}}{\text{investment} + \Delta \text{maintenance}}$$

The pertinent data are summarized below:

	Annualized cost	Δmaintenance	Δuser cost	B/C
1—existing	220	40	500	1.92
2—existing	441	30	800	1.70
3—existing	661	40	900	1.28
4—existing	771	20	1100	1.39

Each of the alternatives has a benefit/cost ratio of greater than 1.0; however, incremental cost/benefit analysis must be undertaken to determine which alternative is the best investment. Alternative 1 is regarded as the first defender, and the incremental costs and benefits are compared for the increasingly expensive challengers.

$$B/C_{i-j} = \frac{\text{incremental benefits}}{\text{incremental costs}}$$

$$= \frac{\Delta \text{user cost}_{i-j}}{\Delta \text{annualized investment}_{i-j} + \Delta \text{maintenance}_{i-j}}$$

for 1–2:
$$B/C_{1-2} = \frac{800 - 500}{(441 - 220) + (30 - 40)} = 1.42$$

Since the incremental benefit/cost ratio is greater than 1.0, alternative 2 represents a better investment than alternative one. It then becomes the defender and is compared with alternative 3.

$$B/C_{2-3} = \frac{100}{(661 - 441) + (40 - 30)} = 0.43$$

Therefore, alternative 3 does not represent an attractive investment over alternative 2. Alternative 2 is now compared with alternative 4:

$$B/C_{2\text{-}4} = \frac{300}{(771 - 441) + (20 - 30)} = 0.94$$

Thus, alternative 2 represents the best investment.

Incremental Rate of Return Analysis

When comparing mutually exclusive alternatives by rate of return, it is necessary to compute the incremental rate of return. The individual rates of return do not form a basis for comparing alternatives. Economically feasible alternatives are ranked in order of ascending cost. The rate of return is computed for the difference in benefits (returns) between one alternative, and the next higher priced alternative, and the incremental cost between the same two alternatives. If this incremental rate of return is larger than the minimum attractive rate of return, the more expensive alternative is preferable to the less expensive one. It would then become the defender against the next higher priced alternative. If the incremental rate of return is smaller than the minimum attractive rate of return, it would remain as the defender against the next higher priced alternative.

In many ways, incremental rate of return analysis is similar to incremental benefit/cost analysis; and if properly applied, will result in the same conclusions being made. It is not, however, as widely used, nor as generally understood, as incremental benefit/cost analysis. Example 7.10 shows application of incremental rate of return analysis to the same transportation improvement problem posed in Example 7.9. Note that both examples reach the same conclusions. This example also illustrates how the economic analysis tables of Appendix B can be used to simplify and streamline the analysis.

EXAMPLE 7.10

Use incremental rate of return analysis to determine the most advantageous transportation improvement project for the alternatives posed in Example 7.9. The pertinent data are given below:

Alternative	Investment	Net annual benefits	P/A	Approximate internal rate of return
1	$2000	$460	4.35	>20%
2	4000	770	5.19	>15%
3	6000	860	6.98	>12%
4	7000	1080	6.48	>12%

Note: useful life = 25 yr; all $ amounts are in thousands of dollars.

The approximate internal rate of return is found by comparing the ratio of investment to net annual benefits with values of the uniform series present worth factor given in Table B.1. For example, for alternative 1, USPWF = P/A = 2000/460 = 4.35. In Table B.1, (USPWF, 20%, 25) = 4.95 and (USPWF, 25%, 25) = 3.985. Therefore, the internal rate of return that equalized discounted costs and benefits is between 20 and

25%; thus, the >20% approximation. There is no need to be more accurate than this, because it is only necessary to show that the internal rate of return is greater than the minimum attractive rate of return (10% for this case). Each of the alternatives under consideration has an internal rate of return greater than 10%, so they are all economically feasible. Alternative 1 has the largest internal rate of return; but as will be demonstrated, it is not the most advantageous alternative.

The incremental investment between alternatives 1 and 2 is 2000×10^3, while the incremental net annual benefits are 310×10^3. Therefore, the incremental value of P/A is 2000/310 = 6.45. Using the USPWF column of Table B.1 for $n = 25$, it can be seen that the incremental rate of return for alternatives 1 and 2 is greater than 15%. This means that alternative 2 is preferable to alternative 1. Alternative 2 will now be used as the defender with alternative 3.

Comparing 2 with 3, the incremental P/A is (6000 − 4000)/(860 − 770) = 22.22. Table B.1 shows that this corresponds with an incremental rate of return of less than 1%. Therefore, alternative 3 is not preferable to alternative 2, and 2 remains as the defender.

Comparing 2 with 4, the incremental ratio of P/A is (7000 − 4000)/(1080 − 770) = 9.68, which corresponds to an incremental rate of return of less than 10%. Therefore, alternative 2 is judged to be the most advantageous.

LINEAR PROGRAMMING AS APPLIED TO ECONOMIC ANALYSIS

Example 7.11 illustrates a combination of linear programming and economic analysis applied to bidding on engineering projects. The objective of the bidder is to maximize the present worth of payments to be received at different times during the duration of the contract. This example has been adapted from an example presented by Stark and Nichols (1972). Further discussion of this type of problem can be found in Gates (1967), Mayer, Stark, and Fitzgerald (1969), and Stark (1968).

A danger inherent in unbalanced bidding is that the proposed and the actual amounts of work may not be the same. If the bidder makes a low bid on a high-cost item such as rock excavating, and the actual amount of work is considerably greater than the proposed, a significant loss can be incurred. A prudent bidder might make actual unit costs the lowest bid that would be submitted, and only unbalance the bid with respect to profit on each item with the objective of increasing total present worth.

EXAMPLE 7.11 ____

Bidding on construction work, especially highway work, is often conducted through unit price proposals. The responsible agency advertises for bids, indicating the quantities of various items (cubic yards of concrete, miles of pavement, etc.) that are expected to be needed. Bidders are invited to submit unit bids ($/unit of work accomplished) as an overall bid on the project. The low overall bidder is generally awarded all of the work.

An overall bid that reflects the actual cost of doing the work plus a reasonable profit can be submitted; however, it may be advantageous for the contractor to submit unit bids that do not reflect the actual cost of accomplishing a particular item of work. This is termed unbalanced bidding. Unbalanced bidding may be advisable, because the time value of money will place a decreased present worth on delayed payment for work to be accomplished later in the project. If high unit bids are submitted for work to be accomplished early in the contract, the payment for this work can help finance work later in the contract.

If the overall objective is to maximize the present worth of all revenues received from a contract, a strategy can be developed for determining unbalanced bids. To simplify the example, it will be assumed that payment for each item of the contract will be made at the unit bid price after that part of the contract has been completed. These payments can be discounted to present worth to determine the total present worth of the project to the bidder. Consider the following data:

Bid information given to the contractor			Contractor balanced bid		
Item	Description	Estimated no. of units	Balanced unit bid	Amount	Estimated completion time (months in the future)
1	Corridor clearing	35,000 yd^2	$4.00	$140,000	2
2	Cut	80,000 yd^3	$7.00	$560,000	6
3	Fill	75,000 yd^3	$2.00	$150,000	8
4	Final grading	35,000 yd^2	$4.00	$140,000	12
			Total bid	$990,000	

If the attractive rate of return is approximately 12%/yr, the payments can be discounted at the rate of 1%/month. Thus, the total present worth (PW) of the contract is

$$\text{PW} = 140,000(1.01)^{-2} + 560,000(1.10)^{-6} + 150,000(1.01)^{-8} + 140,000(1.01)^{-12}$$

$$= 140,000(0.980) + 560,000(0.942) + 150,000(0.923) + 140,000(0.887)$$

$$\text{PW} = \$927,350$$

In unbalanced bidding, the unit bids are treated as variables, and the objective function can be written

$$\max \text{PW} = 35,000(0.980)X_1 + 80,000(0.942)X_2$$

$$+ 75,000(0.923)X_3 + 35,000(0.887)X_4$$

$$= \$34,300X_1 + 75,360X_2 + 69,225X_3 + 31,045X_4$$

in which X_1 through X_4 are the unit bids to be submitted for items 1 through 4. These unit bids are subject to certain constraints. First of all, the total bid price will equal the balanced bid price of $990,000, or

$$35,000X_1 + 80,000X_2 + 75,000X_3 + 35,000X_4 = \$990,000$$

The company making the bid has a policy of not setting unit bids at less than $1.00/ unit, and normally makes the unit bid for cut at least 1.5 times the unit bid for fill operations. Therefore,

$$X_1 \geq 1.0$$

$$X_2 - 1.5X_3 \geq 0$$

$$X_3 \geq 1.0$$

$$X_4 \geq 1.0$$

Based on these constraints and the objective function, the following unit bids should be made:

$$X_1 = 21.71$$

$$X_2 = 1.50$$

$$X_3 = 1.0$$

$$X_4 = 1.0$$

and the maximum present worth would be \$958,100, an increase of approximately 3% over the balanced bid price.

Suppose that the bidder has investigated the quantities of each item contained in the proposal, and has estimated that only 70,000 yd^3 of cut will have to be made, but 90,000 yd^3 of fill will have to be placed. This changes the objective function to

$$\max PW = 35,000(0.980)X_1 + 70,000(0.942)X_2$$
$$+ 90,000(0.923)X_3 + 35,000(0.887)X_4$$
$$= 34,300X_1 + 65,940X_2 + 83,070X_3 + 31,045X_4$$

The constraints will remain as before. For this situation, the unit bids remain the same, but the maximum present worth will drop slightly, to \$957,800 due to the longer completion time for the fill. However, if research reveals that 90,000 cubic yards of cut but only 70,000 yd^3 of fill will be required, the objective function becomes

$$\max PW = 34,300X_1 + 84,780X_2 + 64,610X_3 + 31,045X_4$$

which attains a value of \$1,025,400 for unit bids of $X_1 = 1.00$, $X_2 = 10.56$, $X_3 = 1.00$, and $X_4 = 1.00$. This is an increase of 11% over the present worth of the balanced bid.

SUMMARY

Although the basic equations of engineering economics are easily understood, their applications can become extremely complex. Some costs and benefits are difficult, if not impossible, to estimate. The assumption that the minimum attractive rate of return will remain constant over a period of time may be tenuous. Therefore, it is essential that the engineer be knowledgeable about all aspects and limitations of economic analysis techniques.

Economic analysis will play a crucial role in determining if a project will be undertaken; or, after it has commenced, if the project will be finished. Cost data must be carefully and accurately gathered, then efficiently analyzed by accepted methods. The engineer must be able to interpret the products of cost data analysis and make the proper economic choices. The methods described in this chapter provide a framework to assist the engineer in making rational choices based on the best information available.

APPLICATIONS EXERCISES

7-1. Compute the benefit/cost ratio for the following proposed rapid transit system.

Costs

Item	Initial cost (× $ millions)	Salvage value (× $ millions)	Service life (yr)
Right of way	200	—	100
Main structures	250	25	50
Drainage	70	10	50
Earthwork	50	—	50
Signalization	100	25	25
Road bed	20	—	25
Track	40	8	10

Annual Benefits (for 50-yr Period)

Decreased direct user costs	$40 million/yr
Indirect community benefits	$55 million/yr

(less congestion, increased commercial business, increased development, increased sales tax revenues, etc.)

Use a discount rate of 8%.

7-2. A proposal has been made for the improvement of a downtown area of a small town. The plan calls for banning vehicular traffic on the main street and turning this street into a pedestrian mall with tree plantings and other beautification projects. This plan will involve actual costs of $7,000,000 and, according to its proponents, produce benefits and disbenefits to the town as follows:

Benefits

1. Increased sales tax revenue — $400,000/yr
2. Increased real property taxes — $300,000/yr
3. Benefits due to decreased air pollution — $ 50,000/yr
4. Benefits due to other improvements in the quality of life of people using the area — $ 50,000/yr

Disbenefits

1. Increased maintenance — $150,000/yr

(a) Compute the benefit/cost ratio of this plan based on an 8% cost of capital to the town, and an infinite life for the project. Use $(B - D)/C$.

(b) How does the benefit/cost ratio change for a 20-yr project life and a 10% cost of capital to the town?

7-3. A regional authority has determined that there are two possible courses of action to treat and control hazardous wastes generated within the region.

Plan **A**   Plan A calls for the construction of five hazardous waste treatment plants with the following associated costs and capacities:

Site	Capacity (million gal/yr)	Process cost/ million gal	Capital cost	Transport costs/yr
A	40.0	$ 9,000	6.1×10^6	0
B	61.0	$13,000	7.0×10^6	$1,500
H	21.0	$ 5,000	2.7×10^6	$9,000
P	9.7	$ 2,000	1.9×10^6	$6,000
T	25.7	$ 5,000	2.8×10^6	$7,500

The following additional costs would be associated with this alternative:
1. The cost of accident cleanup is estimated at 1.2×10^6/yr.
2. The cost of increased air pollution caused by incineration is estimated at 1.0×10^6/yr.
3. The loss of agricultural land is estimated at 0.6×10^6/yr.

Plan **B**   A single hazardous waste treatment facility located at H is an alternative. The costs and disbenefits are as follows:

1. Capital cost = 11.0×10^6
2. Operation and maintenance cost = $9,000/million gal
3. Transportation cost = $182,000/yr
4. Cost of accident clean up = 1.8×10^6/yr
5. Increased air pollution = 1.5×10^6/yr
6. Loss of agricultural land = 0.4×10^6/yr

Benefits for either alternative are as follows:

1. Health benefit from prevention of groundwater contamination = 4×10^6/yr
2. Environment, wildlife, health of ecosystem = 1.5×10^6/yr
3. Decreased cost of future landfill (fewer necessary and less risk of future problems because chemicals are treated) = $600,000/yr
4. Reclamation of chemicals for reuse = $200,000/yr

Compute the cost/benefit ratio using $B/(C + D)$ for each alternative and compare alternatives. Use a discount rate of 8 and 10%. The expected life for either alternative is 40 yr. Assume zero salvage values for both.

7-4. A water resources project has been proposed which serves the purposes of flood control, irrigation, and hydroelectric power. The following data apply:

A. Construction costs

1. Dam and reservoir, including $34,900,000
   a. Power penstocks $500,000
   b. Irrigation outlet $300,000
   c. Damsite water outlet $100,000
2. Irrigation canal and distribution system $ 2,500,000
3. Power plant and transmission facilities $ 2,000,000

   Total construction costs $39,400,000

B. Benefits

1. Annual flood control benefits $ 1,000,000
2. Annual irrigation benefits $ 3,500,000
3. Annual power benefits $ 850,000

   Total annual benefits $ 5,350,000

C. Disbenefits (annual)

1. Loss of forest land and hunting grounds $ 425,000
2. Loss of agriculture $ 1,500,000

   Total annual disbenefits $ 1,925,000

D. Annual operation and maintenance costs

1. Dam and reservoir $ 25,000
2. Power plant and transmission facilities $ 150,000
3. Irrigation facilities $ 110,000
4. General expense (administration,
      overhead, etc.) $ 35,000

   Total annual cost $ 320,000

E. Service life and salvage values

Item	Salvage value	Service life (yr)
1. Dam and reservoir	—	100
2. Irrigation canal and distribution system	—	100
3. Power plant and transmission facilities	$500,000	50
Expected service life of the dam		100

Use present worth and annual return computations to calculate the benefit/cost ratio for this project.

7-5. Your construction firm needs to expand its capacity for indoor storage of equipment. Your present storage facility could be expanded to meet your needs at a cost of $500,000. Operation and maintenance costs for the expanded facility would be about $80,000/yr. An entirely new facility could be built for approximately $1,350,000. The new facility would have an estimated annual operation and maintenance cost of $50,000. Also, the present storage facility could be sold for $300,000. Salvage value for the new facility is estimated to be $500,000 after a 25-yr useful life. The expanded facility would have the same useful life and a salvage value of about $200,000. If the prevailing interest rates are around 12%, which alternative is preferable?

7-6. A corporation decides to build a new building of 50,000 ft^2 at a cost of $40/ft^2. Based on their rate of growth, they will need to have an additional 20,000 ft^2 of space 10 yr from now. The corporation currently has a $2600K fund allocated for building. The corporation has two building alternatives:

Alternative **A**   Build 70,000 ft^2 now at $40/ft^2. Since there is only $2600K available, additional money will be borrowed at an interest rate of 20% compounded annually and repaid with equal end-of-the-year payments over a 10-yr period. The additional 20,000 ft^2 built now will cost the corporation $2000/yr in maintenance costs.

Alternative **B**   Build 50,000 ft^2 now and invest the remainder of the presently available $2600K in an account that pays 14% interest compounded quarterly. It is estimated that in 10 yr, it will cost $60/ft^2 to build the additional 20,000 ft^2.

If the corporation uses a minimum attractive rate of return of 15%, which of the two building alternatives is better?

7-7. You have been assigned the task of evaluating the advisability of the state investing in an urban redevelopment project. The project will consist of high rise apartments designed for a mix of tenants from different economic strata. The space scheduled for redevelopment now consists of a number of vacant lots, plus tenements occupied by people from a common ethnic background. List all the costs, benefits, and disbenefits you should consider in computing the benefit/cost ratio for this project. Indicate which of these items would be difficult to quantify, and why.

7-8. The initial cost of a piece of equipment is $300,000. It is expected to bring in a return of $25,000/yr for its expected life of 20 yr. Salvage value at the end of 20 yr should be approximately $25,000. At what rate of return would this investment be attractive?

7-9. A corporation is considering two alternative bulk storage facilities with the following characteristics:

Type	Cost	Life (yr)	Salvage value
Prefab. concrete	$1,600,000	60	−$100,000 (demolition cost)
Clad steel frame	$ 800,000	30	$ 60,000

The concrete building will require no maintenance for the first 15 yr and will cost $10,000/yr thereafter. The steel building will cost $14,000 each year for maintenance.

(a) At a rate of interest of 10%, which will be the most economical choice for 60 yr of service?

(b) Is there any attractive rate of return at which the other alternative will become preferable? If so, what is it?

7-10. An enclosed pedestrian and freight bridge will join two buildings in an industrial complex, allowing exchange of personnel and material without having to utilize the interior elevators and outside streets. Determine how many trips must be made each year for this project to be financially attractive. Use both present worth and annual payment analysis for interest rates of 10, 12, and 15%. The following data are applicable:

$$\text{savings for the first 10 yr} = \$0.80/\text{trip}$$

$$\text{savings for the next 20 yr} = \$0.90/\text{trip}$$

$$\text{savings for the next 20 yr} = \$1.10/\text{trip}$$

Maintenance costs will be $2000/yr for the first 20 yr and $5000/yr in subsequent years.

$$\text{project cost} = \$161,200$$

$$\text{project life} = 50 \text{ yr, with no salvage value}$$

7-11. A company is faced with the decision of selecting either machine A or B, both of which are capable of doing the same kind of job. Machine A costs $1250 with annual maintenance and operating charges of $150 for the first 10 yr and $180 for the following 10 yr. Machine B costs $1500 and operating and maintenance expenses are $100 for the 20-yr period. Both machines have zero salvage value at the end of their economic lives. Determine:

    (a) Which machine is more economical for the company at the attractive rate of interest of 15%

    (b) By how much should the manufacturer of machine A increase or decrease the price of the machine so that it is competitive with machine B?

7-12. As part of a water quality study, it is necessary to develop wastewater treatment cost estimates for two different sized plants. Cost estimates were found in appropriate government publications; however, they were not in the units desired, and they were published several years ago. It is now desired to update the cost estimates to 1983 values, and to convert the cost estimates to the desired units of dollars per pound of pollutant removed. At the present time, the best estimate of the attractive rate of return is 8%. Assume a 20-yr useful life for the plants. Accomplish this conversion if:

    (a) A constant inflation rate of 10% is assumed.

    (b) A cost index with a 1960 value of 100 is used to project costs. Values of the cost index for the years in question are given below.

1977	1978	1979	1980	1981	1982	1983 (est)
361	390	430	495	570	630	662

Other appropriate data are given below.

Treatment plant	Flow rate (10^6 gal/day)	lb/s removed	Capital costs (1977 $)	Operation/maintenance costs (1978 $/yr)
A	6.5	0.106	15×10^6	1.2×10^6
B	13.0	0.213	25×10^6	2.7×10^6

7-13. A state highway department is faced with the problem of deterioration of and unsafe conditions on a winding, seaside highway. Three alternatives have been identified. Alternative A involves resurfacing, minor route realignment, and upgrading of safety barriers and signs. Annual maintenance cost for this alternative will be high because of frequent land slides and unstable foundation conditions. Alternative B is a major realignment of the route slightly inland from the coast. It is 7 mi shorter, but still follows the terrain. The initial cost of this alternative is higher, but soil conditions are more stable and the annual maintenance cost is less. Alternative C is farther inland. It follows a much straighter path, but requires significant cut and fill. All three alternative highways will need resurfacing every 5 yr.

    Highway usage is expected to remain approximately constant over the next 20 yr, which is the planning life for the project. Average speed will vary depending on the type of traffic (heavy truck, light truck, or automobile) and the alternative. Therefore, operation costs will vary. Also, the three alternatives will provide different levels of accident prevention. Estimates are that each accident produces approximately $10,000 in damages.

Pertinent data are given in the tables below. Determine the most attractive alternative at attractive rates of return of 8, 10, and 12%. [This problem was adapted from an example in White, Agee, and Case (1977).]

Route data

	Length (mi)	Initial cost (10^6 \$)	Annual maintenance (\$/mi)	Resurfacing cost (10^6 \$)	Accidents/yr
Alternative A	50	5	20,000	4	250
Alternative B	43	20	10,000	5	180
Alternative C	38	30	12,000	6	120

Traffic volume data

	Vehicles/ day	Percent commercial	Cost of operation (\$/mi)	Time cost for commercial vehicles (\$/hr)
Heavy trucks	500	100	0.50	15
Light trucks	500	75	0.30	12
Automobiles	4000	20	0.20	20

Traffic speed data (mph)

	Automobile	Light trucks	Heavy trucks
Alternative A	40	30	25
Alternative B	45	40	35
Alternative C	50	45	40

7-14. A study is being made to determine the economic feasibility of a college performing arts center. The following costs and benefits have been projected for the center. Determine by using interest rates of 12 and 15% the benefit/cost ratio using both present worth and annual worth. Building life is expected to be 50 yr before renovation. Also, determine the approximate minimum attractive rate of return. Discuss the appropriateness of the cost and benefit data given below, and indicate any additional costs or benefits you think should be included. Give reasons for your answers.

Costs

Construction	\$2,200,000
Operation and maintenance	\$ 20,000/yr
Management of theaters	\$ 20,000/yr

Benefits

Ticket sales for 200 days at $2000	\$ 40,000/yr
Donations toward building	\$1,000,000
Increased enrollment of 30 performing arts majors contributes $2000/ student out of tuition	\$ 60,000/yr

       Entertainment worth to
          students, $50 each for
          2000 students $ 100,000/yr

7-15. The ACME Construction Company owns three tracts of land on Puget Sound in Washington State. Each tract contains a large volume of standing timber. During off-construction periods, ACME plans to use its personnel to harvest the timber. Contracts have already been signed for purchase of the lumber by a manufacturer of containers. ACME now needs to purchase specialized timber harvesting and transportation equipment so it can meet its commitments. Three alternatives have been identified. Pertinent cost data are given below. Find the most attractive alternative if the attractive rate of return is 12%. Use either net present worth or equivalent uniform net annual return.

Plan A

Initial investment	$ 400,000
Expected life of equipment	5 yr
Salvage value after 5 yr	$ 100,000
Estimated cost of replacement of equipment after 5 yr	$ 520,000
Annual operation and maintenance costs	$ 25,000
Estimated annual income from equipment	$ 130,000

Plan B

Initial investment	$ 700,000
Expected life of equipment	10 yr
Salvage value after 10 yr	$ 100,000
Annual operation and maintenance costs	$ 30,000
Estimated annual income from equipment	$ 150,000

Plan C

Initial investment	$1,000,000
Expected life of equipment	10 yr
Salvage value after 10 yr	$ 150,000
Annual operation and maintenance costs	$ 50,000
Estimated annual income from equipment	$ 220,000

7-16. At an attractive rate of return of 10% a certain project has a net present worth of $3292, while at 15% the net present worth is −$1917. Calculate the approximate internal rate of return for this project.

7-17. You are the owner of a growing engineering firm that is in need of larger quarters. Three methods are available for financing the new building. They are:

Alternative 1   Lump sum payment of $250,000.

Alternative 2   Six equal year-end payments of $50,000.

Alternative 3 $50,000 down and $28,000/yr for 10 yr.

Find the most attractive alternative at rates of return of 8 and 12%.

7-18. An engineering firm is considering building a new office building that will accommodate the firm and provide additional income-producing rental space. The costs and benefits considered are:

Costs

Building	$850,000
Sitework	$300,000
Architect	$ 50,000
Engineering	$ 13,000
Legal services	$ 12,000
Land	$350,000
Financing costs	$100,000

Benefits

Rental income	$330,000/yr

(a) Assess the economic feasibility of this project if the expected life is 40 yr, and the attractive rate of return is 15%. The building is expected to have the same value in 40 yr that it does now, and the land is expected to triple in value.

(b) Is there a minimum attractive rate of return at which this project would not be economically feasible? If so, what is it?

(c) Are there any additional costs or benefits that should be considered in this problem? How might they affect the results?

7-19. A community is evaluating a set of projects that will collectively expand its water supply, drainage, flood protection, and wastewater collection systems. Benefit/cost analysis will be undertaken to evaluate the economic feasibility of the project set. Planners have identified the following possible benefits and disbenefits that should be considered in the analysis:

Benefits

Increased water supply

Increased protection from floods

Increase in quality of life

Disbenefits

Destruction of forest and wildlife

Degradation of downstream water quality

(a) Discuss how each of these benefits and disbenefits could be quantitatively estimated.

(b) Are there any of the listed benefits and disbenefits that should not be considered by the community in its analysis? If so, why?

(c) What other benefits and disbenefits should be considered, and how could they be quantified?

(d) Discuss the arguments for and against considering these projects as a set, rather than considering the economic feasibility of each project individually.

7-20. A chemical refinery has to replace approximately 795,000 ft of pipe. Three alternative types of pipe are being considered.

	Useful life (yr)	Initial cost ($/ft)	Annual maintenance cost ($/10,000 ft)	Installation cost ($)
Iron pipe	7	3	50	105,000
Iron alloy pipe	10	5	40	105,000
Chrome–iron alloy pipe	15	6	30	105,000

(a) If the established discount rate is 8%, determine the most attractive alternative.

(b) How would the solution change if pipe costs are expected to increase by 5%/yr, construction costs are expected to rise by 6%/yr, and maintenance costs are expected to increase by 4%/yr?

THE PLANNING/DESIGN PROCESS: IMPLEMENTATION

Chapter 8

Final Planning/Design and Implementation

INTRODUCTION

The preceding chapters of this book have brought the engineer through the various processes involved in preparing a problem for its systematic solution. By this point in a project, an engineering firm would have made a considerable investment in defining the appropriate problem to be solved, in the gathering of data pertinent to that problem and in the analysis of the competing alternatives intended to solve the problem. Recycling and feedback should have been accomplished whenever the conditions that warranted them developed.

The point in the process has now been reached where the engineer should pause and review the project. In order for any benefit to accrue from the preparations that have been completed, the bridge from preliminary planning to project implementation must be spanned. It is now important to undertake postoptimality analysis either to confirm any previous analyses, or to indicate where any changes in conditions might most influence the system's outputs. Not all changes in conditions will be unfavorable, but those that are should be carefully analyzed, and appropriate steps should be taken to minimize their impact. There may be changes in conditions that will produce little change in the output, so they can safely be neglected. Favorable changes in conditions should be noted and their effect on the output estimated.

Since few outcomes in real engineering problems are ever 100% certain, it is important for the engineer analyzing alternative systems with differing outputs to weigh the probability that a projected output would actually occur.

A decision must now be made concerning which of the alternative courses of action should be taken. Since it is often someone other than the engineer who has the authority for making the final decision, it is essential for the engineer to be able to communicate any pertinent information to this decision maker. Once the decision maker has chosen the alternative to be undertaken, the final planning and design can commence. Essentially, a blueprint for implementation must be devised, which includes detailed plans and specifications. Once this stage has begun, any feedback or recycling becomes time consuming and expensive. With proper planning and attention to detail, such feedback should be minimized; however, unforeseen occurrences or changing conditions could still warrant it. Throughout the implementation of the plan, a manager should carefully monitor the project to assure that its blueprint is being followed and that any changing conditions are being noted, with modifications being made to the plan to accommodate them.

The end product of this process should be a workable solution that meets the needs of or improves conditions for the groups affected by the original problem, without adversely affecting other groups.

POSTOPTIMALITY ANALYSIS

As part of the final decision-making process, the engineer should carefully review all previous analyses and interim decisions that have led to the present conditions. This is the last point where feedback can be accomplished without a significant penalty in either time or money. This is also the point where the decision must be made as to whether or not a workable solution to the problem exists. It is possible that even though a problem has been properly posed and defined, it can not be solved given the constraints on the system.

The engineer who is utilizing systems analysis tools to assist in the systems approach must be cognizant of their limitations, and not attempt to apply them to situations for which they are not valid. In such instances, either the engineer must use the best information available to undertake subjective optimization, or recognize the potential for applying more sophisticated techniques, such as nonlinear programming, dynamic programming, or stochastic programming. Utilization of these techniques at this stage would require feedback; but if they were needed, they would facilitate successful implementation after the reanalysis.

As the engineer reviews the application of the systems approach, particular attention should be paid to the mathematical models and techniques that have been used to add objectivity to the process. The overall goal of using these methods is to maximize the benefit resulting from any problem solution per unit of input toward that solution. However, the results from mathematical models and solution techniques are often subject to outside interpretation and skepticism. The engineer must be prepared to deal with any critics in a calm, objective, and professional manner. This can most easily be done if the engineer has anticipated and prepared for the questions most likely to arise.

To be prepared for these questions, the engineer must check and verify the results of all mathematical models used. This can be accomplished in a number of ways. The experienced engineer should have an intuitive feel for whether or not output is reasonable. A sample hand solution for one example of the problem at hand, or previously completed solutions to similar problems which are known to be correct, can be used to check model output. Simplified analyses can also be useful in establishing reasonable ranges for the model solution variable values.

If the solution variable values are not reasonable, the engineer must investigate several possible problem areas before proceeding with the systems approach. This is, in itself, a form of feedback. If a computer has been used to assist in the analysis, the transformation of the mathematical theory into program code must be checked. Data file entries are also possible sources of error. If a previously developed software package is being utilized, the theory upon which the model is based must be compared to the theoretical solution for the problem at hand. In this manner, it can be determined if they are sufficiently similar for comparative results.

It is also important to analyze the theoretical development of the mathematical model that has been used. In a deterministic or stochastic mathematical model, there must be a reasonable linkage between the dependent and independent variables; or, in the case of a linear programming problem, the linkage should be apparent between the constraints and the objective function. Relationships that may have seemed reasonable during the early stages of the problem may no longer appear that way because of the information that has been developed through subsequent analysis. It may even be found that the problem was improperly posed or defined originally. Care should be taken to ensure that the mathematical model that was used applies to the properly posed problem, and that the model input is not out of the range for which the model was developed and intended.

Data that have been utilized at various points in the application of the systems approach should be rechecked to ensure that they are of sufficient accuracy and precision for the analyses undertaken. Data may have been used to estimate model parameters; such as the objective function cost coefficients, the constraint right-hand-side stipulations, and the coefficients of the constraint variables in the linear programming model; and the coefficients and exponents of other linear or nonlinear mathematical models. Data may have also been used for input to simulation-type models. Sensitivity analysis, as described in the next section, can reveal which parameters and/or types of input data most affect model output. The data related to those parameters or input should be accorded extra attention in determining their accuracy and reliability. An engineer must be confident in the validity of data used in order to have confidence in the validity of model output.

Sensitivity Analysis

Sensitivity analysis has been introduced previously with respect to its application to mathematical models and to linear programming applications. Additional discussion will be presented here, with a detailed discussion for sensitivity analysis of linear programming models. It is important that model sensitivity be considered prior to the final formulation and implementation of a plan or design. Sensitivity analysis allows testing of the reactions of dependent variables to variation in model parameter and/or independent variable values. If the model output is sensitive to a particular perturbation (change in parameter or input variable value), care should be taken to ensure that estimates for the applicable parameters and/or independent variables are as accurate as possible. Lack of confidence in these estimates can indicate the need for recycle back to the problem definition, data gathering, or model formulation stages.

Sensitivity analysis for continuous functions can be undertaken by evaluating the derivative or partial derivatives in the vicinity of the expected values for independent variables. Parameter values can be perturbed, allowing the engineer to observe the resultant dependent variable reaction, or parameters can be treated as variables and derivatives can be taken. Complex system models that do not lend themselves to derivative analysis can be analyzed by perturbing independent variables and parameters in turn, while observing dependent variable reactions. An example of this sort was discussed in Chapter 3.

Sensitivity analysis of linear programming models demonstrates the sensitivity of the optimal solution to variation in the objective function coefficients, right-hand side stipulations, or constraint coefficients. It can be undertaken with information supplied by the optimum simplex tableau. The two most common types of sensitivity analysis are variation of the right-hand side constraint stipulations (b_i), or variation of the objective function cost coefficients (c_j). These two types will be discussed in detail. A third type of sensitivity analysis can be undertaken to evaluate the effects of changes in the constraint coefficients (a_{ij}). This analysis is more difficult, especially when several of the coefficients vary simultaneously. Discussions of constraint coefficient sensitivity analysis can be found in optimization texts such as *Foundations of Optimization* by Wilde and Beightler (1976).

Sensitivity analyses for the b_i and c_j parameters of the linear programming model should be considered together to assess their relative impact on model performance. Also, the process of sensitivity analysis depends on what type of model (minimization or maximization) is being analyzed. Therefore, the process is best described through examples. Equations 6.22 presented the following linear programming model:

$$\left. \begin{array}{l} \max Z = 6X_1 + 7X_2 \\ \text{ST:}\quad \text{constraint 1}\quad 2X_1 + 3X_2 \le 12 \\ \phantom{\text{ST:}}\quad \text{constraint 2}\quad 2X_1 + X_2 \le 8 \end{array} \right\} \qquad (8.1)$$

No units were assigned to this model in Chapter 6; however, it could easily represent a product mix problem. The right-hand side stipulations, $b_1 = 12$ and $b_2 = 8$, could represent the amount of resource of types 1 and 2 respectively, which are available for producing the product. To avoid confusion with the b_i values that appear in the final tableau, the original right-hand-side values will be referred to as RHS$_i$ for sensitivity analysis. Variables X_1 and X_2 represent the number of units of products 1 and 2 that will be produced, and the cost coefficients, $c_1 = 6$ and $c_2 = 7$, indicate how much income or profit can be realized by producing one unit of products 1 or 2. Constraint coefficients, a_{ij}, show how many units of a particular resource have to go toward one unit of a particular product. For example, $a_{12} = 3$ indicates that three units of resource 1 have to go into each unit of product produced.

Prior to finding the optimum solution for this system using the simplex algorithm, slack variables have to be added to make the constraints equalities. In Equations 8.2, slack variable X_3 represents the number of units of resource 1 that are not utilized, while X_4 represents the same for resource 2.

$$\left. \begin{array}{l} \max Z = 6X_1 + 7X_2 + 0X_3 + 0X_4 \\ \text{ST:} \qquad 2X_1 + 3X_2 + \ X_3 \qquad \qquad = 12 \\ \qquad \qquad 2X_1 + \ X_2 \qquad \quad + \ X_4 = 8 \end{array} \right\} \qquad (8.2)$$

The optimum solution after application of the simplex algorithm is given in Table 8.1. This is the same solution found in Table 6.4, with $X_2 = 2$, $X_1 = 3$, and the objective function value $= 32$. Neither of the slack variables is in the final optimal solution; however, both have negative $(c_j - Z_j)$ values. This can be interpreted to mean that there are no alternate optima (no $c_j - Z_j$ for variables not in the solution are zero), but there would be an improvement in the objective function value if the constraints associated with slack variables X_3 and X_4 could be relaxed.

To relax less than or equal constraints, the right-hand side stipulation would have to be increased. In this problem, additional units of resource would have to be located to relax the constraints. If the cost of the additional units of resource is different from the cost of the originally available units, or the selling price of additional units of product varies, then a new linear model has to be formulated, and a new optimum solution found.

If one additional unit of resource depicted by the first constraint (associated with X_3) could be acquired, the objective function value would be improved by two units, to 34. Similarly, one additional unit of resource 2 would improve the objective function value by one unit.

Sensitivity analysis of the constraint right-hand-side stipulations can be used to determine the range of RHS_i values for which this linear relationship between RHS_i and Z will hold true. The information required to undertake this sensitivity analysis is contained in the final (optimal) tableau. If the amount of resource one available is changed by a small amount, ΔRHS_1, it amounts to changing the corresponding slack variable, X_3, by an amount $-\Delta\text{RHS}_1$. It can be shown that such perturbations in the RHS_i will not change the optimality of a solution provided that the feasibility of the solution is maintained. The final tableau of Table 8.1 represents the following set of equations:

TABLE 8.1 OPTIMUM SOLUTION FOR MAXIMIZATION PROBLEM

c_j			0	0	6	7
	Solution variables	B	X_3	X_4	X_1	X_2
7	X_2	2	1/2	$-1/2$	0	1
6	X_1	3	$-1/4$	3/4	1	0
	Z_j	32	2	1	6	7
	$c_j - Z_j$		-2	-1	0	0

$$\left. \begin{array}{l} \tfrac{1}{2}X_3 - \tfrac{1}{2}X_4 \qquad\quad + X_2 = 2 \\ -\tfrac{1}{4}X_3 + \tfrac{3}{4}X_4 + X_1 \qquad = 3 \end{array} \right\} \tag{8.3}$$

When the perturbation $-\Delta\mathrm{RHS}_1$ is added to X_3, Equations 8.3 become

$$\left. \begin{array}{l} \tfrac{1}{2}X_3 - \tfrac{1}{2}\,\Delta\mathrm{RHS}_1 - \tfrac{1}{2}X_4 \qquad\quad + X_2 = 2 \\ -\tfrac{1}{4}X_3 + \tfrac{1}{4}\,\Delta\mathrm{RHS}_1 + \tfrac{3}{4}X_4 + X_1 \qquad = 3 \end{array} \right\} \tag{8.4}$$

Since X_3 and X_4 are not in the final solution, Equations 8.4 reduce to

$$\left. \begin{array}{l} -\tfrac{1}{2}\,\Delta\mathrm{RHS}_1 \qquad\quad + X_2 = 2 \\ \tfrac{1}{4}\,\Delta\mathrm{RHS}_1 + X_1 \qquad = 3 \end{array} \right\} \tag{8.5}$$

Variables X_1 and X_2 must remain positive if the solution is to remain feasible. From Equations 8.5, it can be seen that

$$\left. \begin{array}{l} X_2 < 0 \quad \text{if} \quad \Delta\mathrm{RHS}_1 \leq -4 \quad (|\Delta\mathrm{RHS}_1| \geq 4) \\ X_1 < 0 \quad \text{if} \quad \Delta\mathrm{RHS}_1 \geq 12 \end{array} \right. \tag{8.6}$$

and RHS_1 can vary in the range

$$\begin{array}{c} 12 - 4 \leq \mathrm{RHS}_1 \leq 12 + 12 \\ 8 \leq \mathrm{RHS}_1 \leq 24 \end{array} \tag{8.7}$$

and still maintain feasibility, and optimality of the solution. Each unit of RHS_1 within this range will add or subtract two units from the objective function value. It also must be understood that the values of X_1 and X_2 will also change as RHS_1 increases or decreases, but X_1 and X_2 will remain in the solution. For instance, when $\mathrm{RHS}_1 = 24$, $X_1 = 0$, $X_2 = 8$, and $Z = 56$, an increase of 24 units over the previous optimal value of 32.

Figure 8.1 shows the geometric interpretation of varying RHS_1. When $\mathrm{RHS}_1 > 24$, constraint 1 can be met by making X_3 enter the solution, and there will still be a feasible solution. The variables in the solution, however, will be different, and the linear change of Z with RHS_1 will no longer hold. For example, if RHS_1 is set equal to 30, X_2 will equal 8 and X_3 will equal 6 when the optimum solution is found. However, the value of Z for this optimal solution will be 56, which is the same value as when RHS_1 was set equal to 24. The reason is apparent from an examination of Figure 8.1. When $\mathrm{RHS}_1 > 24$, constraint 1 has no point of concurrence with constraint 2. Since both are $\leq$ constraints, constraint 2 governs for this situation, and X_2 cannot exceed 8.

Similar analysis for constraint 2 will give the following results:

$$\left. \begin{array}{l} \tfrac{1}{2}\,\Delta\mathrm{RHS}_2 \qquad\quad + X_2 = 2 \\ -\tfrac{3}{4}\,\Delta\mathrm{RHS}_2 + X_1 \qquad = 3 \end{array} \right\} \tag{8.8}$$

Therefore $\qquad\qquad \left. \begin{array}{l} X_2 < 0 \quad \text{if} \quad \Delta\mathrm{RHS}_2 \geq 4 \quad \text{and} \\ X_1 < 0 \quad \text{if} \quad \Delta\mathrm{RHS}_2 \leq -4 \end{array} \right\} \tag{8.9}$

The sensitivity range for RHS_2 is thus $4 \leq \mathrm{RHS}_2 \leq 12$, and each unit change in RHS_2 within this range will result in a one-unit change in the objective function value.

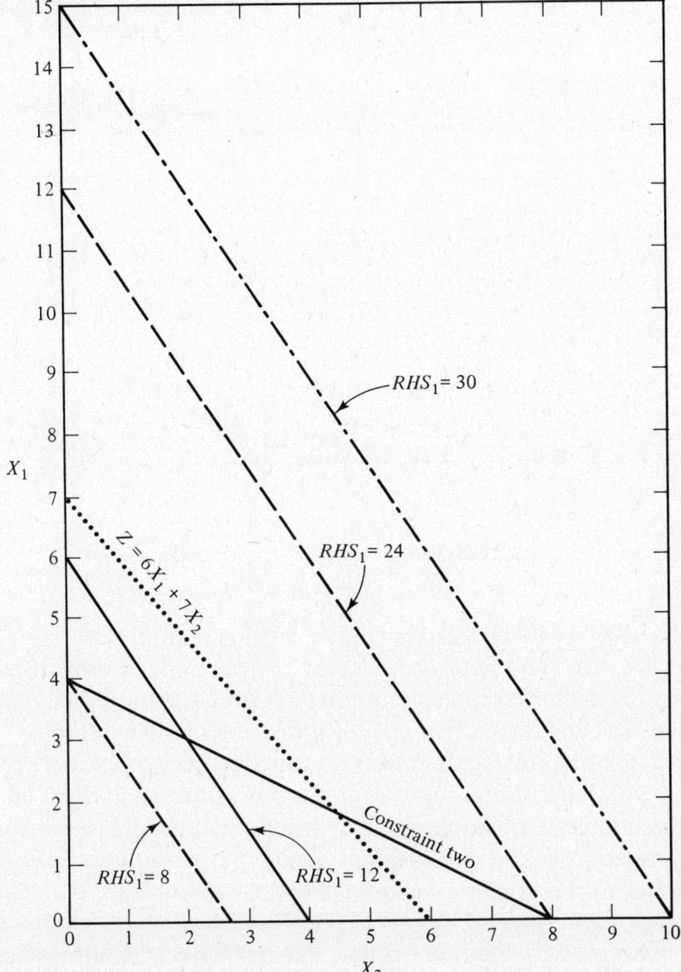

Figure 8.1 Variation of RHS_1.

It can be concluded from this sensitivity analysis that the best decision is to acquire additional units of resource 1, rather than resource 2. Resource 1 not only has a larger range of choices ($8 \leq RHS_1 \leq 24$), but, it also can realize a larger marginal gain. However, it should be remembered that other factors may enter into the decision that can not be included in the mathematical analysis.

A direct method for finding the sensitivity range for a right-hand side stipulation can be developed. The b_i value in each row of the final tableau is divided by the coefficient, a_{ij}, under the slack or surplus variable column corresponding to the constraint whose right-hand side stipulation sensitivity is being tested. The division is accomplished for all of the applicable a_{ij} values that are not equal to zero. In equation form

$$(\Delta RHS_i)_j = -\left(\frac{b_i}{a_{ij}}\right) \quad \text{for} \quad \leq \text{constraints}$$

$$(\Delta RHS_i)_j = \frac{b_i}{a_{ij}} \quad \text{for} \quad \geq \text{constraints} \qquad (8.10)$$

where i = the identification number for the row of the tableau for which ΔRHS_i is being calculated,

j = the variable indentification number corresponding to the slack or surplus variable for the constraint being tested, and

ΔRHS_i = the sensitivity quotient for row i of the tableau

The sensitivity range is given by the positive and negative quotients that are smallest in absolute value (closest to zero on either side). For example, if ΔRHS_i values of -10, 8, -3, 4, and 6 had been found for a problem with five constraints, the sensitivity range would be -3 to $+4$. Application of Equation 8.10 to Table 8.1 would give

$$(\Delta RHS_1)_3 = -\frac{b_1}{a_{13}} = -\frac{2}{1/2} = -4 \quad \text{and}$$

$$(\Delta RHS_2)_3 = -\frac{b_2}{a_{23}} = \frac{-3}{-1/4} = +12$$

which gives the same sensitivity range as was found using Equations 8.6 and 8.7. Similar analysis could be carried out for the second constraint.

Sensitivity analysis for the objective function cost coefficients (c_j) can also be undertaken with information supplied in the optimal tableau. Since all $c_j - Z_j$ values in the optimal tableau are nonpositive, it follows that the cost coefficient (c_j) for any variable (primary, slack, or surplus) not in the optimal solution can increase by an amount equal to the absolute value of the corresponding $c_j - Z_j$. In the optimal solution previously discussed, slack variables X_3 and X_4 are not in the solution. However, they have $c_j - Z_j$ values of -2 and -1, respectively. Cost coefficients for both X_3 and X_4 are presently set at zero because they are slack variables. The sensitivity ranges for their cost coefficients have little practical meaning in this case, unless a net return could be realized by not using units of some resource. For instance, if c_3 could be increased to 3, X_3 would enter the solution with $X_1 = 4.0$, $X_3 = 4.0$, and $Z = 36$, an increase of 4 over the previous objective function value. Cost coefficient sensitivity for variables, other than slack or surplus variables, which are not in the final solution will give an estimate of the relative change in c_j required to bring X_j into the solution. Careful attention should be paid to accurate estimation of cost coefficients that can change the final solution with only small changes in c_j.

Changing the objective function cost coefficient for variables in the solution can have an effect on several of the $c_j - Z_j$ values. For instance, if c_1 was changed from 6 to 8, then $c_3 - Z_3$ in the optimum tableau would change from -2 to -1.5, and $c_4 - Z_4$ would go from -1 to -2.5. Provided that these changes do not produce any positive $c_j - Z_j$ values, the solution will remain optimal and the same variables will remain in the solution at a constant value; however, the value of the objective function will vary because of the variation in the cost coefficients, c_j. If any of the $c_j - Z_j$ do become positive, then the objective function could be improved by bringing those variables with positive $c_j - Z_j$ into the solution.

Therefore, the limiting case for either increasing or decreasing a particular cost coefficient is when one of the $c_j - Z_j$ values is driven to zero. Increasing a cost coefficient will cause $c_j - Z_j$ values corresponding to negative a_{ij} coefficients to approach zero. Decreasing a cost coefficient will cause $c_j - Z_j$ values corresponding to positive a_{ij} coefficients to approach zero. If the variable whose cost coefficient is being perturbed is identified as X_k, then the limiting case for the change in c_k, called Δc_k, can be found by dividing the appropriate $c_j - Z_j$ values by the corresponding a_{ij} values. For this analysis, the subscript k will identify the variable whose cost coefficient sensitivity is being investigated. Thus, it will identify the row of the tableau to be used; however, it is independent of the i value usually associated with a row of the tableau. For example, to find the permissible range of c_1 for the problem described by Table 8.1, X_1 is associated with the second row, and $i = 2$ while $k = 1$. In equation form,

$$\left.\begin{aligned}
\text{negative } \Delta c_k &= \left| \frac{(c_j - Z_j)}{a_{ij}} \right| \quad \text{for} \quad \min a_{ij} > 0 \\
\text{and} \qquad \text{positive } \Delta c_k &= \left| \frac{(c_j - Z_j)}{a_{ij}} \right| \quad \text{for} \quad \min |a_{ij} < 0|
\end{aligned}\right\} \tag{8.11}$$

The $\min|a_{ij} < 0|$ for positive Δc_k means that the negative a_{ij} value closest to zero should be chosen.

In Table 8.1, c_1 is presently set at 6. Looking across the row corresponding to X_1 in the tableau, $a_{23} = -\frac{1}{4}$, $a_{24} = \frac{3}{4}$, $a_{21} = 1$, and $a_{22} = 0$. Coefficient a_{22} is not considered because it is zero. This will always be the case for variables in the solution other than the variable being tested. However, a_{ij} may also be zero for variables not in the solution. Coefficient a_{21} is also not considered, because it corresponds with X_1 and will provide no useful information. The remaining a_{ij} values can be considered for the analysis. The minimum $a_{ij} > 0$ is $\frac{3}{4}$ for a_{24}, so

$$\text{negative } \Delta c_1 = \left| \frac{c_4 - Z_4}{a_{24}} \right| = \left| \frac{-1}{3/4} \right| = \frac{4}{3}$$

similarly, $\min|a_{ij} < 0|$ is $\frac{1}{4}$ for a_{23}, and

$$\text{positive } \Delta c_1 = \left| \frac{c_3 - Z_3}{a_{13}} \right| = \left| \frac{-2}{-1/4} \right| = 8$$

Therefore, $\qquad\qquad\qquad\qquad -\frac{4}{3} \leq \Delta c_1 \leq 8 \tag{8.12}$

This means that c_1 can vary between 4.67 and 14 without changing variables (or their values) in the optimal solution. However, the objective function value will change. Similarly, for c_2

$$-4 \leq \Delta c_2 \leq 2 \tag{8.13}$$

and c_2 can vary within the limits of 3 and 9 without changing the optimal solution.

Figure 8.2 shows the relative sensitivity ranges for the profits from products 1

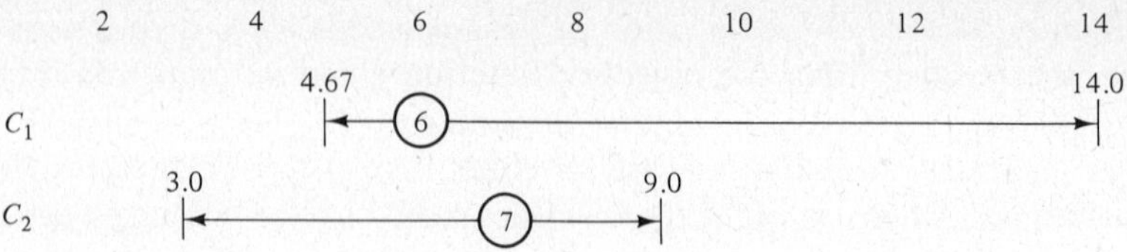

Figure 8.2 Sensitivity ranges for cost coefficients.

and 2. Product 1 is less sensitive to changes in c_1 on the high side of the present value, while product 2 is less sensitive to changes on the low side. Figure 8.3 shows the effect of changing cost coefficients on the slope of the objective function. In the limiting cases, the objective function coincides with one of the constraints. At this point, alternate optima exist. It can be seen that if Δc_j for the cost coefficient being varied is increased further, the optimum solution would switch to either point a or point b.

The sensitivity analysis for larger systems would proceed in the same manner; however, the relationships could not be visualized as they were for the two-dimensional case. The procedures developed above will now be applied to a minimization model with all three types of constraints. This is the second model used to illustrate the simplex algorithm in Chapter 6. Equations 6.28 are repeated here to define the model. Table 8.2 shows the optimal tableau, with X_3 being the slack variable for constraint

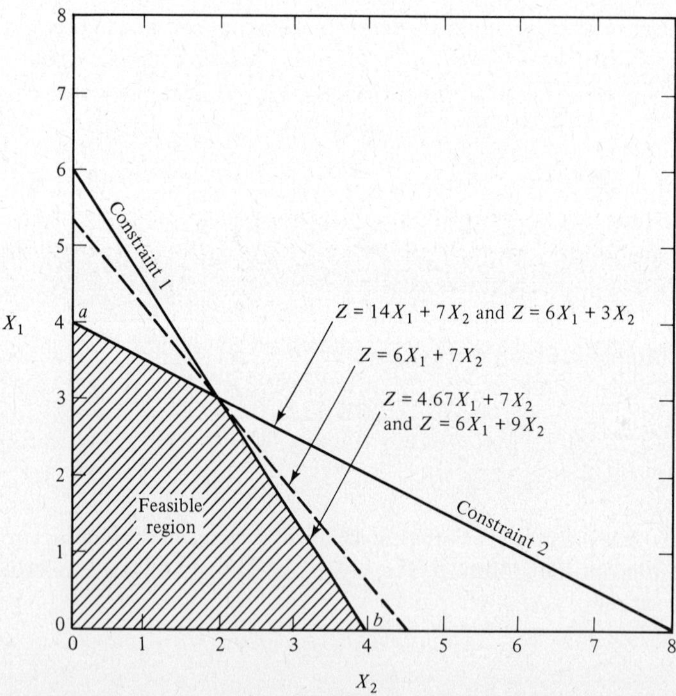

Figure 8.3 Variation of objective function cost coefficients.

TABLE 8.2 OPTIMAL TABLEAU

c_j			0	$-M$	$-M$	-2	-8	0
	Solution mix	B	X_3	X_5	X_6	X_1	X_2	X_4
0	X_3	18	1	2	$-2/10$	0	0	-2
-8	X_2	14	0	1	0	0	1	-1
-2	X_1	2	0	-2	$2/10$	1	0	2
	Z_j	-116	0	-4	$-4/10$	-2	-8	4
	$c_j - Z_j$		0	$-M + 4$	$-M + 4/10$	0	0	-4

1 and X_4 the surplus variable for constraint 2. Variables X_5 and X_6 are artificial variables for constraints 2 and 3 respectively.

$$\min Z = 2X_1 + 8X_2$$

$$
\left.
\begin{array}{l}
X_1 = \text{number of truckloads of mix 1 used} \\
X_2 = \text{number of truckloads of mix 2 used} \\
\text{ST:} \quad X_1 \le 20 \text{ (upper limit on use of mix 1)} \\
\quad\quad\quad X_2 \ge 14 \text{ (lower limit on use of mix 2)} \\
\quad\quad\quad 5X_1 + 10X_2 = 150 \text{ yd}^3 \text{ of gravel required}
\end{array}
\right\} \quad (8.14)
$$

The artificial variables, X_5 and X_6, can be ignored in the postoptimality analysis. Primary variables in the final solution are $X_1 = 2$, $X_2 = 14$, while slack variable $X_3 = 18$ is also in the solution, indicating that 18 units of mix 1 were not utilized. Surplus variable X_4 is not in the solution; however, it has a $c_j - Z_j$ value of -4, indicating that the optimal objective functional value of 116 could be improved by 4 if constraint 2 could be relaxed by one unit. Since constraint 2 is a $\ge$ constraint, relaxing the constraints means reducing the requirement for mix 2 to 13 units. Also, because this is a minimization problem, the objective function value would be reduced to 112 with a one-unit relaxation of constraint 2.

Perturbing a greater than or equal constraint by an amount ΔRHS_i is the same as adding the same amount to the corresponding surplus variable. Incorporating ΔRHS_2 into the equations represented by Table 8.2, while eliminating the artificial variables, results in the following system:

$$
\left.
\begin{array}{rcl}
X_3 \quad\quad - 2X_4 - 2\,\Delta \text{RHS}_2 &=& 18 \\
X_2 - X_4 - \Delta \text{RHS}_2 &=& 14 \\
X_1 \quad + 2X_4 + 2\,\Delta \text{RHS}_2 &=& 2
\end{array}
\right\} \quad (8.15)
$$

Since $X_4 = 0$ (is not in the solution),

$$\left.\begin{array}{llll} X_3 < 0 & \text{if} & \Delta\text{RHS}_2 \leq -9 \ (|\Delta\text{RHS}_2| \geq 9) \\ X_2 < 0 & \text{if} & \Delta\text{RHS}_2 \leq -14 \ (|\Delta\text{RHS}_2| \geq 14) \\ X_1 < 0 & \text{if} & \Delta\text{RHS}_2 \geq 1 \end{array}\right\} \qquad (8.16)$$

Taking the minimum absolute positive and negative ΔRHS_2, the range of the constraint 2 stipulation becomes

$$14 - 9 \leq \text{RHS}_2 \leq 14 + 1$$
$$5 \leq \text{RHS}_2 \leq 15 \qquad (8.17)$$

and each unit change in RHS_2 between these limits will change the objective function value by four units. If RHS_2 could be lowered to 5, the objective function would have a value of 80, with $X_1 = 20.0$, and $X_2 = 5.0$. Equation 8.10 could be used to confirm these results.

It would appear that insufficient information is present in the optimal tableau to conduct a sensitivity analysis of constraint 3, the equality constraint. A slight alteration of the model would provide this information. If constraint 3 is changed into a greater than or equal constraint, the optimal solution will not change. Since the objective function calls for minimization, the delivery of extra gravel would be prevented. Sensitivity analysis would indicate that within the range $140 \leq \text{RHS}_3 \leq 240$, each unit of constraint 3 will change the objective function value by 0.4 units. This shadow price of 0.4 units could be verified by solving the new system with the modified constraint 3 by the simplex algorithm. However, there is actually enough information available in the final tableau to provide all of these answers. Even though the artificial variables are not included in the sensitivity analysis, the a_{ij} coefficients for X_6, the artificial variable associated with constraint 3, are the negative of what the coefficients would be for the surplus variable. Also, the real part of the $c_j - Z_j$ value for X_6 ($\frac{4}{10}$) is the negative of what the $c_j - Z_j$ value would be for the surplus variable. Thus, the range of RHS_3 values would be

$$150 - \frac{2}{2/10} \leq \text{RHS}_3 \leq 150 + \frac{18}{2/10} \qquad (8.18)$$

or
$$140 \leq \text{RHS}_3 \leq 240$$

which is the same result found with the modified model. A more general approach to the treatment of equality constraints is to replace each one with two constraints, one $\leq$ and one $\geq$. Carrying this modified model through the simplex algorithm will produce sufficient information to conduct all desired sensitivity analyses.

Sensitivity analysis for the cost coefficients can now be undertaken. Surplus variable X_4 is not in the final solution. Examination of the tableau indicates that X_4 would enter the solution if it had a $c_4 = 4$. However, in the real model, this would correspond to a -4, due to the sign inversion for minimization. A negative cost coefficient would require that money could actually be saved by using excess mix 2, which would be unlikely for the problem depicted by this model.

Inspection of the model shows that c_2 could increase without bound, and not change the solution variables. However, if c_2 decreased to less than 4, it would be more economical to provide 15 truckloads of mix 2. Cost coefficient c_2 is presently 8.0. The range, $4 \le c_2 \le \infty$ corresponds to $-4 \le \Delta c_2 \le \infty$. Comparison of this range with line 2 of the tableau shows that the a_{ij} values considered in each part of Equation 8.11 must be reversed for minimization problems. Values of $a_{ij} < 0$ are used on the left side, and $a_{ij} > 0$ are used on the right. Therefore, for minimization problems

$$
\left.
\begin{array}{ll}
\text{negative } \Delta c_k = \left| \dfrac{c_j - Z_j}{a_{ij}} \right| & \text{for} \quad \min |a_{ij} < 0| \\[1em]
\text{and} \qquad \text{positive } \Delta c_k = \left| \dfrac{c_j - Z_j}{a_{ij}} \right| & \text{for} \quad \min a_{ij} > 0
\end{array}
\right\}
\tag{8.19}
$$

From Equation 8.19, the value c_3 for slack variable X_3 can vary between the limits $-2 \le c_3 \le \infty$, without changing the solution. To evaluate the sensitivity range for c_1, it is best to consider the augmented model with constraints 3 transformed into a greater than or equal constraint. Otherwise, only an upper bound limitation on Δc_1 will be evident from the optimal tableau. Using Equation 8.19 and the expanded solution, $-2 \le \Delta c_2 \le 2$ is calculated, and the range of c_1 is $0 \le c_1 \le 4$. If c_1 increased beyond 4, one more unit of mix 2 will be utilized, with no units of mix 1.

All of the foregoing analysis indicates that this is a rather restrictive model, with little latitude or range of choice. The effects of the various changes in the constraint stipulations and the objective function cost coefficients could also be illustrated through graphs similar to Figures 8.1 and 8.3.

Table 8.3 presents a compilation of the various formulas and decision rules for

---

TABLE 8.3   SUMMARY OF SENSITIVITY ANALYSIS RULES

Constraint right-hand-side stipulations

$$
(\Delta \text{RHS}_i)_j = -\left(\frac{b_i}{a_{ij}} \right) \quad \text{for} \quad \le \text{ constraints}
\qquad
(\Delta \text{RHS}_i)_j = \frac{b_i}{a_{ij}} \quad \text{for} \quad \ge \text{ constraints}
$$

where $i =$ the identification number for the row of the tableau for which ΔRHS_i is being calculated,
$j =$ the variable identification number corresponding to the slack or surplus variable for the constraint being tested, and
$\Delta \text{RHS}_i =$ the sensitivity quotient for row i of the tableau

The sensitivity range is given by the positive and negative quotients that are smallest in absolute value (closest to zero on either side).

Cost coefficients: maximization problem

$$
\text{negative } \Delta c_k = \left| \frac{(c_j - Z_j)}{a_{ij}} \right| \quad \text{for} \quad \min a_{ij} > 0
$$

$$
\text{positive } \Delta c_k = \left| \frac{(c_j - Z_j)}{a_{ij}} \right| \quad \text{for} \quad \min |a_{ij} < 0|
$$

$\min a_{ij} < 0$ means that the negative a_{ij} value closest to zero should be chosen.

Cost coefficients: minimization problem

$$
\text{negative } \Delta c_k = \left| \frac{c_j - Z_j}{a_{ij}} \right| \quad \text{for} \quad \min |a_{ij} < 0|
$$

$$
\text{positive } \Delta c_k = \left| \frac{c_j - Z_j}{a_{ij}} \right| \quad \text{for} \quad \min a_{ij} > 0
$$

---

sensitivity analysis. Most computer packages for linear programming solutions provide options for sensitivity analysis.

CHOICE OF ALTERNATIVE

After completion of postoptimality analysis, it is important for the engineer to check again for the need to recycle to one of the previous steps. Additional knowledge may have been gained through the analysis, or the lack of knowledge about some key aspect may have become evident. Changes in baseline conditions may also have been recognized. The whole systems approach must be treated as an adaptive or learning process in which newly acquired data permit the engineer to refine the system and adapt to a changing environment. When the engineer is sufficiently satisfied that the learning process has progressed to an acceptable level, the alternative choices can be presented to the decision maker.

The engineer must now assemble all available objective information concerning the viable alternatives and present this information to the decision maker in a concise, understandable, format. When there is a range of alternatives, each viable, but each carrying a different amount of risk or uncertainty, the engineer should supply information concerning the advisability of choosing one over the other. The various methods which assist in decision making under risk and uncertainty will be presented in Chapter 10. These methods will aid the engineer in preparing the final presentation.

Although the final results must be presented in the most objective manner possible, the engineer should be prepared for the actual decision to be made on more subjective grounds. If the engineer presents only those alternatives which would be acceptable or feasible and which could each be implemented with confidence in the final results, then, a choice based on political considerations or personal preference would still be viable.

IMPLEMENTATION

In order for a good plan or design to realize its goals, meet the needs it is designed to meet, and have a high probability of maximizing benefits, it must be properly implemented. The improper implementation of even the most carefully planned project can result not only in the loss of any anticipated benefits, but, it can also prove to be detrimental to the group that it was meant to benefit.

Final planning and design can begin as soon as the decision maker chooses one of the alternatives presented by the engineer. In the case of a design problem such as a high rise building or a freeway interchange, final design would involve detailed design of all subsystems and their coordination to make the overall system function as intended. Although a considerable amount of preliminary design work has already been completed, if the rule of least commitment was followed during the preliminary analysis stages, detailed design and development of material and construction specifications should not begin until a decision has been made concerning which alternative should be implemented.

Implementation of a design project will also involve considerable planning. Implementation schedules have to be formulated, and plans made for material acquisition. Equipment and personnel should be scheduled so that enough of each is present when needed, but excesses are avoided at other times.

Scheduling of work and construction activities, in conjunction with equipment and personnel resources, and keeping track of project status during implementation are best accomplished using network models such as the Critical Path Method (CPM) or the Program Evaluation and Review Technique (PERT). Both of these methods will be discussed in detail in Chapter 9.

The implementation of both plans and designs involves similar strategy. In implementing plans, however, the development of institutional systems may replace the construction of physical systems. There may be more political play involved in implementing a plan, and the benefit resulting from the implementation of the system may be more difficult to measure. If the engineer carefully defines the measures of effectiveness as part of the implementation of the plan, misunderstanding of the final results should be minimized. The implementation of a plan frequently leads to the design of physical subsystems, and, thus, to a corresponding application of the systems approach to those subsystems.

SUMMARY

If the engineer has carefully followed the systems approach, applying systems analysis techniques where appropriate, various viable alternatives should now be available for presentation to the decision maker. The engineer should have gained important information through postoptimality, sensitivity, and decision analyses, and should have narrowed the list of viable alternatives to those that are most nearly optimal. These alternatives, together with any objective information needed to evaluate their relative merits, should provide the decision maker with sufficient basis for making a final decision.

However, the engineer should be prepared to have the final decision made more on a basis of political, or other subjective, constraints. If the engineer has taken care to present only feasible alternatives, then the chosen alternative will still be implementable, even though it may not technically be the optimum alternative.

The implementation of plans or designs is one of the most difficult steps involved in meeting the goals developed in the problem definition stage. A great deal of management skill and attention to detail must be applied. Judicious application of systems analysis management and project planning tools will be of considerable assistance toward successful implementation.

APPLICATIONS EXERCISES

8-1. Change the third constraint of Equations 8.14 into a $\geq$ constraint, and resolve the problem to show that the sensitivity analysis described for that constraint is correct.

8-2. Demonstrate graphically the complete sensitivity analysis undertaken for Equations 8.14. Explain what is illustrated by each graph.

8-3. Accomplish sensitivity analysis for assigned problems from Chapter 6. Explain the meaning and implications of all aspects of the sensitivity analysis.

8-4. Given the following linear programming model:

$$\min Z = 1000X_1 + 800X_2$$

$$\begin{aligned}
\text{ST:} \quad X_1 & & & \geq 10 \\
0.8X_1 + & X_2 & & \geq 16 \\
X_1 & & & \leq 20 \\
& X_2 & & \leq 14
\end{aligned}$$

After addition of slack, surplus, and artificial variables, the model becomes

$$\min Z = 1000X_1 + 800X_2 + 0X_3 + MX_4 + 0X_5 + MX_6 + 0X_7 + 0X_8$$

$$\begin{aligned}
\text{ST:} \quad X_1 & & - X_3 + & X_4 & & & & = 10 \\
0.8X_1 + & X_2 & & & - X_5 + & X_6 & & = 16 \\
X_1 & & & & & & + X_7 & = 20 \\
& X_2 & & & & & + X_8 & = 14
\end{aligned}$$

The final, optimal, tableau for this model is

c_j			-1000	-800	0	^-M	0	$-M$	0	0
	Solution variables	B	X_1	X_2	X_3	X_4	X_5	X_6	X_7	X_8
-1000	X_1	10	1	0	-1	1	0	0	0	0
-800	X_2	8	0	1	0.8	-0.8	-1	1	0	0
0	X_7	10	0	0	1	-1	0	0	1	0
0	X_8	6	0	0	-0.8	0.8	1	-1	0	1
	Z_j	$-16,400$	-1000	-800	360	-360	800	-800	0	0
	$c_j - Z_j$		0	0	-360	$-M$ $+360$	-800	$-M$ $+800$	0	0

Do a complete sensitivity analysis for the right-hand side constraint values, and the cost coefficients for variables X_1 and X_2. Write descriptive statements for the sensitivity analysis. Illustrate your analysis graphically.

Management Techniques

INTRODUCTION

This chapter introduces the student to some effective techniques that can be used in managing and controlling many types of engineering projects and studies. These techniques are derived from systems concepts and should be in the repertoire of any project manager.

The size and complexity of many contemporary projects dictate the necessity for these project management techniques. No longer can the manager mentally keep track of project status and resource management. As more emphasis is placed on bringing projects in on time and within budgetary constraints, close control of project progression becomes essential.

Before a project begins, a manager must be able to schedule and sequence the various project activities in order to optimally utilize available resources and to finish the project in a reasonable amount of time. It is necessary to establish certain controls that will enable the manager to assess whether a project is behind, on, or ahead of schedule. It is equally as important to establish an information system which will facilitate the gathering of information necessary to evaluate the project status at any given moment.

During execution of the project, project status must be monitored, and frequently updated estimates made of the estimated project completion time. If any problems arise, corrective action must be initiated to ensure smooth progression of the project. The objective of these measures is to maximize project productivity by minimizing costly delays and idleness of labor and equipment.

Labor, equipment, materials, cash, and information are the project resources that the project manager is concerned with. The availability of required labor, equipment, and materials is a necessary physical prerequisite before any activity on the project can begin. Cash resources will determine, within permissible bounds, how activities should be scheduled, and may also determine how quickly and effectively unforeseen difficulties can be overcome. The information resource (gathered through the information system) is what is necessary to monitor and adjust the progress of the project.

Program Evaluation and Review Technique (PERT) and the Critical Path Method (CPM) are two management techniques that, when used properly, can assist the manager in efficiently utilizing resources toward the completion of a project. PERT was developed for the Navy in 1957 to help in management of the Polaris missile project. Three thousand different contracting agencies were working on this project, and PERT was used for planning, scheduling, and controlling its many diverse activities. As a result, this project was completed ahead of schedule, whereas, previous Navy contract projects had experienced substantial delays. CPM was developed at about the same time by E.I. DuPont Co., to meet its construction project management needs. Although the methods were developed independently, they are surprisingly similar. The major difference is in the determination of activity duration estimates. CPM utilizes a single expected completion time, while PERT time estimates are based on a weighted average of the most optimistic completion time, most probable completion time, and most pessimistic completion time. The latter method allows the estimation of the variance of activity completion times. Certain probability inferences can be made from these, as will be demonstrated in Chapter 10. Development of the method in the present chapter will utilize a single time estimate, and will be referred to as PERT/CPM.

PERT/CPM is most effective when used to control nonrepetitive projects. Through proper application of PERT/CPM, the manager can:

1. Define all activities pertinent to the project.
2. Define all dependencies among these project activities.
3. Schedule the occurrence of all events as well as the performance of all activities.
4. Determine all critical events and activities (those which control target dates).
5. Analyze, modify, and refine the work plan.
6. Allocate resources in an optimal manner.
7. Determine project status at any point during execution.
8. Establish optimal time–cost trade-offs.

The term "events" as it is used above refers to certain milestones in the project, or to the completion of a series of key activities. The activities themselves are the actual work tasks associated with the project.

Any project manager should bring a considerable amount of experience to the job. The PERT/CPM method can augment, but not replace, that experience. The method will allow the manager to organize more effectively and will facilitate management by exception, allowing the manager to concentrate on the critical activities of the project that govern project completion time. The actual mechanics of calculation in PERT/CPM are relatively simple. It is the ability to properly define activities and activity dependencies which is most difficult and which requires the expertise of an experienced manager.

Both the definition of activities and the determination of activity interdependencies are usually best accomplished by a group. Once the activities and interdependencies are established, a directed linear graph network can be developed which depicts the interdependencies and project flow. The project completion time will be controlled by the longest path through this network. This path is called the critical path. Every network must have one or more critical paths. There is no requirement that the critical path be unique; however, all critical paths for a given network will have the same completion time. Finding the critical path is a specialized assignment problem, which is in turn a specialized transportation problem, which is in turn a specialized linear programming problem. Therefore, this longest path problem could be formulated as a linear programming problem and solved by the simplex algorithm. However, because of the special characteristics of the directed path network, it is much more efficient to use a combinatorial approach to find the longest path or paths. Longest path problems could also be solved through dynamic programming, which will be discussed in Chapter 11.

The manager's task is not finished after the network has been defined and the critical path found. The good manager will study the network to find places where it could be changed to improve project work flow or to reduce overall project duration. Resolving resource conflicts, interchanging resources, and changing arrangement of activities are three types of network replanning and adjustment.

As the project progresses, the manager must update the network by introducing the latest available information into it. This will facilitate the planning and programming of the remaining portion of the project. If problems arise that may prevent completion of the project on time, it may be necessary to compress the length of time required to complete certain activities on the critical path. Although this can generally be done by supplying additional resources to an activity, doing so will likely incur an economic penalty. Time–cost optimization will indicate the most cost-efficient way to compress the project duration by a given amount. Each of these PERT/CPM concepts will be discussed in detail.

There has been reluctance on the part of some managers to use PERT/CPM. This probably stems from their failure to recognize the true simplicity of the method and their misconception that scheduling specialists and computers are necessary for its use. They have continued to attempt to plan and control their projects with bar (Gantt) charts, or have delegated the planning, scheduling, and controlling of project activities to a technician and a computer. This is unfortunate because bar charts do not adequately depict the interdependencies among the project activities, and because schedules which do not reflect the thinking of those actually engaged in carrying out the activities can seldom be followed. It is hoped that through increased understanding, most project managers will see the value of using PERT/CPM concepts.

PERT/CPM CONCEPTS

Activity Definition

The activities that make up a project must be carefully defined before any management strategy can be formulated. Activities are work tasks that utilize some of the types of resources available to the project. The number of activities defined will depend on the size of the project, the level of detail required, and the intended management use of the model developed from these activities. In a large project, there may be a general model of overall activities, while each of these general activities may be further broken up into networks of more detailed activities.

The overall model would be useful to higher level managers, while the more detailed models would be useful for the managers responsible for those levels of project accomplishment. Continuation of activity on the overall network would be controlled by completion of work on the appropriate predecessor subnetworks.

In any case, the definition of activities is an exacting process, which is best done through a group effort. Use of a work breakdown schedule may assist in defining the relevant activities that would have to be efficiently scheduled to ensure project completion in a reasonable amount of time. The work breakdown schedule starts with the completion of the project and works backward to define ever more detailed activities that contribute to project completion. It acts as a hierarchy of activities, but it does not explicitly point out where interactions and dependencies occur among activities. The definition of those are left up to the judgment and experience of the project managers. Figure 9.1 illustrates such a work breakdown schedule.

It is also necessary to make time estimates for activity durations. CPM analysis is done using a single estimate for activity duration. This estimate is based on experience, judgment, and possibly on unit production factors from a handbook. The best estimates of activity duration can usually be obtained from the individuals responsible for performing the work effort on the activity. However, as will be discussed later, there are many exogenous factors, which must be taken into account in the management strategy, and which can influence the actual activity duration. At this point a best initial estimate is satisfactory. PERT uses three time estimates. These will be discussed in a later chapter.

Along with the time estimate, the project planner must evaluate and enumerate the resources (material, equipment, labor, finance, etc.) that are necessary to complete each activity. The technological methods to accomplish each activity have to be selected, and the necessary conditions for beginning each activity must be determined.

Activity Interactions

Once individual activities are defined, they have to be combined into a coherent structure that can be used in management decision making. Establishing the project objective is the first step in developing the management model. This objective is the final event, or milestone, and all branches of the project network model have to converge to this point. A project starting event, from which all branches of the network model will originate, can also be defined. To help define interactions among the various activities in a project, it is useful to establish immediate predecessor and successor activities for each activity on the work breakdown schedule. Predecessor activities must be completed before the activity in question can be started. Successor activities are those activities that depend on completion of the activity in question. Also, activities that could possibly be undertaken concurrently with the activity in question should be identified.

The interactions and interdependencies among activities will determine their relative position within the project work flow. This coordination of the timing of

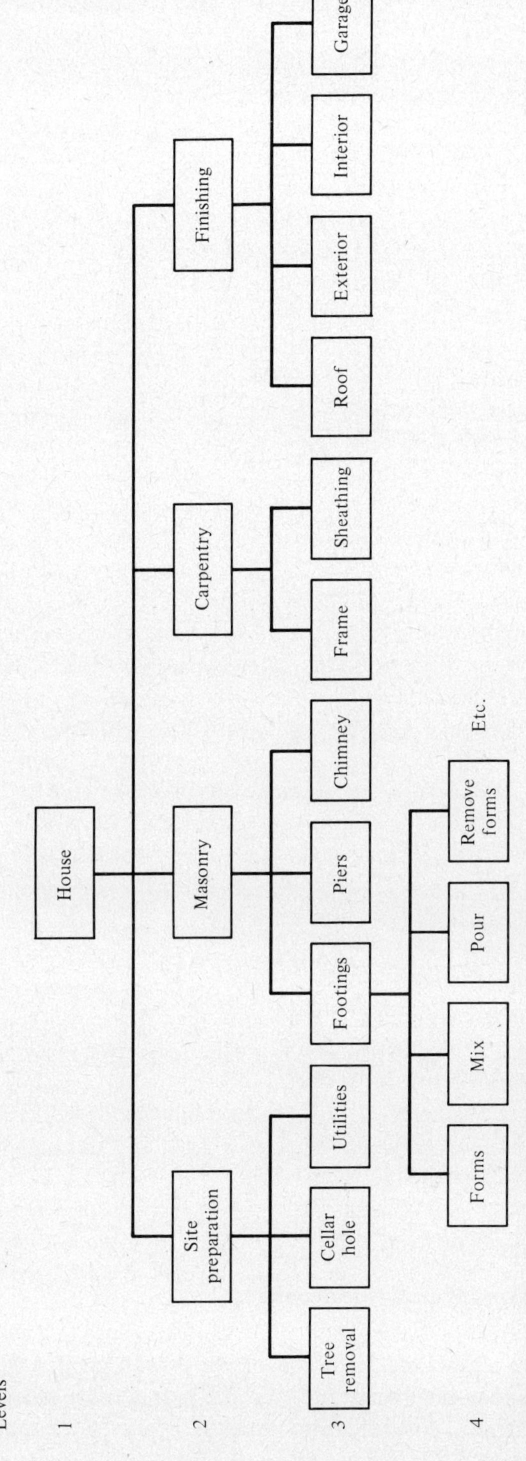

Figure 9.1 Work breakdown schedule for house construction.

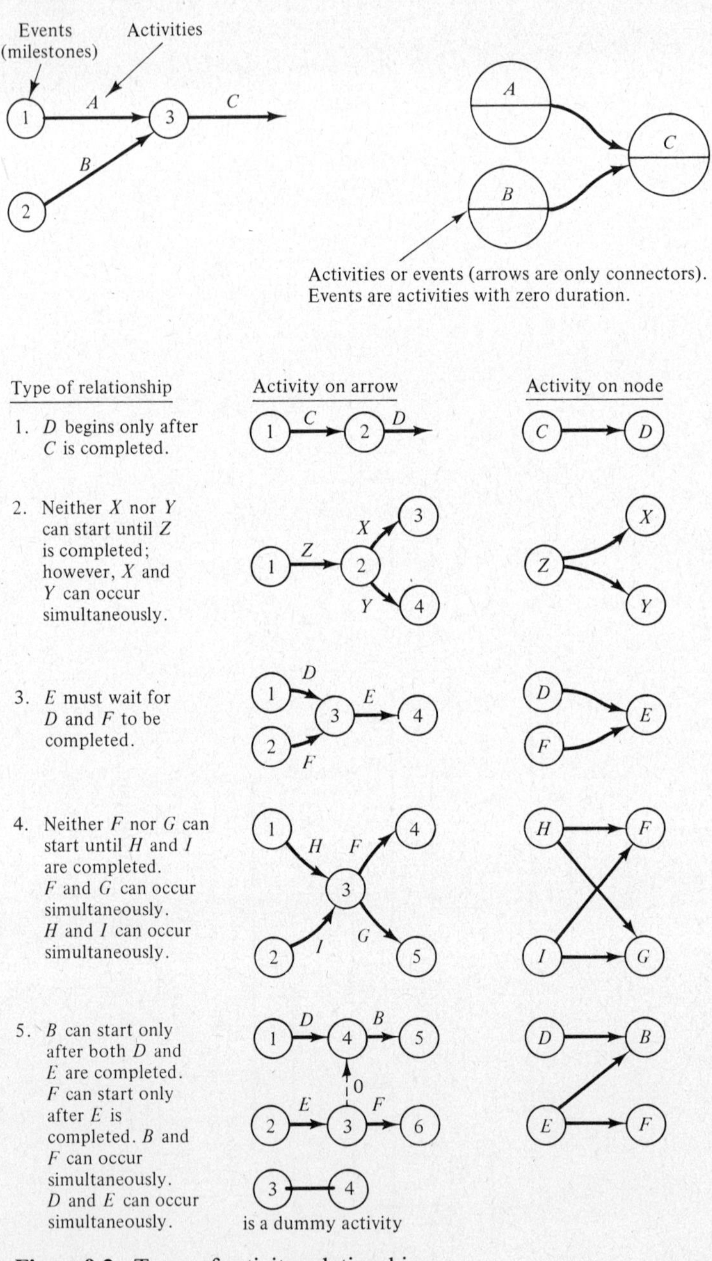

Figure 9.2   Types of activity relationships.

activities will allow determination of the overall project time. The scheduling order for activities may not be unique, and the best activity sequencing may only be determined after careful consideration of the alternatives. Therefore, this initial scheduling

6. Series to parallel conversion

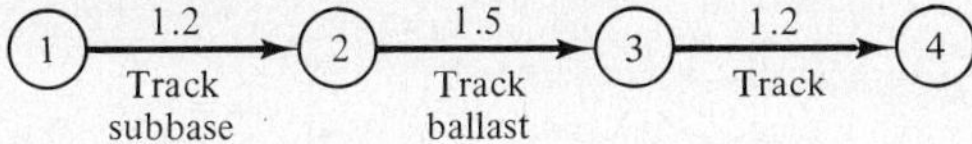

Three series-connected activities are shown below:

However, track ballast can begin to be laid 0.3 units of time
after the track subbase is started, and track can begin to be
laid 0.4 units after ballast is started.
Therefore, a better representation of the relationship
between these activities would be:

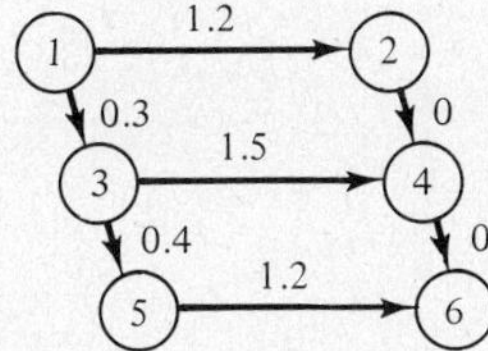

Figure 9.2 *(Continued)*

model should be considered as a feasible solution, but it may not be the optimum
feasible solution.

Proper sequencing of activities leads to the directed path network models common
to PERT/CPM analysis. These directed path networks are helpful in scheduling the
procurement and efficient disbursement of the resources necessary to complete a
project.

Two types of networks, which will each depict the overall relationships among
activities, can be found. Activity on arrow networks represent the activities as directed
arrows. The nodes that join the arrows together depict events or milestones in the
network flow. Activity on node diagrams depict both activities and events as nodes,
while the arrows indicate connectivity of project flow among activities. Figure 9.2
shows examples of activity relationships and how they would be represented under
both schemes. Note the use of a dummy activity in the fifth example. Dummy activities
have zero duration, and they are used to establish predecessor–successor relationships
when it would be incorrect to bring the preceding activity directly into the node (node
4, in this case). Activity E cannot end at node 4 because activity F is only dependent
on the completion of activity E, not both E and D.

It is important for users of PERT/CPM to understand both types of networks,
because both are used in practice. The activity on arrow network is slightly easier to
solve by hand, while the activity on node network provides added information. With
practice, one type of network can easily be transformed into the other type. It should
also be noted that many computerized PERT/CPM applications packages accept input
as activity on arrow, then provide output as activity on node.

Figure 9.3 shows examples of both types of networks that could be developed
directly from the activity relationships given. The relationships provide the instructions

Relationships

Activity A is first activity

Activities B and C can be accomplished simultaneously,
   but each depends on completion of activity A.

Activities D and E can be started after completion of
   activity B. Activities D and E could be accomplished
   simultaneously.

Activity F depends on completion of activities C and D.

Activities E and F could be done simultaneously and are
   the last activities in the project.

Activity on arrow network

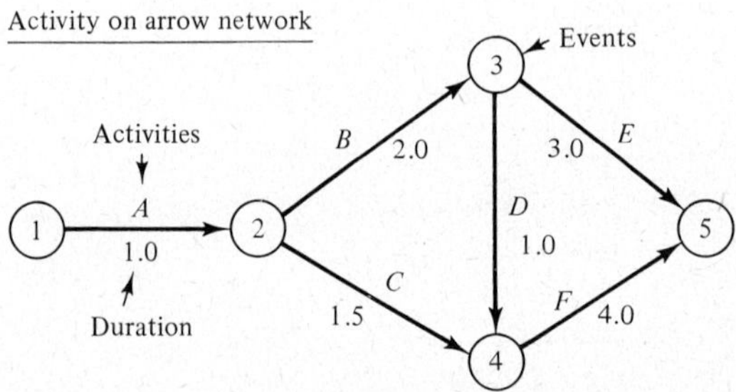

Activity on node network

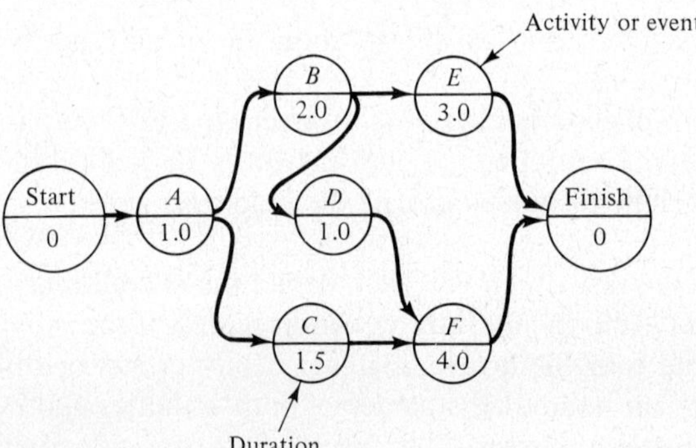

Figure 9.3 Example networks.

for constructing the connectivity of the network. Note also the method of indicating activity duration on each type of network, and the use of start and finish events in the activity on node network. The solution of both types of networks will be discussed in detail.

Management and Scheduling Aids

Bar (Gantt) Charts   During World War I, Henry L. Gantt developed a graphical display for production control. It was basically a bar chart with time lines superimposed upon it. The activities were represented by bars of length proportional to their duration and were placed on the time lines at the point where each would be undertaken. Figure

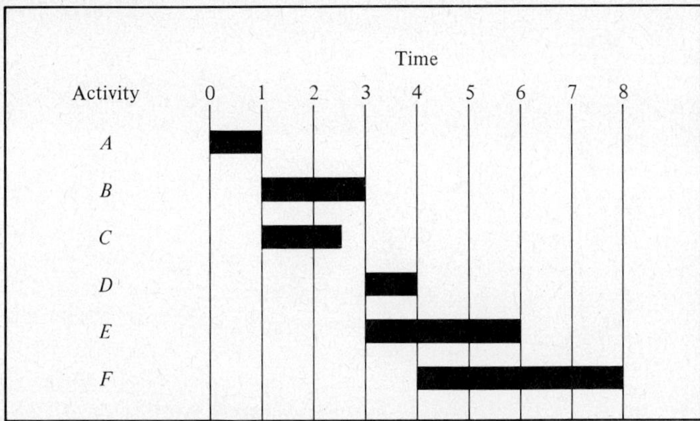

Figure 9.4 Bar (Gantt) chart.

9.4 shows a bar chart for the networks illustrated in Figure 9.3. Note that the expected project duration is shown by the end of the bar representing activity *F*.

For many years the Gantt chart was the primary tool used to control projects, and it is still in widespread use today. The bar chart is an excellent tool for determining resource conflicts and for leveling resource commitment. Its main weakness is that it does not adequately reflect interactions among activities. Thus, it is difficult to use bar charts to monitor project completion rates and make updated estimates of project completion time. This weakness led to the development of PERT/CPM methods. It should be emphasized that even with PERT/CPM, bar charts are still excellent tools and can illustrate information that networks cannot.

Activity on Arrow Network PERT/CPM uses directed path networks to represent the interactions among activities. The first of these is the activity on arrow network, in which activities are represented by directed arrows, and events are represented as nodes. Figure 9.5(a) shows a typical activity on arrow diagram. Activity durations are given in units appropriate to the project at hand, usually in days, weeks, or months. Activities can either be labeled with letters, or identified by the node numbers on either end of the activity; activity 1–2, activity 2–5, and so on. Note the dummy activities 2–3 and 6–7.

Since each link in this network represents an activity duration, and all activities have to be completed before the project is completed, the longest path through the network will set the minimum expected project completion time. This is the critical path. If any of the activities on this path are delayed, then the project completion time will also be delayed. Often a project will have multiple critical paths.

Determination of project completion time and the critical path(s) is a three-step process. The first step is to go forward through the network and estimate the earliest expected completion time (T_E) for each event (node) of the network. T_E is the estimated time required to complete work up to a certain point in the network. Since the critical path is always the longest path, it is necessary to study the completion times of all of the paths that enter each node. The T_E for that node would be the maximum estimated

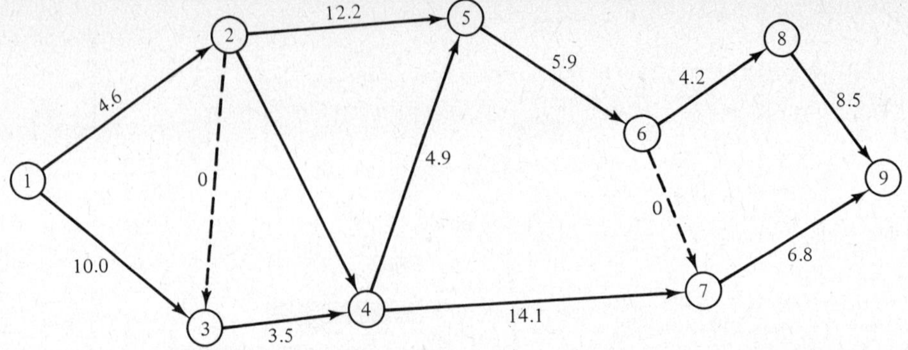

(a) Basic network (durations are in weeks).

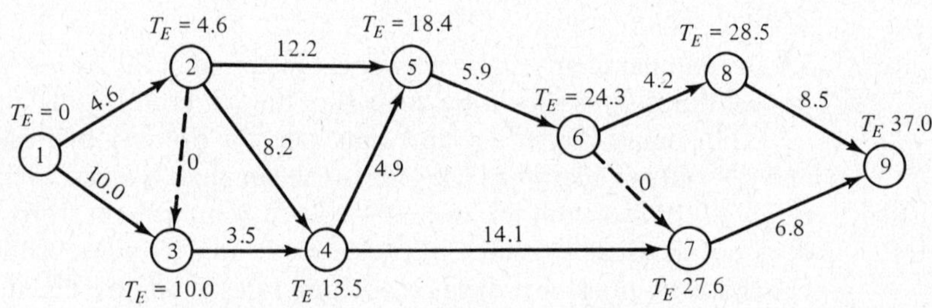

(b) Network with T_E times added.

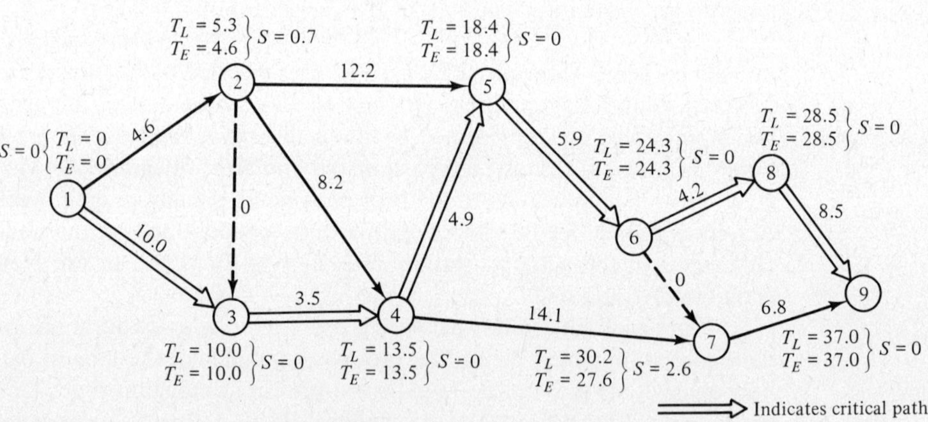

(c) Network with T_L and slack times added.

Figure 9.5 Activity on arrow network.

completion time among all activities entering the node. Completion time for a particular activity entering a node is determined by adding the duration of that activity to the T_E for the node from which the activity originates. For example, in Figure 9.5(b), at node 5 the choice is between the expected completion of activity 2–5 (12.2

$+ 4.6 = 16.8$) and the expected completion time of activity 4–5 ($4.9 + 13.5 = 18.4$). The T_E for node 5 would be 18.4 weeks. Other T_E values given in Figure 9.5(b) can be found in a similar manner.

At each node on the forward pass, attention should focus on the entering activities. T_E for the ending event of the network is the estimated project completion time. Collectively, the T_E estimates for the network provide a schedule of when each event should be finished and enable the manager to determine if the project is on, behind, or ahead of schedule, at any point during project execution.

After completing the forward pass through the network, a pass is made in the other direction (from finish to start) to determine the latest allowable finish time (T_L) for each event. T_L is the greatest amount of time that can elapse between the network beginning event and any event in question without delaying the scheduled completion of the project (T_E for final event). By definition, T_L for the ending event is equal to the estimated project completion time (T_E for ending event). The T_L value for each earlier event is the difference between the length of the longest path from that event to the ending event and the project completion time. However, as before, the computations can proceed from node to node. T_L is the minimum value of T_L for events directly successor to the event in question minus the intervening project duration. Figure 9.5(c) shows the network of 9.5(a) with T_L times added. At node 8, T_L is equal to T_L for node 9 (37.0 weeks) minus the intervening activity duration for activity 8–9 (8.5 weeks). At node 6, T_L is the minimum between T_L of node 8 (28.5 weeks) minus activity 6–8 (4.2 weeks), and T_L of node 7 (30.2 weeks) minus activity 6–7 (0 weeks). The backward pass would continue back to the starting node.

The difference between T_E and T_L for any event is the slack time (S) for that event. Slack time is the amount of time a particular event could be delayed without delaying the project completion time. Critical paths have zero slack time. Once the critical paths have been established, the manager can undertake further analysis to eliminate any resource conflicts that might exist or to even out resource commitment levels over the duration of the project.

Activity on Node Networks The major advantage of the activity on node network solution is that it provides information on slack time for individual activities, not events. This will be helpful in considering network replanning and adjustment. Figure 9.6(a) shows the network of Figure 9.5 redrawn in activity on node representation. Note that this representation eliminates the need for dummy activities.

Forward and backward passes are again made through the network. Slightly different definitions are given to the times found because the times are referring to activities instead of events. However, it should be noted that there can also be events in the activity on node network. Events in this instance would be activities with zero duration, and would depict milestones in the project. The starting and ending events of Figure 9.6(a) are examples of milestones.

On the forward pass, the following definitions are used:

EST: earliest start time of an activity = maximum EFT of operations immediately preceding activity in question.

EFT: earliest finish time of an activity = EST + activity duration.

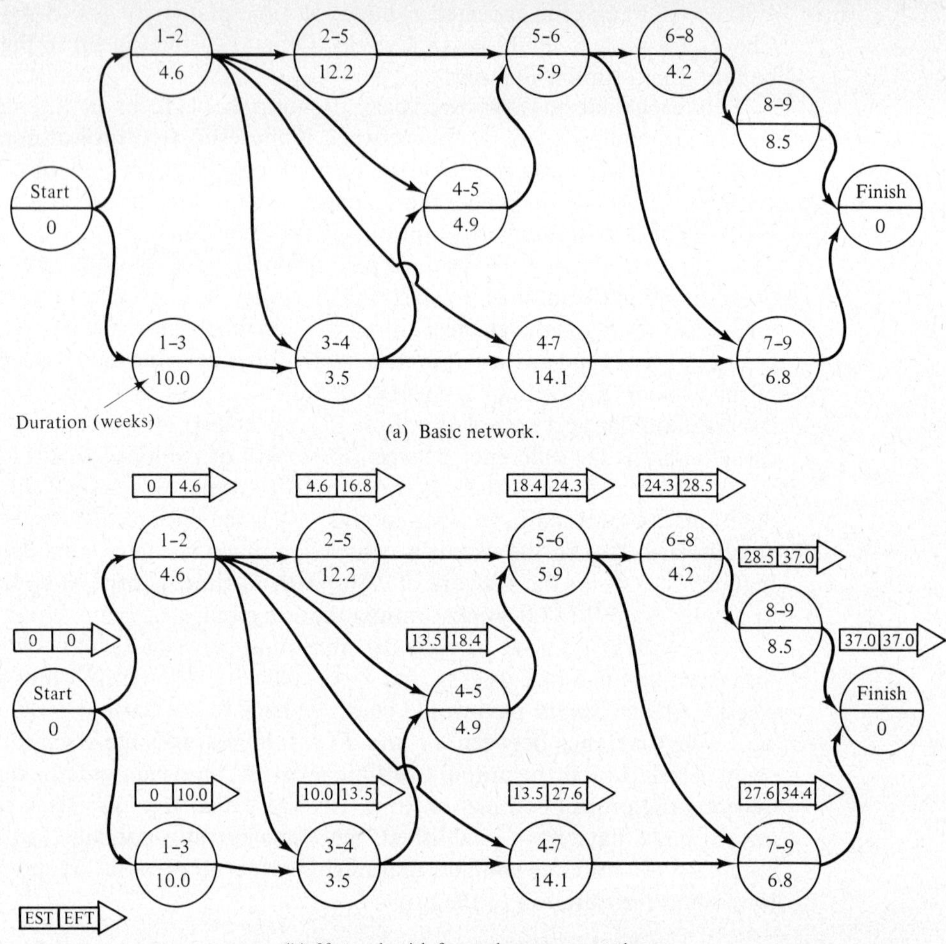

Figure 9.6   Activity on node network.

The EFT is similar to T_E for the activity on arrow network. Forward pass times are usually identified by an arrow pointing to the right $\boxed{\text{EST}\ \text{EFT}}\!\!>$ with the EST and EFT times placed in the indicated boxes. Figure 9.6(b) shows network 9.6(a) with forward pass data recorded on it. The EFT for the ending event is 37.0 weeks, as before. Note how decisions are made for activities with more than one arrow leading into them.

As with the activity on arrow network, the backward pass for the activity on node network will establish latest times, but now they will be times for activities. The following definitions will be used:

LFT: latest finish time of activity = minimum LST of operations immediately following activity in question.

LST: latest start time of an activity = LFT − duration.

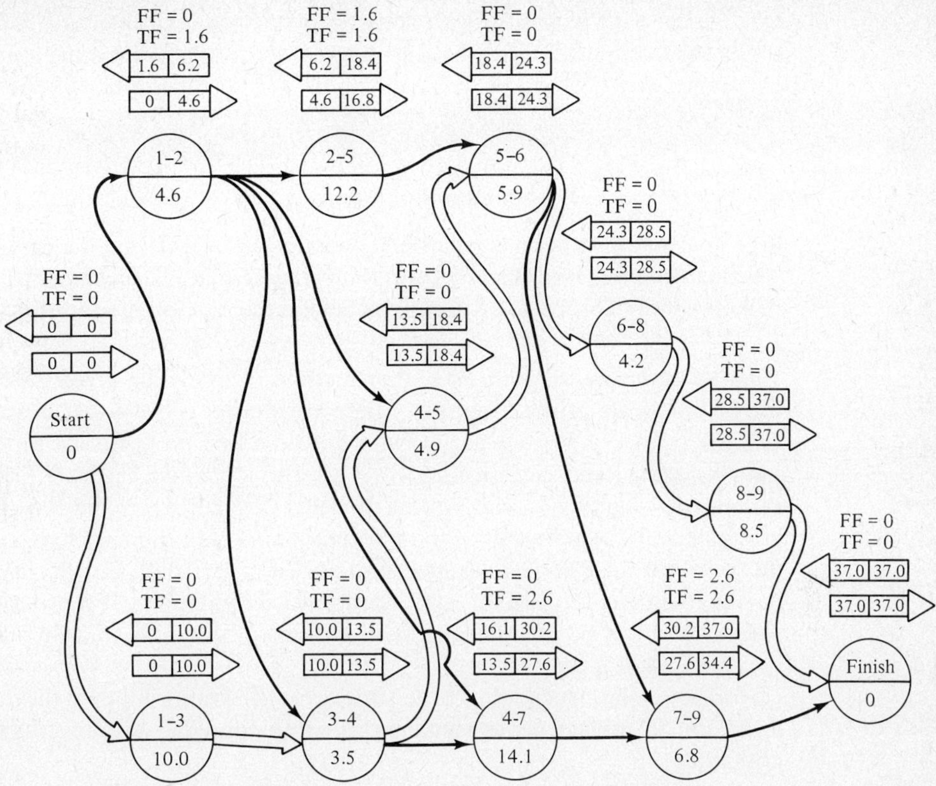

(c) Network with backward pass and float computations.

Figure 9.6 *(Continued)*

The LST is the latest time at which an activity may be started if the project is to be completed as scheduled. Similarly, the LFT is the latest time at which an activity may be finished without delaying project completion. The LFT is similar to T_L for the activity on arrow network. Figure 9.6(c) shows the network of 9.6(a) with backward pass calculations added. Note how decisions are made for activities with more than one arrow leaving them. For example, the LFT for activity 1–2 is the minimum of the LSTs for activities 2–5, 4–5, 4–7, and 3–4. Backward pass times are usually identified by an arrow pointing to the left, ◁ LST | LFT , with the LST and LFT times placed in the indicated boxes.

After completion of the backward pass, slack times for the activities can be calculated. There are three types of activity slack: total float, free float, and interfering float. Total float (TF) is the amount of time a particular activity could be delayed without delaying project completion

$$TF = LST - EST = LFT - EFT \qquad (9.1)$$

The critical path passes through activities with zero total float. Total float is similar to the slack of the activity on arrow diagram. Free float (FF) is the amount of time a particular activity could be delayed without delaying any other activity.

$$FF = \left\{ \begin{array}{l} \text{minimum EST of activities} \\ \text{immediately following} \\ \text{activity in question} \end{array} \right\} - \left\{ \begin{array}{l} \text{EFT of} \\ \text{activity in} \\ \text{question} \end{array} \right\}$$

Free float will always be $\leq$ total float. Interfering float (IF) is the difference between total float and free float. Figure 9.6(c) shows the TF and FF calculations for the network of 9.6(a). Only TF or slack will be indicated in the remaining examples of the chapter.

Time-Scale Arrow Diagram Another method of modeling project work flow is the use of time-scale arrow diagrams. These are a cross between bar charts and PERT/CPM networks. Activities are drawn on a time line proportional to their duration, but predecessor and successor activities are also identified. Figure 9.7 shows the time-scale arrow diagram for the project modeled in Figures 9.5 and 9.6. The time-scale arrow diagram contains all of the information of both the bar chart and the activity on node PERT/CPM network, plus it shows the critical path and activity slack without having to do any forward and backward pass computations. It shows all relationships visually and can be used effectively in arriving at the most efficient way to schedule the activities of a project.

The major criticisms of time-scale arrow diagrams are they they are difficult to draw and to update. Advancements in computer-aided drafting may negate these criticisms.

EXAMPLE 9.1 ____

You are an engineer/manager in charge of constructing a 4-mile segment of an interstate highway. You have completed the breakdown schedule with estimated activity durations shown in Figure 9.8. You now have to formulate the PERT/CPM network and make estimates of completion times.

After consulting with your fellow engineers, you decide that the following relationships and dependencies will exist among the activities of the project.

1. Clearing the corridor will be the initial activity.
2. Road cut and fill and interchange cut and fill can be undertaken simultaneously, but will follow completion of corridor clearing.
3. Bridge substructure (supporting columns, etc.) can be started after roadway cut and fill.
4. Construction of bridge abutments can commence after completion of the interchange cut and fill.
5. Slope stabilization should be started after completion of cut and fill and finished before any final surfacing.
6. Interchange subbase preparation can commence after interchange cut and fill.

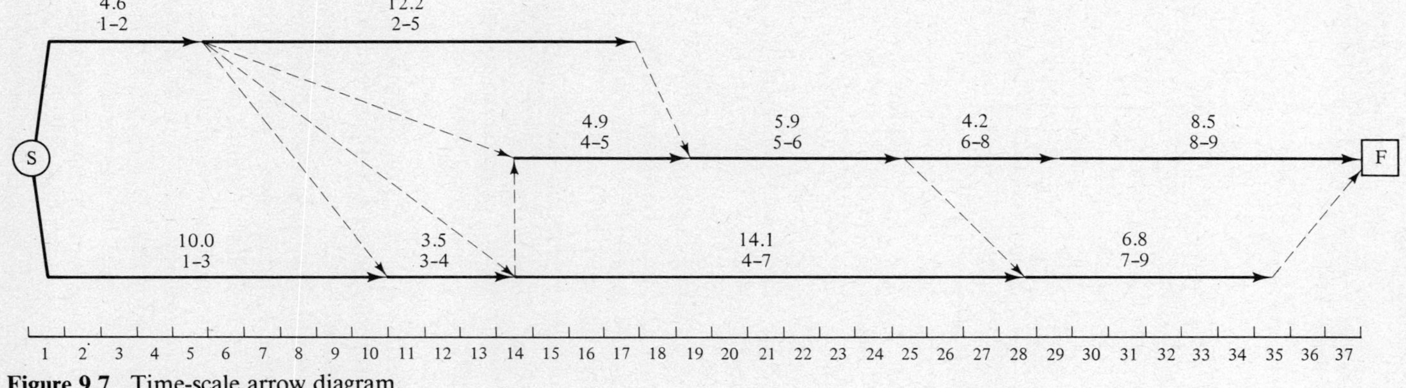

Figure 9.7 Time-scale arrow diagram.

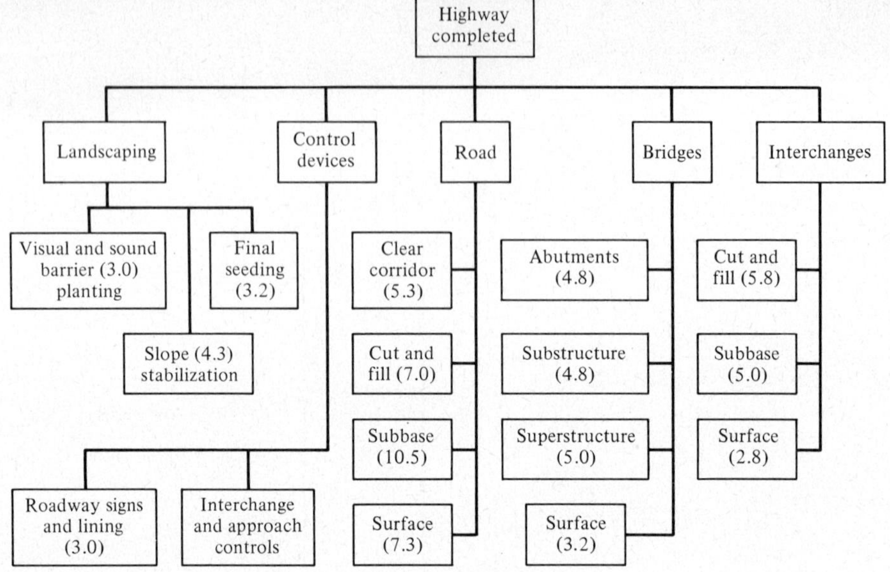

(Time in parentheses are in weeks)

Figure 9.8   Work breakdown schedule for Example 9.1.

7. The sound barrier can be constructed after all cut and fill operations are completed, and must be finished before final seeding.
8. The activities described in 3 through 7 above could all be undertaken simultaneously.
9. Road subbase preparation should start after completion of bridge substructure.
10. Bridge superstructure depends on completion of bridge substructure and abutments, but could be undertaken simultaneously with road subbase preparation.
11. Bridge surfacing depends on completion of bridge superstructure.
12. Road surfacing should not be started until completion of the road subbase, bridge superstructure, and slope stabilization, but could be done simultaneously with bridge surfacing and/or interchange surfacing.
13. Interchange surfacing should follow bridge surfacing, interchange subbase preparation, and slope stabilization.
14. Road signs should follow road surfacing.
15. Interchange and approach controls should be installed after completion of interchange surfacing.
16. Final seeding could be accomplished simultaneously with roadway signs and interchange controls, but should follow all surfacing and installation of the sound barrier.
17. Completion of 14, 15, and 16 finishes the project.

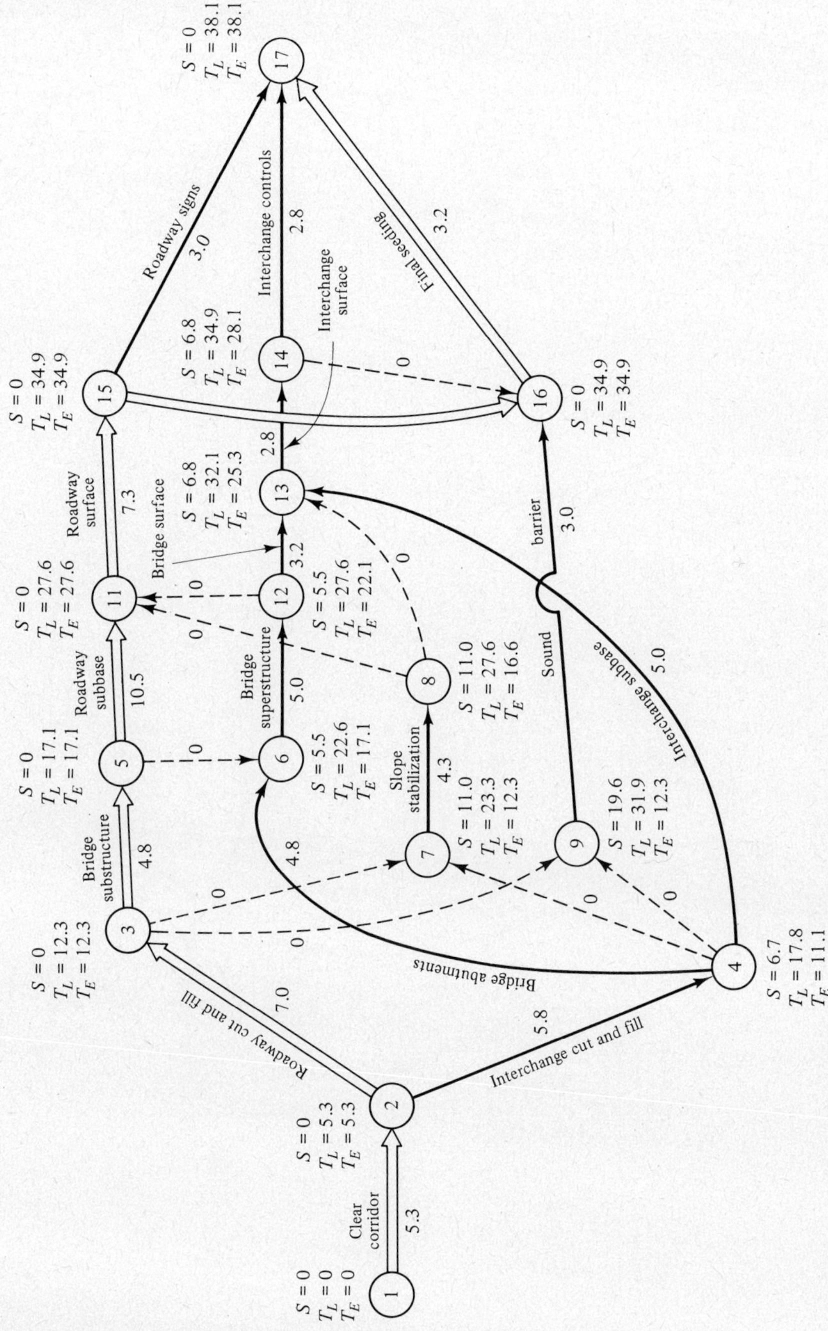

Figure 9.9 Activity on arrow network for Example 9.1.

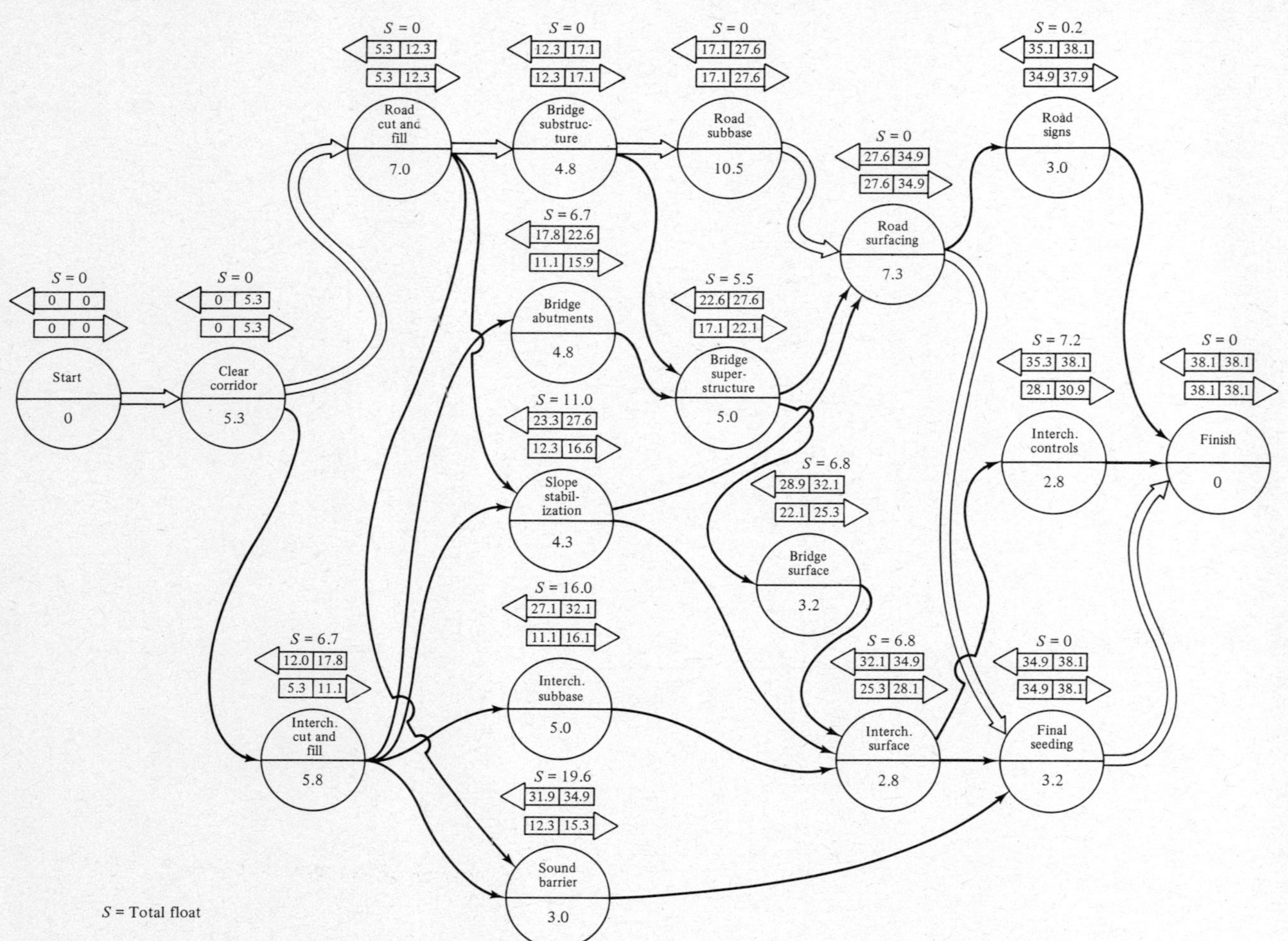

Figure 9.10   Activity on node network for Example 9.1.

These relationships show that activities from one branch of the work breakdown schedule may depend on completion of an activity from another branch. Also, the relationships developed assume that there are ample resources to complete all activities. This assumption must be checked during network replanning and adjustment.

Figure 9.9 gives the activity on arrow network that could be developed from the relationships defined. The expected project duration is 38.1 weeks. The critical path is shown by double arrows. Figure 9.10 shows the corresponding activity on node network with calculated data. Figure 9.11 is the equivalent time-scale arrow diagram for the project.

Network Replanning and Adjustment

The initial modeling of the project work flow is only the first step in the planning of a project. Once a network, bar chart, or arrow diagram is set up, the manager must then check the scheduling of activities for any resource conflicts that might arise. It may be desirable to replan the scheduling so that resource commitment is evened out during the progress of the project. Planning labor resources in such a way as to avoid excessive hiring and layoffs of workers is especially important.

Furthermore, after the project has begun, it is necessary to update the project scheduling model in order to check the status of the project. Various occurrences can affect the rate of accomplishment of project activities. Among these are:

1. Availability of labor and other resources.
2. Crew efficiency.
3. Weather and other factors of nature.

If one or more of these contribute to activity delays that threaten to extend the finishing time of the project, then decisions have to be made as to the best course of action to take. Will it be better to let the project run its natural course, or should efforts be made to speed up critical activities so that the project can be finished on or near the original completion date? Succeeding sections will show how the basic scheduling models can assist in these project refinements. Some of the methods to be discussed may not be applicable to a certain project, but all should be considered and used whenever appropriate. It is also necessary to consider the interdependencies among various methods and to realize that the order of application may vary from one project to another.

RESOURCE REALLOCATION

Resource reallocation involves moving resources from activities not on the critical path to critical path activities. The purpose of this is to reduce project duration, and to reduce slack time in the network. In order to reallocate resources between activities, the resource to be reallocated has to be compatible with both activities. For example, bricklayers would not contribute to shortening an activity that requires plumbers, whereas, if each activity requires plumbers, reallocation would speed up the receiving

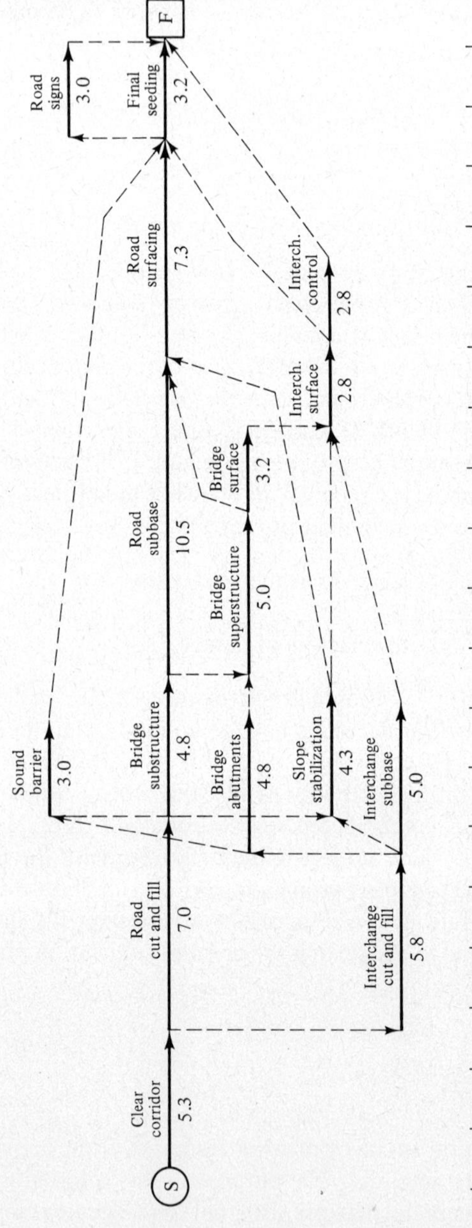

Figure 9.11 Time-scale arrow diagram for Example 9.1.

activity, with a proportional slowdown in completion of the donor activity. Resources could also be effectively interchanged between two construction activities that each require a number of cranes.

Several things have to be taken into account when considering the reallocation of resources. The activities have to be occurring simultaneously. If they are not, the same resource could be used for each and there is actually no resource to reallocate. Reallocating resources will shorten the duration of an activity on the critical path while lengthening the duration of an activity off the critical path. This trade-off can only go so far, however, before all of the slack for the off critical path activity is used up, and an additional critical path is created. Further reallocation will lengthen project duration. Also, transfer of resources from one activity to another may not result in a proportional reduction in activity duration for the receiving activity if the added resource cannot be used efficiently. For example, if 4 laborers could do a certain job in 10 days, there is no guarantee that 10 laborers could do the same job in 4 days because they might interfere with each other.

If the two activities between which resources are to be switched both require the same level of resource, the activity off the critical path can be increased in duration by a time equal to half the slack time for that activity, while the receiving activity on the critical path can be reduced by an equal amount. If there are additional activities in progress at the same time, their slack also has to be examined to be certain that the reallocation does not overshoot and create a different critical path. If more than one activity could contribute resources, the one with the most slack is usually the best one to pick as the resource donor.

If the activities do not require the same level of resource, then appropriate scaling factors have to be used between the donor and receiver activities. The following set of simultaneous equations can be solved to estimate donor activity duration extension and receiver activity duration reduction.

$$AX = BY \qquad\qquad (9.2)$$

$$X + Y = C \qquad\qquad (9.3)$$

where A = normal level of resource for activity on critical path,
B = normal level of resource for activity off critical path,
C = amount of slack time available to activity off critical path,
X = amount of time by which activity on critical path is shortened, and
Y = amount of time by which activity off critical path is extended

A, B, and C will be given by the problem, and X and Y can be solved for. This is a mathematical balancing of slack and does not account for any diminishing productivity. It also does not account for competing slack along parallel paths.

EXAMPLE 9.2 ⎯⎯⎯⎯⎯⎯⎯⎯⎯⎯⎯⎯⎯⎯⎯⎯⎯⎯⎯⎯⎯⎯⎯⎯⎯⎯⎯⎯⎯⎯⎯⎯⎯

Suppose that the network shown in Figure 9.12(a) has been developed for a project. Activities B, C, and D utilize a compatible labor resource. Activities B and C each require 10 workers and D requires 8 workers. Also, activities J and K have compatible labor resources, J requiring 12 workers and K, 12 workers. Any possible resource reallocation should be made to reduce the overall project duration.

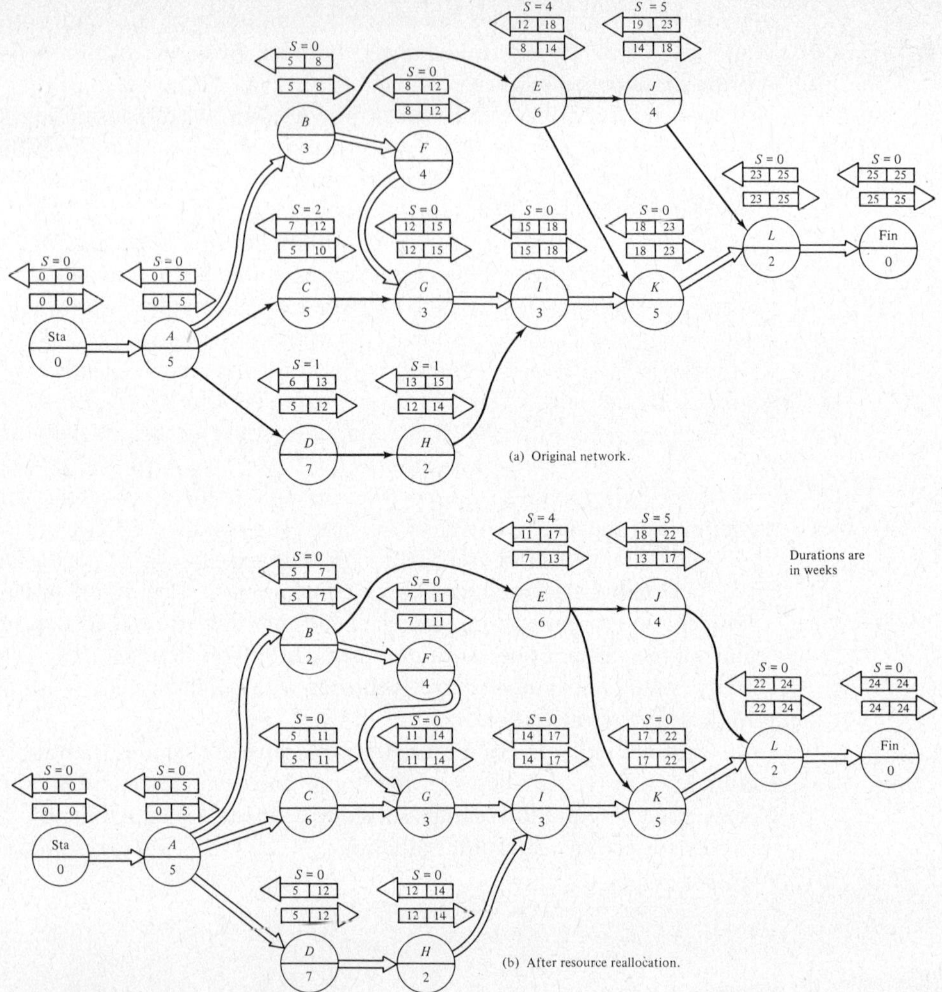

Figure 9.12 Resource reallocation.

The earlier activities should be examined first. Activities B, C, and D all have time overlap, so resources could be reallocated among them. Activity C has 2 weeks of slack while activity D has only 1 week of slack. It will probably be preferable to shift laborers from activity C to activity B. Since they each require the same level of resource, logic says that activity B could be shortened by 1 week while activity C would be lengthened by 1 week. Application of Equations 9.2 and 9.3 will give the same result.

$$10X = 10Y; \quad X + Y = 2; \quad X = Y = 1$$

Since there are three simultaneous activities taking place, the slack for activity D has to be considered. Since it has 1 week of slack available, the reduction in the duration of B by 1 week should just balance off the slack of D. Also, it would be expected that the total project duration would be reduced by exactly 1 week. Figure 9.12(b) confirms this. Note the three-branch critical path going through activities *B*, *C*, and *D*. If the original slack in activity *D* had 0.5 week instead of 1 week, the total project duration could not be shortened more than 0.5 week, even if the full reallocation had been made.

Now consideration can be given to reallocation of resources between activities *J* and *K*. Since there is no expected overlap in these activities (EFT for *J* is 17 weeks and EST for *K* is 17 weeks), there is nothing to be gained by attempting reallocation. Activity *J* could be delayed so *J* and *K* would overlap, but that would be counterproductive because of the large fluctuations in labor resource required.

RESOURCE CONFLICTS

There may be limitations on the amounts of certain resources that are readily available for commitment to a project. If additional resources are necessary, obtaining them will incur an added expense. If these necessary resources are not acquired, there may be conflicts in the project schedule that will have to be resolved by resequencing activities. Such resequencing may cause an increase in total project duration. The decision maker must evaluate the most economical course of action between acquiring additional resources, or allowing the project duration to extend and absorbing any penalties or increased carrying costs involved.

Resource limitations are essentially constraints on the conduct of the project. A contractor may have a limited number of cranes or other equipment that could be used. There may be limitations on the number of workers available who have critical skills. Limitations may be caused by shortages of material supplies, or, they may be imposed by financial resources. A self-imposed limitation may be the desire to keep the level of resource usage nearly constant for the duration of its use on the project. This is especially applicable to labor resources. Leveling of commitment avoids rapid fluctuation in the work force and minimizes hiring and firing.

A modified bar chart is helpful in identifying where resource conflicts may occur and how they might be resolved. Figure 9.13(a) shows the network from Figure 9.12 after resource reallocation. Figure 9.13(b) shows a modified bar chart for a portion of the activities of this network. Note the addition of the EST, LST, EFT, and LFT to the chart. The solid portion of the bar is the expected duration of the activity. Slack time is shown by the open portion of the bar. Activities on the critical path have no open bar.

Each of the activities shown in Figure 9.13(b) requires one or more cranes for completion. The numbers required are:

Activity	Number of cranes required
E	1
I	1
J	2
G	1

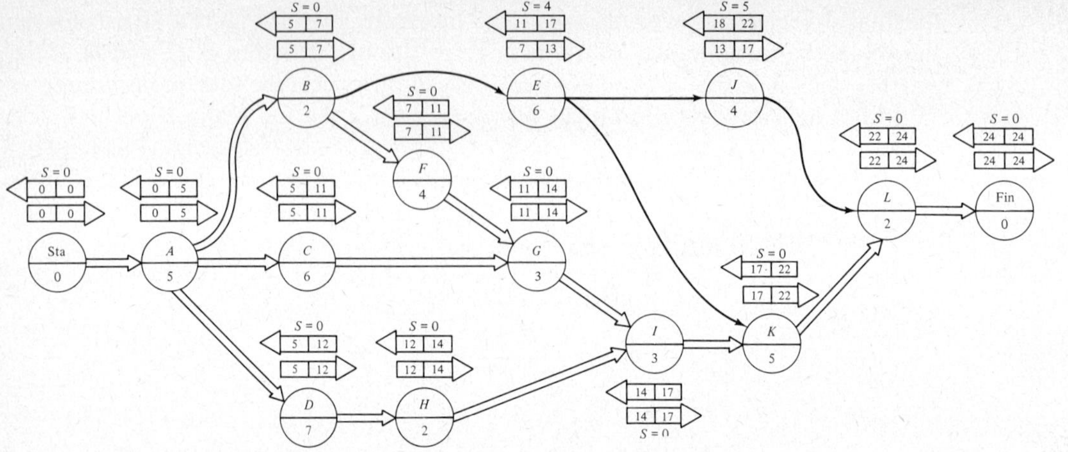

(a) Network from Figure 9.12 after resource reallocation.

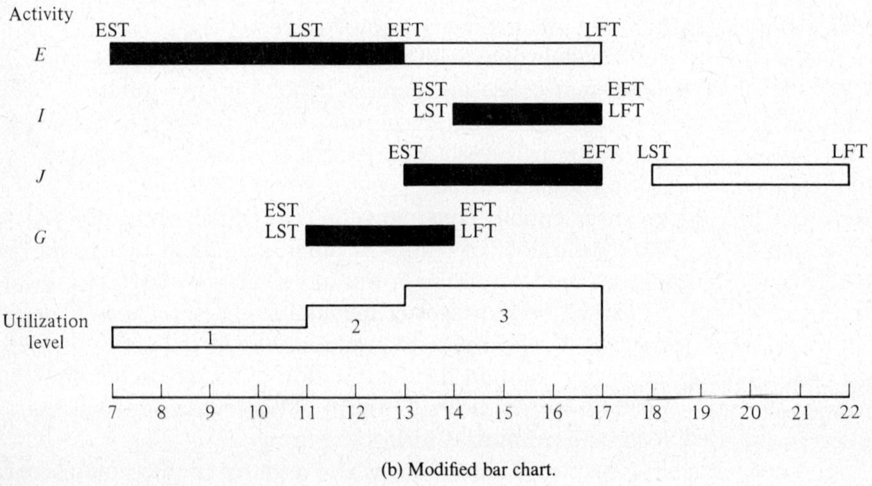

(b) Modified bar chart.

Figure 9.13   Resolving resource conflicts.

However, the contractor for this project has only 2 cranes available for commitment to the project. The resource utilization level graph at the bottom of Figure 9.13(b) shows that during the 14th, 15th, 16th, and 17th weeks, three cranes are required. Careful examination of the bar chart can reveal how to resolve the conflict with minimum increase in the project duration. Concentrate on the time frame during which there is a resource conflict. This shows that activities G, I, and J contribute to the conflict. From these activities, it is necessary to find the activity with the minimum EFT (activity G has EFT of 14 weeks) and the activity with maximum LST (activity J has LST of 18 weeks.) If J is resequenced after activity G, minimum increase in project duration should result, while at least part of the conflict will be resolved. The increase in project duration (IPD) will be given by

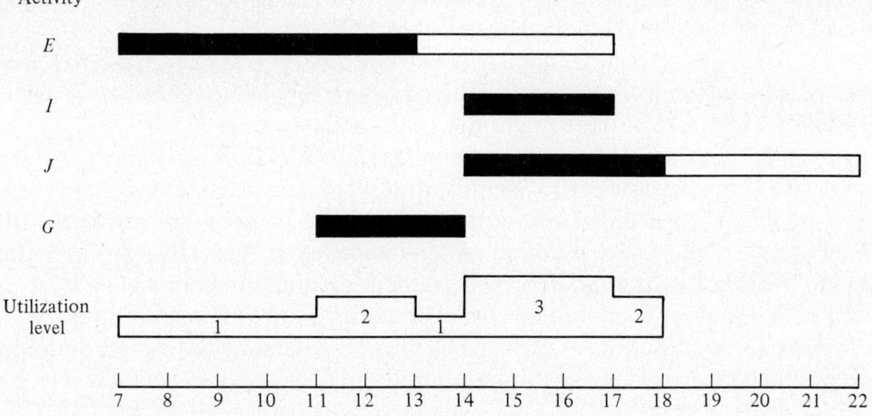

(c) Bar chart after rescheduling of *J* after *G*.

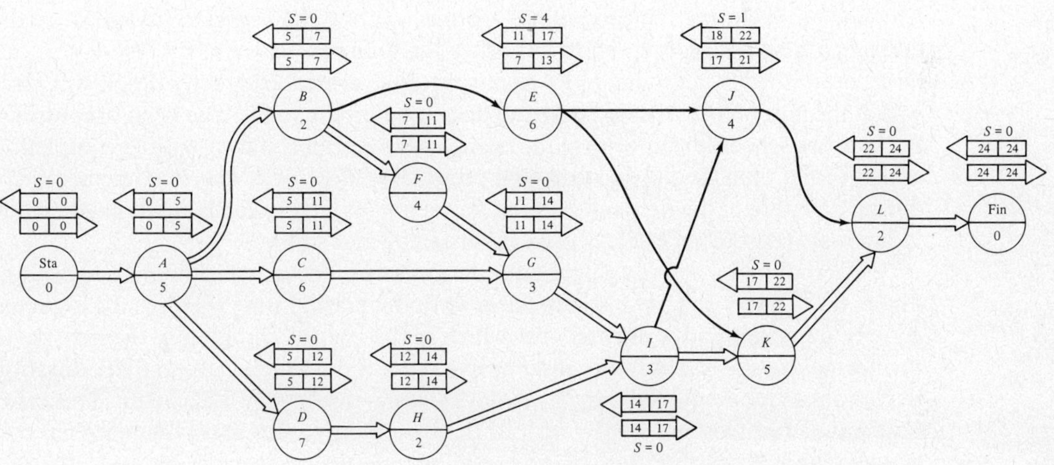

(d) Network rescheduled to eliminate resource conflict.

Figure 9.13 *(Continued)*

$$IPD = EFT_{min} - LST_{max}$$
$$= 14 - 18 = -4 \tag{9.4}$$

If IPD is negative or zero, there is no increase in project duration. It is quite possible that the same activity will have minimum EFT and maximum LST. In that case, the activity should be compared with the second largest LST activity.

Figure 9.13(c) shows the bar chart depicting activity *J* rescheduled after activity *G*. Note that this removes the conflict during the 14th week, but not during the 15th, 16th, and 17th weeks. Activities *I* and *J* have to be compared to eliminate the remaining conflict. Activity *I* has the minimum EFT of 17 weeks and *J* has the maximum LST of 18 weeks. Rescheduling *J* after *I* will eliminate the conflict and still have an IPD of zero. Since *I* is already dependent on *G*, only one new arrow from *I* to *J* has to be drawn. Figure 9.13(d) shows the revised network. Note that the project duration is still 24 weeks, but that the slack for activity J has been reduced to 1 week.

The utilization level chart and the bar chart can also be used to even out resource commitments by showing which activities could be rescheduled to better utilize available resources, even though there might not be an absolute limit.

Essentially, elimination of resource conflicts involves construction of the bar chart and utilization level chart to identify if any conflicts exist. If conflicts exist, activities should be rescheduled two at a time until the conflicts are resolved with minimum increase in project duration. After rescheduling, the network has to be modified to include the appropriate new dependencies, and all network data have to be recomputed. Finally, the additional cost of procuring more resources from external sources versus the additional costs incurred by possibly extending the project duration must be evaluated to arrive at the most economically advantageous action.

TIME–COST OPTIMIZATION

At some point during the progress of a project, it may be necessary to speed up the critical path activities in an effort to reduce the total project duration. While this can be accomplished by applying more resources to the critical path activities, it will carry with it the burden of increased cost and decreased productivity. The need to compress the network schedule may arise at the beginning of a project if the required completion date is earlier than the normal expected project duration, or, it may become necessary during updating of the network because key activities have fallen behind schedule due to labor strikes, delivery delays, natural disasters, and so on.

An individual activity might have a time versus cost curve similar to curve A of Figure 9.14. Every activity has associated with it a normal time which is the expected activity duration, and a normal cost which is the cost of completing the activity in the normal time. Most activities also have a crash time, which is the activity duration if all of the stops are pulled out (seven days a week, three shifts a day, etc.). This crash time has a crash cost associated with it. Between the two extremes, the marginal cost of reducing activity duration by one unit of time would increase as the activity is

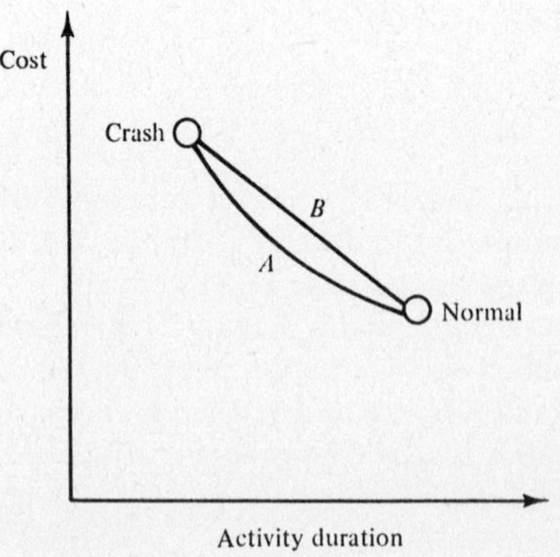

Figure 9.14 Variation of cost with activity duration.

TABLE 9.1 NORMAL AND CRASH DATA FOR PROJECT

Activity	Normal time (weeks)	Crash time (weeks)	Normal cost	Crash cost	Marginal cost to crash ($/week)
A	3	2	$20,000	$22,000	2,000
B	5	3	30,000	36,000	3,000
C	4	2	25,000	27,000	1,000
D	3	2	10,000	11,500	1,500
E	5	4	15,000	17,500	2,500
F	4	4	20,000	20,000	—
G	7	6	12,000	12,500	500
H	5	3	15,000	22,000	3,500
I	6	4	10,000	18,000	4,000
J	4	3	8,000	9,000	1,000

progressively compressed, because of decreasing returns on investment. However, to simplify computations, it is generally assumed that cost varies linearly with duration reduction between the normal and crash times (line B of Figure 9.14). The slope of line B is referred to as the marginal cost to crash (the incremental cost of reducing the activity duration by one unit of time).

The objective of time–cost optimization is to reduce the project completion time to the new deadline with the minimum increase in project cost. The additional cost can then be compared with the cost of delaying project completion beyond the deadline. First it has to be determined if it will be possible to reduce project duration to the new deadline. This can be accomplished by calculating the minimum possible project duration if all activities are crashed.

Table 9.1 shows the normal and crash data for the network given in Figure 9.15. Normal completion time for this project would be 23 weeks, and the normal cost would be $165,000. The marginal cost to crash is computed as follows:

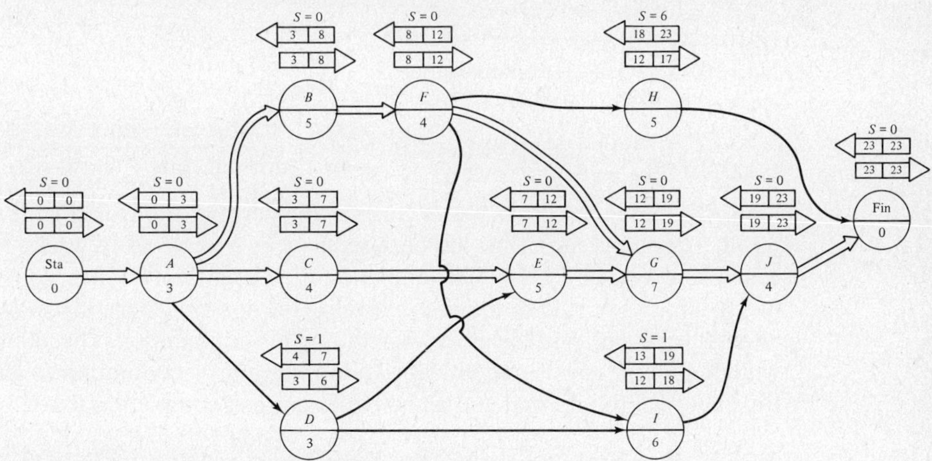

Figure 9.15 Network for time–cost optimization.

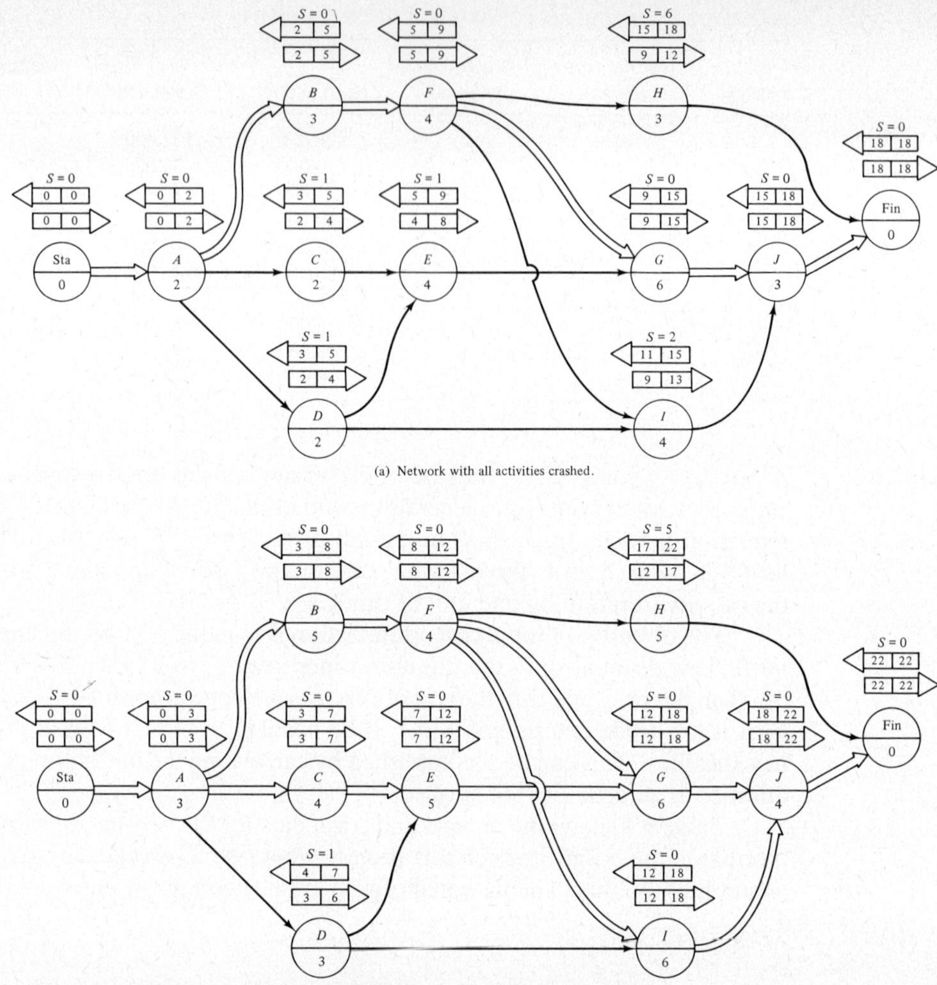

Figure 9.16   Time–cost optimization.

$$\text{marginal cost to crash} = \frac{\Delta c}{\Delta t} = \frac{\text{crash cost} - \text{normal cost}}{\text{normal time} - \text{crash time}} \tag{9.5}$$

If all activities in the project were crashed, the network shown in Figure 9.16(a) would result. The completion time has been reduced by 5 weeks to 18 weeks, but the project cost would now be $195,500. Completion time could not be reduced below 18 weeks unless some way could be found to rearrange the network. Generally, however, crashing all activities is not an efficient way of reducing project duration. The following approach will identify the most economical way to reduce project duration to any completion time between the normal and all activities crashed completion times.

1. Find the critical path or paths.
2. Find the least expensive activity on the critical path. When multiple critical

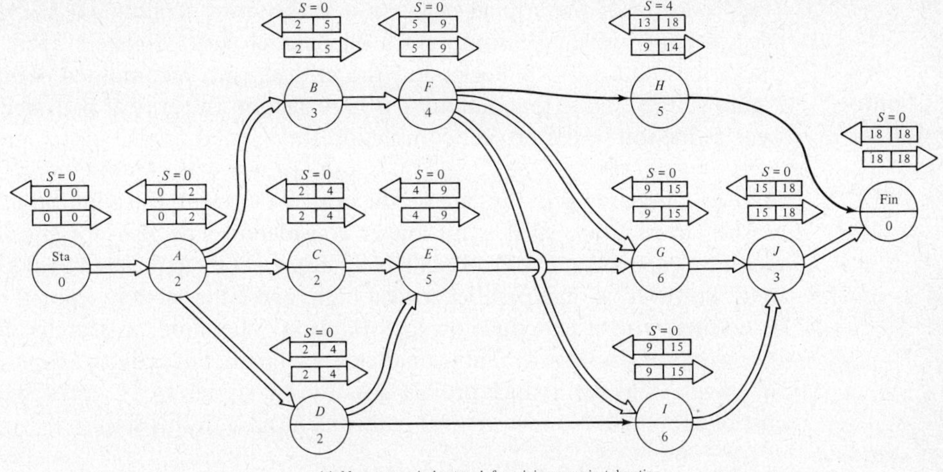

(c) Most economical network for minimum project duration.

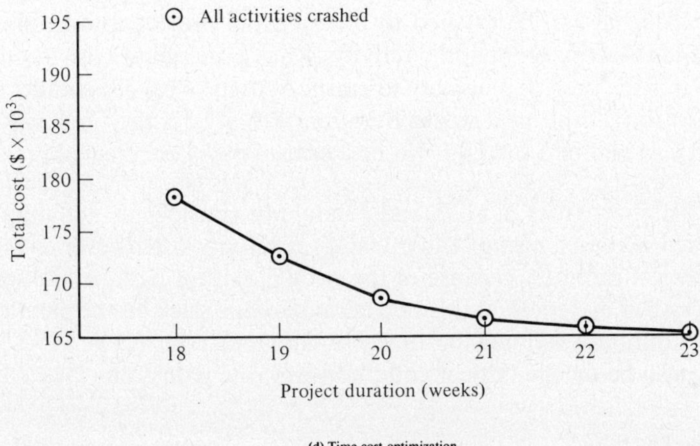

(d) Time-cost optimization.

Figure 9.16 *(Continued)*

paths exist, it may be most economical to crash a combination of two activities on parallel critical paths.

3. Crash the most economical activity or activities by an amount equal to the minimum slack on parallel paths, but not beyond its crash time.

4. Recompute network data and confirm critical path or paths.

5. Repeat steps 2, 3, and 4 until desired completion time is reached.

If the objective of network compression for the project depicted in Figure 9.15 was to reduce project duration to 18 weeks in the most economical manner, the following steps would be required.

Step 1. Activities on the critical paths are A, B, C, E, F, G, and J. There are multiple critical paths at this step. Activity F cannot be crashed (fixed time to complete) so it will not be considered. Activity G has the lowest marginal cost to crash and is on the single portion of the critical path, so it is the most economical to crash. It can be crashed by 1 week. Minimum slack on parallel paths is also 1 week. Therefore, G should be reduced to 6 weeks and the critical path recomputed. This is shown in Figure 9.16(b). Note that activity I is now also on the critical path because of the 1-week reduction in the project completion time. Total project cost is now $165,500.

Step 2. The same activities are on the critical path with the addition of I. G has been crashed to its limit, so it is no longer considered. F is also not considered. Of the remaining activities, C and J both have marginal costs to crash of $1,000/week. Activity C, however, is on a parallel critical path, and either activity B or F would have to be crashed with it to reduce project duration. Therefore, it is preferable to crash J. Five weeks of slack are available on the parallel path, but activity J can only be crashed by 1 week. This will reduce project duration by 1 week to 21 weeks, with a new total cost of $166,500. No new critical path was formed in this step.

Step 3. Next, activities A, B, C, E, and I should be considered. Activity C has the minimum marginal cost to crash; however, it is on a parallel critical path, so both B and C would have to be crashed to produce any project compression. This would require $4,000/week reduction. Activity A has a marginal cost to crash of $2,000/week, so it is the preferred activity to crash. Activity A has no parallel paths, so it can be crashed to its limit of 2 weeks duration. Project duration will now be 20 weeks, and total cost will be $168,500. No new critical paths are created.

Step 4. Next, activities B, C, E, and I should be considered. Activities B and C have a combined marginal cost of $4,000, as does activity I. However, crashing I will not reduce project duration because of the parallel critical path. Activities B and C can only be crashed by 1 week at this step because of the slack on the parallel path through D. Project duration will now be 19 weeks, and total cost will be $172,500. Activity D will now also be on the critical path; however, the path from D to I is not a critical path.

Step 5. Finally, activities B, C, D, E, and I should be considered. B and C can still be considered because they have not been crashed to their limit. Combinations of B, C, and D, or B and E could be crashed. The total marginal cost of each combination is $5,500. The choice would depend on project conditions and ease of crashing. In this case, B, C, and D will be crashed by 1 week, with a resulting project duration of 18 weeks and a total cost of $178,000. Additional crashing will not result in any further reduction in project duration. Figure 9.16(c) shows the final network, while Figure 9.16(d) shows a plot of project duration versus total cost. This plot could be used to determine the best way to reduce project duration to any time between the normal and crash limits. Note also that the total cost to reach the minimum possible project duration is $178,000 rather than the $195,500 required if all activities are crashed.

EXAMPLE 9.3 ———————————————————————————————

You have formulated the activity schedules for the project described in Example 9.1. Now, you want to discern if there is any need for network replanning and adjustment.

Road surfacing and interchange surfacing each require two paving machines, and your company has two machines. Figure 9.17 reveals that there is a conflict during 0.5 week. Interchange surfacing has the minimum EFT of 28.1 weeks, but it also has the maximum LST of 32.1. Rescheduling road surfacing after interchange surfacing will result in an increased project duration of 0.5 week, and will also produce a new critical path. The critical path will now be: clear corridor–roadway cut and fill—bridge substructure—bridge superstructure—roadway surfacing—final seeding.

The next step would be to consider whether any reallocation of resources would reduce the project duration to or below the original estimated completion time of 38.1 weeks. Roadway cut and fill will require five earthmovers and interchange cut and fill will require seven. There will be sufficient earthmovers available, and they can be freely interchanged among the activities. Roadway cut and fill is on the critical path, while interchange cut and fill is not. Interchange cut and fill has 1.2 weeks of slack. Using Equations 9.2 and 9.3,

$$5X - 7Y = 0$$
$$X + Y = 1.2$$
$$X = 0.7$$
$$Y = 0.5$$

Assuming that there are no diminishing returns, roadway cut and fill can be reduced by 0.7 week, with a corresponding reduction in project duration. The new duration for roadway cut and fill would be 6.3 weeks, the same time required for interchange cut and fill. The total estimated project duration will now be 37.9 weeks, 0.2 week shorter than the previous expected completion time.

When reducing activity duration by resource reallocation, caution must be exercised not to attempt to reduce an activity duration below its crash time. Resource reallocation could not realistically do that.

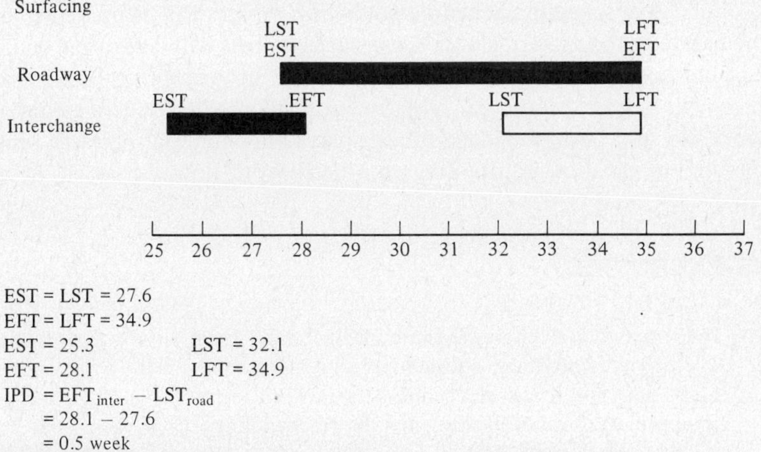

$$\text{IPD} = \text{EFT}_{inter} - \text{LST}_{road}$$
$$= 28.1 - 27.6$$
$$= 0.5 \text{ week}$$

Figure 9.17 Bar chart, resource conflicts, Example 9.3.

If it is not possible to reach the desired completion time by the methods used in this example, either additional equipment rental, or time–cost optimization could be undertaken. If this is required, the decision maker will have to evaluate the relative economics of different courses of action. Rental of additional equipment will cost money, as will crashing of activities. Extending the project duration may cause two types of additional cost. One of these is the carrying cost, which represents the normal overhead that the firm executing the project would incur for each week's extension. Additionally, many projects either have a bonus for finishing on or before a certain deadline, or a penalty for finishing late.

SUMMARY

PERT/CPM can be an effective tool in the management and scheduling of nonrepetitive types of projects. Project activities can be linked together in a directed arrow diagram indicating activity interdependencies and project work flow. Various techniques can be utilized to reduce project duration or to efficiently distribute resource commitment levels during execution of the project.

As with most tools, the user must be skilled to ensure PERT/CPM's proper use. All competing aspects of the project must be considered in developing the initial scheduling strategy. After initial scheduling, the user must treat the dynamics of project execution. Seldom does any project progress exactly as it was planned. Therefore, it is necessary to update PERT/CPM at regular intervals during the project to plan for the efficient scheduling and completion of the remaining activities.

Numerous computer applications packages are available to help in implementation of PERT/CPM. These packages do not generate networks, nor do they optimize activity scheduling. They can, however, facilitate network analysis and produce useful reports that can assist the user in making decisions about the project. Among these reports are schedule status and trouble area reports, which are necessary to assess the current degree of project completion and to indicate whether the project is behind, on, or ahead of schedule. Schedule prediction reports will estimate project completion time based on the present status. Cost status reports will determine actual costs versus budgeted costs. These reports can be utilized in generating cost prediction reports which will extrapolate actual costs in order to predict cost overruns or underruns. Manpower and equipment loading displays and material delivery schedules can be generated to assist in maintaining an orderly work flow.

APPLICATIONS EXERCISES

9-1. In the network given in the figure for Exercise 9-1, activity B requires one crane, activity C requires two cranes, and activity D requires one crane. Three cranes are available. Show how you would reschedule this network to resolve any conflicts and indicate the total project duration before and after rescheduling.

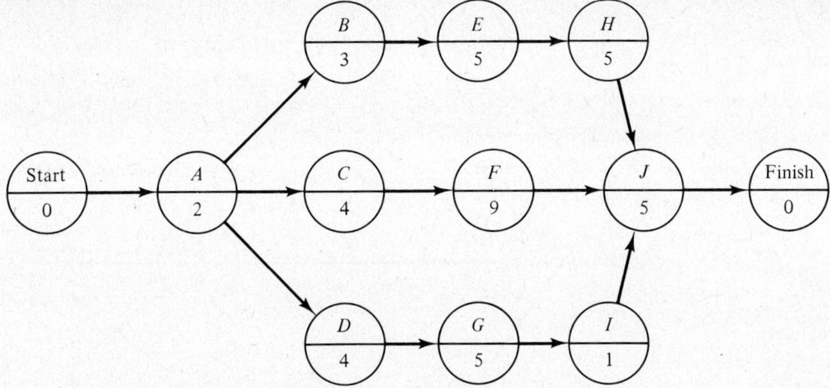

9-2. Again, use the network given in Exercise 9-1 (before rescheduling). Activities *F* and *G* are labor compatible; however, activity *F* requires 9 people while activity *G* requires 6 people. How much can you reduce the total project duration by reallocating resources between *F* and *G*?

9-3. Formulate the following problem into a PERT/CPM network. Use both activity on node and activity on arrow.

A, *B*, and *C* can be accomplished simultaneously and are the first activities in the network.

D and *E* can be accomplished simultaneously and follow *A*.

H and *I* can be accomplished simultaneously and follow *E*.

G follows *C*, *F*, and *H*.

F follows *B* and *C*.

J and *K* can be accomplished simultaneously and follow *B*, *D*, *G*, and *I*.

L and *M* can be accomplished simultaneously and follow *J*.

N follows *L*.

O follows *K* and *M*.

N and *O* can be accomplished simultaneously and are the last activities in the network.

Note: This is the same project description given in Exercise 4-9. You should compare your present solution to what you developed previously.

9-4. (a) Transform the network in the figure for Exercise 9-4 on p. 354 from activity on node to activity on arrow.

(b) For the original network given in (a). *E*, *F*, and *G* each require one crane. You have only two cranes. Is there a resource conflict? If so, how would you best reschedule the network to eliminate it? How much will the project duration be extended?

(c) Again, using the original network of (a), suppose *H* and *I* require 8 and 4 units of compatible labor input, respectively. How would you reallocate this labor resource to get the most reduction in project duration? What is the new project duration?

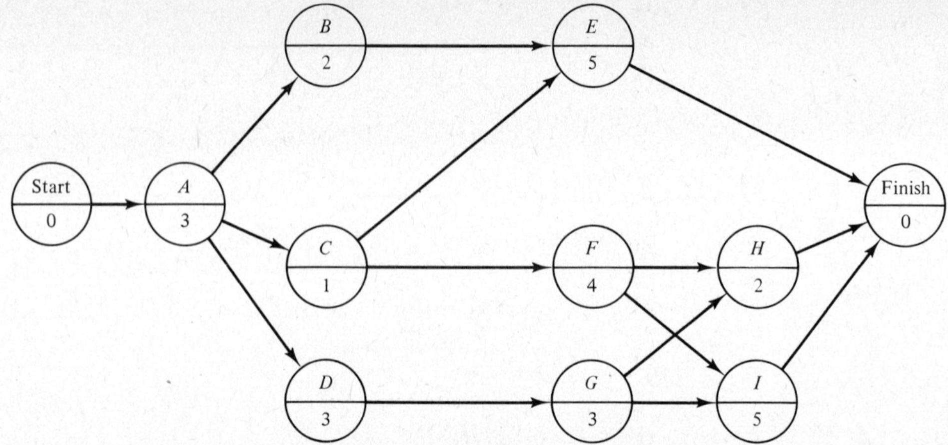

9-5. Given the activity on node network in the figure for Exercise 9-5 (all activity durations in weeks), and using the original network for each following question:

   (a) What is the expected project duration? Indicate the critical path and float for each activity.

   (b) Transform the network to an activity on arrow network and recompute the estimated project duration.

   (c) If activities D, E, and F each require one crane, and there are two cranes available:

      1. Do you have a resource conflict?

      2. If you do, how could it be resolved by rescheduling? What is the increase in project duration?

      3. If it costs $10,000 each week of project delay and you can rent an extra crane for $7,000/week, should you reschedule or rent another crane? If you rent a crane, how long should you rent it for?

   (d) The following activities can be crashed:

Activity	Normal time	Crash time	Marginal crash cost ($/week)
L	4	2	4,000
H	1	0.5	1,000
J	2	1	2,000
G	2	1	1,500

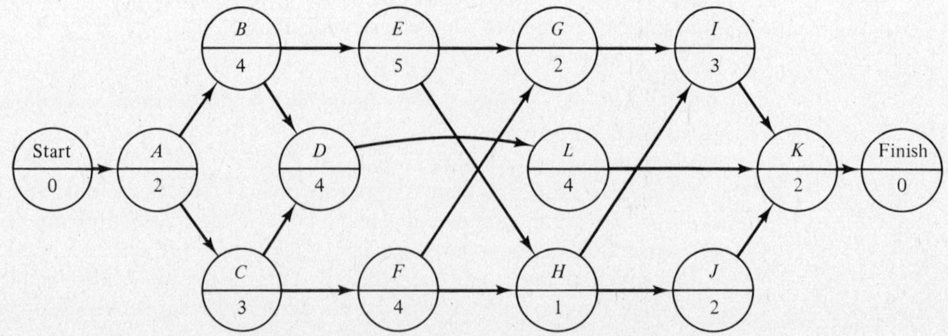

    1. Is it possible to reduce the estimated project duration by 2 weeks by crashing selected activities?

    2. If it is, how much extra will it cost you? If it is not possible, why not?

9-6. Given the network in the figure for Exercise 9-6, plus the information in the table, find the following:

(a) The expected completion time.

(b) The minimum possible project duration.

(c) The minimum cost to reach the minimum time found in (b). (Plot time–cost curve.)

Activity	Normal time (weeks)	Crash time (weeks)	Normal cost ($)	Crash cost ($)
1–2	1	1	20,000	20,000
2–3	3	2	30,000	35,000
2–4	4	3	25,000	28,000
2–6	2	1	20,000	30,000
3–5	9	7	10,000	15,000
3–4	0	0	—	—
4–5	7	5	20,000	28,000
4–7	6	4	15,000	20,000
6–7	7	6	10,000	12,000
5–7	3	2	12,000	13,000
5–8	5	3	18,000	21,000
7–8	2	2	20,000	20,000

Network:

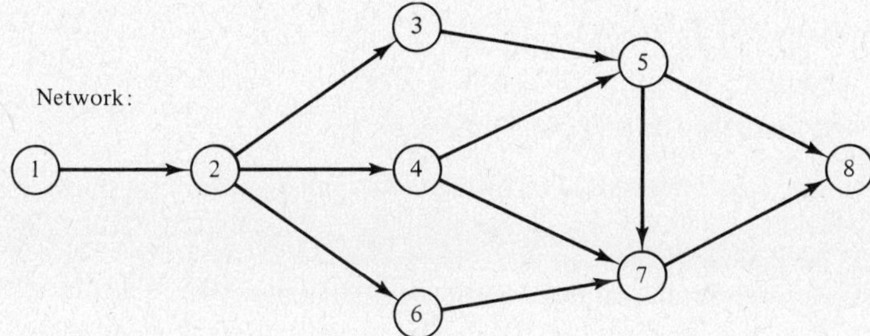

9-7. Two activities (A and B) have compatible resources. Activity A is on the critical path and requires 5 units of resource. Activity B is off the critical path and requires 10 units of resource. Activity B has 5 weeks of slack associated with it.

(a) How much, and in which direction, should A and B be adjusted; and how much will the project duration be reduced?

(b) Assume that A can only be reduced by 1 week without going below its crash time. Under these conditions, how much should B be adjusted and in which direction? How much slack would remain in activity B and how much would the project duration be reduced?

9-8. You are an engineer in charge of a large project which you have broken down into the following activities:

Activity	Duration (weeks)	Normal cost ($)	Crash time (weeks)	Crash cost ($)
A	2.00	5,000	1.0	7,000
B	4.00	12,000	3.0	15,000
C	3.67	8,000	3.0	9,000
D	2.17	6,000	1.5	8,000
E	5.83	15,000	4.0	20,000
F	3.00	10,000	2.0	13,000
G	4.17	13,000	3.0	18,000
H	7.33	14,000	5.0	20,000
I	6.50	13,000	5.0	16,000
J	8.00	11,000	6.0	14,000
K	10.33	14,000	7.0	21,000
L	9.67	12,000	6.0	16,000

From you work breakdown schedule, you have decided the activities should be sequenced as follows:

A is the first activity.

B, C, and D can be done simultaneously but only after A is completed.

E follows B.

F follows C.

G follows D.

H and J can be accomplished simultaneously, but only after F and G are completed.

I follows E and H.

K follows I.

L follows J.

K and L are the last activities in the network.

You are to accomplish the following tasks so that you can get the project completed on schedule:

(a) Formulate the network.

(b) Accomplish the initial critical path computations.

(c) Activity F requires two cranes while activities B and G require one each. Unfortunately, you only have three cranes available. Use an activity bar chart to resolve any conflict and reschedule the activities, if need be. Indicate the critical path and expected project duration. If a penalty of $5,000 is assessed for each week that the project extends beyond 35 weeks, and an additional crane can be rented for $2,500/week, should you rent an additional crane to solve any resource conflict rather than rescheduling?

(d) Manpower resources are freely interchangeable among activities E, H, and J; however, E and J require 8 men while H requires 16 men. Using the network as modified in (c), make further modifications to reduce slack time as much as possible. Again, indicate critical path(s) and expected project duration. No activity can be reduced below its crash level during reallocation.

(e) What would be the time savings and extra cost if you were to crash all activities for the network found in (d)? Indicate the critical path(s).

(f) Can the same time savings be accomplished more economically? If so, how? Indicate critical path(s) and additional and total project costs.

    (g) From the results of (f), what is the most economical way to compress project duration by 3 weeks?

9-9. Set up the network for the following relationships; find the critical path, and the expected project completion time. Indicate slack time for all activities.

A is the first activity.

M is the final activity.

B, C, and D can be done simultaneously and follow A.

M depends on L, K, and J.

F and E can be done simultaneously and depend on B.

G and H can be done simultaneously and depend on B and C.

J, K, and L can be done simultaneously.

L follows E.

I and J follow D and H.

I and J can be done simultaneously.

K depends on F, G, and I.

Activity	Duration
A	5
B	4
C	3
D	2
E	7
F	6
G	4
H	5
I	2
J	6
K	8
L	3
M	2

9-10. **(a)** Activities K, L, and M form part of a network, as is depicted in the figure for Exercise 9-10 on p. 358.

      K requires 1 crane.

      L requires 2 cranes.

      M requires 1 crane.

      Three cranes are available.

Is there a resource conflict? If so, how would you alleviate it? Assume that no additional cranes can be rented. What is the increase in project duration?

    (b) Use the diagram from (a) before any modifications are made. This is a new problem without the crane requirements. If activity L requires 8 people, and activity M requires 12 people, and these people can be freely exchanged between the activities, how should L and M be adjusted to optimize resource allocation? What reduction would you expect in the project duration?

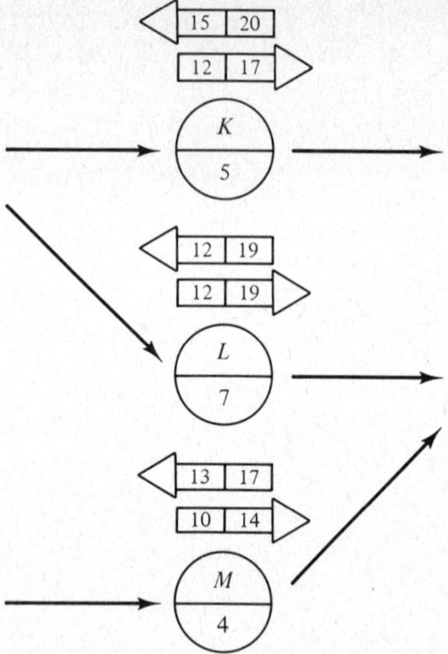

9-11. The following table gives the activity durations for the project described in Example 9.1, after the network modifications described in Example 9.3 have been made. (Times are in weeks.)

Activity	Normal duration	Normal cost ($)	Crash time	Crash cost ($)
1. Landscaping				
a. Visual and sound barrier planting	3.0	300,000	2.0	350,000
b. Slope stabilization	4.3	200,000	3.0	300,000
c. Final seeding	3.2	60,000	2.5	90,000
2. Control devices				
a. Roadway signs	3.0	50,000	2.5	60,000
b. Interchange and approach signs	2.8	50,000	2.3	60,000
3. Roadway				
a. Clear corridor	5.3	500,000	4.0	700,000
b. Cut and fill	6.3	750,000	5.0	950,000
c. Subbase	10.5	800,000	8.0	1,100,000
d. Surface	7.3	400,000	6.5	500,000
4. Bridges				
a. Abutments	4.8	400,000	4.5	450,000
b. Substructure	4.8	200,000	4.0	280,000
c. Superstructure	5.0	300,000	4.0	350,000
d. Surface	3.2	150,000	3.0	170,000
5. Interchanges				
a. Cut and fill	6.3	800,000	5.0	900,000
b. Subbase	5.0	600,000	4.0	650,000
c. Surfacing	2.8	250,000	2.5	280,000

The estimated project duration of 37.9 weeks and the associated critical path(s) can be found using the activity durations given above.

(a) Use time–cost optimization to determine how much extra it would cost to reduce the project duration to 34 weeks. Indicate which activities should be crashed, and by how much each should be crashed.

(b) Continue the time–cost optimization to find the minimum cost method of reducing the project to its minimum possible duration. Plot the time–cost curve and indicate which activity or activities are associated with each segment of the curve.

(c) If the required time of finishing the project is 36 weeks, and the penalty per week additional spent is $100,000, what is the best course of action to take? What additional costs are incurred above normal completion costs?

9-12. Construction activity data for the pedestrian/freight bridge of Exercise 7-10 is given in the following chart:

Activity number	Activity name	Dependencies	Duration (days)
0	Start		
1	Pier excavation	0	6
2	Cut in door no. 1	0	3
3	Demolish stairs	2	1.5
4	Install new stair	3	1.5
5	Relocate duct bank	1, 4, 15, 22	3
6	Construct piers and bollards	1, 4, 15, 22	3
7	Install roof and floor deck	5, 6, 16	4.5
8	Erect structural steel	5, 6, 16	3
9	Rough heating	7	4.5
10	Rough electric	7	2
11	Install windows	10, 27	2.5
12	Pour decks	8, 17, 24	2.5
13	Install metal siding	12, 18	3.5
14	Finish heat	9, 11, 13, 19	3
15	Temporary fire egress	2	3
16	Cut in openings	1, 4, 15, 22	3
17	Remove window/block opening	5, 6, 16	3
18	Corridor	8, 17, 24	3
19	Install roofing and flashing	12, 18	3.5
20	Finish electric	9, 19, 11, 13	3
21	Interior painting	14, 20	6
22	Rebar shop drawings	0	6
23	Blockwork	1, 4, 15, 22	3
24	Erect stairs	23	3
25	Painting and caulking	21	6
26	Touch up	25	3
27	Roof drain piping	12,18	1
28	Finish	26	

Using the given data:

(a) Construct an activity on arrow network and an activity on node network.

(b) Using the activity on arrow network, determine the critical path and the project length. Check your results using the computer.

(c) If all labor resources are compatible, use the given labor requirements to reallocate resources and shorten the project duration, if possible.

Activity	Number of workers required
5	2
6	6
16	2
23	8

(d) Activities 11, 13, and 19 require 1, 2, and 2 aerial platforms, respectively. The construction company has only two aerial platforms available to devote to this project initially; however, a third platform will become available on the 18th day. Resolve any resource conflicts that exist, and determine any increased project duration that will result.

(e) Determine the minimum project completion time and the accompanying cost if the appropriate activities are judiciously crashed. Assume the conflict of (d) has been resolved without rescheduling. Plot the time–cost optimization curve. Use the following normal and crash times and costs in determining your answer.

Activity	Normal time (days)	Crash time (days)	Normal cost ($)	Crash cost ($)
1	6	4.5	3,000	4,500
2	3	2.5	1,000	1,250
3	1.5	1.5	1,500	—
4	1.5	1	2,000	2,500
5	3	2	2,000	3,000
6	3	2.5	10,000	13,000
7	4.5	3.5	2,000	3,000
8	3	3	43,500	—
9	4.5	4	5,000	6,000
10	2	1.5	1,500	1,750
11	2.5	2	20,000	22,000
12	2.5	2	1,000	1,200
13	3.5	2.5	30,000	31,000
14	3	2	1,300	1,900
15	3	3	0	—
16	3	2.5	2,860	3,110
17	3	2.5	1,375	1,500
18	3	2	1,000	1,500
19	3.5	2	5,000	6,500
20	3	2	665	965
21	6	3	12,000	13,500
22	6	4	0	500
23	3	2	10,000	12,500
24	3	2	1,000	1,500
25	6	4	1,500	2,000
26	3	3	0	—
27	1	1	2,000	—

9-13. (a) Determine the critical path for the activity on node network given in the figure for Exercise 9-13.

(b) Transform the network (a) into an activity on arrow diagram.

(c) Develop a time–cost optimization curve for the network in (b). What is the minimum possible project duration, and the most economical way to reach this duration? Use the time–cost data given below. All times are in weeks.

Activity	Normal time	Crash time	Normal cost ($)	Crash cost ($)
A	5	3	10,000	11,000
B	4	4	15,000	15,000
C	3	2	6,000	7,000
D	5	4	8,000	9,500
E	7	5	8,000	11,000
F	5	4	6,000	8,000

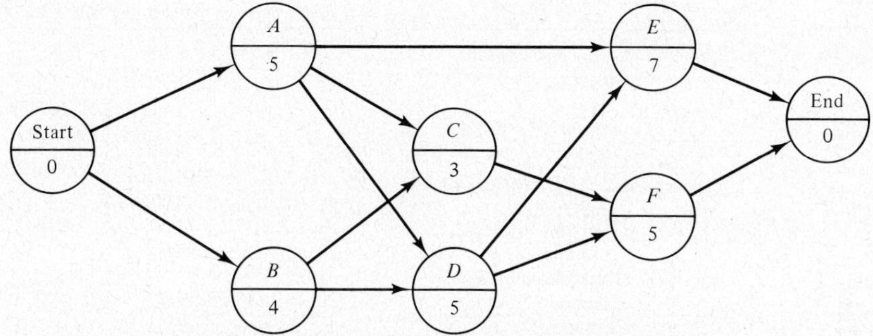

9-14 (a) Smooth out resource commitment for the activities shown in the figure for Exercise 9-14, so that no more than 30 units of resource are assigned to the project at any one time. Indicate how much the project duration would be expected to increase.

(b) How would the rescheduling change if the maximum resource commitment is to be 25 units at any one time? What is the expected increase in project duration for this case?

Activity	Resource required
A	10
B	15
C	10
D	5

9-15. A construction firm is preparing a bid for building a new service station. The specifications state that construction must be completed in 45 work days (9 weeks). A $2,000/day penalty will be assessed for each working day beyond 45 required for completion. This is the last project of the construction season, so no additional return can be expected by shortening the duration of the project. Activity descriptions, durations, and costs are shown in the table at the end of this exercise. The costs given reflect the actual cost plus the overhead and profit margins. Additional information is given below. What is the lowest bid that the firm should submit for the project?

Additional Information

1. Prefabricated walls rest on the floor.
2. The site is already cleared.
3. Excavations must be back-filled prior to the erection of walls.
4. Hydraulic lift, grease equipment, and air compressor must be installed inside the building after the building is completely closed in.
5. Only one activity involving concrete can be accomplished at any particular time. The activities that involve concrete work are H, K, S, Z, AA, and BB.

Activity	Description	Normal time (days)	Normal cost ($)	Crash time (days)	Crash cost ($)
A	Layout and excavate	2	6,000	1	9,600
B	Order and deliver signs	15	14,000	10	17,000
C	Order and deliver fuel tanks	20	16,000	14	22,000
D	Order and deliver hydraulic lift	30	13,000	13	19,000
E	Order and deliver air compressor	10	9,000	10	—
F	Order and deliver fuel pumps	25	16,000	20	20,000
G	Erect fence	3	8,000	2	9,000
H	Construct sign foundations	1	2,000	1	—
I	Install piping to tanks	5	4,000	3	6,000
J	Construct building foundations	3	15,000	2	19,000
K	Construct lift foundation	1	1,000	1	—
L	Install signs	2	3,000	1	5,000
M	Install fuel tanks	2	4,000	2	—
N	Install piping between tanks	4	4,000	4	—
O	Back fill tanks and grade	2	6,000	2	—
P	Install building piping	3	8,000	2	9,000
Q	Install hydraulic lift	3	4,000	2	6,000
R	Back fill building and foundation	1	2,000	1	—
S	Form and pour floor	2	9,000	2	—
T	Erect building walls	3	25,000	2	29,000
U	Erect building roof	10	23,000	6	28,000
V	Glaze windows	1	5,000	1	—
W	Install doors	2	4,000	2	—
X	Install air compressor	4	2,000	2	3,000
Y	Install grease equipment	4	1,000	2	2,000
Z	Construct pump islands	2	3,000	1	5,000
AA	Pour sidewalks	7	6,000	2	7,000
BB	Pave service and parking areas	5	25,000	3	30,000
CC	Install fuel pumps	2	2,000	1	3,000
DD	Order and deliver grease equipment	20	<u>10,000</u> $250,000	10	15,000

9-16. A new bus station is to be built for a medium-sized city. The activities associated with this project, and their inter-relationships, are given below:

Activity	Description	Normal Duration (weeks)	Normal Cost ($)	Crash Duration (weeks)	Crash Cost ($)
A	Remove interfering utilities	2	10,000	1.5	15,000
B	Survey lot	1	3,000	0.5	6,000
C	Clear and site excavation	2	15,000	1	20,000
D	Dig footings	2	5,000	1	10,000
E	Construct forms and pour footings	3	50,000	2	60,000
F	Electric service	1	15,000	1	15,000
G	Water service	0.5	6,000	0.5	6,000
H	Sanitary service	2	9,000	0.5	12,000
I	Steam service	1.5	7,000	1	9,000
J	Pour floor slab	5	90,000	4	110,000
K	Support frame	20	500,000	16	700,000
L	Roof	4	100,000	3	125,000
M	Interior frame	14	200,000	12	300,000
N	Exterior (block) walls	12	170,000	9	250,000
O	Interior walls	12	175,000	10	255,000
P	Air conditioning	10	300,000	8	400,000
Q	Plumbing	7	60,000	6	80,000
R	Safety	5	50,000	4	65,000
S	Electric	12	150,000	10	225,000
T	Painting	6	40,000	5	50,000
U	Floor covering	4	40,000	3	50,000
V	Seating	3	25,000	2	30,000
W	Exterior finish	10	150,000	7	225,000
X	Landscape	3	10,000	2	15,000
Y	Patios (entrances)	4	20,000	3	25,000

Interrelationships

A and B can be done simultaneously and are the first activities in the network.

C follows both A and B.

E follows D, which follows C and can be done simultaneously with H and I.

J follows H, I, and E.

K follows J.

F, G, M, N, and L follow K.

S and R can be done simultaneously and follow F and M.

Q follows G and M.

P and O can be done simultaneously and follow M.

T follows S, R, Q, P, and O.

V follows U, which follows O.

W and Y can be done simultaneously and follow N.

X follows W and Y.

T, V, X, and L are the last activities in the network.

(a) Develop an activity on node network for this project and find the expected project duration and critical path(s).

(b) Comment on the flexibility of this scheduling network. Where might opportunities exist for resource reallocation? Where might resource conflicts exist? Will resources be committed smoothly over the duration of the project? Suggest alternative schedules that might be better, and explain why they are better.

(c) Find the absolute minimum project duration for the original network.

(d) Determine the optimal time–cost trade-off curve to reach this minimum duration, or any duration between it and the normal duration.

9-17. For the CPM network given in the figure for Exercise 9-17 (numbers indicate working days):

(a) Find the critical path.

(b) Indicate the free float for activity Ⓒ–Ⓖ.

(c) The following table shows the cost per day of activity duration reduction. Assuming that any activity can be crashed to 50% of its original duration, how could $150 best be spent to maximize reduction in project duration?

Costs per day of savings ($)	
AB	200
BC	50
BD	200
BE	100
CG	30
DH	60
EJ	20
FH	—
GJ	100
HI	120
IJ	150
DE	150
DF	80

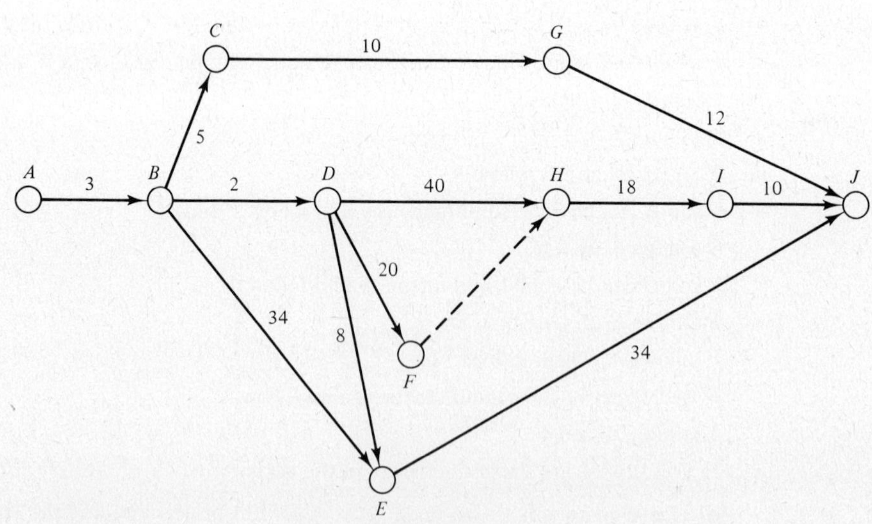

9-18. A contractor has taken out plans and specifications for a sewer contract in a small town in the Northeast. The contract consists of three projects which are detailed below.

A proposed bid, covering costs, is required of each contractor who applies for the job. One of the contract contingencies is that the work be done as soon as possible. The engineering firm overseeing the contract has estimated a completion time of 150 days. If the contract is finished before this, the contractor will realize a bonus of $2,000/day up to $30,000.

The contracting firm has decided to submit a proposal; but first, an estimate must be made concerning how long it should take the firm to complete the project given the contractor's contingencies and costs.

Based on the contracting firm's data, shown below, the contractor wants to estimate the firm's labor costs for the contract.

Contract Data

Project A	Project B	Project C
10-house services	12-house services	20-house services
4000 ft of pipe	5000 ft of pipe	6000 ft of pipe
4 manholes	4 manholes	3 manholes
Restoration	Restoration	Restoration
Shallowest	Shallowest	Shallowest
excavation = 15 ft	excavation = 19 ft	excavation = 12 ft
Deepest	Deepest	Deepest
excavation = 19 ft	excavation = 25 ft	excavation = 12 ft

Soil borings show the soil to be acceptable for use as backfill over the pipe. Therefore, no virgin soil will have to be trucked to the site. The contractor owns two backhoes, one of which can dig up to a depth of 25 ft, while the other can only dig to a 15-ft depth.

The contractor is required by state safety codes to have two trench boxes (one on top of the other) for any excavation over 20 ft deep. Also, the contractor is required to have at least one trench box up to 20 ft deep. The contractor at this time owns two trench boxes.

After meeting with the contracting firm's superintendents and foremen, visiting the job sites and reviewing the pertinent data, the contractor has come up with the following time estimates for each project. The contractor also realizes a cost of $250/day for the crew that works with the 25-ft backhoe and a cost of $200/day for the crew that works with the 15-ft backhoe.

Project A	Project B	Project C
$\frac{1}{2}$ day per lateral	$\frac{1}{2}$ day/lateral	$\frac{1}{2}$ day/lateral
75 ft of pipe/day	100 ft of pipe/day	100 ft of pipe/day
$\frac{1}{2}$ day/manhole	$\frac{1}{2}$ day/manhole	$\frac{1}{2}$ day/manhole
12 days restoration	10 days restoration	15 days restoration
Total—61 days for work	Total—58 days for work	Total—72 days for work
12 days for restoration	10 days for restoration	15 days for restoration

Use a bar chart and/or CPM network to establish how the job should be scheduled, what the expected completion time will be, and what the expected cost will be. Should the contractor expect to be eligible for the bonus offered? If so, how much of a bonus could be expected? What other costs besides labor costs should be considered in formulating a bid?

9-19. (a) The network in the figure for Exercise 9-19 describes a project using the activity on arrow method. Draw a Gantt chart for this network.

(b) The table below also shows those activities which require a diesel compressor. You have two compressors available. When does a resource conflict occur? What do you recommend? Explain.

Activity	Time (days)	Compressor required?
A	1	Yes
B	3	No
C	4	Yes
D	2	Yes
E	1	Yes
F	2	No
G	3	No
H	7	Yes

(c) Activity *B* may be compressed 2 days at a cost of $200/day saved, *C* 3 days at a cost of $400/day saved, and *E* 1 day at a cost of $50. No other activities can be compressed. You are offered a bonus of $400 for every day by which you beat the nominal schedule. What do you do?

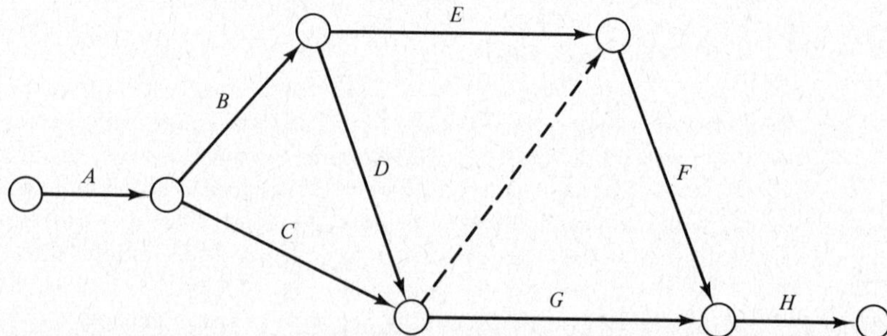

9-20. The network in the figure for Exercise 9-20 describes the construction of a reinforced concrete culvert using the activity on arrow convention. The times and costs for each operation for "normal" and "crashed" times are shown in the table on the next page.

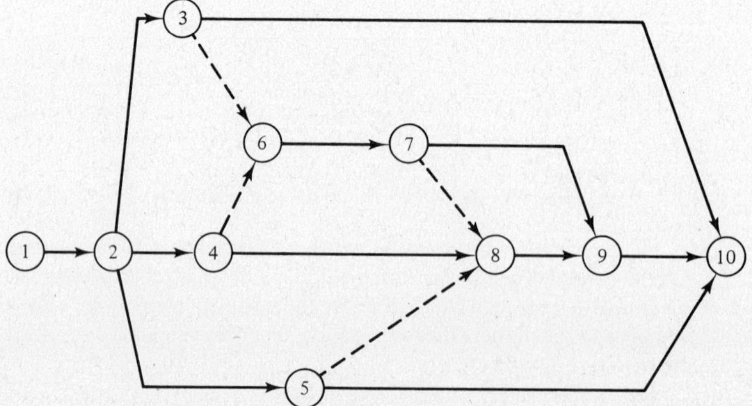

Operation	Activity	Normal		Fully crashed		Labor required
		Time (days)	Cost ($)	Time (days)	Cost ($)	
1–2	Base slab	4	1,500	3	1,750	5
2–3	North apron	4	750	3	850	7
2–4	North side walls	7	1,200	5	1,400	6
2–5	South apron	4	750	3	850	7
3–10	North approaches	15	4,100	10	6,000	8
4–8	South side walls	7	1,200	5	1,400	6
5–10	South approaches	10	2,400	10	2,400	8
6–7	North wing walls	6	1,500	5	1,700	4
7–9	North roof	14	2,000	14	2,000	4
8–9	South wing walls	7	1,600	6	1,900	4
9–10	South roof	14	2,000	14	2,000	5

(a) Construct a Gantt, or bar, chart for the project, showing the normal completion date of the project and the floats, where they are appropriate.

(b) Develop a chart or graph showing the number of laborers required, assuming that each task is started at its earliest start time. Show if it is possible to smooth out the labor requirements without delaying the project.

(c) If a bonus of $150/day for completion ahead of the "normal" time is offered, what is the optimum strategy? (Assume that labor resources are not a problem.)

9-21. A one-story commercial building is to be erected on the site of an abandoned house. The exterior and interior walls will be concrete block. The roof will be steel joists and steel decking, covered with rigid insulation and built-up roofing material. The floor will be a concrete slab on grade with an asphalt tile finish. Interior finish on all walls will be paint. The project has been broken down into the 18 activities given below. Expected activity durations are also given.

1. *Demolition.* Including the demolition of the present structure, removal of debris, and rough grading of the site. Time: 2 days.

2. *Foundations.* The construction of the foundation for the new structure. Time: 3 days.

3. *Underground Services.* The installation of water and sewer service from mains in the street. Time: 1 day.

4. *Exterior Walls.* The construction of the exterior block walls. Time: 6 days.

5. *Interior Walls.* The construction of the interior block walls. Time: 3 days.

6. *Roof Steel.* Includes the installation of the long-span joists, the bar joists, and the steel roof deck. Time: 2 days.

7. *Roof Finish.* Includes the installation of the rigid insulation on the steel deck and the installation of the built-up roofing. Time: 2 days.

8. *Floor Slab.* Assumed to include fine grading, sand fill and compaction, membrane waterproofing, casting and finishing of the floor slab. Time: 3 days.

9. *Floor Finish.* Includes the installation of asphalt tile and base mold. Time: 2 days.

10. *Rough Plumbing and Heating.* Includes the setting of the heating unit and rough ductwork, the installation of rough plumbing above the floor slab, vents, etc. Time: 3 days.

11. *Finish Plumbing and Heating.* Includes the final installation of the heating radiators, controls, and so on, and the installation of the sink and water closet. Time: 4 days.

12. *Rough Electrical.* The installation of conduit, service inlet and meter box, etc. Time: 3 days.

13. *Finish Electrical.* The installation of wire and fixtures. Time: 3 days.

14. *Rough Carpentry.* The installation of rough door frames, display window framing, etc. Time: 2 days.

15. *Finish Carpentry.* Installation of door trim, hanging of doors, etc. Time: 4 days.

16. *Ceiling.* The installation of the suspended acoustic ceiling. Time: 3 days.

17. *Display Windows.* The installation of the display window glass and metal trim. Time: 1 day.

18. *Paint.* Painting of interior walls and trim. Time: 3 days.

(a) Draw the best network relating these activities, and estimate the project duration. Indicate activities on the critical path and any slack time in the network.

(b) Use the network established for (a) to complete the following analyses:

1. Floor finishing, rough carpentry, finish carpentry, and finishing plumbing and heating each require one radial saw. Two saws are available.
   a. Will there be a resource conflict?
   b. If there is, what is the best way to eliminate it?
   c. When might it be more economical to rent another saw rather than rescheduling?
2. Use the network as modified by step 1. If rough carpentry, floor finishing, and finish carpentry take six, four, and three carpenters, respectively, will reallocation of these carpenters result in a time savings. If so, how much?
3. The following table indicates which activities could be crashed:

Activity	Normal time (days)	Crash time (days)	Normal cost ($)	Crash cost ($)
Demolition	2	2	10,000	—
Foundations	3	2	5,000	7,000
Underground services	1	1	1,000	—
Exterior walls	6	5	5,000	6,000
Interior walls	3	3	3,000	
Roof steel	2	1	4,000	5,000
Roof finish	2	1.5	2,000	2,500
Floor slab	3	2	2,000	3,000
Rough plumbing and heating	3	1.5	2,000	3,500
Finish plumbing and heating	4	3	3,000	4,000
Rough electrical	3	2.5	1,500	1,750
Finish electrical	3	2	2,000	3,000
Rough carpentry	2	2	1,200	—
Finish carpentry	4	4	3,000	—
Ceiling	3	3	2,500	—
Display windows	1	1	2,000	—
Paint	3	2.5	1,500	2,000

   a. What is the normal cost for the building?
   b. How much extra would it cost to finish the building 2 days early?
   c. What is the minimum duration of the project if all appropriate activities are crashed? What is the most economical way of reaching this minimum time?
   d. What is the total cost if all activities are crashed?

9-22. Shown in the figure for Exercise 9-22 is a time-scale network diagram for a bridge construction project. Reformulate this network into an activity on node network, and accomplish the necessary computations to establish the critical path and slack. Compare the two methods of representing project activity scheduling, discussing the advantages and disadvantages of each.

9-23. A building project consists of the following activities and activity durations:

Activity	Estimated duration (days)
Layout	22
Excavation	30
Pile work	48
Foundation	48
Structure	280
Facade	217
Interior Core	292
Interior Work	288
Site Work	120

The relationships among the activities are:

Excavation cannot start until the layout is finished.

When half-finished with excavation, pile driving can start.

When three-fourths finished with piles, foundation can be poured.

When two-thirds done with foundation, structure can start.

When half-finished with structure, facade can start.

When seven-eighths finished with structure, interior core can start.

When one-fourth finished with interior core, interior work can start.

When one-third finished with interior finish work, site work can start.

(a) Draw activity on arrow and time scaled arrow networks for this project. Determine the critical path(s) and activity slack.

(b) Develop a bar chart for this project.

(c) The following contracts are due to expire on the dates shown:

Laborers	4/30/83
Carpenters	7/31/83
Cement masons	8/31/83

The project is due to start on 1/15/83. Discuss ways of dealing with the possibilities of the various workers not resolving their contract disputes promptly (i.e., rescheduling activities so as to make use of the cheaper costs of labor before the contract expiration dates, or so the work can progress even if there are strikes).

9-24. A one-story (3200 ft^2) control building with office space and restroom facilities is being constructed. An activity on node network for the project is shown in the figure for Exercise 9-24.

Pertinent data for the activities are shown below:

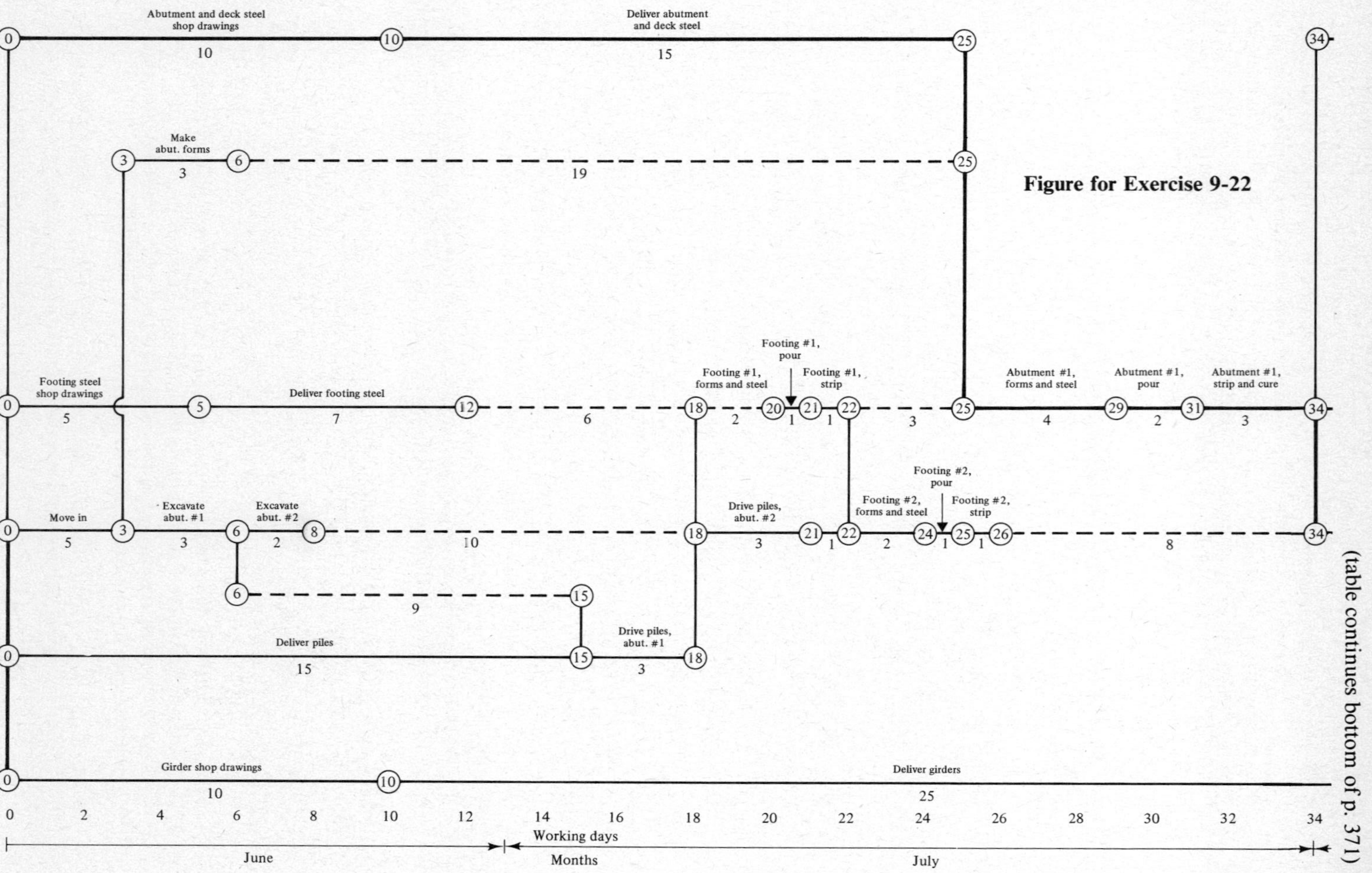

Figure for Exercise 9-22

(table continues bottom of p. 371)

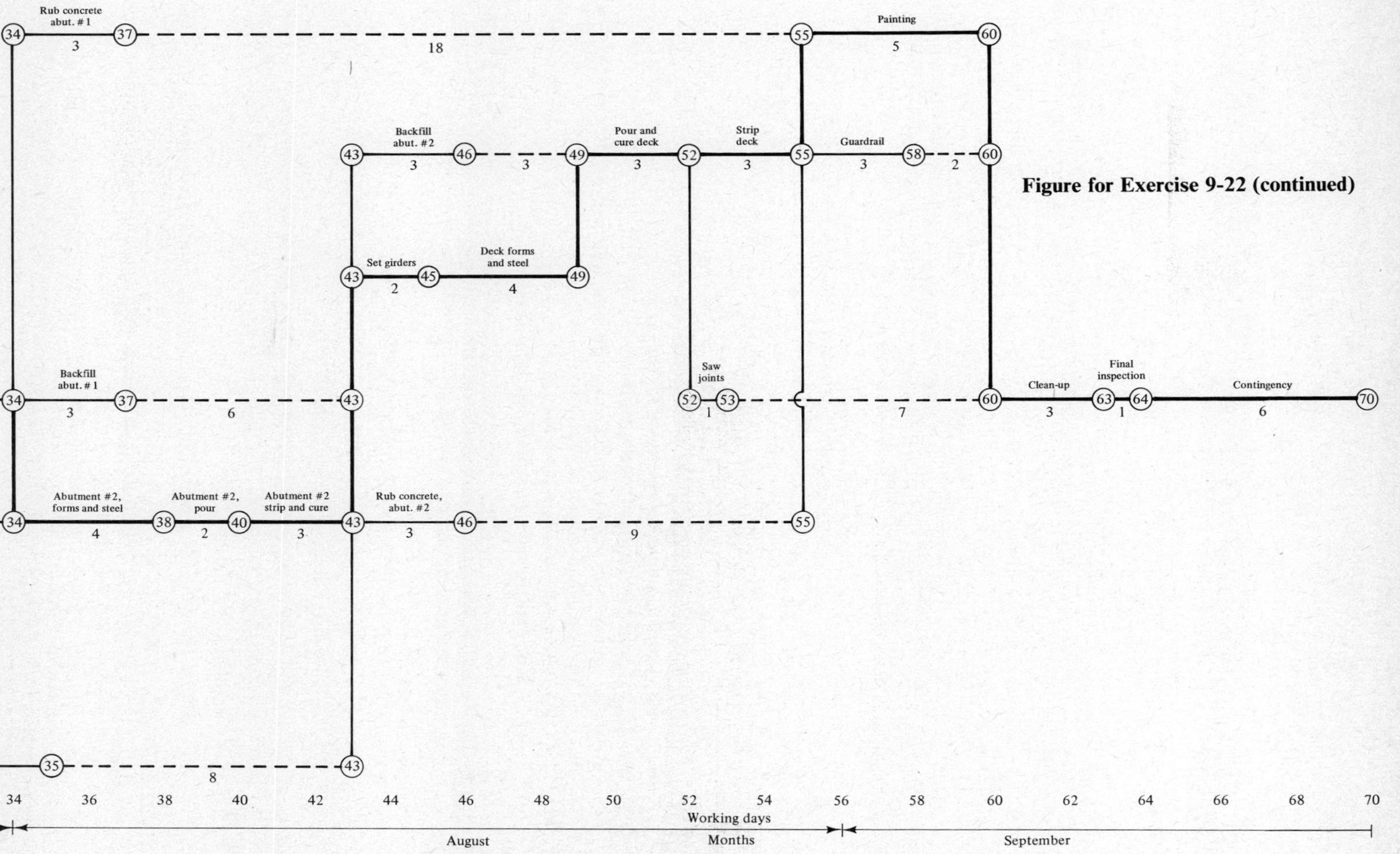

Figure for Exercise 9-22 (continued)

Identifier	Activity description	Duration (weeks)
A	Excavation	1
B	Footings	1
C	Concrete walls (foundation)	3
D	Backfill	2
E	Subgrade preparation	2
F	Structural steel and metal deck	3
G	Exterior masonry	4
H	Interior masonry	5
I	Interior slab on grade	2
J	Roofing	3
K	Exterior door frames	1
L	Interior door frames	1
M	Doors and hardware	2
N	Finishing and building fixtures	6
O	Painting and finishing	7
P	Ceiling system	3
Q	Flooring	3
R	Underground electrical	2
S	Electrical system	7
T	Plumbing service attachment	2
U	Interior plumbing	3
V	Heating and air conditioning	11

(a) Work backward to develop the work breakdown schedule for this project.

(b) Develop a time-scale network for the project.

(c) Develop a bar chart for the project.

(d) Compare the activity on node network, time scale network, and bar charts as project management tools.

(e) Indicate any changes you would make in the sequencing of activities for this project, and explain why you would make the changes.

9-25. A construction manager has developed the following relationships for the activities of a proposed project.

A is the first activity for the construction job.

B, C, and D can be done simultaneously, but only after A is done.

E follows B; F follows C; G follows D.

H and J can be done together, but only after F and G are completed.

I follows E and H.

K follows I; L follows J.

K and L are the last activities in the project, and can be accomplished simultaneously.

You are the assistant to the construction manager, and have been asked to develop a management strategy for the project. From information on past construction projects you have estimated normal and crash activity durations and costs. These are listed in the following table.

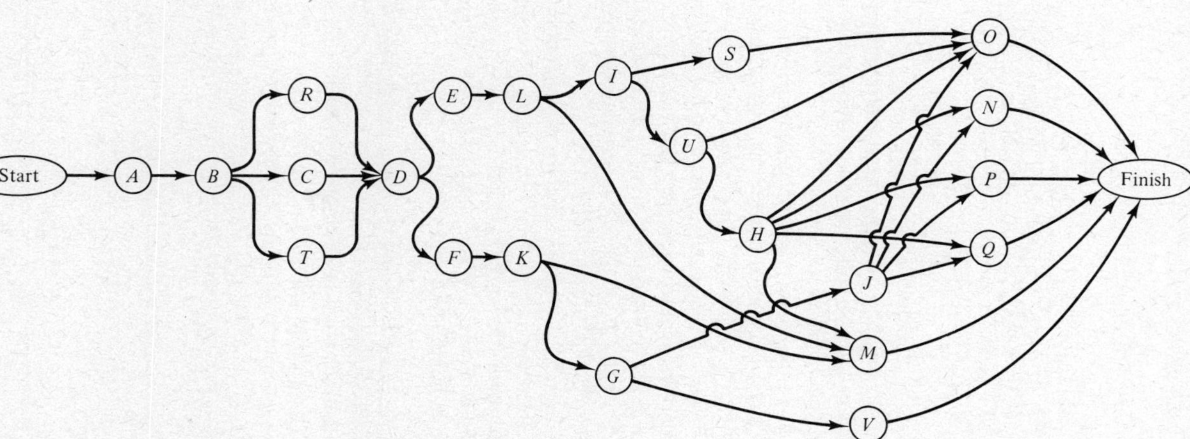

Figure for Exercise 9-24

Activity	Normal duration (weeks)	Normal cost ($)	Crash time (weeks)	Crash cost ($)
A	2.00	$ 9,000	1.0	$11,000
B	4.00	14,000	3.0	17,000
C	3.67	8,000	3.0	9,000
D	2.17	9,000	1.5	12,000
E	5.83	16,000	4.0	21,000
F	3.00	12,000	2.0	15,000
G	4.17	15,000	3.0	20,000
H	7.33	14,000	5.0	20,000
I	6.50	14,000	5.0	18,000
J	8.00	15,000	6.0	19,000
K	10.33	14,000	7.0	21,000
L	9.67	10,000	6.0	14,000

Complete the following steps to help generate information upon which to make decisions.

(a) Develop an activity on node network and find the estimated project duration, critical path, and activity slack.

(b) The project requires two cranes for activity F, while B and G require one each. Your firm only has two cranes available. The job is scheduled to be completed in 35 weeks or less, with a $5,000/week fine for each week beyond that. A crane can be rented for $2,500/week. Resolve any conflicts in the most cost-efficient manner.

(c) The nature of this construction job is that you may interchange laborers from activities E, H, and J. E and J require 8 laborers each while H needs 16 laborers. You are to consider shifting laborers around, to reduce your overall completion time and to reduce slack time as much as possible.

(d) As a contingency, you want to find the most economical way of reducing the project duration from the normal time to any time between it and the minimum possible project duration. It is quite possible that the expected project duration will have to be reduced by 3 weeks. What activities should be crashed, and what is the extra cost of achieving this time reduction?

9-26. The figure for Exercise 9-26 and the table that follows describe a job using the activity on arrow convention.

(a) Using the information in the table, construct a Gantt, or bar, chart for the project, showing the floats as appropriate and the expected duration of the job.

(b) Develop a chart or graph showing the resource needs under the condition that each task is started at its earliest start time. What is the effect on the project duration if the absolute maximum available number of units of resource Z is three (3)?

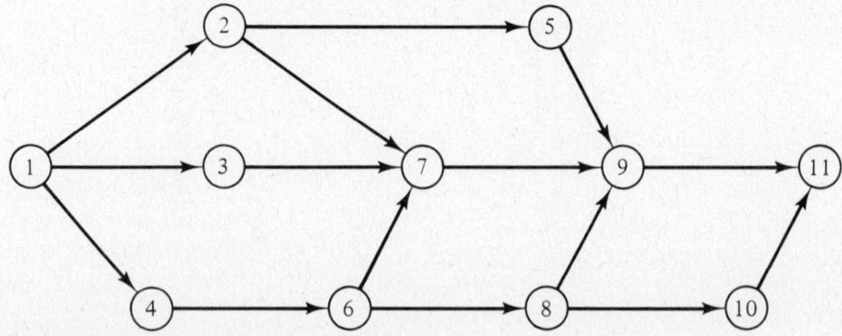

(c) If penalties and indirect costs are \$200/day, and it costs \$2,500 to obtain the necessary quantities of resource Z, what should the manager do?

(d) If the network represents a small construction job in which a road roller is required continuously for activities 1–4, 5–9, 2–7, 8–10, and 9–11; is it possible to schedule the program for the minimum time and use only one roller?

Activity	Duration (days)	Resource	Units of resource needed
1–4	3	X	2
4–6	10	Z	2
1–2	2	X	3
2–7	8	Z	1
1–3	4	Y	2
6–7	6	Z	2
6–8	5	Y	2
3–7	4	Z	1
2–5	6	Y	3
7–9	7	X	4
8–9	7	X	2
5–9	2	X	4
8–10	3	Y	3
9–11	8	Y	2
10–11	8	Z	2

9-27. A construction project consists of the activities listed below. Expected durations are also given.

Activity	Duration (weeks)
A	2
B	4
C	3
D	6
E	8
F	4
G	10
H	6
I	2
J	7
K	5
L	7
M	8
N	3
O	8

A, B, and C may be started immediately and done simultaneously. D and E may only start when A is finished and I must come after D. H and K must follow G, which can itself only be started after C. F follows B. J follows E, F, and H. K must precede L and N. O follows N. M cannot start until I, J, and L are finished. The job is complete when M and O are finished.

(a) Using the above information, construct a CPM diagram using the expected durations.

(b) Develop a Gantt chart for the project, showing the expected earliest start and latest finish times and the slacks.

(c) Develop a chart or graph showing the needs for cranes throughout the project, assuming that every activity is started at its earliest start time. Activities G, H, and O require two cranes, and activities E and F require one.

(d) Rental of a crane is by the month (i.e., in 4-week periods) at \$2,000/month. Penalties and indirect costs are \$800/week. What crane rental policy do you recommend?

SPECIAL TOPICS

Chapter 10

Decision Making Under Risk and Uncertainty

INTRODUCTION

Although this chapter is included in the Special Topics section of the text, parts of it have direct applicability to final planning/design and implementation. The concepts included in the chapter are grouped together here because they all deal with the risk and/or uncertainty involved in most engineering decision making. The statistical methods described in Chapter 3 are frequently helpful in quantifying the degree of uncertainty or risk involved in a decision. It will, however, usually be the engineer who must take the responsibility for quantifying and describing the degrees of uncertainty present in the various alternative decisions which are presented to the decision maker.

RISK AND UNCERTAINTY

The term "uncertainty" implies doubt, or lack of certainty, concerning the outcome produced by a given input to a system or component. System or component output may not be dependable or reliable; it may change or vary with apparent arbitrariness. The degree of uncertainty can range from merely a lack of absolute sureness, as in the case of the yield strength of a particular batch of steel, to such vagueness as to permit estimates based only on educated guesswork, such as what the state of the economy might be in 20 years.

Some degree of uncertainty is involved in virtually every decision that is made. The engineer may have a good idea of what the outcome produced by any given decision will be; however, there is always some finite probability that the outcome will not be the expected one. Changing conditions, unforeseen developments or uncontrollable chance occurrences may alter a predicted outcome.

Examples of uncertainty abound in all areas of engineering decision making. For example, a certain construction activity, which is part of a larger construction project, may under normal conditions take 12 weeks to finish. However, if everything goes extremely well, the weather is ideal, suppliers are able to deliver materials ahead of time, and sufficient equipment and labor are available, then experience has shown that the activity could be completed in 10 weeks. Conversely, if conditions are not favorable, the weather is bad, suppliers experience delays in material delivery, equipment breaks down, or labor strikes occur, then the activity can be expected to take approximately 16 weeks to finish. Note that the upper and lower bounds on these completion times are not symmetrical about the normal completion time. There is also a finite probability that the activity could take longer than 16 weeks or could possibly be finished in less than 10 weeks.

The difficulty involved in predicting hydrologic events is a prime example of uncertainty in applied engineering problems. The discipline of statistical hydrology involves the attempt to predict future events based on past experience. Although contributing factors such as time of the year, geographic location, and soil type can be identified, hydrologic variables, such as streamflow, exhibit sufficient variability to prevent their concise prediction. This uncertainty in prediction is easily understood when considering that rainfall is the primary input to the hydrologic system. Rainfall, although not a purely random occurrence, defies accurate prediction despite the sophisticated equipment utilized by meteorologists.

Risk is a logical extension of uncertainty. It involves estimating the effects of a probabilistic outcome. Under a given set of conditions, it should be possible to estimate the chances of injury, damage, or loss that could result from probabilistic system output. One type of risk, or decision, analysis involves minimizing the expected monetary loss, or maximizing the expected return, for controllable systems and components which have probabilistic outcomes. For example, there is always some risk involved in building in an area that is susceptible to flooding. There is an estimable probability of a certain level of flooding during any one year, and a risk, or monetary, loss can be associated with each flooding condition. Furthermore, certain preventive measures can be taken either to reduce the probability of flooding, or to reduce the damages from a certain degree of flooding. Decision analysis can be used to assess which level of protection will minimize the expected cost of damage from flooding. However, with all probabilistic analysis, the result obtained gives only the set of conditions that over a period of time should produce less damage than other sets of conditions. There is still a finite probability that the worst outcome could occur. Conversely, no flooding or damage is a possible outcome; or any other outcome between the extremes could occur.

DECISION ANALYSIS

The type of decision analysis described in this text involves estimating the expected monetary value (EMV) of a set of probabilistic outcomes. As stated previously, it will identify the alternative that has the best chance of minimizing cost or maximizing return. A tree network can be formed which reflects all the possible final outcomes that can result when decisions are made at key points within the decision network. These decision and outcome points can be organized, along with probabilities, into a tree network leading from the final outcomes at the branch tips of the tree, to the first decision at the base of the tree. The decision, chance, and outcome points can be represented as nodes, connected by links to form the tree.

Decision analysis is based on the calculation of an EMV for each outcome. The EMV for an outcome is its probability of occurrence, $P(X_i)$, times its expected value, X_i.

$$\text{EMV}_{\text{outcome}} = X_i P(X_i) \tag{10.1}$$

When n (two or more) outcomes are combined at a chance node, the EMV for the chance node is given by the summation of the EMVs of the outcomes.

$$\text{EMV}_{\text{chance node}} = \sum_{i=1}^{n} X_i P(X_i) \tag{10.2}$$

A useful property at the chance nodes is that the summation of the probabilities for the n outcomes has to add up to 1.0.

$$\sum_{i=1}^{n} P(X_i) = 1.0 \tag{10.3}$$

Application of the method is best illustrated through a practical application.

A consulting firm is designing a temporary access bridge to be constructed in an area of known seismic risk. The decision makers responsible for the bridge must be informed as to what degree of earthquake resistance incorporated into the design should minimize total costs associated with seismic design and damages.

Records for the past 50 years show that earthquakes have occurred in 40 of the 50 years. Of the 40 earthquakes, 24 had a magnitude of 0 to 4, 12 had a magnitude of 4 to 7, and 4 had a magnitude greater than 7.

The bridge can be made resistant to increasing earthquake magnitudes, with an associated increased cost. Cost versus earthquake resistance data are as follows:

Earthquake magnitude	Cost to be resistant to damage ($)
0–4	1,000
4–7	3,000
>7	6,000

If no earthquake resistance is incorporated into the design, then damage is dependent on the magnitude of the quake. If a certain level of earthquake resistance is incorporated, damage can still occur if an earthquake of larger magnitude is experienced. Table 10.1 shows the expected damages depending on the earthquake magnitude and the level of seismic resistance incorporated into the bridge design and construction.

A decision tree will be used to assist in determining which degree of earthquake resistance the bridge should be designed for. In this problem, there will be only one level of decision making: what magnitude of earthquake to design for. Therefore, the tree will start from this base decision, as shown in Figure 10.1(a). Regardless of what level of resistance the bridge is designed for, there are 40 chances out of 50, or a 0.80 probability that there will be an earthquake of some magnitude. Conversely, there is a 0.20 probability that there will not be an earthquake.

Thus, a chance node must precede each alternate design level leading into the decision node. These are shown as chance nodes B, C, D, and E of Figure 10.1(b). If no earthquake occurs, that is an outcome, and the total loss for that outcome will be the cost for the level of protection chosen. On the other hand, if an earthquake does occur, there will be another level of probability related to how severe the earthquake is, and a second chance node on that branch of the tree. When an earthquake does occur, the probability of its magnitude being between 0 and 4 is 24/40 or 0.60. Similarly, the probability that a quake will measure between 4 and 7, given that there is a quake, will be 0.30, and the probability of a quake of magnitude >7 will be 0.10. Note again that the probabilities add up to 1.0. The magnitudes of the earthquake, given that there is a quake, form outcomes, and lead into chance nodes F, G, H, and I. Figure 10.1(c) shows the addition of the second level of chance nodes, along with the expected values for the outcomes. The expected values for the earthquake outcomes are found by adding the cost of the designed protection to the additional cost for damages inflicted by an earthquake more severe than the one protected against. For instance, the sixth expected value represents $1,000 for protection against an earthquake of magnitude 0–4, plus $7,000 additional damage caused by an actual earthquake of magnitude greater than 7. The negative signs indicate that the dollar amounts represent costs, rather than returns.

Once the decision tree network is established, the EMVs of each node can be determined in the following sequence:

1. Start at the outcomes (ends of tree branches).
2. Compute the EMV for each outcome.

TABLE 10.1 EARTHQUAKE DAMAGE COSTS

Earthquake magnitude bridge designed for	Magnitude of earthquake incurred	Cost of damages ($)
None	0–4	3,000
	4–7	5,000
	>7	8,000
0–4	0–4	0
	4–7	3,000
	>7	7,000
4–7	0–4	0
	4–7	0
	>7	4,000
>7	>7	0

3. Compute the EMV for chance nodes closest to the outcomes. Note: If one chance node leads into another, use the EMV for the first chance node as the expected value for the next branch.
4. At the first level of decision nodes:
   a. Compare the EMVs of the alternatives.
   b. Choose the best.
   c. Mark out the others. (In some way, indicate that the particular branch will not be considered further.)
   d. Assign the EMV of the best alternative to the decision node.
5. Proceed to the next level of chance nodes and repeat 3 and 4.
6. Continue 3, 4, and 5 until the root of the tree is reached.
7. Alternatives that are not marked out give the best decision.

Applying Equation 10.2 to the chance nodes closest to the tips of the tree produces the following EMVs:

$$\begin{aligned}
\text{EMV @ } F &= 0.6(-3{,}000) + 0.30(-5{,}000) + 0.10(-8{,}000) = -\$4{,}100 \\
\text{EMV @ } G &= 0.6(-1{,}000) + 0.30(-4{,}000) + 0.10(-8{,}000) = -\$2{,}600 \\
\text{EMV @ } H &= 0.6(-3{,}000) + 0.30(-3{,}000) + 0.10(-7{,}000) = -\$3{,}400 \\
\text{EMV @ } I &= -\$6{,}000
\end{aligned} \quad (10.4)$$

These then become the expected values for their respective branches leading into the second chance node. Applying Equation 10.2 to the second chance nodes produces

$$\begin{aligned}
\text{EMV @ } B &= 0.80(-4{,}100) + 0.20(0) = -\$3{,}280 \\
\text{EMV @ } C &= 0.80(-2{,}600) + 0.20(-1{,}000) = -\$2{,}280 \leftarrow \text{least} \\
\text{EMV @ } D &= 0.80(-3{,}400) + 0.20(-3{,}000) = -\$3{,}320 \\
\text{EMV @ } E &= 0.80(-6{,}000) + 0.20(-6{,}000) = -\$6{,}000
\end{aligned} \quad (10.5)$$

Since node C has the least EMV, the bridge should be designed to resist earthquakes of magnitude 0 to 4. This conclusion is not obvious from the stated data, even for this simple example.

Again, it must be emphasized that designing the bridge to withstand an earthquake of magnitude 0–4 is no guarantee that the actual monetary cost will be $2,280. The actual costs could vary anywhere from $1,000 to $8,000. Choosing the 0–4 design only provides the best chance of minimizing costs.

Example 10.1 shows application of decision analysis to evaluating the best way to prepare a bid for a project. In this case, some of the outcomes have positive expected values, representing the amount of profit estimated to be received from the successful completion of the project. The decision tree indicates that hiring a consultant to help with the preparation of the bid is the best course of action.

Example 10.2 illustrates the application of decision analysis to a problem involving possible economic and/or physical harm to residents of a flood prone area. A difficult part of this problem is to quantify how much damage would be done by the severe flood described. Possible loss of goods and structures can be quantified, but putting a value on the possible loss of life is extremely difficult.

This example illustrates two levels of decisions. The first level is deciding whether

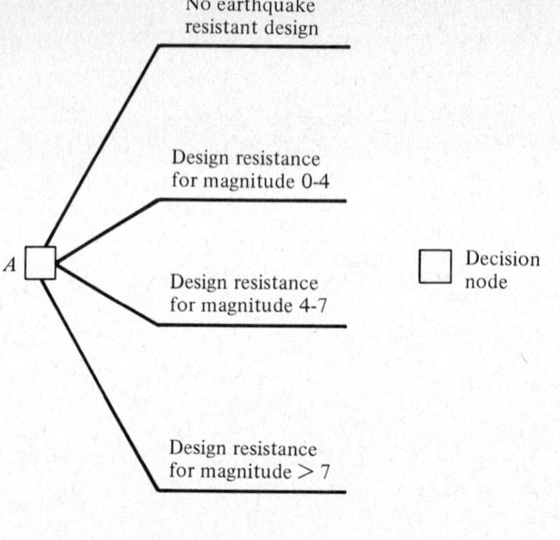

(a) First decision level.

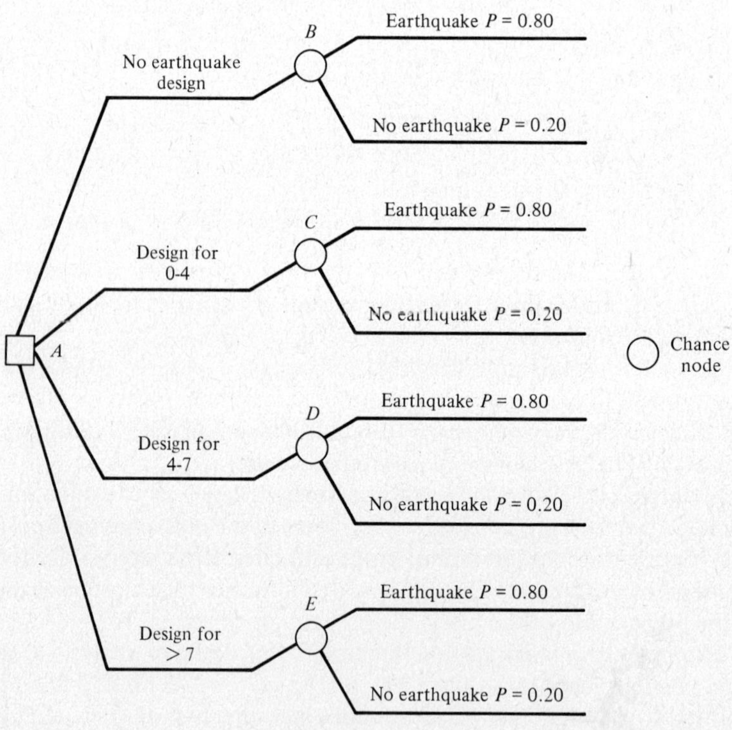

(b) First chance node.

Figure 10.1 (a)–(d) Decision tree for seismic design.

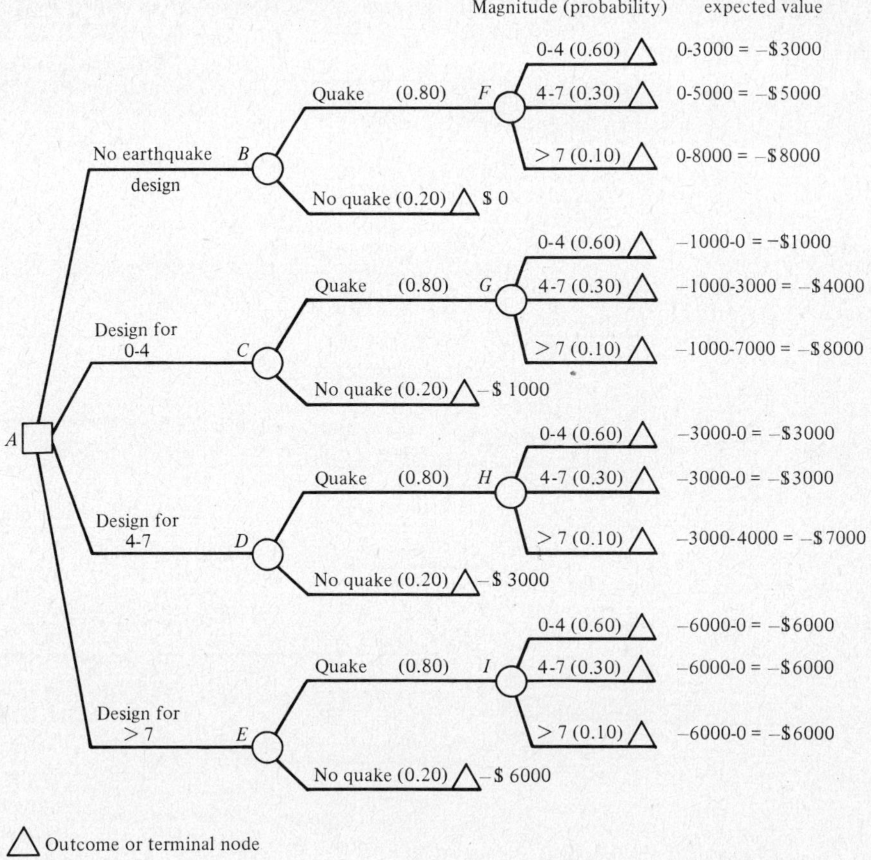

(c) Second chance node.

Figure 10.1 *(Continued)*

or not to institute the flood prediction system. Once it has been determined that the system should be instituted, a second decision must be made based on the predictions that the system provides. The probabilities that are associated with the various branches and outcomes are conditional probabilities, which are described in Appendix C.

It should be noted that the conclusions drawn in this problem might be different if different probabilities or costs were associated with the outcomes. Conditions could easily change enough so that in pure EMV terms, the best alternative might be not to institute the warning system. However, since our society places a high value on life, that alternative would probably not be politically attractive.

EXAMPLE 10.1

A consulting firm has to make a decision concerning bidding on a major project that has a projected profit for the firm of $100,000. Engineers for the firm estimate that they have about a 20% chance of receiving the contract if all the work on the proposed

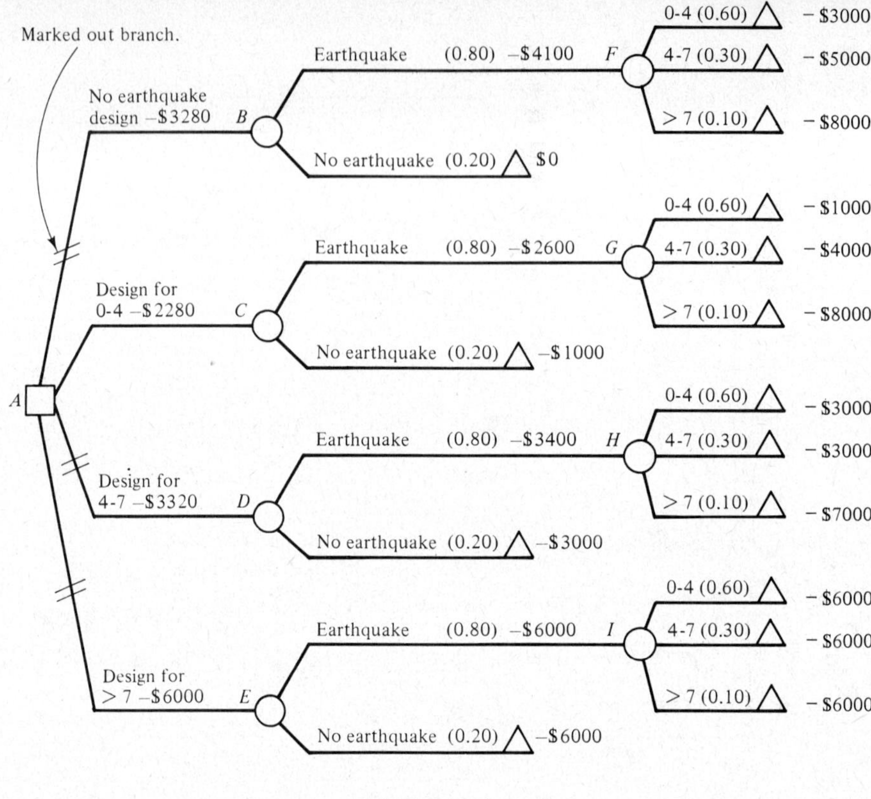

(d) Final decision analysis.

Figure 10.1 *(Continued)*

bid is done in-house. An outside consultant could be hired to help prepare the bid, which would increase the probability of success in being awarded the contract to about 50%. The outside consultant would cost $20,000. The decision tree shown in Figure 10.2 indicates that the firm should hire the outside consultant.

EXAMPLE 10.2 ____

A flood control district desires to minimize the flooding risk for its jurisdiction. Best estimates indicate that there is a 1 in 1000 chance that in the next 10 yr there will be a flood large enough to overtop the flood control dam upstream and cause extensive damage. If such a flood occurs without any warning, the loss in lives and property would have an economic value of approximately $100 million. If an impending flood could be predicted 24 hr in advance, an evacuation could be ordered. The evacuation would cost about $100,000, but would lower the economic loss to approximately $1 million.

A flood prediction system is available to the flood control district from the U.S. Army Corps of Engineers. Unfortunately, this prediction system is not always accurate.

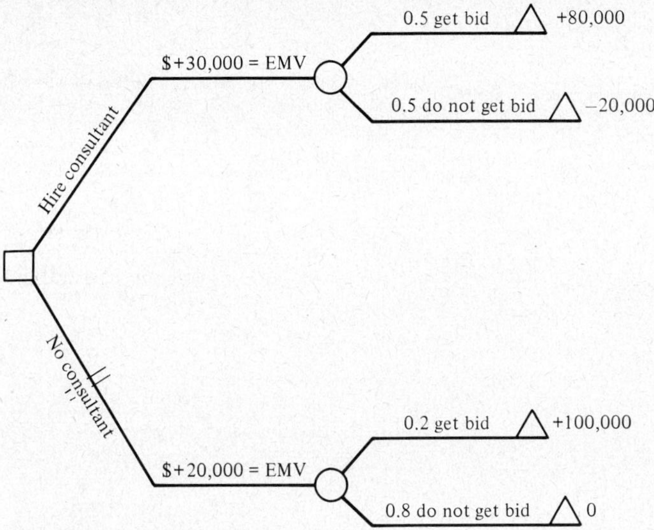

Figure 10.2   Decision tree for bidding.

Its developers have estimated that if actual conditions that will cause the major flood exist 24 hr in advance, the system will have the following reliability in predicting it:

Prediction	Percent of occurrences prediction will be made
Flood	60
Indefinite	25
No flood	15

On the other hand, the system may mistakenly predict the major flood when actual conditions will be less severe. Over the 10-yr period, the chances of this happening are:

Prediction	Probability
Flood	0.01
Indefinite	0.45
No flood	0.54

The flood control board wishes to determine if it would be wise to institute the flood prediction system; and if so, what the EMV of the system would be.

A decision tree is developed in Figure 10.3 which will assist in doing this. The first chance node on the detection system side of the tree involves the probability of one of the three indications being made. Since predictions can be made under two sets of circumstances, the probabilities associated with the indications are compound probabilities. The overall probability that the system will predict a flood when there actually is a flood is the product of the prediction reliability and the probability of a flood.

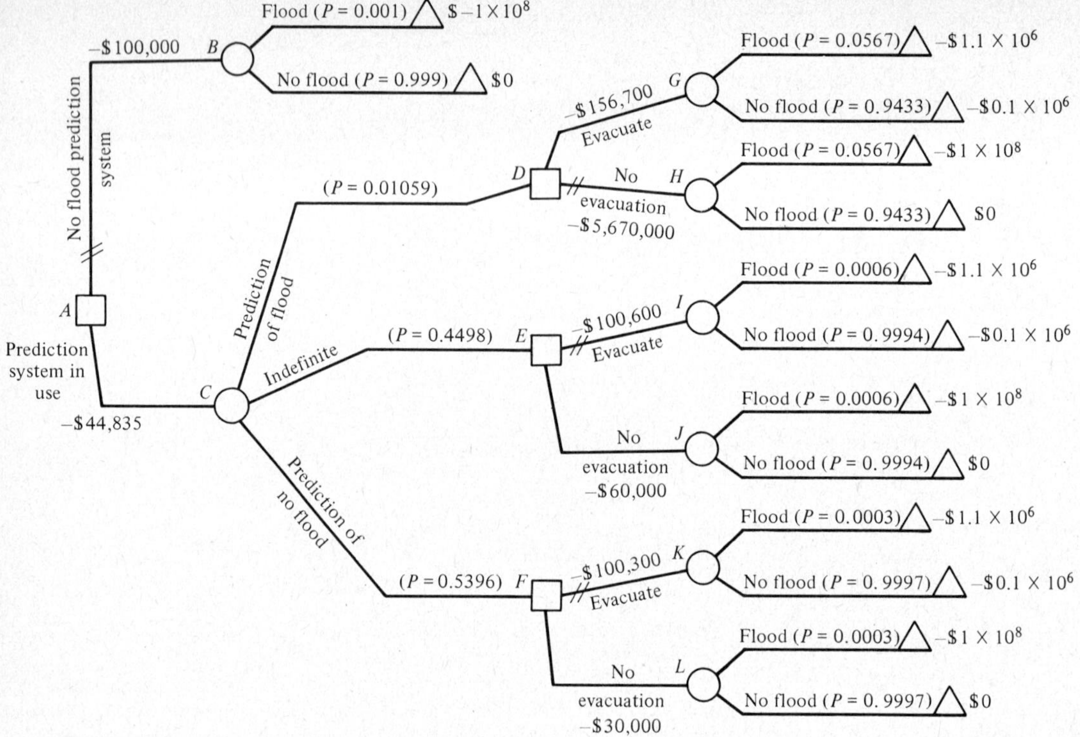

Figure 10.3   Decision tree for flood prediction system.

P(accurately predicting flood)
$= P$(predicting flood given that flood will actually occur)
$\times P$(flood will occur)
$= (0.60)(0.001) = 0.0006$ (10.6)

The total probability of predicting a flood is the sum of the probability of accurately predicting the flood and the probability of erroneously predicting a flood:

P(predicting flood) $= P$(accurately predicting flood)
$+ P$(erroneously predicting flood)
$= (0.60)(0.001) + (0.01)(0.999)$
$= 0.01059$ (10.7)

Similarly, the probability of an indefinite prediction is

P(indefinite prediction) $= (0.25)(0.001) + (0.45)(0.999) = 0.4498$ (10.8)

and the probability of a no flood prediction is

P(predicting no flood) $= (0.15)(0.001) + 0.54(0.999) = 0.5396$ (10.9)

As would be expected, the sum of the probabilities is 1.00.

After one of the three predictions is made, a decision about evacuation must be made. This decision leads to the final chance node of whether or not a flood will actually occur. The probability that a flood will occur when one is predicted is given by the equation

$$P(\text{having flood when flood is predicted}) = \frac{P(\text{accurately predicting flood})}{P(\text{predicting flood})}$$

$$= \frac{(0.60)(0.001)}{0.01059} = 0.0567. \qquad (10.10)$$

Conversely, the probability of not having a flood when one is predicted is 0.9433. This gives the probabilities for the top four outcome branches. The remainder of the outcome probabilities are found in a similar manner. After assigning the appropriate cost to each outcome and determining the EMV of the first-level chance nodes, decison nodes D, E, and F can be evaluated. If a flood is predicted, node D shows that the best course of action is to order an evacuation, while an indefinite or no flood prediction would best be ignored. The EMV for the chance node at C is $-\$44,835$, which is less than the $-\$100,000$ EMV for node B. Thus, it is wise to implement the flood prediction system, and the value of the system is said to be $\$55,165$.

UNCERTAINTY AS APPLIED TO PERT/CPM

When an engineer is applying the PERT/CPM procedures, it is sometimes both desirable and possible to estimate a range of durations for each activity of a project network. If this can be done, probability estimates can be generated for completion of strings of activities along the critical path. Using a range of activity duration estimates in PERT is essentially what differentiates it from CPM.

The completion times for an activity will range from the most optimistic time (a), to the most likely time (m), to the most pessimistic time (b). There is only a small probability of reaching either the most optimistic or the most pessimistic time. Generally, it is assumed that there is only a 1% chance of finishing earlier than the most optimistic time, and a 1% chance of finishing later than the most pessimistic time. There is also only one most likely time.

If sufficient data are available, a probability distribution can be developed to predict the probability of finishing anytime between the most optimistic and most pessimistic times. The distributions for durations of activities have been found to be asymmetrical, or skewed to one side or the other. This type of distribution, termed a beta distribution, appears in Figure 10.4. Frequently, insufficient data exist to explicitly define the probability distribution, so the most likely, most optimistic, and most pessimistic activity durations are estimated by people with a good deal of experience and knowledge as to how long an activity should take, and what factors might influence the completion time.

Once the most optimistic, most likely, and most pessimistic times are established, a single estimate of the expected completion time (t_e) for the activity (its expected value) must be generated. The accepted practice is to accomplish this by taking a weighted average of a, m, and b.

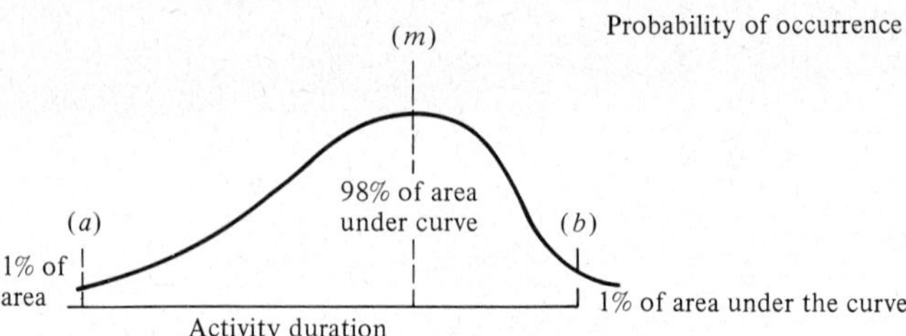

(a) More likely to finish late than early.

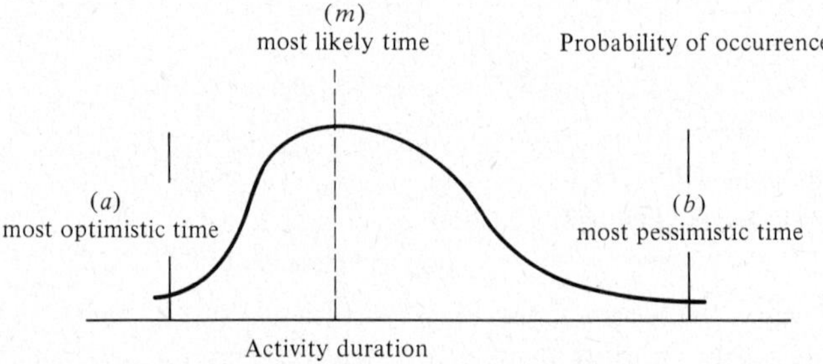

(b) More likely to finish early than late.

Figure 10.4 Beta distribution.

$$t_e = \left(\frac{a + 4m + b}{6} \right) \qquad (10.11)$$

This expected activity duration, t_e, would be used when computing the expected project duration for a CPM or PERT network.

It is generally assumed that the distribution of activity durations is close enough to normal that the normal distribution standard deviation can be used. Furthermore, it is assumed that the interval $b - a$ represents plus or minus three standard deviations ($\pm 3\sigma$) from the expected value. This is not an exact approximation, since $\pm 3\sigma$ represents 99.7% of the area under the normal distribution, while the area between a and b represents 98% of the area under the beta distribution. Considering all of the uncertainty involved in estimating a, b, and m, however, the approximation should be considered valid. This is a good assumption, since it facilitates making probability estimates based on the properties of the standard deviation and the normal probability distribution. If $b - a$ is $\pm 3\sigma$, then

$$\sigma_{\text{activity}} = \frac{b - a}{6} \qquad (10.12)$$

When several activities are connected, each one being dependent on the completion of the previous activity, the overall standard deviation of the string, which is also the standard deviation of the completion time for the series of activities, can be found by combining the standard deviations of the individual activities. Consider the three activities shown in Figure 10.5. The pooled, or overall, standard deviation for a series of activities is best given by

$$\sigma_e = \sqrt{\sigma_{1-2}^2 + \sigma_{2-3}^2 + \cdots + \sigma_{(n-1)-n}^2}$$

$$= \sqrt{\sum_{i=2}^{n} \sigma_{(i-1)-i}^2} \tag{10.13}$$

This is analogous to the estimate of the pooled standard deviation for different samples taken from the same population, given by the equation

$$\sigma_{\text{overall}} = \sqrt{\sum_{i=1}^{n} \frac{(X_i - \bar{X})^2}{n} + \sum_{j=1}^{m} \frac{(Y_j - \bar{Y})^2}{m} + \sum_{k=1}^{p} \frac{(z_k - \bar{Z})^2}{p}}$$

$$= \sqrt{\sigma_n^2 + \sigma_m^2 + \sigma_p^2} \tag{10.14}$$

For the series of activities given in Figure 10.5, the standard deviation for the series, which is also the standard deviation for the ending event, ④, is

$$\sigma_4 = \sqrt{\sigma_{1-2}^2 + \sigma_{2-3}^2 + \sigma_{3-4}^2}$$

$$= \sqrt{2^2 + 2^2 + 1^2}$$

$$= 3.0 \text{ weeks} \tag{10.15}$$

The expected completion time, T_E, for event ④ of 31.3 weeks, coupled with the standard deviation, σ_4, of 3.0 weeks, defines a normal probability distribution curve, which can be used to make probability inferences about completion times other than T_E. As expected, the probability of finishing on or before T_E is 0.50, since T_E is the mean value for the normal curve. Characteristics of the standard deviation and the normal probability distribution are discussed in Chapter 3 and Appendix C. They will be used here to estimate the probability of finishing either before, or after, a given time. A small standard deviation indicates that projected completion times will be tightly grouped around T_E, which means that the estimators are fairly confident of

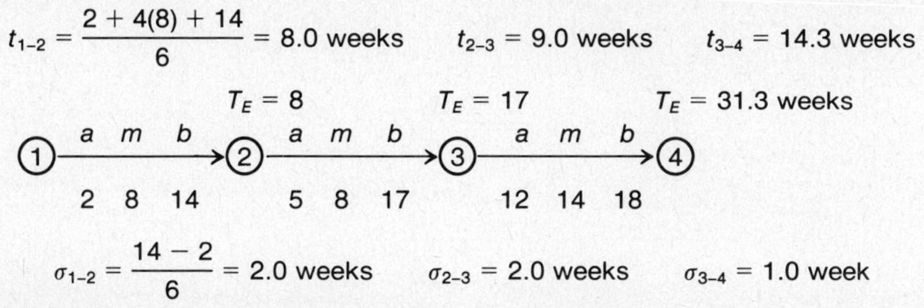

$$t_{1-2} = \frac{2 + 4(8) + 14}{6} = 8.0 \text{ weeks} \qquad t_{2-3} = 9.0 \text{ weeks} \qquad t_{3-4} = 14.3 \text{ weeks}$$

$$\sigma_{1-2} = \frac{14 - 2}{6} = 2.0 \text{ weeks} \qquad \sigma_{2-3} = 2.0 \text{ weeks} \qquad \sigma_{3-4} = 1.0 \text{ week}$$

Figure 10.5   Activities in series.

the accuracy of T_E. A larger standard deviation indicates more uncertainty in the estimate. The probability of finishing on or before a particular required time, T_L, is given by the area under the normal probability curve to the left of the time in question. This is illustrated in Example 10.3. The probability of finishing on, or later than, a given T_L is represented by the area under the curve to the right of the T_L, as shown in Example 10.4. When using Table C.1, it is important to remember that Z is the number of standard deviations the point in question is from the mean. For project networks,

$$Z = \frac{T_L - T_E}{\sigma_{\text{path}}} \tag{10.16}$$

When applying probability analysis to a **PERT** network, it is the standard deviations of the critical paths that must be considered. The standard deviation for an ending event of a network is the pooled standard deviation of the activities on the critical path leading to that event. Activities not on the critical path should not be considered. When multiple critical paths lead to an ending event, that ending event will have more than one standard deviation. When undertaking probability analysis under multiple critical path conditions, the probability associated with each standard deviation should be evaluated, and the worst case taken. When T_L is less than T_E, the smallest pooled standard deviation will govern, while if T_L is greater than T_E, the largest pooled standard deviation will produce the worst probability estimate.

Example 10.5 illustrates application of uncertainty principles to a project activity network for a highway construction project. The network for this project was developed in Example 9.1 and revised in Example 9.3.

EXAMPLE 10.3 ____

Suppose that T_E = 13.0 weeks, T_L = 12.0 weeks, and σ = 1.4 weeks. What is the probability that the project will be finished in 12 weeks or less?

Figure 10.6 shows a definition sketch for $T_L < T_E$. Using Equation 10.16,

$$Z = \frac{T_L - T_E}{\sigma} = \frac{12.0 - 13.0}{1.4} = -0.7143$$

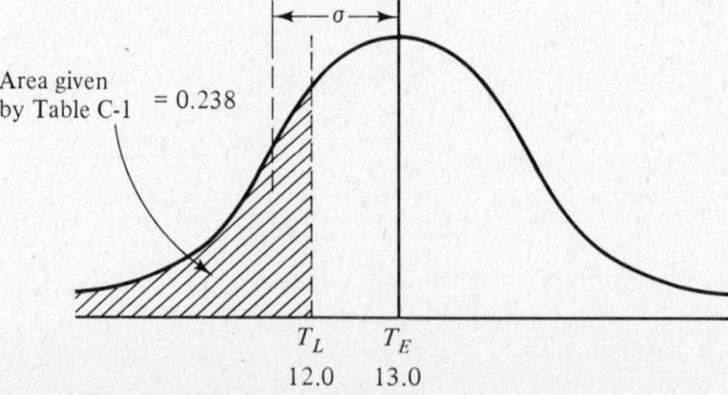

Figure 10.6   Definition sketch for $T_L < T_E$.

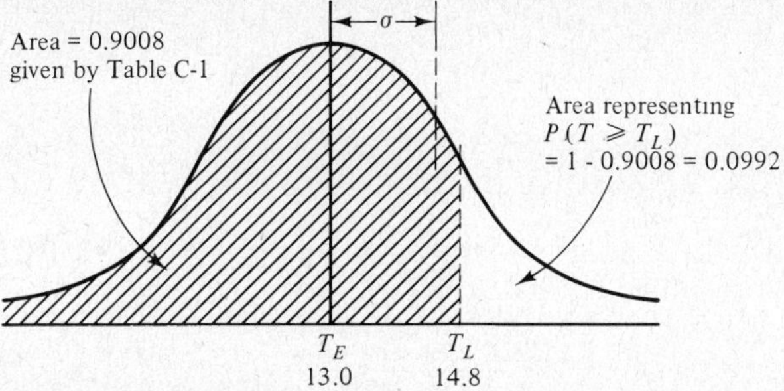

Figure 10.7 Definition sketch for $T_L > T_E$.

Utilizing Table C.1 and linear interpolation, the area under the curve from $-\infty\sigma$ to -0.7143σ is approximately

$$0.2389 - 0.0031\,\frac{(0.7143 - 0.71)}{(0.01)} = 0.2376$$

Therefore, $P(T \le T_L) \sim 0.238$, or the probability that the project will be finished in 13 weeks or less is about 24%.

EXAMPLE 10.4

Suppose $T_E = 13$ weeks, $\sigma = 1.4$ weeks, and T_L (latest allowable completion date) = 14.8 weeks. What is the probability that the project will be completed after T_L?

Figure 10.7 shows the definition sketch for $T_L > T_E$. From Equation 10.16,

$$Z = \frac{T_L - T_E}{\sigma} = \frac{14.8 - 13.0}{1.4} = 1.286$$

Using Table C.1 and linear interpolation, the probability of $T \le T_L$ for 1.286σ is approximately 0.9008. Therefore, the probability that $T \ge T_L$ is given by

$$P(T \ge T_L) = 1 - 0.9008 = 0.0992 \tag{10.17}$$

EXAMPLE 10.5

A project activity network for a highway construction project was developed in Example 9.1. In Example 9.3, this network was modified to resolve any resource conflicts and to accommodate certain resource reallocation. Figure 10.8 shows the network after the adjustments have been made. Activity names have been left off to simplify the diagram; however, they can be ascertained from Table 10.2. The estimated project duration after adjustment is 37.9 weeks.

The activity durations used in the previous examples were derived from estimates of most optimistic, most likely, and most pessimistic completion times for activities. These estimates are given in Table 10.2. It can be assumed that the most optimistic and most pessimistic times will not change with the amount of resource reallocation that is indicated. Therefore, the estimates of the standard deviations for the activities (σ_{activity}) can be found using Equation 10.12. These are also shown in Table 10.2.

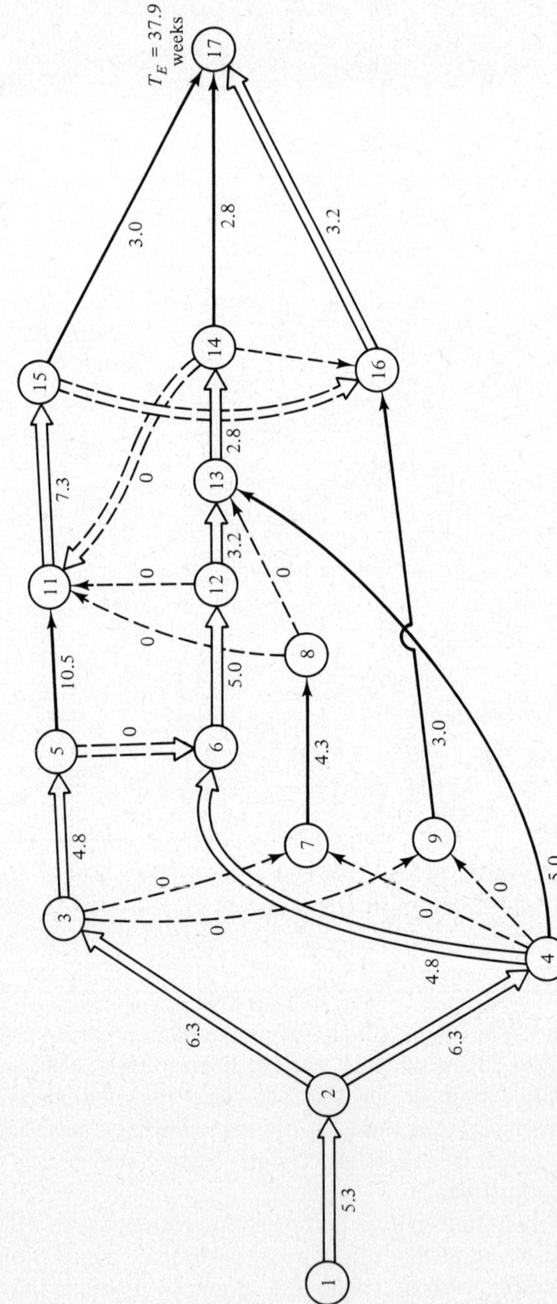

Figure 10.8 Highway construction project schedule after adjustment.

TABLE 10.2   HIGHWAY CONSTRUCTION PROJECT DATA

Activity	Ending nodes	Time (weeks) a	m	b	t_e	Adjusted t_e	$\sigma_{activity}$
1. Landscaping							
a. Visual and sound barrier planting	9–16	2	3	4	3.0		0.333
b. Slope stabilization	7–8	3	4	7	4.3		0.667
c. Final seeding	16–17	2	3	5	3.2		0.500
2. Control devices							
a. Roadway signs	15–17	3	3	3	3.0		0.000
b. Interchange and approach signs	14–17	2	3	3	2.8		0.167
3. Roadway							
a. Clear corridor	1–2	4	5	8	5.3		0.667
b. Cut and fill	2–3	5	7	9	7.0	6.3	0.667
c. Subbase	5–11	8	10	15	10.5		1.167
d. Surface	11–15	6	7	10	7.3		0.667
4. Bridges							
a. Abutments	4–6	3	5	6	4.8		0.500
b. Substructure	3–5	4	5	5	4.8		0.167
c. Superstructure	6–12	4	5	6	5.0		0.333
d. Surface	12–13	3	3	4	3.2		0.167
5. Interchanges							
a. Cut and fill	2–4	4	6	7	5.8	6.3	0.500
b. Subbase	4–13	4	5	6	5.0		0.333
c. Surfacing	13–14	2	3	3	2.8		0.167

After determining that the expected completion time is 37.9 weeks, notification is received that the project has to be completed in no longer than 33 weeks. Two probability estimates can be made for this problem. In the first instance, the probability of finishing in 33 weeks if no further network modifications are made can be determined. Then, assuming that time–cost trade-offs for crashing activities could reduce the project duration to 33 weeks, a further reduction to produce a 90% certainty of finishing in 33 weeks could be made. For this initial estimate it can be assumed that the critical path pooled standard deviation would not change during the project duration reduction.

The first step in this analysis is to determine the pooled standard deviation for the critical path or paths of the network. There are two critical paths in Figure 10.6. Standard deviations for the activities on these paths are given in Table 10.3. Equation 10.13 can be utilized to determine the pooled standard deviation for each critical path.

$$\sigma_{upper} = \sqrt{\sum_{i=1}^{11} \sigma_i^2} = \sqrt{1.779} = 1.334$$

$$\sigma_{lower} = \sqrt{\sum_{j=1}^{10} \sigma_j^2} = \sqrt{1.805} = 1.344$$

(10.18)

TABLE 10.3   STANDARD DEVIATION FOR ACTIVITIES ON
                 CRITICAL PATHS

Upper critical path	$\sigma_{activity}$	Lower critical path	$\sigma_{activity}$
1–2	0.667	1–2	0.667
2–3	0.667	2–4	0.500
3–5	0.167	4–6	0.500
5–6	0	6–12	0.333
6–12	0.333	12–13	0.167
12–13	0.167	13–14	0.167
13–14	0.167	14–11	0
14–11	0	11–15	0.667
11–15	0.667	15–16	0
15–16	0	16–17	0.500
16–17	0.500		

The required reduction in project duration is 4.9 weeks. This represents $-3.67\sigma_{upper}$ and $-3.65\sigma_{lower}$. Table C.1 shows that the area under the normal probability curve from $-\infty$ to -3.67 or from $-\infty$ to -3.65 is less than 0.0002. In other words, there is almost no chance of the project's being finished in 33 weeks or less without crashing some activities. Also, in this case, the difference in standard deviation for the two paths is insignificant.

The next step is to assume that the estimated project duration, T_E, has been reduced to 33 weeks through time–cost optimization of activity crashing. Exercise 9-11 established that it is possible to reduce T_E to 33 weeks. It is also being assumed that the pooled estimates of standard deviation for the critical paths will not change during the crashing of activities and the reduction in project duration.

To find the planned project duration that would provide a 90% chance of finishing in ≤ 33 weeks, the duration of 33 weeks becomes T_L, and Table C.1 is utilized to find a new T_E that will provide the 0.90 probability. A value of $Z = +1.28$ provides a probability of approximately 0.90. Therefore,

$$1.28 = \frac{T_L - T_E}{\sigma}$$

$$= \frac{33.0 - T_E}{\sigma} \tag{10.19}$$

The upper critical path standard deviation will provide an estimate of $T_E = 31.29$ weeks, while σ_{lower} gives an estimate of 31.28 weeks. The difference in this case is inconsequential; however, it was carried through to illustrate the method. The smaller of the two estimates would represent the worst case, because it would require more crashing of activities. Again, time–cost optimization would have to be undertaken to determine if a project duration of 31.28 weeks is possible; and if so, if it is economically justifiable. Exercise 9-11 established that the minimum project duration for this case is 32 weeks, so a T_E of 31.28 weeks could not be reached through crashing of activities.

SUMMARY

The risk and uncertainty that are a part of nearly any ordinary decision-making process also affect most engineering decisions. Just as the layman qualitatively weighs the risks of most normal activities, studying alternatives and possible outcomes before making a judicious choice, so, too, the engineer conducts a program of research into risk and uncertainty before making a recommendation to the project decision maker. The engineer, however, has the responsibility of quantifying the risk and uncertainty in order to make a substantive recommendation.

This chapter discussed two of the many methods available for assessing risk and uncertainty. Decision analysis (a type of risk analysis) and uncertainty as applied to the Program Evaluation and Review Technique (a variation of the Critical Path Method) were covered in some detail. Further useful methods for assessing risk and uncertainty, such as utility theory, can be found in other engineering literature.

APPLICATIONS EXERCISES

10-1. You are the project manager for a project that is under way. At the completion of the 15th week, you find that your expected completion time is 44 weeks, while your contract requires that you have the project finished in 40 weeks. The critical path for the remaining activities is given in the figure for Exercise 10-1. Parallel activities to these have sufficient slack to accommodate your decision-making options. You could crash the remaining activities on the critical path by either 2, 4, or 6 weeks for the following extra costs:

Time reduction (weeks)	Cost increase ($)
2	3,000
4	7,000
6	12,000

If the project is allowed to extend beyond 40 weeks, you will incur a $1,000/week carrying charge. This is essentially the opportunity cost for remaining committed to this project beyond 40 weeks. The maximum carrying charge that you will consider will be $10,000. Also, your contract calls for the following penalty charges to be assessed for tardiness in project completion:

Time for completion (weeks)	Penalty ($)
$T_E \leq 40$ weeks	0
$40 < T_E \leq 42$	2,000
$42 < T_E \leq 44$	4,000
$44 < T_E \leq 46$	6,000
$46 < T_E \leq 48$	8,000
$T_E \geq 48$	10,000

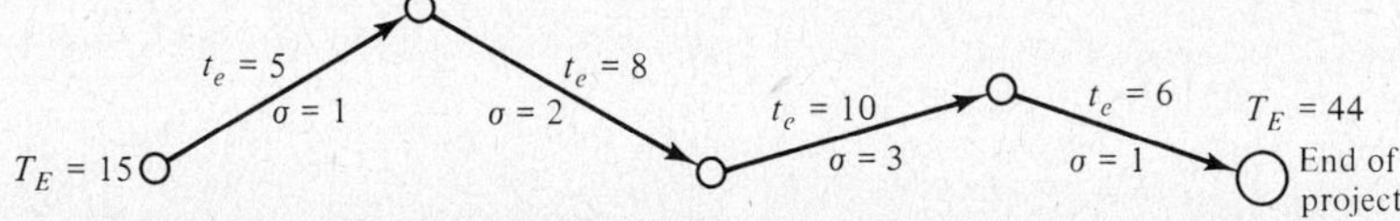

Assuming that the standard deviations of the critical path remain the same after crashing, what is your best management decision at the 15-week point?

10-2. Given the two critical paths shown in the figure for Exercise 10-2:
  (a) What is the expected project duration?
  (b) What is the probability of finishing more than 2 weeks later than the expected project duration?

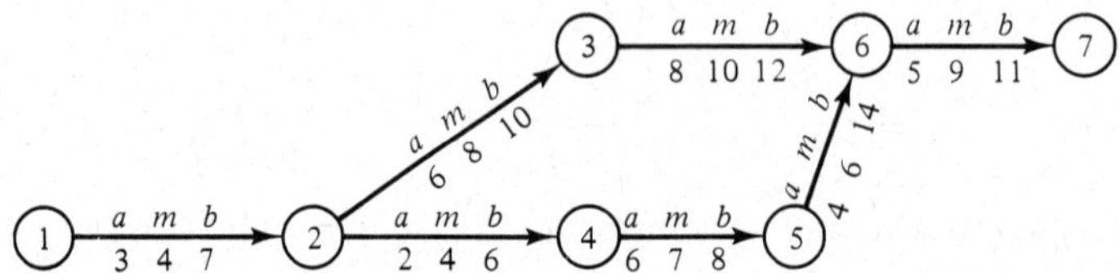

10-3. The activities shown in the table given below form the critical path through a network. All times are in weeks.

Activity	a	m	b	Normal cost ($)	Crash time	Crash cost ($)
1	4	5	8	10,000	4.5	14,000
3	1	6	8	12,000	4.0	16,000
7	5	9	10	15,000	7.0	20,000
9	6	7	9	17,000	6.0	19,000
10	2	5	11	12,000	5.0	17,000

  (a) What is the expected completion time?
  (b) If your latest allowable completion time is 30 weeks, what is the probability that you will make it?
  (c) Considering only the activities on the critical path given above, what will be the extra cost of crashing enough activities to give you a 75% chance of completing work in 30 weeks, assuming that the standard deviations remain the same?

10-4. Problem 9.6 presented a project activity network and associated data. Use the estimates for activity standard deviations given below to estimate the following probabilities:
  (a) The probability of finishing the project on or before the minimum time found in (b) of Exercise 9-6.
  (b) The probability of finishing the project more than two weeks after the expected completion time found in (a) of Exercise 9-6.

Activity	Activity σ
1–2	0.5
2–3	0.5
2–4	1.0
2–6	0.5
3–5	2.0
3–4	0
4–5	1.5
4–7	1.0
6–7	0.5
5–7	0.5
5–8	0.5
7–8	0.5

10-5. The following data were used to estimate the original activity durations used in Exercise 9-11.

		Time (weeks)		
Activity		a	m	b
1. Landscaping				
a. Visual and sound barrier planting		2	3	4
b. Slope stabilization		3	4	7
c. Final seeding		2	3	5
2. Control devices				
a. Roadway signs		3	3	3
b. Interchange and approach signs		2	3	3
3. Roadway				
a. Clear corridor		4	5	8
b. Cut and fill		5	7	9
c. Subbase		8	10	15
d. Surface		6	7	10
4. Bridges				
a. Abutments		3	5	6
b. Substructure		4	5	5
c. Superstructure		4	5	6
d. Surface		3	3	4
5. Interchanges				
a. Cut and fill		4	6	7
b. Subbase		4	5	6
c. Surfacing		2	3	3

Using these data, complete the following analyses:

(a) What is the probability of finishing in the required 36 weeks without crashing any activities?

(b) If enough activities are crashed to reduce the duration to 36 weeks, what is the probability of finishing in 36 weeks?

(c) Is it possible to crash enough activities to be 90% certain of finishing in 36 weeks? Show your work, and indicate how much additional cost would be associated with reaching this degree of certainty.

(d) Reevaluate your decision as to the best course of action to take when the penalty is $100,000/week of delay beyond 36 weeks, in terms of expected monetary value.

10-6. The expected project duration for Exercise 9-4 (after resolving resource conflicts in activities E, F, and G, and reallocating resources between activities H and I) is 13 weeks. The following standard deviations have been estimated for the activities of the project:

Activity	Standard deviation (σ)
A	1.0
B	0.5
C	0.33
D	0.5
E	0.75
F	0.0
G	0.75
H	0.5
I	0.67

(a) What is the probability of finishing the project in 16 weeks or less?

(b) If activities B, C, and D are finished at their earliest finish times (EFT), what is the probability, at that point, of finishing the project in 11 weeks or less?

(c) If activity D is delayed by 2 weeks because of a strike, what is the probability, at that point in time, of finishing on or before the original expected project duration of 13 weeks?

10-7. This problem refers back to Exercise 9-8. Assuming that resource reallocation has not changed the standard deviation of activities E, H, or J, estimate the standard deviation of the network ending event. What is the probability that the project will be completed before the required completion time of 35 weeks? Now, suppose that Typhoon Person hits and halts all activity on activities E, H, and J for 2 weeks. What is your probability of finishing in 35 weeks? If your probability of finishing is low, how might you improve it?

The original estimates of most optimistic, most likely, and most pessimistic activity times are given below:

Activity	Time (weeks)		
	a	m	b
A	1	1.5	5
B	3	3.5	7
C	1	4	5
D	1	2	4
E	3	6	8
F	2	3	4
G	2	4	7
H	5	7	11
I	2	7	9
J	6	8	10
K	6	11	12
L	6	10	12

10-8. Utilize the activity on arrow network developed for Exercise 9-5, and assume that the event signaling completion of activities E and F is reached in 11 weeks. Assume also that the standard deviation for all actitivies is 1.0. What is the probability of finishing the project in 19 weeks or less?

10-9. If it is assumed that the standard deviation for all activities associated with the pedestrian/freight bridge of Exercise 9-12 is 1.0 day, determine which activities would have to be crashed, and by how much, to be 95% sure of finishing on or before the expected project duration (after resource reallocation and resolving of resource conflicts). How much extra would this degree of certainty cost?

10-10. Your construction firm has just been awarded a contract for improvement of public facilities in a residential area. The project includes removing and replacing all of the curbs and sidewalks, improving the drainage system, and resurfacing the streets. You have developed a list of the activities involved, and their relationships with each other. A key factor is that all new grades will be taken off the new curb; therefore, the curb work must be completed to get the correct grade for the new sidewalk, drainage system improvements, and street surfaces. Pertinent data are listed in the table below:

Activity	Normal time (weeks)			Normal cost ($)	Crash time (weeks)	Crash cost ($)
	a	m	b			
A—old curb removal	3.0	4.8	7.8	10,000	3.5	13,500
B—pour new curb	2.5	5.2	6.7	16,000	4.3	21,000
C—root cutting	1.4	2.5	4.8	1,100	1.7	2,000
D—tree removal	0.8	2.0	2.6	1,800	1.1	2,500
E—old sidewalk removal	4.8	5.7	8.4	12,000	4.1	16,550
F—catch basins	1.0	1.4	2.4	2,500	1.0	3,100
G—manhole construction	2.1	4.7	4.9	3,000	2.3	5,400
H—drainage pipe	3.2	4.4	9.2	8,500	2.5	14,600
I—road subbase around curb	4.3	6.0	6.5	1,000	4.7	1,320
J—road subbase over drainage	0.8	1.2	2.2	800	0.9	1,100
K—pour new sidewalk	3.0	5.9	15.4	25,000	5.8	37,500
L—sod and seeding	1.5	4.5	5.1	4,000	3.1	6,700
M—surface road	1.9	3.0	5.3	7,000	3.0	8,000

1. Excavation of the old curb and tree root cutting are the initial activities.
2. Any trees that are dangerous can be removed after the roots have been cut.
3. The new curb pouring and the removal of the old sidewalk can be done simultaneously but only after the old curb has been excavated.
4. The catch basins, manholes, and drainage pipe installation begins after the new curb is in.
5. After the curb installation is completed, the subbase of the road around the curb can be prepared.
6. Subbase of the road over the new drainage can be prepared after the catch basins, manholes, and drainage pipe installation are completed.
7. The new sidewalk can be poured after the old sidewalk is removed and must be completed before final sod and seeding starts.
8. Final road surfacing can occur simultaneously with the sod and seeding but only after the subbase of the road over the drainage and around the curb is completed.
9. Road surfacing and sod and seeding finish the project.

(a) Formulate the activities into a PERT/CPM network using activity on node. Determine the critical path and the project duration.

(b) Determine the minimum project duration and the critical path.

(c) Your firm has received the contract to a similar job in an adjacent town. There is a stipulation stating that the job must begin by the first working day in August. A penalty of $1,000 will be assessed for each day after this date that the project is not begun. The job that you are planning now will not be started until March 15. August 2, the first working day, is 20 weeks away. Will you make the deadline? What is the probability that you will complete the job in time?

(d) You decide that you want to be 75% sure that you will be finished in 20 weeks. How much will the project duration have to be reduced and what is the cheapest way to achieve this?

(e) Because your firm is small and not well equipped to do the drainage work, you decide to subcontract out this work. You won the bid on the job for $124,500. The cheapest bid that you received for the drainage job was $20,000. If you maintain your 75% assurance level of finishing in 20 weeks, how much profit will you make? The drainage work subcontracted includes the catch basins, manholes, drainage pipe, and subbase of the road.

(f) Because you are working in a residential area, there is a possibility that you could do some private jobs, such as driveways and walks, to earn some extra money. If you did do private work, you estimate that you could make $7,000; but, it would take you 1.2 weeks to complete the jobs. Is it worth the risk to do the private work?

(g) List some of the problems that might be encountered in this job and any possible ways you could work to avoid them.

10-11. Using the network of Exercise 9-21, as modified by steps 1 and 2 of that exercise, answer the following questions:

(a) What is the probability of finishing at your estimated project duration?

(b) What is the probability of finishing 2 days early?

(c) What is the probability of finishing at least 1 day late?

Use the following table for estimating your standard deviations. All activities other than those listed are estimated to have zero standard deviations.

	Time estimates (days)	
	a	b
Foundations	2	5
Rough plumbing and heating	1.5	4
Exterior walls	5	9
Floor slab	2	7
Roof steel	1	3
Roof finish	1.5	4
Rough electrical	2.5	4
Finish electrical	2	4
Finish plumbing and heating	3	6
Paint	2.5	8

10-12. Using the network in the figure for Exercise 10-12, plus the table below, answer the following questions:

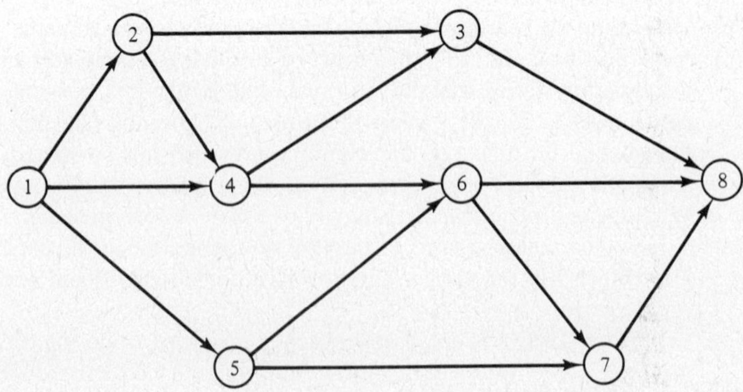

(a) What is the estimated project duration?
(b) What is the estimated probability of finishing within 2 weeks of the estimated project duration?
(c) What is the probability of finishing in less than the estimated project duration?
(d) What is the probability of finishing in less than or equal to the project duration plus 4 weeks?

Activity	a	m	b
1–2	2	3	5
1–4	1	4	5
1–5	2	4	6
2–4	3	5	6
2–3	4	5	8
4–3	3	8	10
3–8	2	3	5
4–6	4	6	8
5–6	4	5	7
5–7	5	7	8
6–7	3	5	6
7–8	2	5	7
6–8	2	4	7

10-13. There is a 1 in 800 chance that an earthquake will cause a large earthen dam to fail within the next 10 yr. Probable damage caused by dam failure with no advance warning would be $50,000,000. If an impending quake of magnitude likely to cause failure of the dam could be predicted 12 hr in advance, mitigating measures could be taken. These would reduce the expected damage to $1,000,000, but would cost $80,000 to implement. An earthquake prediction system is available, but its performance is not perfect. When subjected to test conditions that would cause dam failure, the system predicts failure 70% of the time. Also, there is a 0.5% chance that the system will predict a failure when actual conditions would not cause one.
(a) What is the value of the prediction system, using EMV criteria?
(b) If the prediction system is in place, what action should be taken for each system indication?
(c) How much would you be willing to pay for insurance against losses with, and without, the prediction system in place?

10-14. Redo Exercise 10-13 with the following added information. If conditions 12 hr in advance indicate a quake of failure magnitude is imminent, there is a 60% chance that such a quake will actually occur. There is also a 1×10^{-5}% chance that a failure quake could occur during any 12-hr period when the indicative conditions are not present at the beginning of the period.

10-15. Suppose in Exercise 10-13 that the predictions of the system are reliable, but that the predictions are expressed in terms of probability of occurrence; that is, 0, 20, 40, 60, 80, 100% chance of failure quake occurrence. At what level of prediction should mitigation measures be instituted?

10-16. Redo Exercise 10-15 with the added conditions given in Exercise 10-14.

10-17. A construction firm needs to use an old generator on a job. It would cost $1,000 to overhaul the generator before using it. If the generator fails on the job, it would cost $2,000 for repairs and added job costs. There is a 60% chance that the generator will be reliable without overhaul, and a 95% chance if it is overhauled. The manager could spend $150 on an evaluation of the generator which would indicate whether or not the generator could be expected to fail with 90% reliability. (90% reliability means there is a 10% chance that an actually good generator will test out as bad, and vice versa.) What should the manager do?

10-18. A contractor is performing some work in a river flats area that has been subjected to high water conditions and occasional destructive flooding in the past. There is a four-month period during which the contractor has no use for the equipment either on this job or on others. The equipment can be kept on this job in the river flats, or moved out, stored and then moved back, at a total cost of $1,800.

If the equipment is kept in the river flats, the contractor has the option of building a platform for the equipment at a cost of $500, which will protect it against high water but not against a destructive flood. The damage that would be caused by high water amounts to $10,000, if there is no platform. A destructive flood would cause a loss of $60,000, regardless of whether or not the contractor builds a platform. The probability of high water in the four-month period is 0.25; the probability of a destructive flood is 0.02.

What is the contractor's optimum policy?

10-19. Two cities are located in a river floodplain. Some degree of flooding occurs every year due to snow melt runoff. Records for the past 65 yr indicate the following flood frequencies:

Number of occurrences	Flood magnitude (ft^3/s)
27	10,000–35,000
15	36,000–75,000
10	76,000–105,000
8	106,000–140,000
5	141,000–175,000

Channel improvements that would increase the carrying capacity of the stream could be made, with a corresponding decrease in flood damage. Two sites (C and D in the following table) are also available upstream for flood control dams. Site conflicts prevent dams being constructed at both C and D; however, either dam could act in conjunction with channel improvements. Flood damage costs and project costs for the various alternatives are given in the following table. If the cost of capital is 12%, determine the most attractive alternative.

Estimated Costs of Flood Damages for each Flood Magnitude

Flood magnitude (×1000 ft^3/s)	No flood mitigation ($)	Channel improvement alone ($)	Dam at site C alone ($)	Dam at site D alone ($)	Dam at site C with channel improvement ($)	Dam at site D with channel improvement ($)
10–35	500,000	250,000	190,000	125,000	100,000	60,000
36–75	800,000	500,000	400,000	330,000	280,000	250,000
76–105	1,050,000	800,000	650,000	600,000	500,000	460,000
106–140	1,250,000	1,050,000	870,000	825,000	730,000	675,000
141–175	1,500,000	1,250,000	1,050,000	950,000	815,000	750,000

Project Costs and Expected Life

Project	Investment ($)	Annual operation and maintenance costs ($)	Project life (yr)
I. No flood mitigation	0	0	—
II. Channel improvement alone	500,000	100,000	25
III. Dam at site C alone	3,000,000	60,000	100
IV. Dam at site D alone	4,000,000	80,000	100
V. Dam at site C and channel improvement	3,500,000	160,000	100
VI. Dam at site D and channel improvement	4,500,000	180,000	100

10-20. An international corporation wishes to expand by constructing a 60 million dollar addition to existing research and development facilities. Activities for one subcontract of this project are given in the following table.

		Time estimates		
Activity	Name	Most optimistic a (weeks)	Most pessimistic b (weeks)	Most likely m (weeks)
Excavate to lay pipe	A	4	8	4.5
Install pipe	B	0.5	4.5	4
Fill pipe excavation	C	2.5	7.5	3.5
Excavate for retaining wall	D	1	3.2	1.2
Install drainage system for retaining wall	E	0.25	0.75	0.5
Construct retaining wall	F	1.4	4.2	1.6
Backfill at retaining wall	G	1.5	3.7	3.2
Excavate for slab on grade	H	0.6	2.2	0.8
Pour concrete slab	I	0.4	0.8	0.6
Masonry work	J	10	24	14
Washdown masonry	K	0.7	1.3	0.7
Repair roads in complex	L	2	6	2.5
Raise scaffold in atrium area (bldg. interior)	M	0.6	1.4	1
Put up studs and drywall in atrium area	N	8	18	11.5
Paint atrium	O	0.8	2.4	2.2
Apply top course to roads	P	1	2	1.5

A and H are the first activities and may be done simultaneously.

A must be completed before work on B can begin (but B depends on A).

E follows B and D.

I depends upon H.

J cannot start before I is completed.

D, K, and M may be performed simultaneously after J is completed.

C is dependent upon *B*.

L follows *K* and *N* follows *M*.

G depends upon *F*, but *E* must be completed before *F* can start.

N is followed by *O*.

P depends on *G*, *L*, *C* and *O*.

P is the last activity.

(a) Develop a CPM network for this subcontract.

(b) You are required to have the subcontract completed within 35 weeks. What is the probability that you will meet this deadline?

(c) Shear failure of a slope stops all work on activities *G* and *L* for 4 weeks. How will this affect the probability of finishing within 35 weeks?

(d) If activity *I* is delayed for 6 weeks by a labor dispute, what will be the probability of finishing later than the 35th week?

10-21. An engineering firm has to design a stormwater overflow station for a sewer project. A problem exists in that the station has to be located near the intersection of two rivers. It is realized that during the springtime this area is often flooded. Records for the past 100 yr show that flooding has occurred in 80 of the 100 yr.

The following data express the severity of each flood, number of occurrences, and the expected cost to build the station to resist a flood of that magnitude.

Using an expected monetary value analysis, decide which magnitude to design the station for.

Flood magnitude (ft)	Number of occurrences in past 100 yr	Cost to resist ($)
0–5	45	8,000
5–10	25	14,000
10–15	9	19,000
15+	1	25,000

Design for	Magnitude	Cost of damage ($)
None	0–5	16,000
	5–10	20,000
	10–15	26,000
	15+	30,000
0–5	0–5	0
	5–10	15,000
	10–15	20,000
	15+	25,000
5–10	0–5	0
	5–10	0
	10–15	19,000
	15+	24,000
10–15	0–5	0
	5–10	0
	10–15	0
	15+	22,000

Design for	Magnitude	Cost of damage ($)	*(Continued)*
15+	0–5	0	
	5–10	0	
	10–15	0	
	15+	0	

10-22. Your firm plans to put on a benefit July 4 picnic for a local charity. All of the food and beverages have been donated. Each person attending will be asked to make a $3 donation.

A decision must be made beforehand whether to hold the party outdoors or in a fieldhouse. The chosen location influences the turnout as shown in the following table:

	Turnout (people)	
	Outdoors	Fieldhouse
No rain	400	160
Rain	80	200

The record of previous July 4 weather is that it has rained 6 times in the last 10 yr.

(a) Where should the party be held?

(b) What is the expected profit?

(c) You are offered a rain-prediction device that will forecast 6 hr in advance. It has had the following record over the past 2 yr:

	Indicated rain	
	Yes	No
Rain	4	2
No rain	1	3

If the final decision as to where to hold the picnic can be delayed until the morning of the event, what decision rules should be applied? Assume that the same turnout data apply as before.

(d) How much is the rain predictor worth?

Chapter 11

Dynamic Programming

INTRODUCTION

Dynamic programming takes advantage of the condition that many complex systems can be modeled as a series of subsystems linked together. The serial structure requires that there be no feedback between subsystem stages; however, feedback can be accommodated within a particular subsystem stage. A serial structure allows the decomposition of complex systems into components small enough for optimization, while at the same time accounting for the interaction of the different parts. Figure 11.1 gives a definition sketch depicting the model structure for dynamic programming.

Richard Bellman (1957) advanced the concept of dynamic programming through his postulate that "an optimal set of decisions has the property that whatever the first decision is, the remaining decisions must be optimal with respect to the state which results from the first decision." In other words, the decisions to be made in subsequent stages will be dependent on the decisions made in the present stage, and the decisions made in the present stage are dependent on the decisions made in previous stages. However, decisions made in previous stages were not dependent on decisions made in the present stage. Time is a prime example of the serial structure required for dynamic programming. Decisions made on Wednesday will influence decisions to be made on Thursday; however, decisions made on either Wednesday or Thursday have no influence over decisions made on Tuesday. The key is that the analyst does not have to keep track of all past decisions, but only needs to know what the current system state is. The current stage embodies all relevant information needed for making optimal decisions from the current stage on.

There is no standard mathematical technique for dynamic programming problems. Dynamic programming is an approach to problem solving, and particular mathematical equations have to be developed to fit each situation. An understanding of

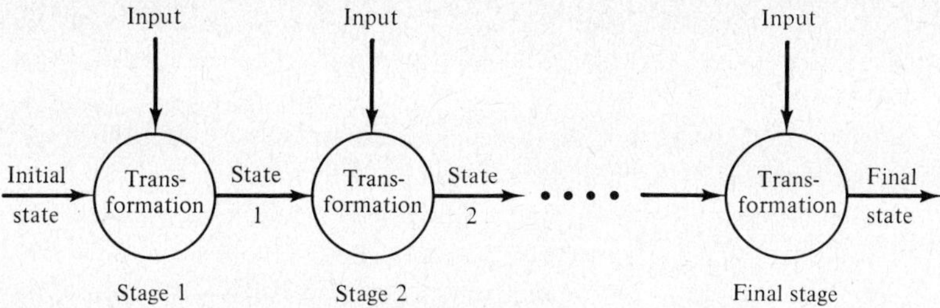

Figure 11.1   Dynamic programming structure.

the general structure of dynamic programming problems is essential in determining if a particular problem fits the dynamic programming mold; and, if it does, how to apply the methods to it.

Four features characterize the problems to which dynamic programming can be successfully applied. The problem must be able to be divided into stages, with a decision required at each stage. The stages may represent different points in space (selecting the route for a new expressway) or different points in time (determining the optimum storage versus time for controlling stormwater runoff). Each stage of the problem must have associated with it a finite number of states that reflect the possible conditions of the system. The volume of stormwater stored at a particular time could represent the state of the system. System state is defined by the values of the state variables at each stage. State variables were previously defined in Chapter 1. Decisions made at each stage transform the present state of the system to the state of the subsequent stage, through transformation functions. Continuous transformation functions can be examined at discrete points to give a finite number of states. The volume of stormwater released during a stage will affect the volume stored for the next stage. It is a continuous function, but the effect can be studied at appropriate discrete points. There must be some way of measuring the effectiveness of the transformation from one state to the next, usually in terms of benefit or cost.

Since the decisions to be made in subsequent stages are dependent on the decisions made at the present stage, many dynamic programming problems proceed in a reverse direction, from the final stage to the initial stage. A particular decision-making path from the present stage to the final stage is considered only if that path has a chance of being optimal. The procedure continues backstepping until the initial stage is reached, and the overall optimal decision path identified. Examples in this text will be of the backward dynamic programming type. There are other problems that lend themselves to forward dynamic programming. Examples of this type can be found in dynamic programming texts, such as the one by Dreyfus and Law (1977).

The efficiency in dynamic programming arises because not all possible decision paths are considered. Normally, the number of computations required to consider all alternatives rises geometrically with the number of variables, whereas, they rise approximately linearly with the number of stages for the dynamic programming formulation. Dynamic programming is generally more efficient than linear programming for solving problems to which either would be applicable. Example 11.1 will show the computational efficiency of dynamic programming over consideration of all possible alternatives, while Examples 11.2 and 11.3 will demonstrate the efficiency of dynamic programming over the simplex algorithm solution of the linear programming model for the same problem.

Dynamic programming is not limited to linear systems, however. One of its major strengths is that it can accommodate deterministic and stochastic systems, continuous and discontinuous systems, linear and nonlinear systems, or any mix of these, provided that the serial structure is maintained.

A drawback to dynamic programming is the difficulty in handling problems with more than three state variables at each stage. Research work has been done on such problems; however, the solution techniques are often inefficient and cumbersome.

The best way to become familiar with the basic application of dynamic programming is to develop specific examples. Examples in this chapter will use linear transformations for each stage, to compare dynamic programming solutions with other methods of solution. It should be obvious, especially in Example 11.3, that the transformation functions could be nonlinear without changing the solution procedure.

NETWORK PROBLEMS

Many graphical network problems can be solved by utilizing dynamic programming. Longest and shortest paths between pairs of locations in a transportation network can be ascertained. Branching networks, such as sewer systems, can be optimized using dynamic programming, if proper conditions are imposed. Decision analysis, using the decision tree network, is a type of dynamic programming.

Example 11.1 will illustrate the application of dynamic programming to find the shortest path through a transportation network. The same network that was developed for Example 5.10 will be used. In Example 5.10, the simplex algorithm was used to solve the linear programming model for the shortest path through the network. This network was again utilized in Chapter 6 to illustrate how the assignment algorithm could be used to find the shortest path between two points of a network. The assignment algorithm provided a computationally more efficient way of finding the shortest path. Dynamic programming will provide an even more efficient, and certainly a more physically appealing, way of finding this shortest path.

Although there are more efficient ways of solving this problem (Dreyfus and Law, 1977), applying dynamic programming to it provides a good example of developing the serial structure when it may not be obvious at first glance. Examples 11.2 and 11.3 will demonstrate problems with a more obvious serial structure.

EXAMPLE 11.1 ————————————————————————————————————

Figure 11.2 shows the transportation network of Figure 6.5, which has been used previously to illustrate development of a linear programming model and application of the assignment algorithm. The present objective is to apply dynamic programming to find the shortest path through this network between locations 1 and 6.

To apply dynamic programming, the network has to be divided into stages. The state variables for each stage are the minimum distances from each node within that stage to the final destination. For this problem, it is more efficient to have only

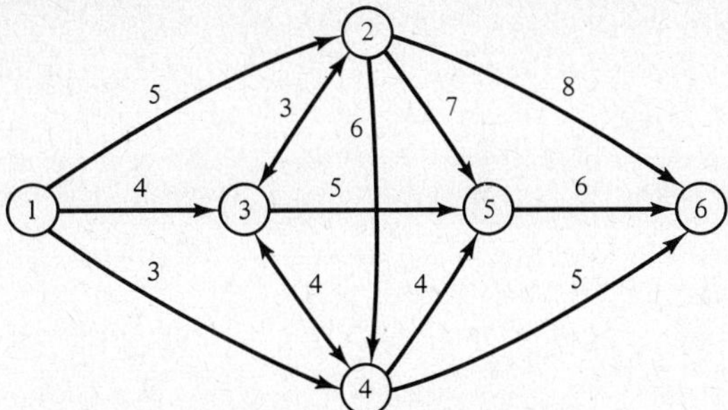

Figure 11.2　Transportation network.

one node within each stage. Since minimum distances from subsequent nodes to the final destination will be known, the total number of possible paths that have to be considered is reduced. This illustrates the utility of dynamic programming. Figure 11.3 shows the network with the dynamic programming stages defined.

Working backward from node 6, stage 1 shows that node 5 is the only node that must be investigated, and the only possible path is ⑤–⑥. At stage 2, node 4 is evaluated. Node 2 was not included at this stage because it has two paths leading into node 4, while there is only one possible path from node 4 to node 2. Therefore, evaluating 4 first should save some computations. There are five possible paths from node 4 to node 6, the shortest being the path directly from 4 to 6. The other paths are enumerated in Table 11.1. Path ④–⑥, with a length of five, is the only one that has to be considered in subsequent stages. From node 2, (stage 3) there are seven possible paths to node 6, but only five of these need to be considered because of the previous stages. The direct path from node 2 to node 6 is again the shortest; however, it is essential to enumerate the other paths to confirm this. At stage 4, the number of paths that have to be enumerated is reduced significantly because of the previous stages. Path ③–④–⑥ is

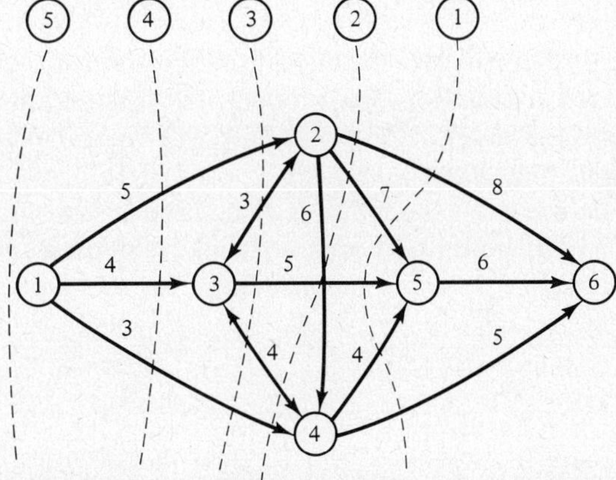

Figure 11.3　Transportation network with stages defined.

TABLE 11.1   PATH LENGTHS FOR DYNAMIC PROGRAMMING

Number of stages from node 6	Distances to node 6			Least distance to node 6	
		From node	Distance	Distance	Path
1		5	6	6	5–6
2	4	4–6	5	5	4–6
		4–5–6	$4 + 6 = 10$		
		4–3–5–6	$4 + 5 + 6 = 15$		
		4–3–2–6	$4 + 3 + 8 = 15$		
		4–3–2–5–6	$4 + 3 + 7 + 6 = 20$		
3	2	2–6	8	8	2–6
		2–5–6	$7 + 6 = 13$		
		2–4–6	$6 + 5 = 11$		
		2–3–4–6	$3 + 4 + 5 = 12$		
		2–3–5–6	$3 + 5 + 6 = 14$		
4	3	3–2–6	$3 + 8 = 11$	9	3–4–6
		3–4–6	$4 + 5 = 9$		
		3–5–6	$5 + 6 = 11$		
5	1	1–2–6	$5 + 8 = 13$	8	1–4–6
		1–3–4–6	$4 + 9 = 13$		
		1–4–6	$3 + 5 = 8$		

the shortest of the three considered in Table 11.1. Only three paths have to be considered for the fifth, and final stage, also. The shortest of these is path ①–④–⑥ with a length of 8 units, which is the same result found previously.

If all paths from node 1 to node 6 were considered, a total of 20 paths would have to be enumerated, requiring a total of approximately 50 additions. All of the dynamic programming computations required only 20 additions.

PRODUCTION SCHEDULING PROBLEMS

Production scheduling problems involve the allocation of resources over several time periods to meet the projected demand for some product. The time periods will represent the stages for dynamic programming. No opportunity will be available for feedback since going back in time is impossible.

Example 11.2 will demonstrate the solution of the production scheduling model developed for Example 5.12 by both linear programming techniques and by dynamic programming. The efficiency of dynamic programming for this type of problem is evident.

EXAMPLE 11.2

Example 5.12 presents a production scheduling problem in which a manager desires to minimize the cost of producing concrete sewer pipe over the next 5 months, while meeting the projected demand each month. Constraints are placed on the amount of processing time available each month at standard wage rates, and on the amount of raw material available each month. Two thousand hours of regular processing time are available each month. The amount of material available for processing varies each month because of seasonal variations in construction activity. If material is not used one month, it cannot be carried over to the next month.

Any product on hand at the beginning of the first month has already been subtracted from the first month demand. It is assumed that no product will be left over at the end of the fifth month.

It takes 75 hr of processing time to produce one unit of pipe, and costs approximately $20 to store one unit of pipe from one month to the next. Figure 11.4 depicts the material and product flow over this 5-month period. It also includes other pertinent problem data. The variable XS_i represents the units of product carried over from month i to month $i + 1$, and is the state variable for dynamic programming.

In Example 5.12, the variable XR_i was used to represent the number of units of pipe produced on regular time during month i, and XO_i, the number of units of pipe produced on overtime during month i. It was assumed that sufficient overtime was available to meet the needs in any particular month. A linear programming model was developed, and is repeated here, as Equation 11.1, with the meaning of each constraint indicated. Twelve iterations and approximately 1500 calculations were required to find the minimum value solution shown in Figure 11.5. The total cost for this alternative was $24,520.00.

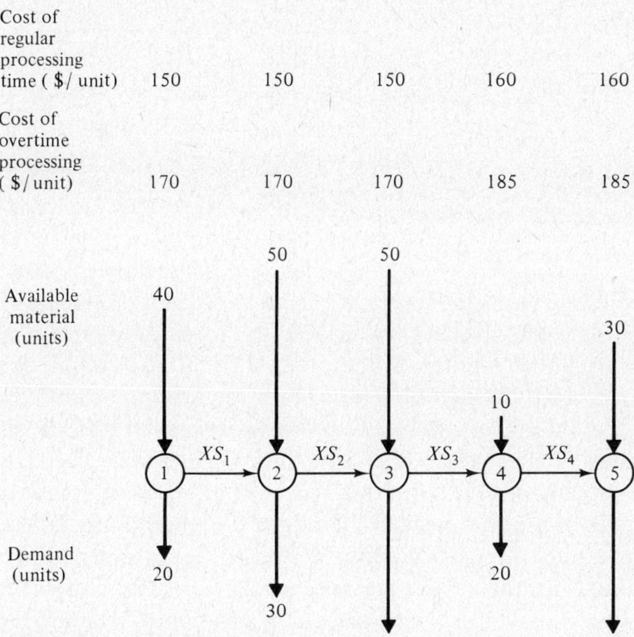

Figure 11.4 Production scheduling flow.

$$\min Z = 150XR_1 + 150XR_2 + 150XR_3 + 160XR_4 + 160XR_5 + 170XO_1$$
$$+ 170XO_2 + 170XO_3 + 185XO_4 + 185XO_5$$
$$+ 20XS_1 + 20XS_2 + 20XS_3 + 20XS_4$$

ST: $75XR_1 \le 2000$ regular production time available during month 1

$XR_1 + XO_1 \le 40$ resource available during month 1

$75XR_2 \le 2000$ regular production time available during month 2

$XR_2 + XO_2 \le 50$ resource available during month 2

$75XR_3 \le 2000$ regular production time available during month 3

$XR_3 + XO_3 \le 50$ resource available during month 3

$75XR_4 \le 2000$ regular production time available during month 4

$XR_4 + XO_4 \le 10$ resource available during month 4

$75XR_5 \le 2000$ regular production time available during month 5

$XR_5 + XO_5 \le 30$ resource available during month 5

$XR_1 + XO_1 - XS_1 \qquad = 20$ production continuity during month 1

$XR_2 + XO_2 + XS_1 - XS_2 = 30$ production continuity during month 2

$XR_3 + XO_3 + XS_2 - XS_3 = 40$ production continuity during month 3

$XR_4 + XO_4 + XS_3 - XS_4 = 20$ production continuity during month 4

$XR_5 + XO_5 + XS_4 \qquad = 40$ production continuity during month 5

(11.1)

Dynamic programming will now be used to solve the problem. During month 5, 40 units of pipe are required, and material is available for producing 30. Examination of the problem data reveals that the cheapest way to produce pipe in any month is by regular time during that month. Furthermore, if the 2000 available production hours per month are divided by the number of hours required to produce one unit of pipe (75 hr), then 26.67 units/month can be produced at regular time. This reduces somewhat the number of alternatives that have to be examined at any particular stage.

There are really only two options that need to be examined in the fifth month. Since more units of pipe are needed than there are units of material available, some amount of pipe must be carried forward from the previous month. The choice is whether or not to use all of the material available in the fifth month by utilizing overtime. Option A of Table 11.2(a) uses all of the material available, while option B uses only the portion available to regular time processing. The total cost attributed to month 5 is shown for each option, and will be carried forward to subsequent computational stages.

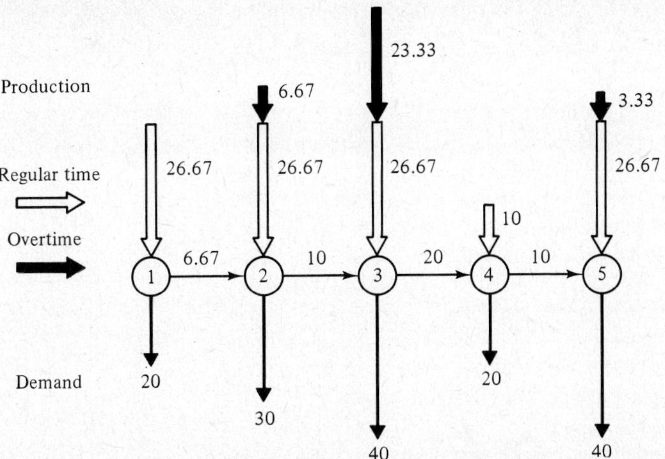

Figure 11.5 Optimum solution for production scheduling problem.

In month 4, there is again less available material than product demand, and there is also less material than could be used during regular time processing. Therefore, available options are again limited, and are governed by the options developed for month 5. For either option, an additional 10 units have to be added to the state variable, XS_3, to meet the excess demand during month 4. Table 11.2(b) shows the resulting computations at month 4, and the combination with the fifth month.

There is excess material available during month 3, so a wider range of options is available. The total demand for month three is 40 units plus the value assigned to XS_3. Therefore, the total demand for option A is 60 units, while it is 63.33 units for option B. Table 11.2(c) shows the suboptions, within the options, chosen for evaluation. The first three combinations for both option A and option B were chosen so that the corresponding XS_2 values would be equal. This allows the elimination of some suboptions from consideration in subsequent stages. The suboptions of this stage become the options of the subsequent stage. Options B_1, B_2, and B_3 can be eliminated in this manner, because each is more expensive than the A option with the same carryover from month 2 (XS_2). Options A_1, A_2, A_3, and B_4 will be carried forward.

Table 11.2(d) shows the suboptions considered in the second month within the options retained from the third month. Option A1a provides the minimum cost alternative with a value of 13.33 for the state variable, XS_1. Similarly, options A1b, A1c, and B4d provide the minimum cost for XS_1 values of 6.67, 0, and 20.0, respectively. Note that fewer suboptions were considered for the last two options of Table 11.2(d) because of limitations on the material available, and the total number of units that could be carried forward from month 1.

Only one option is available during month 1 to meet each of the state variable values, XS_1, for the selected options in month 2. These options are enumerated in Table 11.2(e). The total combined cost can be used to determine the optimum strategy. However, for this problem, there are two options with the same total cost (accurate to four significant figures). Options A1b and A1c both have a total cost of \$24,520.00, which is the same amount found with the linear programming model. Option A1b is the same solution as the linear programming model; however, examination of the

(a) Fifth month

Option A XS_4 = 10

26.67 units on regular time	@ \$160/unit	
3.33 units on overtime	@ \$185/unit	\$5,083.25
10 units carried over from month 4	@ \$ 20/unit	

Option B XS_4 = 13.33

26.67 units on regular time	@ \$160/unit	
13.33 units carried over from month 4	@ \$ 20/unit	\$4,533.80

(b) Fourth month

Option A (XS_4 = 10, XS_3 = 20)

10 units on regular time	@ \$160/unit	
20 units carried over from month 3	@ \$ 20/unit	\$2,000.00

Option B (XS_4 = 13.33, XS_3 = 23.33)

10 units on regular time	@ \$160/unit	
23.33 units carried over from month 3	@ \$ 20/unit	\$2,066.60

Combination with fifth month
Option A: \$7083.25
Option B: \$6600.40

Combined with
subsequent steps

(c) Third month

Option A (XS_3 = 20) Total demand = 60

Option $A$1:	26.67 units on regular time (RT)	@ \$150/unit		
	23.33 units on overtime (OT)	@ \$170/unit	\$8,166.60	\$15,249.85
(XS_2 = 10)	10-unit carryover (CO) from month 2	@ \$ 20/unit		
Option $A$2:	26.67 RT	@ \$150/unit		
	15 OT	@ \$170/unit	\$6,917.10	\$14,000.35
(XS_2 = 18.33)	18.33 CO	@ \$ 20/unit		
Option $A$3	26.67 RT	@ \$150/unit		
	0 OT	@ \$170/unit	\$4,667.10	\$11,750.35
(XS_2 = 33.33)	33.33 CO	@ \$ 20/unit		

Option B (XS_3 = 23.33) Total demand = 63.33

Option $B$1:	26.67 RT	@ \$150/unit		
	23.33 OT	@ \$170/unit	\$8,233.20	\$14,833.60
(XS_2 = 13.33)	13.33 CO	@ \$ 20/unit		
Option $B$2:	26.67 RT	@ \$150/unit		
	18.33 OT	@ \$170/unit	\$7,483.20	\$14,083.60
(XS_2 = 18.33)	18.33 CO	@ \$ 20/unit		
Option $B$3:	26.67 RT	@ \$150/unit		
	3.33 OT	@ \$170/unit	\$5,233.20	\$11,833.60
(XS_2 = 33.33)	33.33 CO	@ \$ 20/unit		
Option $B$4:	26.67 RT	@ \$150/unit		
	0 OT	@ \$170/unit	\$4,733.70	\$11,334.10
(XS_2 = 36.66)	36.66 CO	@ \$ 20/unit		

	(d) Second month			Combined with subsequent steps
Option $A1$ (XS_2 = 10) Total demand = 40				
Option A1a	26.67 RT	@ \$150/unit		
	0 OT	@ \$170/unit	\$4,267.10	\$19,515.95
(XS_1 = 13.33)	13.33 CO	@ \$ 20/unit		
Option A1b	26.67 RT	@ \$150/unit		
	6.67 OT	@ \$170/unit	\$5,267.80	\$20,517.65
(XS_1 = 6.67)	6.67 CO	@ \$ 20/unit		
Option A1c	26.67 RT	@ \$150/unit		
	13.33 OT	@ \$170/unit	\$6,266.60	\$21,516.45
(XS_1 = 0)	0 CO	@ \$ 20/unit		
Option $A2$ (XS_2 = 18.33) Total demand = 48.33				
Option A2a	26.67 RT	@ \$150/unit		
	8.33 OT	@ \$170/unit	\$5,683.20	\$19,683.55
(XS_1 = 13.33)	13.33 CO	@ \$ 20/unit		
Option A2b	26.67 RT	@ \$150/unit		
	15.00 OT	@ \$170/unit	\$6,683.90	\$20,684.25
(XS_1 = 6.67)	6.67 CO	@ \$ 20/unit		
Option A2c	26.67 RT	@ \$150/unit		
	21.67 OT	@ \$170/unit	\$7,684.40	\$21,684.75
(XS_1 = 0)	0 CO	@ \$ 20/unit		
Option $A3$ (XS_2 = 33.33) Total demand = 63.33				
Option A3a	26.67 RT	@ \$150/unit		
	23.33 OT	@ \$170/unit	\$8,233.20	\$19,983.55
(XS_1 = 13.33)	13.33 CO	@ \$ 20/unit		
Option A3d	26.67 RT	@ \$150/unit		
	16.67 OT	@ \$170/unit	\$7,234.40	\$18,984.75
(XS_1 = 20.00)	20.00 CO	@ \$ 20/unit		
Option $B4$ (XS_2 = 36.66) Total demand = 66.66				
Option B4d	26.67 RT	@ \$150/unit		
	20.00 OT	@ \$170/unit	\$7,800.50	\$19,134.60
(XS_1 = 20.00)	20.00 CO	@ \$ 20/unit		
	(e) First month			
Option A1a (XS_1 = 13.33) Total demand = 33.33				
	26.67 RT	@ \$150/unit		
	6.66 OT	@ \$170/unit	\$5,132.70	\$24,648.65
Option A1b (XS_1 = 6.67) Total demand = 26.67				
	26.67 RT	@ \$150/unit		
	0 OT	@ \$170/unit	\$4,000.50	\$24,519.15
Option A1c (XS_1 = 0) Total demand = 20.00				
	20.00 RT	@ \$150/unit		
	0 OT	@ \$170/unit	\$3,000.00	\$24,516.45
Option A3d (XS_1 = 20.00) Total demand = 40.00				
	26.67 RT	@ \$150/unit		
	13.33 OT	@ \$170/unit	\$6,266.60	\$25,251.35

final tableau of the linear programming model reveals the existence of an alternate optimum. This alternate optimum is option $A1c$, which has zero units carried over from month 1 to month 2. Approximately 125 calculations were required for the dynamic programming solution, compared to 1500 calculations for the linear programming solution.

WATER RESOURCES PROBLEMS

Many water resources problems follow a serial structure, and can be efficiently optimized using dynamic programming techniques. Furthermore, many of the transfer functions within these stages are more accurately represented by nonlinear functions. Some examples include finding the optimal size and location of detention facilities for stormwater control, optimizing a multiple reservoir system's operation, and making a least-cost design for urban drainage networks.

A regional management plan for water supply and wastewater disposal will be used to demonstrate the methodology. This is the problem developed in Example 5.5. Use of the simplex algorithm will be contrasted with dynamic programming techniques for finding the optimum solution for the model developed.

EXAMPLE 11.3 ――――――――――――――――――――――――――――――――

The three cities described in Example 5.5 discharge their treated liquid wastes into a common river. Additional pollution enters the river above all three cities in the form of leachate from an unsealed landfill. Cities A and B draw their water supply from the river, and treat it prior to delivering it to their customers. Figure 11.6 shows the general situation, and Table 11.3 defines the variables and parameters. The regional authority is attempting to minimize the total cost of treatment at six locations so that water supply quantity and quality requirements can be met, while maintaining required water quality levels at all points in the river. A general linear model was developed for this in Example 5.5; however, some modifications of this model need to be accomplished before optimization can be undertaken. These modifications will be made, then the simplex algorithm will be used to find an optimum solution to the model. After that, the problem will be reinvestigated utilizing dynamic programming.

The model of Example 5.5 has to be expanded so that continuity, mixing, and quality constraints can be satisfied simultaneously. In addition, all constant parameters of the model need to be identified and evaluated. Also, the constraints must be rearranged so that all constants are on the right-hand side, and all variables on the left-hand side of the equality or inequality. The model, with all constraints identified, is given below as Equations 11.2.

$$\min Z = C_L T_L + C_{A_0} T_{A_0} + C_A T_A + C_{B_0} T_{B_0} + C_B T_B + C_C T_C$$

$$\text{ST:} \quad \frac{B_G}{Q} - \frac{T_L}{Q} = L_{A_0} \qquad \text{Concentration in river after discharge from landfill treatment}$$

$$L_{A_0} \le L_L \qquad \text{Water quality limitation in river at discharge from landfill}$$

$$L_{A0} - \frac{T_{A0}}{Q_A} = L'_{A0}$$ Concentration in water supply after treatment at City A

$$L'_{A0} \leq L_D$$ Water supply quality limitation after treatment at City A

$$L'_{A0} + L_A - \frac{T_A}{Q_A} = L'_A$$ Concentration in effluent after waste treatment at City A

$$\left(\frac{Q_A}{Q}\right)L'_A + L_{A0} - \left(\frac{Q_A}{Q}\right)L_{A0} \leq L_L$$ Water quality limitation in river at discharge from City A

$$Q_A L'_A + Q L_{A0} - Q_A L_{A0} = \left(\frac{Q}{0.80}\right)L_{B0}$$ Water quality at intake for water supply, City B*

$$L_{B0} - \frac{T_{B0}}{Q_B} = L'_{B0}$$ Concentration in water supply after treatment at City B

$$L'_{B0} \leq L_D$$ Water supply quality limitation after treatment at City B

$$L'_{B0} + L_B - \frac{T_B}{Q_B} = L'_B$$ Concentration in effluent after waste treatment at City B

$$\left(\frac{Q_B}{Q}\right)L'_B + L_{B0} - \left(\frac{Q_B}{Q}\right)L_{B0} \leq L_L$$ Water quality limitation in river at discharge from City B

$$Q_B L'_B + Q L_{B0} - Q_B L_{B0} = \left(\frac{Q}{0.70}\right)L_{C0}$$ Water quality in river just upstream from waste outfall for City C*

$$L_C - \frac{T_C}{Q_C} = L'_C$$ Concentration in effluent after waste treatment at City C

$$\left(\frac{Q_C}{Q_C + Q}\right)L'_C + \left(\frac{Q}{Q_C + Q}\right)L_{C0} \leq L_L$$ Water quality limitation in river at discharge from City C

$$(11.2)$$

Identification and quantification of constant parameters will proceed as follows. Pollutant concentrations will be expressed during computations in terms of pounds per cubic foot (lb/ft^3), rather than in milligrams per liter, in order to avoid having to switch units back and forth.

The given flow requirements are

$$Q_A = 10 \ ft^3/s$$
$$Q_B = 20 \ ft^3/s$$
$$Q_C = 10 \ ft^3/s$$

---

* 0.80 and 0.70 are fractions of pollutant not removed between Cities A and B and Cities B and C, respectively.

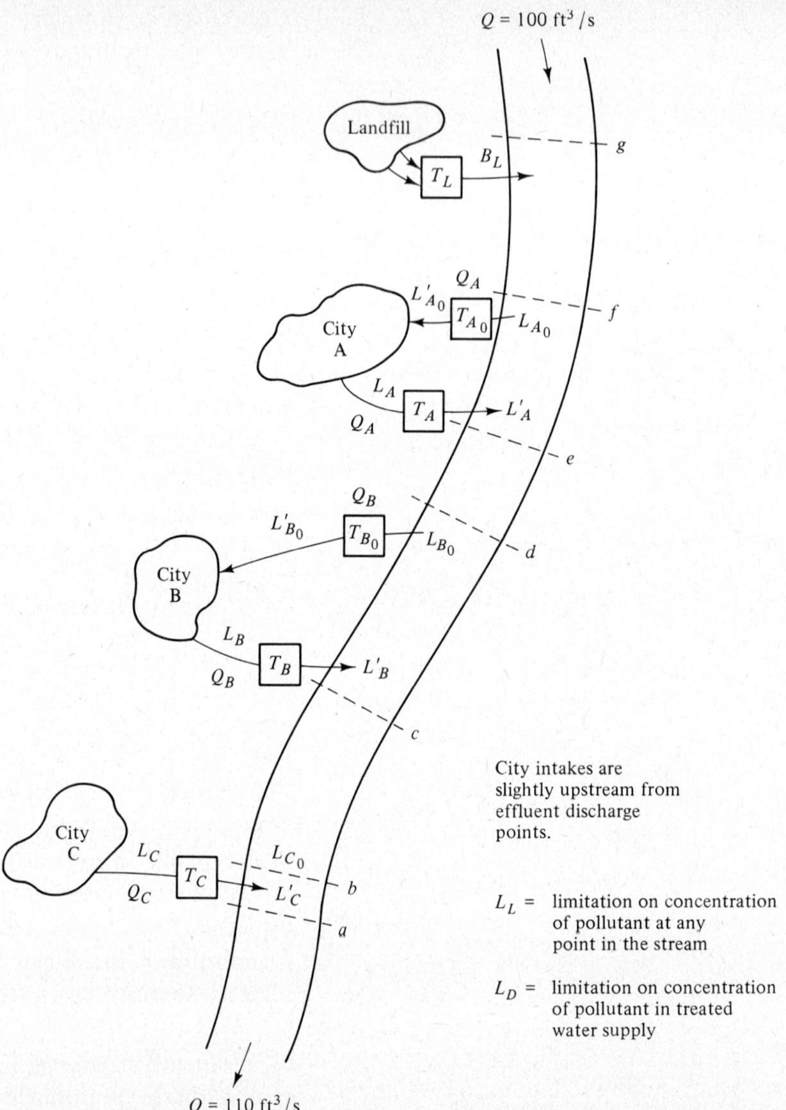

Figure 11.6 Areawide water supply, waste treatment plan, with stage delineation for dynamic programming.

Limitations on pollutant concentration in the water supplies, and in the river, are set by the state, and expressed in milligrams per liter. These, however, can easily be converted to pounds per cubic foot as

$$L_L = 5 \text{ mg/l} = 3.12 \times 10^{-4} \text{ lb/ft}^3$$

and
$$L_D = 0.5 \text{ mg/l} = 3.12 \times 10^{-5} \text{ lb/ft}^3$$

Measurements have shown that each city contributes approximately 200 mg/l of pollutant L to the water supply as the water is utilized for various purposes, and is then

TABLE 11.3 PERTINENT SYSTEM VARIABLES FOR TREATMENT OPTIONS

Location	Flowrate (ft^3/s)	Pollutant load before treatment	Design variable (lb removed)	Pollutant load after treatment	Treatment costs ($/lb removed)
Landfill	~ 0	B_G lb/s	T_L lb/sec	B_L lb/sec	C_L
Water supply A	Q_A	L_{A0} lb/ft^3	T_{A0}	L'_{A0} lb/ft^3	C_{A0}
Waste treatment A	Q_A	$L'_{A0} + L_A$ (lb/ft^3)	T_A	L'_A	C_A
Water supply B	Q_B	L_{B0} lb/ft^3	T_{B0}	L'_{B0}	C_{B0}
Waste treatment B	Q_B	$L'_{B0} + L_B$ (lb/ft^3)	T_B	L'_B	C_B
Waste treatment C	Q_C	L_C lb/ft^3	T_C	L'_C	C_C

NOTE: L_L = limitation on concentration of pollutant at any point in the stream.

L_D = limitation on concentration of pollutant in treated water supply.

recollected for treatment at one of the waste treatment plants. This equates to 1.25×10^{-2} lb/ft^3. Pollutant production from the landfill has been estimated to be 0.02 lb/s.

Treatment costs were developed from U.S. Environmental Protection Agency manuals which provide estimates of capital and operation and maintenance costs for various types of treatment. These estimates were converted to equivalent annual series of payments, then expressed in terms of dollars per pound removed. A 20-yr life was assumed. The methodology for developing these cost estimates has been included as an applications exercise in Chapter 7 (Exercise 7-12). The estimates are given in Table 11.4. These values can now be substituted into the constraints and the constraints arranged in the linear programming format. Parameter values assigned by the above discussion, but not yet listed, are given below:

$$B_G = 0.02 \text{ lb/s}$$
$$L_A = L_B = L_C = 1.25 \times 10^{-2} \text{ lb/ft}^3$$

Equations 11.3 show the linear programming model. All of the constraints have been multiplied by 10^3 to simplify the presentation. This will not affect the final values determined for the variables. Also, the computer package used to find the optimum solution to this model would accept no more than three decimal places for input data. Actually, the theory of linear algebra establishes that not all of the equations need to

TABLE 11.4 TREATMENT COSTS

Location	Variable identifier	Cost ($/lb removed)
Landfill	C_L	4.00
Water supply City A	C_{A0}	1.70
Waste treatment City A	C_A	0.85
Water supply City B	C_{B0}	1.60
Waste treatment City B	C_B	0.75
Waste treatment City C	C_C	0.85

be multiplied by the same factor. They are all multiplied by the same factor here for clarity of presentation.

$$
\left.\begin{aligned}
\min Z = 4.00T_L &+ 1.70T_{A0} + 0.85T_A + 1.60T_{B0} + 0.75T_B + 0.85T_C \\
\text{ST:} \quad 1000L_{A0} &+ 10T_L = 0.200 \\
1000L_{A0} &\leq 0.312 \\
1000L'_{A0} - 1000L_{A0} &+ 100T_{A0} = 0 \\
1000L'_{A0} &\leq 0.0312 \\
-1000L'_{A0} + 100T_A &+ 1000L'_A = 12.5 \\
100L'_A &+ 900L_{A0} \leq 0.312 \\
125{,}000L_{B0} - 10{,}000L'_A &- 90{,}000L_{A0} = 0 \\
1000L'_{B0} - 1000L_{B0} &+ 50T_{B0} = 0 \\
1000L'_{B0} &\leq 0.0312 \\
-1000L'_{B0} + 50T_B &+ 1000L'_B = 12.5 \\
200L'_B &+ 800L_{B0} \leq 0.312 \\
143{,}000L_{C0} - 20{,}000L'_B &- 80{,}000L_{B0} = 0 \\
100T_C &+ 1000L'_C = 12.5 \\
90.9L'_C &+ 909.1L_{C0} \leq 0.312
\end{aligned}\right\} \qquad (11.3)
$$

Solution of this model produces the following variable values at the optimum point:

$$
\begin{aligned}
T_L &= 0.0 \text{ lb/s} \\
L_{A0} &= 0.20 \times 10^{-3} \text{ lb/ft}^3 \\
T_{A0} &= 1.69 \times 10^{-3} \text{ lb/s} \\
L'_{A0} &= 0.03 \times 10^{-3} \text{ lb/ft}^3 \\
L'_A &= 1.32 \times 10^{-3} \text{ lb/ft}^3 \\
T_A &= 112 \times 10^{-3} \text{ lb/s} \\
L_{B0} &= 0.25 \times 10^{-3} \text{ lb/ft}^3 \\
T_{B0} &= 4.37 \times 10^{-3} \text{ lb/s} \\
L'_{B0} &= 0.03 \times 10^{-3} \text{ lb/ft}^3 \\
L'_B &= 0.56 \times 10^{-3} \text{ lb/ft}^3 \\
T_B &= 239 \times 10^{-3} \text{ lb/s} \\
L_{C0} &= 0.22 \times 10^{-3} \text{ lb/ft}^3 \\
L'_C &= 1.25 \times 10^{-3} \text{ lb/ft}^3 \\
T_C &= 113 \times 10^{-3} \text{ lb/s}
\end{aligned}
$$

The objective function value for this solution is $0.381/s. The units involved in this solution may seem unusual, but they are consistent with the rest of the model, and actually represent annualized costs when multiplied by the number of seconds in a year. Thus, the equivalent uniform annual cost is $11,980,000/yr. It took 13 iterations and approximately 1600 computations to arrive at the final solution. The most economical solution is not to treat any of the landfill leachate. Treatment must be provided at the other five locations. Examination of the solution variable values shows that the carrying capacity of the river is used by the model to reduce the amount of treatment necessary. This is why no treatment is required at the landfill. When the effluent mixes with the river, the concentration in the river is below the allowable level. The concentration in the river after mixing with the effluent from each waste treatment plant is at the maximum allowable level. If this had not been the case, it should be suspected that something is wrong with the model. Sensitivity analysis of the optimum solution reveals the following:

1. T_L will enter the solution if its cost coefficient is decreased to 0.935. Currently it is set at 4.0.
2. Cost coefficient for T_{A0} may be varied in the range 0.85 to 32.35 without affecting the final solution. Currently this is set at 1.7.
3. Cost coefficient for T_A may be varied in the range 0.736 to 1.7 without affecting the final solution. Currently this is set at 0.85.
4. Cost coefficient for T_{B0} may be varied in the range 0.75 to 2.313 without affecting the final solution. Currently this is set at 1.6.
5. Cost coefficient for T_B may be varied in the range 0.594 to 0.928 without affecting the final solution. Currently this is set at 0.75.
6. Cost coefficient for T_C may be varied in the range 0 to 1.072 without affecting the final solution. Currently this is set at 0.85.
7. Within the range of 0 to 0.2, each unit change in the right-hand side value for the fourth constraint (water supply quality limitation after treatment at City A) will change the objective function value by 0.001. The right-hand side value is presently set at 0.0312.
8. Within the range of 0.180 to 0.488, each unit change in the right-hand side value for the sixth constraint (water quality limitation in the river at discharge from City A) will change the objective function value by 0.011. The right-hand side value is presently set at 0.312.
9. Within the range of 0 to 0.250, each unit change in the right-hand side value for the ninth constraint (water supply quality limitation after treatment at City B) will change the objective function value by 0.002. The right-hand side value is presently set at 0.0312.
10. Within the range of 0.200 to 0.491, each unit change in the right-hand side value for the eleventh constraint (water quality limitation in the river at discharge from City B) will change the objective function value by 0.016. The right-hand side value is presently set at 0.312.
11. Within the range of 0.198 to 1.33, each unit change in the right-hand side value for the 14th constraint (water quality limitation in the river at discharge from City C) will change the objective function value by 0.094. Currently the right-hand side value is set at 0.312.

Analysis of the cost coefficient sensitivity shows that the solution is most sensitive to the costs of city wastewater treatment. Therefore, these are the cost estimates the analyst should be the most sure of. Treatment at the landfill will only become economical if its cost is reduced by about 75%. Right-hand side value sensitivity analysis for the constraints reveals that the most gain could be made in the objective function value if the water quality requirements in the river could be relaxed.

The optimization will now proceed using dynamic programming techniques. Analysis will commence at the most downstream point, and work upstream toward the landfill. Pollutant concentration in the stream will be the state variable for each stage. Letters have been placed on Figure 11.6 to delineate the stages for dynamic programming.

Stage 1. This is the section of the river between stations a and b, and includes the contribution to river flow and pollution provided by City C. Figure 11.7 shows a schematic of the section. If the carrying capacity of the river is to be used to its fullest extent, then the concentration in the river just below City C should be at the maximum permissible level. This will also lead to the minimum amount of treatment that must be accomplished. Therefore, the continuity equation for mixing at the outfall from City C will be

$$LR_a = \frac{10L'_C + 100L_{C_0}}{110} = 0.312 \times 10^{-3} \text{ lb/ft}^3 \qquad (11.4)$$

No more pollutant can be treated at City C than is produced by the city. Therefore,

$$T_C \leq 0.125 \text{ lb/s} \qquad (11.5)$$

Also, the upper limit on the concentration in the river just above the discharge point from City C is seven-tenths of what the maximum concentration can be just below the discharge point for City B, because of the natural reduction in pollution within the stream. Therefore,

$$L_{C_0} \leq 0.7(0.312 \times 10^{-3}) = 0.22 \times 10^{-3} \qquad (11.6)$$

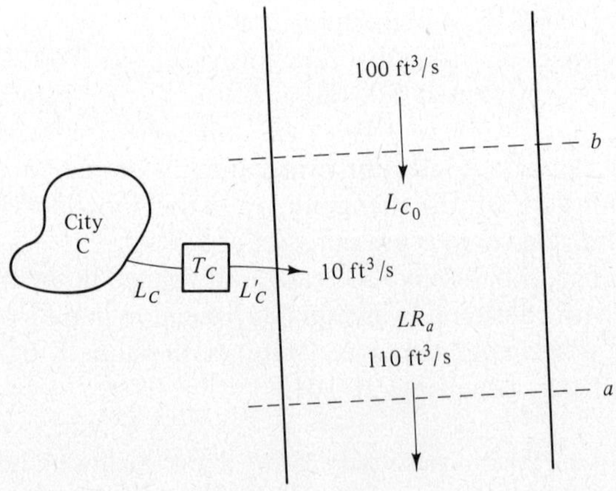

LR_a = concentration of pollutant in river after mixing below
      City C

Figure 11.7 First stage for dynamic programming.

The mass rate of pollutant transport in the discharge line from the treatment plant at C is related to the rate of pollution production by the city and the rate of removal at the treatment plant by Equation 11.7.

$$T_C + 10L'_C = 0.125 \tag{11.7}$$

Equations 11.7 and 11.4 can be combined to form the transfer function for the stage,

$$0.125 - T_C + 100L_{C_0} = 0.0343 \tag{11.8}$$

which can be further transformed into

$$T_C = 100L_{C_0} + 0.0907 \tag{11.9}$$

Utilizing the constraints imposed by Equations 11.5 and 11.6, a range of values for the state variable L_{C_0} can be chosen, and corresponding values of T_C calculated. Table 11.5 shows the results.

Stage 2. Stage two incorporates the segment of the river from just below the discharge point for City B, after mixing has taken place, to just above the discharge point for City C. There are no lateral inputs to this stage; however, there is removal of pollutant through natural processes. The state variable at the downstream end of the stage (station b on Figure 11.6) will be seven-tenths of the corresponding state variable at the upstream end (station c). If LR_c is the concentration at station c, then Table 11.6 shows the values of the state variables at c and b.

Stage 3. Stage 3 is established to include all of City B, from just above the intake for treatment (station d in Figure 11.6) to just below the waste treatment plant effluent

TABLE 11.5   ALTERNATIVES FOR STAGE 1

L_{C_0} (lb/ft^3 × 10^3)	T_C (lb/s)	$/s for treatment @ $0.85/lb removed
0.22	0.113	0.0956
0.20	0.111	0.0941
0.15	0.106	0.0899
0.10	0.101	0.0856
0.0	0.091	0.0771

TABLE 11.6   STATE VARIABLES FOR STAGE 2

L_{C_0} (lb/ft^3 × 10^3)	LR_c (lb/ft^3 × 10^3)
0.220	0.312
0.200	0.285
0.150	0.214
0.100	0.143
0.00	0.0

discharge point (station c). This is done because of the splitting of flows for water supply at City B, then the rejoining of flows after waste treatment. These would be difficult to incorporate into a serial dynamic programming structure. However, with one reasonable stipulation, the whole city can be treated as a single stage, which will make the problem more manageable. Figure 11.8 shows a definition sketch for this stage.

The stipulation is made that if the concentration of pollutant in the intake of the water treatment plant is greater than the permissible level in the water supply, sufficient treatment will be provided to bring the concentration down to the permissible level, but no further. This stipulation is reasonable. The concentration has to meet the quality requirements. Also, since the cost per pound of pollutant removed is greater for the water supply treatment plant than it is for the wastewater treatment plant, it is not economical to provide excess treatment at the supply end to take some of the load off the waste treatment end.

Since 20% of the pollutant in the river below City A is removed by natural means prior to the intake for water supply at City B, the maximum concentration that can be expected at the intake side of the water supply treatment plant is eight-tenths of L_L, or

$$L_{B_0} \le (0.8)(0.312 \times 10^{-3}) = 0.250 \times 10^{-3} \tag{11.10}$$

Since no more pollutant can be removed than is present in the raw water supply, the upper limit on T_{B0} is

$$T_{B_0} \le 20(0.250 \times 10^{-3}) = 4.99 \times 10^{-3} \text{ lb/s} \tag{11.11}$$

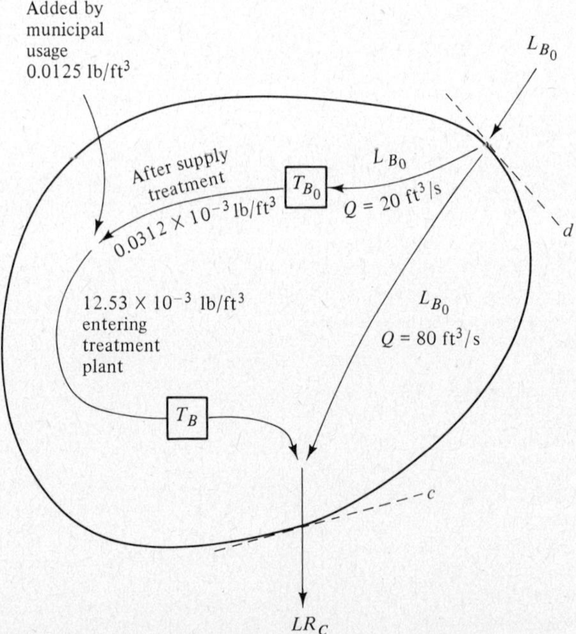

Figure 11.8   Third stage for dynamic programming.

A constant 0.0125 lb/ft^3 is added to the water supply through usage, so the maximum load entering the waste treatment plant, and thus the upper limit for treatment, would be

$$T_B \leq 20(0.0312 \times 10^{-3} + 0.0125) = 0.251 \text{ lb/s} \tag{11.12}$$

There are two transfer functions for this stage, one for each treatment plant. Using the assumption that excess pollutant would not be removed at the water supply treatment plant, the transfer functions become

$$\left. \begin{array}{l} \text{at water supply treatment plant} \quad 20L_{B_0} - T_{B_0} = 20(0.0312 \times 10^{-3}) \\[2ex] \text{at wastewater treatment plant} \quad \dfrac{20(0.01253) - T_B + 80L_{B_0}}{100} = LR_c \end{array} \right\} \tag{11.13}$$

which can be rearranged to give

$$\left. \begin{array}{l} T_{B_0} = 20L_{B_0} - 0.624 \times 10^{-3} \\[1ex] T_B = 80L_{B_0} - 100LR_c + 0.251 \end{array} \right\} \tag{11.14}$$

TABLE 11.7 EVALUATION OF ALTERNATIVES AT STAGE 3

LR_c (lb/ft^3 $\times 10^3$)	\$/s from stage 1	L_{B_0} (lb/ft^3 $\times 10^3$)	T_{B_0} (lb/s $\times 10^3$)	\$/s for treatment at B_0 @\$1.60/lb	T_B (lb/s $\times 10^3$)	\$/s for treatment at B @\$0.75/lb	Total \$/s for City B	Total \$/s stages 1–3
0.312	0.0956	0.250	4.38	0.00700	239.3	0.179	0.186	0.282
		0.200	3.38	0.00540	235.3	0.176	0.182	0.278
		0.150	2.38	0.00380	231.3	0.173	0.177	0.273
		0.100	1.38	0.00220	227.3	0.170	0.173	0.268
		0	0	0	219.3	0.164	0.164	0.260
0.285	0.0941	0.250	4.38	0.00700	242	0.182	0.189	0.283
		0.200	3.38	0.00540	238	0.179	0.184	0.278
		0.150	2.38	0.00380	234	0.176	0.179	0.273
		0.100	1.38	0.00220	230	0.173	0.175	0.269
		0	0	0	222	0.167	0.167	0.261
0.214	0.0899	0.250	4.38	0.00700	249.1	0.187	0.194	0.284
		0.200	3.38	0.00540	245.1	0.184	0.189	0.279
		0.150	2.38	0.00380	241.1	0.181	0.185	0.275
		0.100	1.38	0.00220	237.1	0.178	0.180	0.270
		0	0	0	229.1	0.172	0.172	0.262
0.143	0.0856	0.250	4.38	0.00700	256.2	0.192	0.199	0.285
		0.200	3.38	0.00540	252.2	0.189	0.195	0.280
		0.150	2.38	0.00380	248.2	0.186	0.190	0.276
		0.100	1.38	0.00220	244.2	0.183	0.185	0.271
		0	0	0	236.2	0.177	0.178	0.263
0	0.0771	0.250	4.38	0.00700	270.50	0.203	0.210	0.287
		0.200	3.38	0.00540	266.50	0.200	0.205	0.282
		0.150	2.38	0.00380	262.50	0.197	0.201	0.278
		0.100	1.38	0.00220	258.50	0.194	0.196	0.273
		0	0	0	250.50	0.188	0.188	0.265

TABLE 11.8 STATE VARIABLES FOR STAGE 4

Station d, L_{B_0} (lb/ft^3 × 10^3)	Station e, LR_e (lb/ft^3 × 10^3)
0.250	0.312
0.200	0.250
0.150	0.188
0.100	0.125
0.0	0.0

Values for LR_c are set from the previous stage. A reasonable range of values for the state variable, L_{B_0}, at the upstream end of the stage will be used to generate alternatives. Values for T_{B_0} and T_B will be calculated from corresponding values of LR_c and L_{B_0}, using Equations 11.14, and the constraints of Equations 11.10 through 11.12. Table 11.7 gives the alternatives evaluated and the resulting costs for stages 1 through 3. Note that the same set of L_{B_0} values were used for each value of LR_c. This allows the elimination of some alternatives from further consideration. For example, if L_{B_0} was 0.25×10^{-3} lb/ft^3, the minimum total cost for stages 1–3 occurs when $LR_c = 0.312 \times 10^{-3}$ lb/ft^3. All of the other alternatives with $L_{B_0} = 0.25 \times 10^{-3}$ lb/ft^3 can be ignored in subsequent stages. The question to be asked is, if L_{B_0} is a certain value, what is the least cost way of reaching the end point of the problem? This is the essence of Bellman's theory of optimality described in the introduction to this chapter. Each of the other

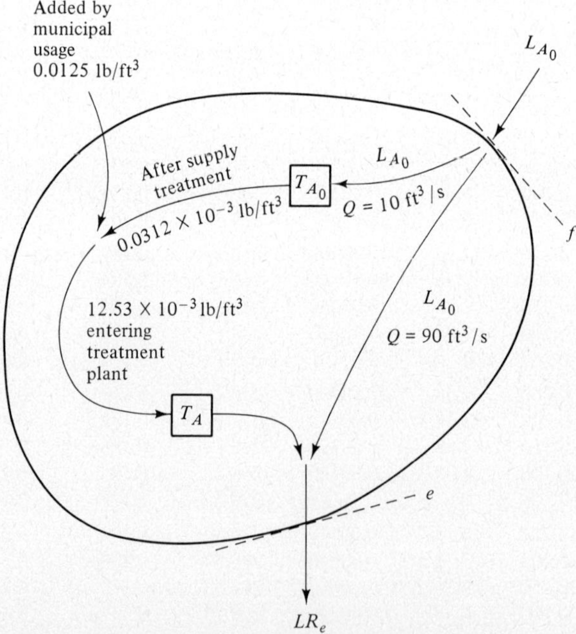

Figure 11.9 Fifth stage for dynamic programming.

TABLE 11.9 EVALUATION OF ALTERNATIVES AT STAGE 5

LR_e (lb/ft^3 × 10^3) ------- Cost stages 1–3	L_{A_0} (lb/ft^3 × 10^3)	T_{A_0} (lb/s × 10^3)	\$/s for treatment at A_0 @\$1.70/lb	T_A (lb/s × 10^3)	\$/s for treatment at A @\$0.85/lb	Total \$/s City A	Total \$/s stages 1–5
0.312	0.312	2.81	0.00478	122.18	0.104	0.109	0.391
	0.250	2.19	0.00372	116.60	0.0991	0.103	0.385
--------	0.200	1.69	0.00287	112.10	0.0953	0.0982	0.380
	0.150	1.19	0.00202	107.60	0.0915	0.0935	0.376
\$0.282	0.100	0.69	0.00117	103.10	0.0876	0.0888	0.371
	0.0	0	0	94.10	0.0800	0.0800	0.362
0.250	0.312	2.81	0.00478	128.38	0.109	0.114	0.391
	0.250	2.19	0.00372	122.80	0.104	0.108	0.386
--------	0.200	1.69	0.00287	118.30	0.101	0.103	0.381
	0.150	1.19	0.00202	113.80	0.0967	0.0988	0.376
\$0.278	0.100	0.69	0.00117	109.30	0.0929	0.0941	0.372
	0.0	0	0	100.3	0.0853	0.0853	0.363
0.188	0.312	2.81	0.00478	134.58	0.114	0.119	0.392
	0.250	2.19	0.00372	129.00	0.110	0.113	0.386
--------	0.200	1.69	0.00287	124.50	0.106	0.109	0.382
	0.150	1.19	0.00202	120.00	0.102	0.104	0.377
\$0.273	0.100	0.69	0.00117	115.50	0.0982	0.0994	0.372
	0.0	0	0	106.50	0.0905	0.0905	0.363

values of L_{B_0} also has a minimum total cost when LR_c is 0.312×10^{-3} lb/ft^3, which is reasonable because this value of LR_c utilizes the full carrying capacity of the river at point c.

Stage 4. Stage 4 is similar to stage 2; the transfer function involves removal of pollutant through natural processes. Twenty percent of the pollutant concentration in the river at station e (of Figure 11.6) is removed by the time the flow gets to station d. Table 11.8 shows what the concentrations will be at station e for the previously selected values of L_{B_0} at station d.

Stage 5. Stage 5 for City A is similar to stage 3, with the exception that $Q_A = 10$ ft^3/s, whereas, $Q_B = 20$ ft^3/s. All of the assumptions and development of transfer functions are the same. The pertinent expressions are given below, and Figure 11.9 shows the definition sketch.

$$\left.\begin{array}{l} L_{A_0} \le 0.312 \times 10^{-3} \text{ lb/ft}^3 \\ T_{B_0} \le 3.12 \times 10^{-3} \text{ lb/s} \\ T_B \le 0.125 \text{ lb/s} \\ T_{A_0} = 10L_{A_0} - 0.312 \times 10^{-3} \\ T_A = 90L_{A_0} - 100LR_e + 0.125 \end{array}\right\} \qquad (11.15)$$

Table 11.9 shows the evaluation of total costs for various values of L_{A_0} assigned for each value of LR_e. The pollutant concentration at station c, LR_c, for each of these alternatives is 0.312×10^{-3} lb/ft^3, as established in stage 3. Again, the options that

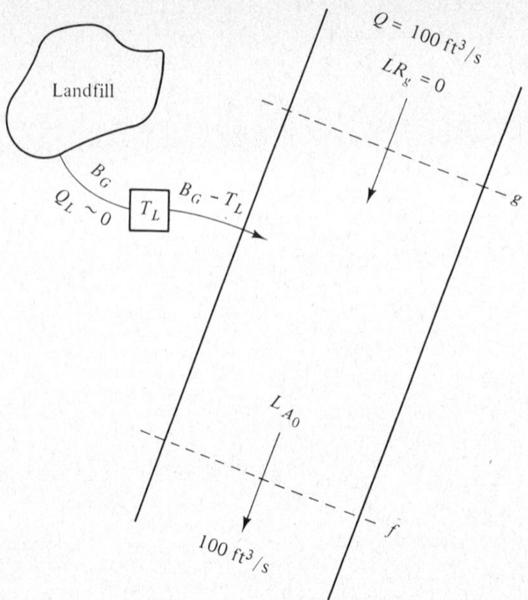

Figure 11.10 Sixth stage for dynamic programming.

fully utilize the carrying capacity of the river, $(LR_e = 0.312 \times 10^{-3} \text{ lb/ft}^3)$, are the most economical.

Stage 6. Figure 11.10 shows the conditions for the final stage of the analysis. The landfill produces 0.02 lb/s, which, if mixed with the river flow, will produce a concentration of $0.200 \times 10^{-3} \text{ lb/ft}^3$. Therefore, the first two alternatives of Table 11.9 $(L_{A_0} = 0.312 \times 10^{-3} \text{ lb/ft}^3$ and $L_{A_0} = 0.250 \times 10^{-3} \text{ lb/ft}^3)$ do not need to be considered because L_{A_0} cannot be that high. It can also be seen that no treatment is required; however, the economic efficiency of providing it should be investigated.

If
$$\frac{B_G - T_L}{100} = L_{A_0} \tag{11.16}$$

then
$$T_L = 0.02 - 100 L_{A_0} \tag{11.17}$$

TABLE 11.10 EVALUATION OF ALTERNATIVES AT STAGE 6

L_{A_0} (lb/ft^3 × 10^3)	Total \$/s stages 1–5	T_L (lb/s × 10^3)	\$/s for treatment at landfill @\$4.00/lb	Total \$/s stages 1–6
0.200	0.380	0.00	0.00	0.380
0.150	0.376	5.00	0.020	0.396
0.100	0.371	10.0	0.040	0.411
0.0	0.362	20.0	0.080	0.442

Table 11.10 shows the evaluation of Equation 11.17 for the appropriate values of L_{A_0}, with the total costs enumerated. The overall minimum cost is \$0.380/s, and if the variable values are traced back through for this alternative, it will be found that they are the same as for the linear programming solution. The total number of computations for the dynamic programming solution was approximately 600, as opposed to 1600 for the simplex method.

Thus, dynamic programming identified the optimum policy for this regional wastewater treatment authority with reference to one pollutant. Other pollutants of concern would have to be similarly evaluated, and final policy decisions made based on all of the information.

SUMMARY

Dynamic programming provides a computationally efficient and readily understood method of finding the optimal operation of systems containing serially structured stages. Furthermore, the transfer function at each stage can be either linear, or nonlinear, and can change from one stage to the next. At each stage, a family of solutions for discrete points within the range of possible production is generated. This represents a distinct advantage of dynamic programming over other optimization techniques, such as linear programming, which produce just one solution.

There is no standard mathematical technique for dynamic programming problems. Dynamic programming is an approach to problem solving, and particular mathematical equations have to be developed to fit each situation. An understanding of the general structure of dynamic programming problems is essential in determining if a particular problem fits the dynamic programming mold, and, if it does, how to apply the methods to it. Application of the method to three problems previously solved by linear programming techniques demonstrates the utility of dynamic programming when the system model exhibits the characteristic serial structure.

APPLICATIONS EXERCISES

Use dynamic programming techniques to find the optimal solutions to the following exercises: 5-9 and 5-32.

Chapter 12

Integer Programming

INTRODUCTION

Problem conditions may require that some or all of the solution variables in a linear programming model take on only integer values. Transportation problems involving loads that cannot be subdivided, problems involving allocation of resources such as people or buildings, and production problems that require the manufacturing or fabrication of whole units are examples. In pure integer problems, all variables are required to have integer values, while mixed integer problems require only certain variables to have integer values.

The simplest approach toward integer solutions is to solve the linear programming problem by the simplex algorithm and hope the appropriate variables do turn out to have integer values. Certain linear programming models will produce integer solutions because of their structure. The assignment problems, illustrated by Example 5.13, fall into this category because of their requirement of a one-to-one correspondence between resources and assignments. Frequently, linear programming models that have all integer constraint stipulations (right-hand side values) will produce integer optimal solutions; but this is not guaranteed. The steel fabrication model of Example 5.11 does result in an integer optimum solution; however, the concrete sewer pipe production model of Example 5.12 does not produce an integer optimal solution even though all of the constraint stipulations, as well as all of the constraint and cost coefficients, have integer values.

A second approach to achieving integer solutions is to round off noninteger values for solution variables. Although this is easy, it may not result in the optimum integer solution, and may even result in an infeasible solution. Consider the model

$$\max Z = 2X_1 + 3X_2$$
$$\left. \begin{array}{l} \text{ST:} \quad 6X_1 + 3.23X_2 \le 30 \\ \phantom{\text{ST:}} \quad 1.39X_1 + 10X_2 \le 31 \end{array} \right\} \qquad (12.1)$$

Solution of this model graphically, or by using the simplex algorithm, gives optimum values for X_1 and X_2 of 3.6 and 2.6, respectively, and an optimal Z value of 15. If X_1 and X_2 are rounded to 4 and 3, the optimal Z value appears to increase to 17; however, examination of Figure 12.1 shows that the point (4, 3) is actually outside the feasible region. Rounding may be justified if the affected constraints are flexible enough to be relaxed to accommodate the new point. Of the integer points within the present feasible region, $X_1 = 3$, $X_2 = 2$, $Z = 12$ provides the largest objective function value. Rounding down variable values (truncating) for maximization problems bounded by $\leq$ constraints may reduce the chance of an integer solution's being outside the feasible region, but does not eliminate it. Neither does it ensure that an optimal integer solution will be found. For the model depicted, rounding down would eliminate the infeasible point, and would result in the optimal feasible integer solution being found. However, if the constraint

$$1.39X_1 + 10X_2 \leq 31 \tag{12.2}$$

had a larger negative slope while still passing through (3.6, 2.6), the point (2, 3) could be included within the feasible region. In that case, rounding down would still give (3, 2), while the optimum integer solution would be at (2, 3).

It should be evident that linear programming models with even one equal constraint will not have any integer solutions unless the equality constraint passes through one or more integer points. As the number of equality constraints increases, the chance of finding an integer solution becomes very small. Another characteristic of integer solutions illustrated by Equations 12.1 and Figure 12.1 is that unless the optimal continuous solution is integer, the optimal integer solution will have a different objective function value from the optimum continuous solution. The integer Z value would be less for maximization problems and greater for minimization problems.

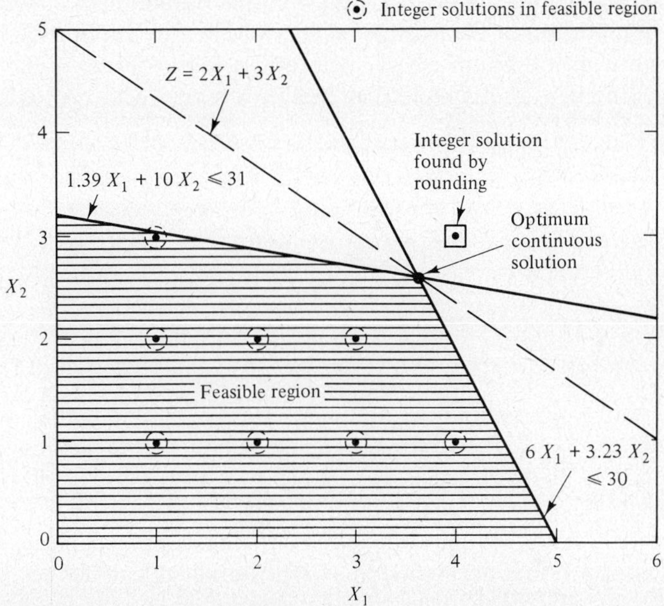

Figure 12.1 Infeasible solution caused by rounding.

Problems with two variables could always be graphed to readily identify the optimal integer solution. However, for problems involving more variables, analytical procedures are needed to find the best integer solution. Integer linear programming algorithms use a series of continuous solutions of the simplex algorithm until the solutions converge on an optimal integer solution. Computer programs for implementing these algorithms are available at most computer centers. However, since successive applications of the simplex algorithm are involved, integer solutions for large models may be excessively consumptive of computer time. Two common techniques for solving integer linear programming problems are the cutting plane algorithm and the branch and bound algorithm. The cutting plane method will be illustrated here because it is intuitively more understandable. Constraints which shrink the feasible region until an optimum integer extreme point is located are added to the initial model. When many integer variables are involved, a large number of cutting plane constraints may have to be added to find this optimal integer solution. Therefore, branch and bound algorithms may be more efficient solution techniques for problems with more than about 30 integer variables. These algorithms use the concept of a tree of solutions, with the nodes representing a subset of the set of all feasible solutions, and the branches of the tree representing directed line segments joining the nodes. The bounding portion of the algorithm provides a means of rejecting certain sequences of branches (paths), thereby, reducing the effort in finding an optimal integer solution. Further information on both methods can be found in Agin (1966), Bradley, et al. (1977), Garfinkel and Nemhauser (1972), and Greenberg (1971).

INTEGER SOLUTION BY CUTTING PLANES

Cutting plane algorithms involve the successive determination of new constraints that will shave off a portion of the feasible region, followed by application of the simplex algorithm, until an optimum integer solution is found. Finding the cutting plane constraints is the key, and is best illustrated by a model with two variables that can also be graphed.

Consider the model

$$\begin{aligned}
\max Z &= 26X_1 + 13X_2 \\
\text{ST:} \quad -2X_1 + 4X_2 &\leq 12 \\
2X_1 + 5X_2 &\geq 10 \\
X_1 + X_2 &\leq 7 \\
5X_1 - 2X_2 &\leq 16.8
\end{aligned} \right\} \tag{12.3}$$

Figure 12.2 shows the optimal continuous solution for this model, with $X_1 = 4.4$, $X_2 = 2.6$, and $Z = 148.2$. Possible integer solution points are also shown. The optimal integer solution is observed to be $X_1 = 4$, $X_2 = 3$, and $Z = 143$. For this problem, rounding would have produced the optimum integer solution.

Tables 12.1(a) and (b) show the initial and final simplex tableaux for Equations 12.3. Note, however, that the second line of the initial tableau has been multiplied by -1 and the artificial variable that would be expected to accompany a greater than or

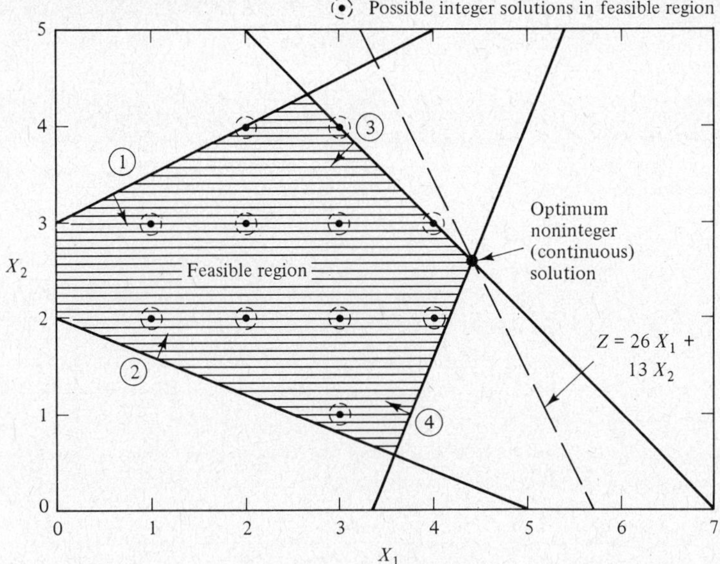

Figure 12.2 Optimal continuous solution.

equal constraint is not included in the matrix. This is because most cutting plane techniques require all constraints to be of the less than or equal type. Greater than or equal constraints can be changed to this form by multiplying by -1 and switching the direction of the inequality. The restriction that all constraint stipulations be ≥ 0 can be neglected in this case. Therefore, Equations 12.3 are changed as follows to allow application of techniques to find optimal integer solutions for X_1 and X_2:

$$\left.\begin{array}{rl}
\max Z = 26X_1 + 13X_2 \\
\text{ST:} \quad -2X_1 + 4X_2 \leq 12 \\
-2X_1 - 5X_2 \leq -10 \\
X_1 + X_2 \leq 7 \\
5X_1 - 2X_2 \leq 16.8 \\
\\
X_1, X_2 = 0, 1, 2, \cdots
\end{array}\right\} \quad (12.4)$$

When slack variables are added, the model becomes

$$\left.\begin{array}{rl}
\max Z = 26X_1 + 13X_2 + 0X_3 + 0X_4 + 0X_5 + 0X_6 \\
\text{ST:} \quad -2X_1 + 4X_2 + X_3 \qquad\qquad\qquad = 12 \\
-2X_1 - 5X_2 \qquad + X_4 \qquad\qquad = -10 \\
X_1 + X_2 \qquad\qquad + X_5 \qquad = 7 \\
5X_1 - 2X_2 \qquad\qquad\qquad + X_6 = 16.8 \\
X_1, X_2, X_3, X_4, X_5, X_6 \geq 0 \\
X_1, X_2 = 0, 1, 2, \cdots
\end{array}\right\} \quad (12.5)$$

Comparison will show that the solution to this continuous model is identical to the solution for Equations 12.3.

TABLE 12.1 CONTINUOUS SOLUTION

(a) Initial tableau

c_j			26	13	0	0	0	0
	Solution variable	B	X_1	X_2	X_3	X_4	X_5	X_6
0	3	12	−2	4	1	0	0	0
0	4	−10	−2	−5	0	1	0	0
0	5	7	1	1	0	0	1	0
0	6	16.80	5	−2	0	0	0	1
	Z_j	0	0	0	0	0	0	0
	$c_j - Z_j$	0	0	26	13	0	0	0	0

(b) Final tableau

c_j			26	13	0	0	0	0
	Solution variable	B	X_1	X_2	X_3	X_4	X_5	X_6
0	3	10.40	0	0	1	0	−2.29	0.86
0	4	11.80	0	0	0	1	4.14	−0.43
13	2	2.60	0	1	0	0	0.71	−0.14
26	1	4.40	1	0	0	0	0.29	0.14
	Z_j	148.2	26	13	0	0	16.7	1.86
	$c_j - Z_j$		0	0	0	0	−16.7	−1.86

The final tableau of Table 12.1(b) represents the following set of equations:

$$\left.\begin{array}{l}
Z = 148.2 \ - 16.7X_5 - 1.86X_6 \\
X_3 = \ \ 10.40 + 2.29X_5 - 0.86X_6 \\
X_4 = \ \ 11.80 - 4.14X_5 + 0.43X_6 \\
X_2 = \ \ \ \ 2.60 - 0.71X_5 + 0.14X_6 \\
X_1 = \ \ \ \ 4.40 - 0.29X_5 - 0.14X_6
\end{array}\right\} \qquad (12.6)$$

Neither X_1 nor X_2 has integer values for this solution; therefore, a cutting plane constraint needs to be developed to remove a noninteger portion of the feasible region.

There is no particular protocol; but commonly, the largest fractional remainder for any of the desired integer variables decides which constraint should be used first to find a cutting plane. In this case, X_1 and X_2 are the desired integer variables. X_1 has a value of 4.40 and X_2 has a value of 2.60 in the optimal continuous solution. Since 0.60 is the largest fractional remainder, the equation

$$X_2 = 2.60 - 0.71X_5 + 0.14X_6 \tag{12.7}$$

with a remainder of 0.60 will be chosen. Equation 12.7 may be rewritten as

$$X_2 + 0.71X_5 + (-1 + 0.86)X_6 = 2 + 0.60 \tag{12.8}$$

and
$$0.71X_5 + 0.86X_6 = 0.60 + 2 + X_6 - X_2 \tag{12.9}$$

which depicts each noninteger coefficient as a sum of an integer and a positive fraction. Equation 12.9 is a manipulation of an original constraint; therefore, any solution to the model, including any integer solution, must still satisfy 12.9. For a linear programming solution, X_5 and X_6 both must be ≥ 0; therefore, the left-hand side of Equation 12.9 must also be ≥ 0. If the left-hand side is ≥ 0, then the right-hand side of Equation 12.9 must also be ≥ 0. As before, X_2 and X_6 must both be ≥ 0. If X_2 is limited to integer values, the minimum value the right-hand side can take on, and still remain greater than or equal to zero, is 0.60. This is regardless of the value of X_6. Consequently, the minimum value the left-hand side of Equation 12.9 can take on is 0.60, inferring that

$$0.71X_5 + 0.86X_6 \geq 0.60 \tag{12.10}$$

Equation 12.10 could be inserted directly into the model after being converted into a less than or equal constraint, and the augmented model solved using the simplex algorithm. However, to illustrate the cutting plane generation, it is advantageous to express Equation 12.10 in terms of the original problem variables, X_1 and X_2. That way the new constraint can be shown graphically. From Equations 12.5

$$\left. \begin{array}{l} X_5 = 7 - X_1 - X_2 \\ X_6 = 16.8 - 5X_1 + 2X_2 \end{array} \right\} \tag{12.11}$$

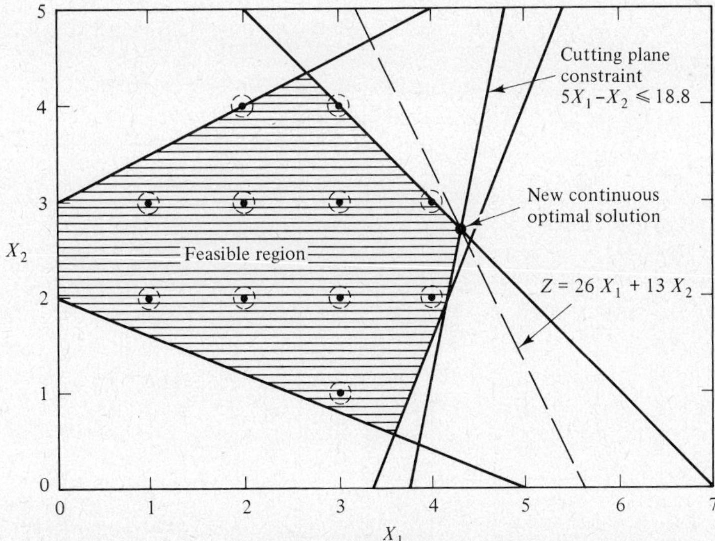

Figure 12.3 Solution after application of first cutting plane constraint.

Substituting Equations 12.11 into 12.10 gives

$$0.71(7 - X_1 - X_2) + 0.86(16.8 - 5X_1 + 2X_2) \geq 0.60$$
$$4.97 - 0.71X_1 - 0.71X_2 + 14.45 - 4.3X_1 + 1.72X_2 \geq 0.60 \qquad (12.12)$$

which reduces to

$$-5X_1 + X_2 \geq -18.8$$

or
$$5X_1 - X_2 \leq 18.8 \qquad (12.13)$$

Figure 12.3 shows the feasible region and optimum continuous solution after applying this additional constraint. It can be seen that a corner has been shaved off of the feasible region, cutting out the current solution, and a new optimal solution has been formed. The final tableau of the simplex algorithm for the modified model is shown in Table 12.2. The objective function value is now 146.9, as opposed to the original value of 148.2. Variables X_1 and X_2 now have values of 4.30 and 2.70, which are closer to integer solutions than the first solution, but are not there yet.

Table 12.2 represents the equations

$$\left.\begin{aligned}
Z &= 146.9 - 15.2X_5 - 2.17X_7 \\
X_3 &= 9.80 + 3X_5 - X_7 \\
X_4 &= 12.10 - 4.5X_5 + 0.5X_7 \\
X_6 &= 0.70 - 0.83X_5 + 1.17X_7 \\
X_1 &= 4.30 - 0.17X_5 - 0.17X_7 \\
X_2 &= 2.70 - 0.83X_5 + 0.17X_7
\end{aligned}\right\} \qquad (12.14)$$

Variable X_2 has the largest remainder, so it will be considered for cutting plane generation. As before,

TABLE 12.2 OPTIMAL SIMPLEX TABLEAU AFTER ADDITION OF FIRST CUTTING PLANE

c_j			26	13	0	0	0	0	0
	Solution variable	B	X_1	X_2	X_3	X_4	X_5	X_6	X_7
0	3	9.80	0	0	1	0	−3.00	0	1.00
0	4	12.10	0	0	0	1	4.50	0	−0.50
0	6	0.70	0	0	0	0	0.83	1	−1.17
26	1	4.30	1	0	0	0	0.17	0	0.17
13	2	2.70	0	1	0	0	0.83	0	−0.17
	Z_j	146.9	26	13	0	0	15.2	0	2.17
	$c_j - Z_j$		0	0	0	0	−15.2	0	−2.17

$$X_2 + 0.83X_5 + (-1 + 0.83)X_7 = 2 + 0.70$$

and
$$0.83X_5 + 0.83X_7 = 0.70 + 2 + X_7 - X_2 \qquad (12.15)$$

which implies that

$$0.83X_5 + 0.83X_7 \geq 0.70 \qquad (12.16)$$

Substituting from Equations 12.14,

$$\left. \begin{array}{l} X_5 = 7 - X_1 - X_2 \\ X_7 = 18.8 - 5X_1 + X_2 \end{array} \right\} \qquad (12.17)$$

into 12.16, gives

$$0.83(7 - X_1 - X_2) + 0.83(18.8 - 5X_1 + X_2) \geq 0.70$$
$$-4.98X_1 \geq -20.71 \qquad (12.18)$$
$$X_1 \leq 4.16$$

However, since X_1 is limited to integer values, Equation 12.18 reduces to

$$X_1 \leq 4 \qquad (12.19)$$

Application of this cutting plane is shown in Figure 12.4 and Table 12.3, both of which show that the optimum integer solution has been found, with $X_1 = 4$, $X_2 = 3$, and $Z = 143$.

　　　If, after the first cutting plane, the constraint for X_1 in Equations 12.14 had been chosen instead of the constraint for X_2, the end result would have been the same. The equation

$$X_1 + 0.17X_5 + 0.17X_7 = 4 + 0.30 \qquad (12.20)$$

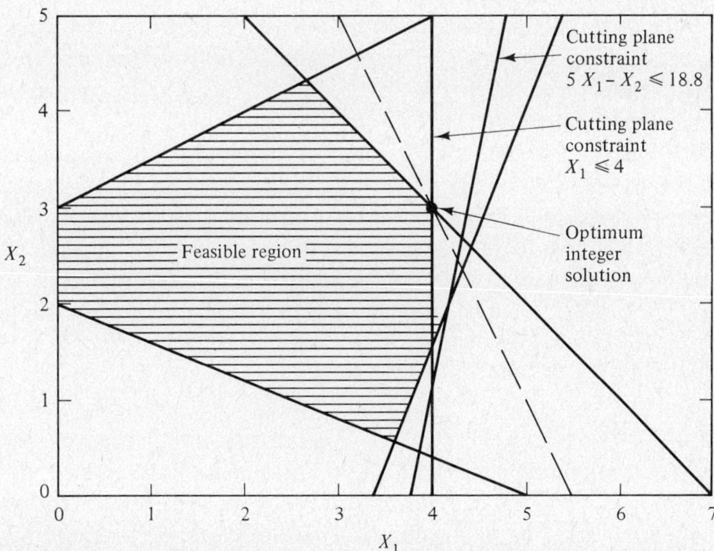

Figure 12.4　Solution after application of second cutting plane constraint.

TABLE 12.3   OPTIMAL SIMPLEX TABLEAU AFTER ADDITION OF SECOND CUTTING PLANE

c_j			26	13	0	0	0	0	0	0
	Solution variable	B	X_1	X_2	X_3	X_4	X_5	X_6	X_7	X_8
0	3	8.00	0	0	1	0	−4.00	0	0	6.00
0	4	13.00	0	0	0	1	5.00	0	0	−3.00
0	6	2.80	0	0	0	0	2.00	1	0	−7.00
26	1	4.00	1	0	0	0	0	0	0	1.00
0	7	1.80	0	0	0	0	1.00	0	1	−6.00
13	2	3.00	0	1	0	0	1.00	0	0	−1.00
	Z_j	143.0	26	13	0	0	13.0	0	0	13.0
	$c_j - Z_j$		0	0	0	0	−13.0	0	0	−13.0

implies that

$$0.17X_5 + 0.17X_7 \geq 0.30 \tag{12.21}$$

Substituting Equations 12.17 into 12.21 gives

$$0.17(7 - X_1 - X_2) + 0.17(18.8 - 5X_1 + X_2) \geq 0.30 \tag{12.22}$$

which simplifies to

$$-1.02X_1 \geq -4.09 \tag{12.23}$$

or

$$X_1 \leq 4 \tag{12.24}$$

which is the same cutting plane as Equation 12.19. In fact, if the equation for X_1 had been chosen when selecting the first cutting plane from Equations 12.6, Equation 12.24 would have been the first cutting plane derived, and the optimal integer solution would have been found one step sooner.

This two-dimensional example serves to demonstrate the physical principles of cutting plane generation for finding optimal integer solutions. It can also be deduced that larger problems will require generation of numerous cutting planes, and significant enlargement of the constraint space. It is quite possible for the size of the constraint space to exceed the available storage space on many computers.

APPLICATIONS OF INTEGER PROGRAMMING

Restrictions placed on a system may require that either input resources or output products, or both, be used in whole units. These restrictions create the need for integer programming procedures, one example of which was described in the previous section. The restrictions may be institutional in nature, such as a requirement to use even blocks of labor resource, or physical, such as a requirement to buy whole lots of a material resource. This section will describe some typical applications.

EXAMPLE 12.1 ———————————————————————————————————

In Examples 5.12 and 11.2, sewer pipe production was optimized over a five-month period. Demand for the pipe and available material varied from month to month. Costs for processing each unit of pipe on regular time and overtime were constant over the three months, but then increased because of new labor contracts. Two thousand hours of regular time processing were available each month, while there was no practical limit placed on the amount of overtime that could be scheduled. Seventy-five hours of processing are required for each unit of pipe produced; therefore, no more than 26.67 units of pipe could be produced in regular time each month. Any units above this limit would have to be produced on overtime. Equations 11.1 show the linear

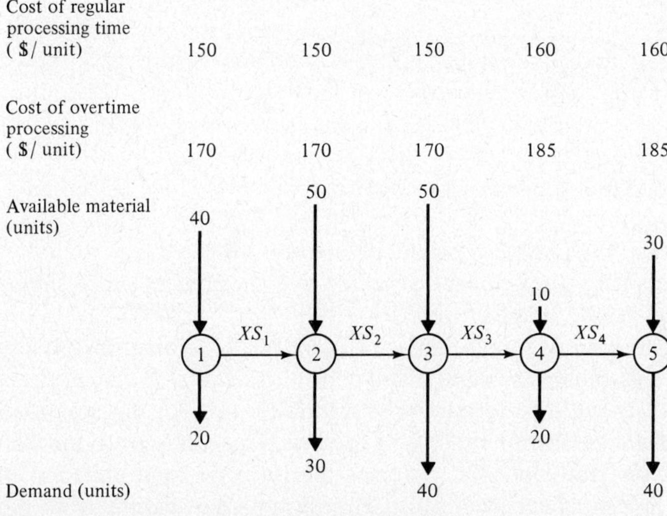

Variables:
XR_i – number of units produced on regular time during month i
XO_i – number of units produced on overtime during month i
XS_i – number of units stored from month i to month $i+1$

Figure 12.5 Pertinent system data, concrete sewer pipe production problem.

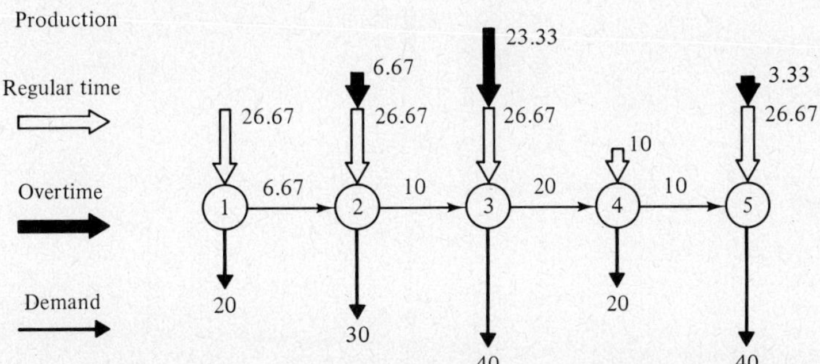

Figure 12.6 Optimum continuous solution.

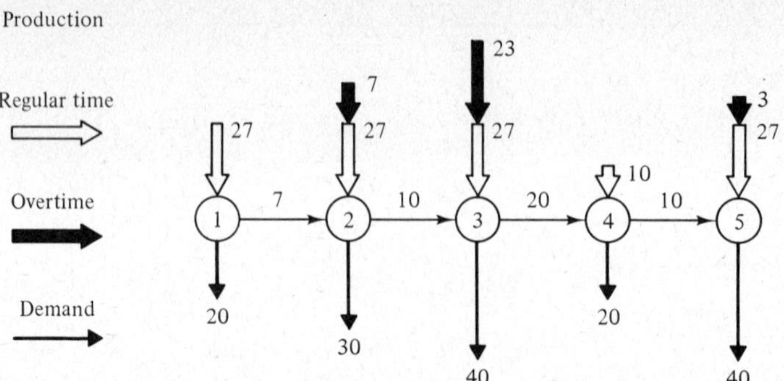

Figure 12.7   Integer solution formed by rounding.

programming model developed for this problem. Figure 12.5 shows the pertinent input data for the problem. Solution of the model by either the simplex algorithm or dynamic programming produced the solution shown in Figure 12.6, and a minimum cost of $24,520.00.

Suppose that the following additional conditions are placed on the problem:

1. Sewer pipe must be produced in even units.
2. Labor resources must be applied in even units of processing time.

These two limitations imply that all variables have to assume only integer values. If the option was chosen to round off all continuous variable values, the solution shown in Figure 12.7 would be formed. The minimum cost for this solution is $24,665.00. However, the regular time processing used each month exceeds the 2000 hr available. Also, flow continuity is violated during month 2: 41 total units are either produced during the month, or carried forward from the previous month, while the total demand is 40 units, 30 used during the month and 10 carred over to month three. It is possible that 25 additional hours of processing time could be scheduled each month, and the discontinuity during month 2 could be alleviated by scheduling 6 units of overtime, rather than 7.

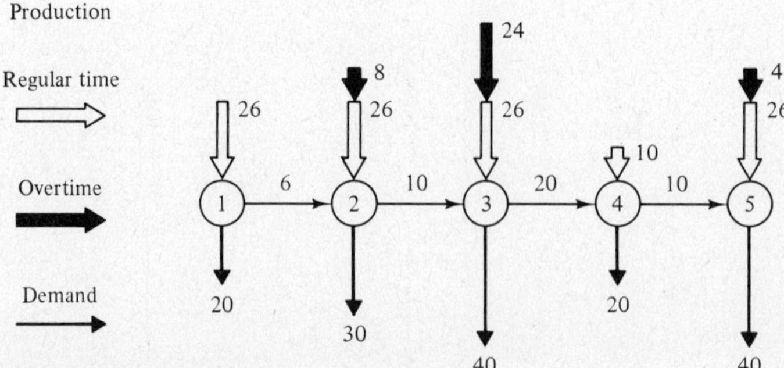

Figure 12.8   Optimum integer solution.

If these adaptations are not possible, then an integer programming computer package could be used to find an optimal integer solution that satisfies all of the constraints. Such a solution is shown in Figure 12.8. The minimum cost for this solution is $24,560.00, a savings of $105 over the roundoff solution, and only $40 more than the continuous solution.

EXAMPLE 12.2

A construction firm desires to build some combination of one-, two-, and three-bedroom apartments. Each type of apartment requires certain amounts of lumber, concrete, and insulation, which are each limited in the total amount available. Also, each type of apartment will produce a different level of rental income. Pertinent data are given in Table 12.4. In addition, local ordinances require at least one three-bedroom apartment for every three one- and two-bedroom apartments. The objective is to maximize the rental income subject to the given resource constraints.

A linear programming model for the problem is presented as Equations 12.25.

$$
\begin{aligned}
\max Z = {}& 250X_1 + 300X_2 + 325X_3 \\
\text{ST:} \quad 1000X_1 + {}& 1400X_2 + 1600X_3 \le 100{,}000 \\
10X_1 + {}& 12X_2 + 13X_3 \le 1{,}200 \\
18X_1 + {}& 24X_2 + 30X_3 \le 2{,}500 \\
0.33X_1 + {}& 0.33X_2 - X_3 \le 0 \\
& X_1, X_2, X_3 \ge 0
\end{aligned}
\tag{12.25}
$$

where X_1 = number of one-bedroom apartments
X_2 = number of two-bedroom apartments
X_3 = number of three-bedroom apartments

The first three constraints reflect the limited supply of lumber, concrete, and insulation available, while the fourth constraint represents the limitation that the sum of one- and two-bedroom apartments cannot be more than three times the number of three-bedroom apartments.

A continuous solution for this model will show that 65.45 one-bedroom apartments and 21.60 three-bedroom apartments will provide a maximum rental income

TABLE 12.4   APARTMENT DATA

Number of bedrooms	Lumber required (board ft)	Concrete required (yd³)	Insulation required (rolls)	Rental income ($/month)
One	1,000	10	18	250
Two	1,400	12	24	300
Three	1,600	13	30	325
Total amount available	100,000	1200	2500	

of \$23,380/month. However, it is obvious that part of an apartment would not provide any rental income, so the variable values should be restricted to integers. Rounding off would suggest that 65 one-bedroom apartments and 22 three-bedroom apartments be built, which would produce rental income of \$23,400/month, which is greater than the continuous solution. Therefore, it should be suspected that the solution violates one or more of the constraints. Inspection shows that the first constraint is the only one violated.

$$1000X_1 + 1400X_2 + 1600X_3 = 100,200 \tag{12.26}$$

If 200 additional board feet of lumber could be allocated, this solution would be possible. Otherwise, one less one-bedroom apartment could be constructed, and all constraints would be satisfied.

When this model was input to an integer programming package, the package was unable to find an optimum solution after more than 600 additional constraints were added. This is not to say that a more sophisticated integer programming package might not have found a solution. However, it does illustrate that even small problems can produce extremely large arrays in the quest for an optimal integer solution.

EXAMPLE 12.3 ———————————————————————————————

Phillips et al. (1982) have reported on application of a mixed integer programming screening model to assist in developing a regional wastewater management plan for Nassau County, New York State. Their approach and results are summarized here as a case study.

Locations of wastewater treatment plants in Nassau County are shown in Figure 12.9. Dates of construction for the plants shown vary from 1927 to 1974. Many of the older plants use less efficient treatment processes than the newer plants; however, all plants provide secondary levels of treatment which remove up to about 85% of most pollutants. It is possible that the oldest of these plants could be phased out, and their flows diverted to more centrally located treatment plants. The objective is lower overall treatment costs because of economies of scale; however, additional costs are incurred because of diversion. The objective function is

$$\min Z = \sum \text{diversion costs} + \sum \text{treatment costs}$$

$$= \sum_{i=1}^{n} \sum_{j=1}^{n} d_{ij}X_{ij} + \sum_{j=1}^{n} (c_j y_j + t_j Q_j) \tag{12.27}$$

where d_{ij} = cost (present worth dollars) for diversion from location i to location j,

   X_{ij} = 0, 1 integer variable indicating whether or not diversion link ij is used (1 = link is used),

   c_j = fixed cost (present worth dollars) of building or upgrading plant j,

   y_j = 0, 1 integer variable indicating whether or not a plant is constructed or upgraded (1 = yes),

   t_j = treatment cost per unit flow at plant j,

   Q_j = total flow entering plant, and

   n = number of possible plants (1 per area).

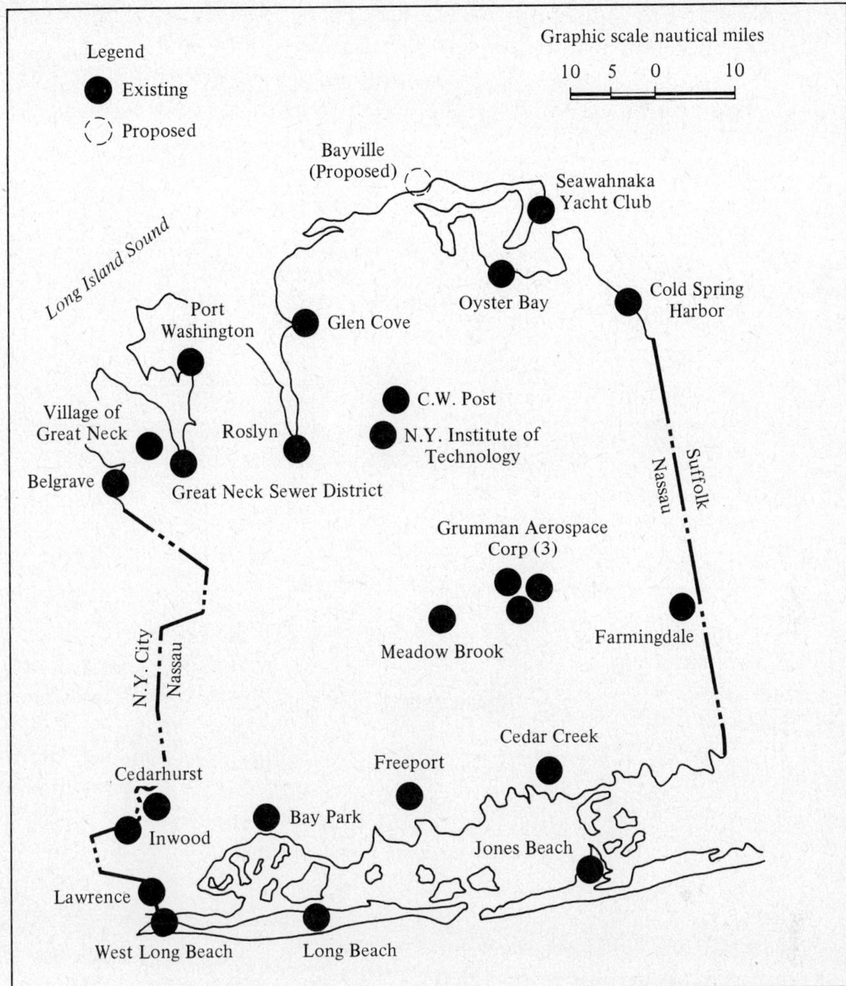

Figure 12.9 Existing wastewater treatment plants in Nassau County. From Kevin J. Phillips et al., *Journal, Water Pollution Control Federation,* Vol. 54, No. 1, January, 1982, Figure 1, p. 88. Used with permission.

Constraints are placed on the system to maintain physical continuity. Each treatment plant must be accounted for, and each plant's flow can only be diverted to one other plant. Therefore,

$$\sum_{j=1}^{n} X_{ij} = 1 \quad \text{for all } i \tag{12.28}$$

when $i = j$, indicates that the flow is not diverted away from plant i.

The flow to be treated at each plant equals the sum of the flows diverted to the plant plus the original flow entering the plant:

$$\sum_{i=1}^{n} q_i X_{ij} - Q_j = 0 \quad \text{for all } j \tag{12.29}$$

where q_i = flow generated by the area tributary to plant i.

If a plant is to be phased out, then it cannot accept any further flow from diversions. This is accomplished by two constraints

$$X_{jj} - y_j = 0 \quad \text{for all } j \tag{12.30}$$

which ensures that if flows from area j are not diverted, then the plant for j is constructed or upgraded, and

$$Q_j - 10^6 y_j \le 0 \quad \text{for all } j \tag{12.31}$$

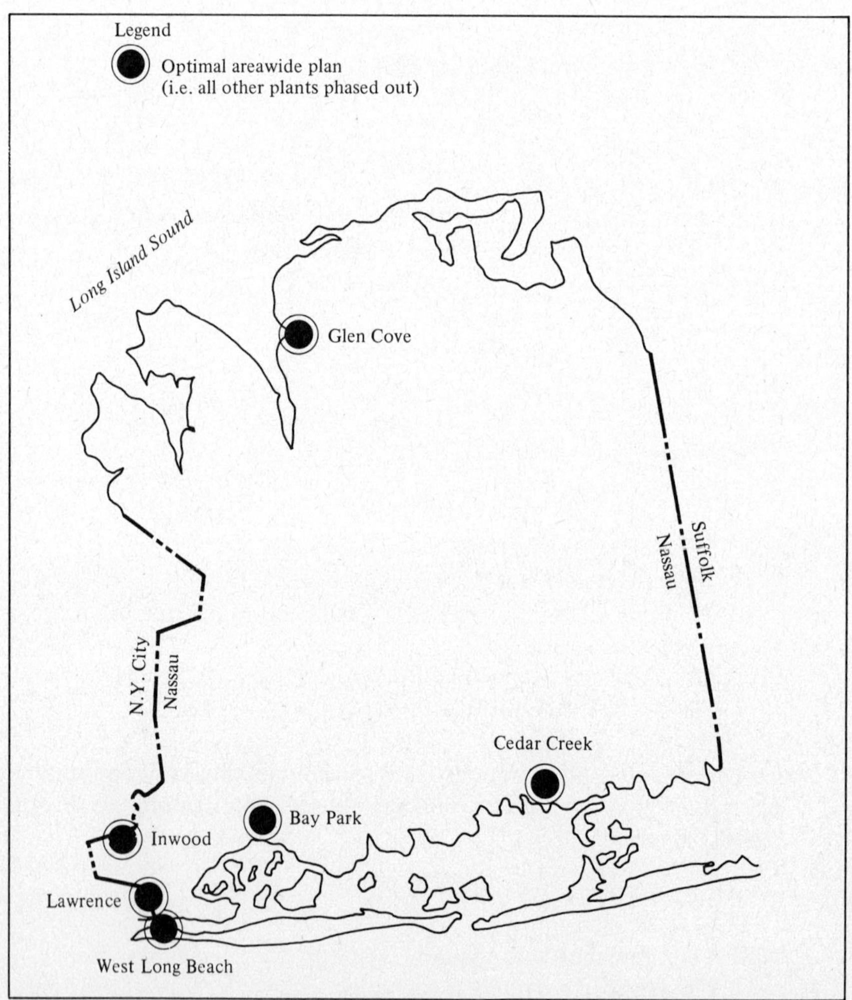

Figure 12.10 Optimal areawide plan for wastewater treatment plants in Nassau County. From Kevin J. Phillips et al., *Journal, Water Pollution Control Federation*, Vol. 54, No. 1, January, 1982, Figure 4, p. 92. Used with permission.

which ensures that the flowrate entering a plant is zero if flows are to be diverted away from that plant. The coefficient 10^6 is much larger than any value expected for Q_j, which is given in units of million liters per day.

The costs for diversion (d_{ij}) were developed from capital and operation and maintenance costs for interceptor sewers, force mains, and pumping stations from area i to area j. Only technically feasible diversions were considered.

The treatment costs were divided into fixed and variable costs. The fixed costs (c_j) were estimated from the zero treatment intercept on the cost axis of a linear regression fit of cost versus treatment capacity. The variable cost is the slope of the linear regression fit. Both the slope and intercept were influenced by the water quality limitations of the body of water receiving the treated effluent. In developing the costs, "due consideration was given to existing treatment plants." It should be pointed out that this particular application does not provide for economies of scale at a particular treatment plant because the variable cost is linear. However, economies of scale between competing plants could be accommodated by using different variable cost slopes and different fixed cost intercepts.

Figure 12.10 shows the optimal areawide plan, which recommends that treatment be consolidated from the 25 locations presently active to six central locations. However, this model does not take into account institutional, social, and political constraints that may prevent implementation of such a plan.

SUMMARY

Certain problems require that some or all variables take on only integer values. The form of the model may produce integer values for the appropriate variables at the optimal solution. If not, specialized linear programming techniques can be used to search for the integer solution which is closest to the optimum continuous solution. However, these integer programming techniques may take an excessive number of iterations to converge to the optimal integer solution. In these cases, approximate solutions, such as rounding continuous optimal solution variable values, may be employed. However, care must be observed to ensure that infeasible solutions are not used.

APPLICATIONS EXERCISES

Use integer programming techniques to find the optimal integer solution to the following exercises: 4-12; 5-1; 5-6; 5-7; 5-8; 5-10; 5-11; 5-13; 5-17; 5-19; 5-21; 5-22; 5-34; 6-1; and 6-3.

Multiobjective Analysis

INTRODUCTION

Traditionally, the primary objective in planning and design has been overall economic efficiency, with the goal of either maximizing benefits or minimizing costs. More recently, the emphasis in certain types of planning, such as water resources planning, has shifted to include all relevant objectives, including environmental quality, social well being, and regional income redistribution. Analytical tools, although they are not perfect, are available for quantifying the objective of economic efficiency. However, few methodologies are available to assist in quantification of social and community goals and objectives.

The purpose of this chapter will be to introduce the student to the basic methodology of multiobjective analysis through two-dimensional models, then present a case study of the application of multiobjective analysis. Additional information on multiobjective analysis can be found in works by Loucks, Stedinger, and Haith (1981); Cohon (1978); and Major (1977). Each of these references also lists other supplemental references.

METHODOLOGY

Multiobjective models are subject to a set of system constraints, either linear or non-linear. However, the constraints have more than one objective function associated with them. For the linear model, the system becomes

$$\left. \begin{array}{l} \text{max or min} \quad Z_1 = c_{11}X_1 + c_{12}X_2 + \cdots + c_{1n}X_n \\[1em] \qquad\qquad\quad Z_2 = c_{21}X_1 + c_{22}X_2 + \cdots + c_{2n}X_n \\[0.5em] \qquad\qquad\quad \vdots \qquad \vdots \qquad \vdots \qquad\quad \vdots \\[0.5em] \qquad\qquad\quad Z_k = c_{k1}X_1 + c_{k2}X_2 + \cdots + c_{kn}X_n \\[1.5em] \text{ST:} \quad a_{11}X_1 + a_{12}X_2 + \cdots + a_{1n}X_n \left(\begin{array}{c} \leq \\ \geq \\ = \end{array}\right) b_1 \\[2em] \qquad\quad a_{21}X_1 + a_{22}X_2 + \cdots ; a_{2n}X_n \left(\begin{array}{c} \leq \\ \geq \\ = \end{array}\right) b_2 \\[1.5em] \qquad\quad \vdots \qquad \vdots \qquad\qquad \vdots \qquad\quad \vdots \\[1.5em] \qquad\quad a_{m1}X_1 + a_{m2}X_2 + \cdots + a_{mn}X_n \left(\begin{array}{c} \leq \\ \geq \\ = \end{array}\right) b_m \end{array} \right\} \quad (13.1)$$

Since the objective function $\mathbf{Z}$ is a vector, it cannot be optimized in the conventional manner. It is not guaranteed that an optimal solution for one objective function will be optimal with respect to any of the others. A methodology must be applied which will quantify the tradeoff among the various competing objectives.

Consider the following two-dimensional model,

$$\left. \begin{array}{l} \text{max } Z_1 = X_1 \\ \text{max } Z_2 = X_2 \\[1em] \text{ST:} \quad \text{①} \;\; X_2 + 0.125X_1 \leq 5 \\ \qquad\;\; \text{②} \;\; X_2 + 0.314X_1 \leq 5.6 \\ \qquad\;\; \text{③} \;\; X_2 + 0.975X_1 \leq 8.83 \\ \qquad\;\; \text{④} \;\; X_2 + 2.857X_1 \leq 21.0 \\ \qquad\;\; \text{⑤} \;\; X_2 + \quad 10X_1 \leq 70 \end{array} \right\} \quad (13.2)$$

Although the two objective functions of this model are unrealistic in terms of real world problems, they do allow careful scrutiny of what is happening in the application of solution techniques. A case study will serve to illustrate the application of the methods to an actual problem.

The constraint space and competing objectives for this model are shown in Figure 13.1. It is desirable to generate a set of noninferior solutions (sometimes referred to as the noninferior frontier), which are solutions that maximize one objective function for a given value of the other objective function. This set will be used, in conjunction with trade-off functions among the objectives, to determine the best multiobjective solution. Two primary methods for developing this set of noninferior solutions are the constraint and weighting techniques. The model of Equations 13.2 will be used

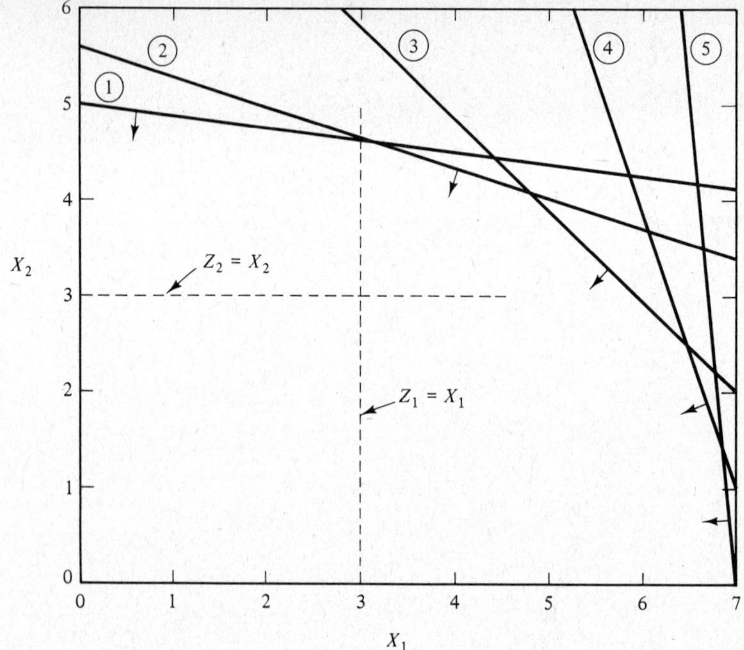

Figure 13.1 Multiobjective model.

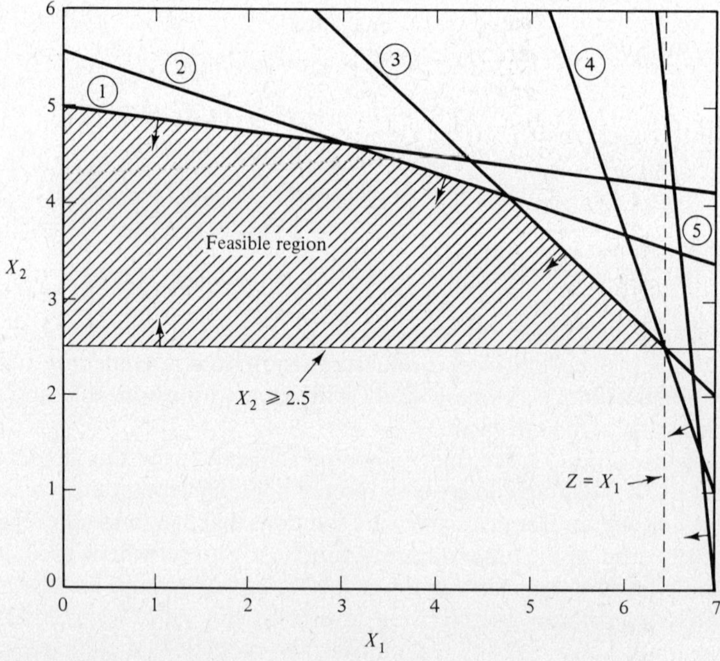

Figure 13.2 Multiobjective analysis by the constraint method.

to illustrate both methods. After generation of the solution set, information can be developed to estimate trade-offs between the competing objectives.

Constraint Method

The constraint method for finding the noninferior solution set involves transforming one of the objective functions into a greater than or equal constraint, and systematically varying the right-hand side value until the boundary of feasibility is reached. If the first objective function of the general model of Equations 13.1 is selected, and the other objective functions are transformed into constraints, the model becomes:

$$\text{maximize or minimize} \quad c_{11}X_1 + c_{12}X_2 + \cdots + c_{1n}X_n$$

$$\text{ST:} \quad Z_2 = c_{21}X_1 + c_{22}X_2 + \cdots + c_{2n}X_n \left(\genfrac{}{}{0pt}{}{\leq}{\geq}\right) B_2$$

$$\vdots$$

$$Z_k = c_{k1}X_1 + c_{k2}X_2 + \cdots + c_{kn}X_n \left(\genfrac{}{}{0pt}{}{\leq}{\geq}\right) B_k$$

(transformed objective functions)

$$a_{11}X_1 + a_{12}X_2 + \cdots + a_{1n}X_n \left(\genfrac{}{}{0pt}{}{\leq}{\geq}{=}\right) b_1$$

$$a_{21}X_1 + a_{22}X_2 + \cdots + a_{2n}X_n \left(\genfrac{}{}{0pt}{}{\leq}{\geq}{=}\right) b_2$$

$$\vdots$$

$$a_{m1}X_1 + a_{m2}X_2 + \cdots + a_{mn}X_n \left(\genfrac{}{}{0pt}{}{\leq}{\geq}{=}\right) b_m$$

(regular constraints)

$$(13.3)$$

For the B_2 to B_k constraints, the greater than or equal sign would be used for maximization objective functions while less than or equal would be used for minimization objective functions.

The two-dimensional model of Equations 13.2 would become

$$\begin{aligned}
&\max Z = X_1 \\
\text{ST:} \quad &X_2 \geq B_1 \\
&X_2 + 0.125X_1 \leq 5 \\
&X_2 + 0.314X_1 \leq 5.6 \\
&X_2 + 0.975X_1 \leq 8.83 \\
&X_2 + 2.857X_1 \leq 21.0 \\
&X_2 + \quad 10X_1 \leq 70
\end{aligned} \qquad (13.4)$$

If B_1 initially has a value of zero, the maximum value of X_1 is 7. As B_1 is increased, the maximum value of X_1 that maintains feasibility will decrease. When $B_1 = 1.40$,

Z_1 will equal approximately 6.86. When B_1 increases to 2.53, Z_1 will be reduced to 6.47. Figure 13.2 shows the constraint space and graphical solution for the latter conditions. Continuing to increase the B_1 value will further shrink the feasible region until B_1 reaches a value of 5, which shrinks the feasible region to a single point. Each value of B_1 represents a value of the objective function $Z_2 = X_2$, and the model optimizes $Z_1 = X_1$ subject to this value of Z_2. A plot of Z_1 versus Z_2 will show the boundary, or noninferior, solutions to the multiobjective model. Any solutions inside this boundary would be feasible, but not as desirable as the boundary solutions. The noninferior solutions maximize the value of one objective function for a given value of the other objective function or functions. Solutions outside the boundary would be infeasible. Figure 13.3 shows the noninferior solution set that would be traced out by the constraint method.

Weighting Method

The second method for finding the noninferior set of solutions is the weighting method. For this method, objective functions are assigned weighting factors and combined into a single, scalar, objective function. One of the weights is assigned a value of 1.0, and is defined as the numeraire. The other weights are systematically varied until the

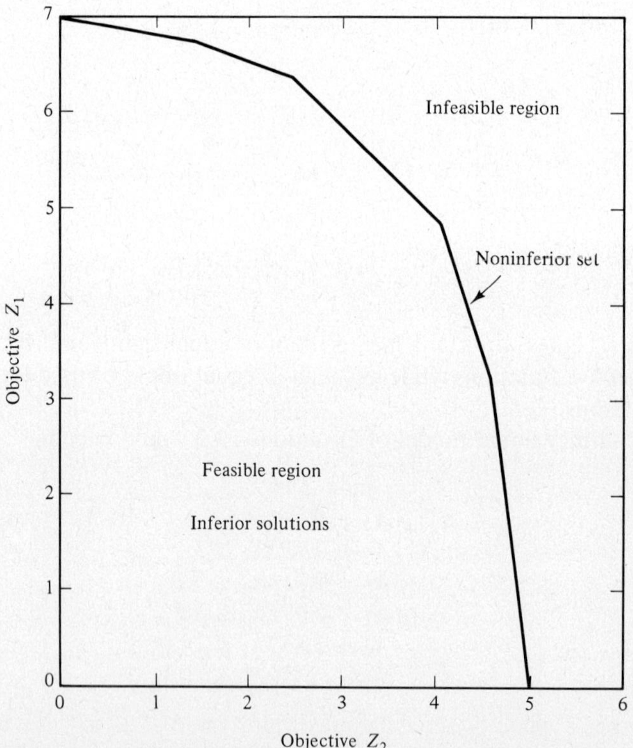

Figure 13.3 Noninferior solution set.

noninferior set is traced out. Transforming the general model of Equations 13.1 into the form necessary for the weighting method would give

$$\text{max or min} \quad \sum_{i=1}^{k} \lambda_i Z_i = \lambda_1 Z_1 + \lambda_2 Z_2 + \cdots + \lambda_k Z_k$$

or,

$$
\left.
\begin{aligned}
\text{max or min} \quad Z = {}& \lambda_1 c_{11} X_1 + \lambda_1 c_{12} X_2 + \cdots + \lambda_1 c_{1n} X_n \\
& + \lambda_2 c_{21} X_1 + \lambda_2 c_{22} X_2 + \cdots + \lambda_2 c_{2n} X_n \\
& \quad \vdots \qquad\qquad \vdots \qquad\qquad\qquad \vdots \\
& + \lambda_k c_{k1} X_1 + \lambda_k c_{k2} X_2 + \cdots + \lambda_k c_{kn} X_n \\[1em]
\text{ST:} \quad a_{11} X_1 + a_{12} X_2 + {}& \cdots + a_{1n} X_n \begin{pmatrix} \geq \\ \leq \\ = \end{pmatrix} b_1 \\[1em]
a_{21} X_1 + a_{22} X_2 + {}& \cdots + a_{2n} X_n \begin{pmatrix} \geq \\ \leq \\ = \end{pmatrix} b_2 \\[1em]
\vdots \qquad\quad \vdots \qquad\quad & \qquad \vdots \qquad\qquad \vdots \\[1em]
a_{m1} X_1 + a_{m2} X_2 + {}& \cdots + a_{mn} X_n \begin{pmatrix} \geq \\ \leq \\ = \end{pmatrix} b_m
\end{aligned}
\right\} \quad (13.5)
$$

The two-dimensional model of Equations 13.2 would become

$$
\left.
\begin{aligned}
\text{max } Z &= \lambda_1 Z_1 + \lambda_2 Z_2 \\
&= \lambda_1 X_1 + \lambda_2 X_2 \\
\text{ST:} \quad X_2 &+ 0.125 X_1 \leq 5 \\
X_2 &+ 0.314 X_1 \leq 5.6 \\
X_2 &+ 0.975 X_1 \leq 8.83 \\
X_2 &+ 2.857 X_1 \leq 21.0 \\
X_2 &+ \quad 10 X_1 \leq 70
\end{aligned}
\right\} \quad (13.6)
$$

If λ_1, and thus Z_1, is chosen as the numeraire, then the objective function becomes

$$\text{max } Z = X_1 + \lambda_2 X_2 \qquad (13.7)$$

As λ_2 is systematically varied, the noninferior set will be traced out. Table 13.1 shows the pertinent values as λ_2 is varied. In effect, what is happening is that as λ_2 increases, the slope of the combined objective function increases in the negative direction. When the slope shifts enough, the optimum solution for the combined objective function shifts to a new extreme point of the feasible region of Figure 13.1. The solutions at these extreme points trace out the noninferior solution set of Figure 13.3.

    The foregoing model is simplistic in terms of the objective functions used; however, it serves well in illustrating the two methods for generating the noninferior solution set. It also aids in understanding the trade-off between the two objective functions. Near the middle of the noninferior set, there is nearly an even trade-off between the two objective functions, whereas, near either end of the range a small change in one objective function produces a large change in the other.

TABLE 13.1   WEIGHTING METHOD VALUES

	Solution variables		Objective function values		
λ_2	X_1	X_2	Z	Z_1	Z_2
0	7	0	7	7	0
0.1	7	0	7	7	0
0.2	6.86	1.40	7.14	6.86	1.40
0.3	6.86	1.40	7.28	6.86	1.40
0.4	6.47	2.53	7.48	6.47	2.53
.	.	.	.	.	.
.	.	.	.	.	.
.	.	.	.	.	.
1.0	6.47	2.53	8.99	6.47	2.53
1.1	4.89	4.07	9.36	4.89	4.07
.	.	.	.	.	.
.	.	.	.	.	.
.	.	.	.	.	.
3.1	4.89	4.07	17.49	4.89	4.07
3.2	3.17	4.60	17.91	3.17	4.60
.	.	.	.	.	.
.	.	.	.	.	.
.	.	.	.	.	.
7.9	3.17	4.60	39.54	3.17	4.60
8.0	0	5.00	40	0	5.00
8.1	0	5.00	40	0	5.00

Nonconvex Noninferior Sets

When the noninferior set forms a convex space, as it does in Figure 13.3, either of the methods can be used to trace out the set. Any linear model will form a convex noninferior set; however, some nonlinear models may form a noninferior set that has some nonconvex regions, as shown in Figure 13.4. For such a model, the weighting method would start with $\lambda_2 = 0$ at point 1, and proceed until $\lambda_2 = T$ at point 2. From there it would jump over to point 4, and continue on to point 5. The entire nonconvex portion of the curve, through points 2, 3, and 4, would not be sensed. The constraint method, however, would trace out the entire noninferior solution set.

Isopreference Curves

Once the noninferior solution set of Figure 13.3 is defined, the next step is to establish isopreference or "indifference" curves. These curves establish the willingness of the interested groups to accept certain levels of one objective function vis-à-vis the other objective function or functions. For example, if one objective is to maximize economic gain and the other is to maximize environmental quality, particular levels of economic gain and environmental quality will produce a certain level of satisfaction or amount of utility. For another economic level, the level of environmental protection that would provide the same overall level of satisfaction would form another point on the isopreference curve. A family of isopreference curves is shown in Figure 13.5, with

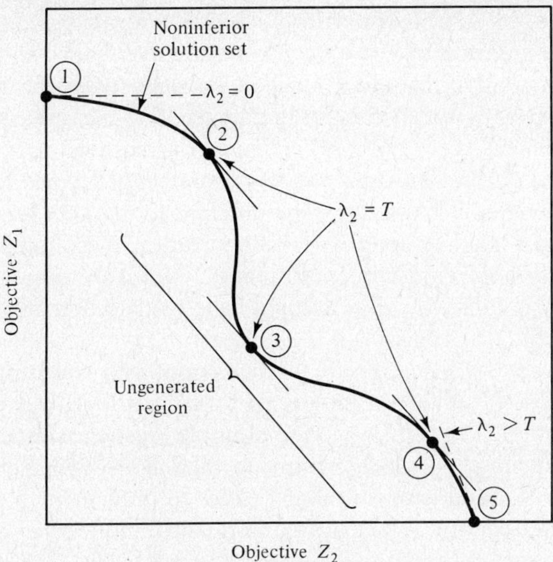

Figure 13.4 Nonconvex noninferior set.

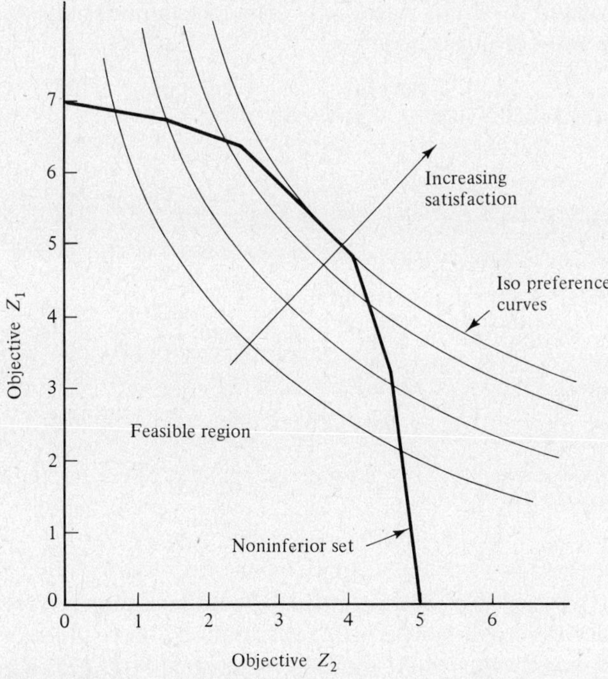

Figure 13.5 Noninferior set with isopreference curves.

the direction of increasing satisfaction shown. The highest satisfaction isopreference curve that touches the noninferior set is the most attractive management strategy.

The difficulty in applying multiobjective analysis effectively lies in defining the isopreference curves. How is satisfaction, or preference, best measured? If both objectives are economic, such as national economic benefit versus regional economic benefits, the trade-offs and preferences can be fairly accurately quantified for a particular group. However, the preferences may be quite different for another group. Also, how is preference measured when one objective is maximization of economic benefits, and another is maximization of environmental quality? Not only may environmental quality be difficult to quantify; but also, it might have vastly different worth to two or more competing groups.

Example 13.1 shows application of the constraint and weighting methods to determine the noninferior set for the model of Equations 6.16, augmented with a second objective function. The solution to the example illustrates some of the advantages of the constraint method over the weighting method. It provides more information about the noninferior set, and is operationally easier to implement. Remember also that the weighting method will not identify and quantify nonconvex regions of the noninferior set.

The next section will present a case study of the application of multiobjective analysis to an actual problem.

EXAMPLE 13.1 ____

Use the constraint and weighting methods to define the noninferior solution boundary for the following two objective model:

$$
\begin{aligned}
\max Z_1 &= X_1 + 2X_2 \\
Z_2 &= 2X_1 + X_2 \\
\text{ST:} \quad -X_1 + 3X_2 &\leq 10 \\
X_1 + X_2 &\leq 6 \\
X_1 - X_2 &\leq 2 \\
X_1 + 3X_2 &\geq 6
\end{aligned}
\tag{13.8}
$$

The constraint method can be applied by transforming one of the objective functions into a constraint. If Z_2 is thus transformed, the model becomes

$$
\begin{aligned}
\max Z_1 &= X_1 + 2X_2 \\
\text{ST:} \quad Z_2 = 2X_1 + X_2 &\geq B_2 \\
-X_1 + 3X_2 &\leq 10 \\
X_1 + X_2 &\leq 6 \\
X_1 - X_2 &\leq 2 \\
X_1 + 3X_2 &\geq 6
\end{aligned}
\tag{13.9}
$$

Figure 13.6 shows the original constraint space with several lines representing $Z_2 = 2X_1 + X_2 \geq B_2$ for different values of B_2 superimposed. For values of B_2 up to and including 8.0, Z_1 has a value of 10 with $X_1 = 2$ and $X_2 = 4$. This corresponds with the objective function $Z_1 = X_1 + 2X_2$ achieving its maximum value of 10 at extreme point ① of the feasible region shown in Figure 13.6. As B_2 increases beyond 8, the

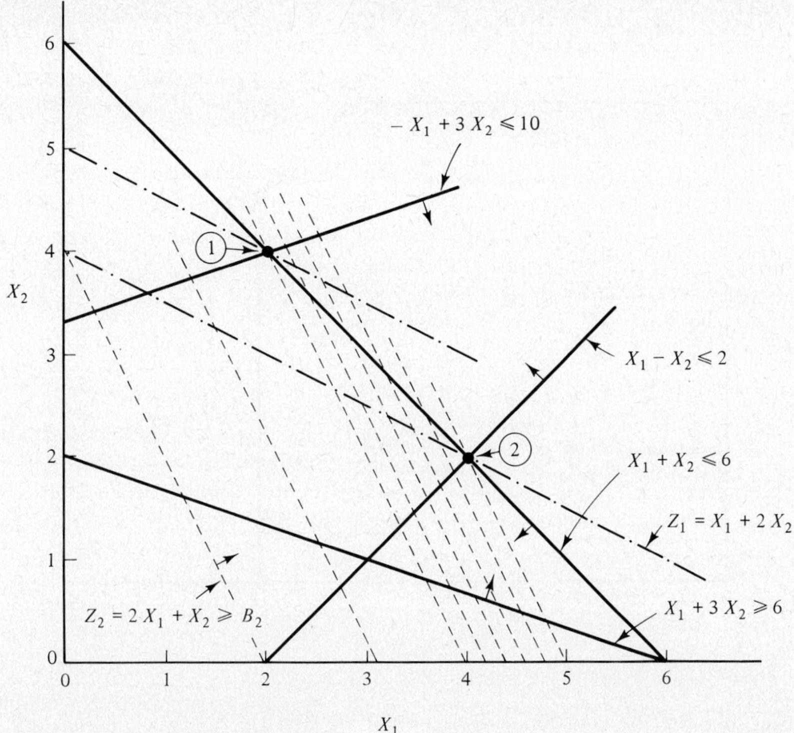

Figure 13.6 Constraint method.

transformed constraint excludes point ① from the feasible range and forces objective function Z_1 to move down to remain in contact with the feasible region. When B_2 reaches 10, the transformed constraint has shrunk the feasible region to a single point (point ② of Figure 13.6) and Z_1 has a value of 8 with $X_1 = 4$ and $X_2 = 2$. Any values of B_2 larger than 10 would make it impossible to find any feasible solutions. Figure 13.7 shows the noninferior set traced out by the constraint method.

In applying the weighting method, Z_2 was chosen as the numeraire, so the objective function would become:

$$\max Z = \sum_{i=1}^{2} \lambda_i Z_i = Z_2 + \lambda_1 Z_1$$

$$= 2X_1 + X_2 + \lambda_1 X_1 + 2\lambda_1 X_2 \tag{13.10}$$

When $\lambda_1 = 0$, $Z_2 = 10$, $X_1 = 4.0$, and $X_2 = 2.0$. These same values for X_1 and X_2 will remain through a value of 1.0 for λ_1. This corresponds with point ② of Figure 13.7. The retransformed Z_1 value for these solutions is 8.0. Larger values of λ_1 will produce a different solution with $X_1 = 2.0$, $X_2 = 4.0$, $Z_1 = 10.0$, and $Z_2 = 8.0$, corresponding with point ① of Figure 13.7. Points ① and ② of Figure 13.7 correspond to solutions at points ① and ② of Figure 13.6. It can be seen that the weighting method skips from point ② to point ① without explicitly defining the intermediate solutions or the non-

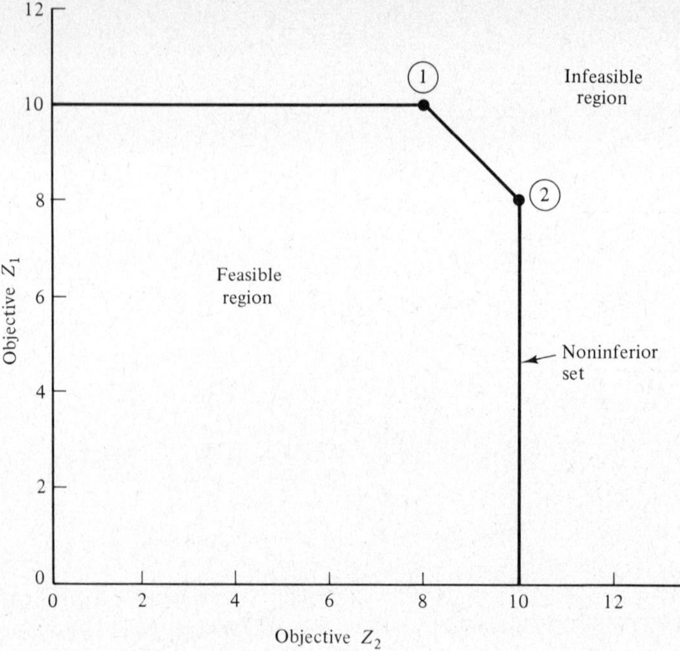

Figure 13.7   Noninferior solution set.

inferior solutions outside the segment ①–②, although these can be deduced from the characteristics of the problem. The shift in solution points occurs because the slope of the combined objective function when $\lambda_1 = 1.0$ is the same as the slope of the constraint $X_1 + X_2 \le 6$.

CASE STUDY

Multiobjective analysis has been used most often in relation to water resources problems, so the case study cited here will be from that area.

This case study is a multiobjective plan for basin-wide water resources management practice for the Santa Ana Region of California. Multiobjective analysis was combined with simulation modeling designed to ascertain the changes in water quality brought about by various usages and by treatment and disposal practices in the basin.

It was recognized that a comprehensive basin-wide plan would have simultaneously to consider objectives relating to water resources allocation, water quality control, and prevention of undesirable effects caused by rapid decline of groundwater levels. After formulating the multiobjective model, the constraint linear programming technique was used to develop the noninferior solution set. The planning study was undertaken for the California Water Quality Control Board (CWQCB), Santa Ana Region, and the Department of Water Resources (DWR), Southern District, and was reported by Louie, Yeh, and Hsu (1984).

For analysis, the basin was subdivided into smaller areas called operational areas, delineated by management districts or other appropriate subdivisions. Within each operational area there were various water use groups seeking allocation of resources, and producing wastewater which required treatment prior to disposal.

The overall water quality objective that was established was the minimization of the deviation between concentration levels estimated by the simulation model and the limiting levels set for the operational areas. The deviation had a positive value if the computed concentration level was above the allowable limit; otherwise, it was set equal to zero. For the test case, the maximum deviation for all subareas was chosen as the measurand, which was to be minimized. Therefore, the water quality objective could be stated as

$$\min Z_{WQ} = \max(D_1, D_2, \ldots, D_n) \tag{13.11}$$

where D_i = deviation for ith operational area
$\quad\quad$ = computed concentration level $-$ allowable concentration for area if result is positive
$\quad\quad$ = zero if result is negative

Two water supply management objectives were identified. The first was to minimize the cost of meeting the water demands in an area. Costs included the initial cost of supplying the water as well as the costs of transporting and treating wastewater. The cost structure was approximated by the linear function:

$$\min Z_{TC} = \sum_{i=1}^{S} \sum_{j=1}^{U} CS_{i,j} QS_{i,j} + \sum_{j=1}^{U} \sum_{k=1}^{T} CT_{j,k} QT_{j,k} + \sum_{k=1}^{T} \sum_{l=1}^{D} CD_{k,l} QD_{k,l} \tag{13.12}$$

where Z_{TC} = total cost of supplying and disposing of water,
$\quad\quad CS_{i,j}$ = cost of supplying one unit of water from source i to user group j,
$\quad\quad QS_{i,j}$ = units of water supplied from source i to user group j,
$\quad\quad CT_{j,k}$ = cost of transporting one unit of water from user j to treatment plant k,
$\quad\quad QT_{j,k}$ = units of water transported from user j to treatment plant k,
$\quad\quad CD_{k,l}$ = cost of transporting one unit of water from treatment plant k to disposal site l,
$\quad\quad QD_{k,l}$ = units of water transported from treatment plant k to disposal site l,
$\quad\quad S$ = number of supply sources,
$\quad\quad U$ = number of water use groups,
$\quad\quad T$ = number of treatment plants, and
$\quad\quad D$ = number of disposal sites

A second water supply management objective was the minimization of overdraft of groundwater supplies

$$\min Z_O = \sum_{i=1}^{S_b} Q_i - \sum_{j=1}^{T_b} Q_j \tag{13.13}$$

where Q_i = volume extracted from ith groundwater source in the basin,
$\quad\quad Q_j$ = volume recharged to groundwater basin from jth treatment plant,

S_b = total number of groundwater sources in the basin, and

T_b = total number of treatment plants recharging to the basin

Clearly, there would have been conflicts if all three objectives were optimized independently, but concurrently. Therefore, they were formulated in the multiobjective context:

$$\min Z = (Z_{WQ}, Z_{TC}, Z_O) \tag{13.14}$$

The solution procedure was broken down into the following steps.

Step 1. A linear programming model was solved to find the optimal solution to the total cost of supplying the water objective function, independent of the water quality and overdraft objective functions.

Step 2. The results of step 1 were input into a simulation model to estimate water quality effects.

Step 3. The water quality objective function was optimized independently of the other two.

Step 4. The overdraft objective function was optimized independently.

Step 5. A payoff table was constructed as per Table 13.2.

Step 6. The original multiobjective problem was converted into a constraint problem, and the values in the payoff table used to determine the range for the right-hand side value for each objective function transformed into a constraint.

Step 7. The constraint problem was solved by a parametric linear programming procedure to produce a noninferior set of solutions.

TABLE 13.2   THREE-OBJECTIVE OPTIMIZATION

Optimal solution for objective function	Objective function value		
	Z_{WQ}	Z_{TC}	Z_O
WQ (water quality)	Z_{WQ} (WQ)	Z_{TC} (WQ)	Z_O (WQ)
TC (total cost)	Z_{WQ} (TC)	Z_{TC} (TC)	Z_O (TC)
O (overdraft)	Z_{WQ} (O)	Z_{TC} (O)	Z_O (O)

TABLE 13.3   PAYOFF TABLE FOR PROBLEM 1

Optimal solution for objective function	Objective function value	
	Z_{WQ}	Z_{TC}
WQ (water quality)	182 mg/l	$\$1.96 \times 10^8$
TC (total cost)	162 mg/l	$\$2.77 \times 10^8$

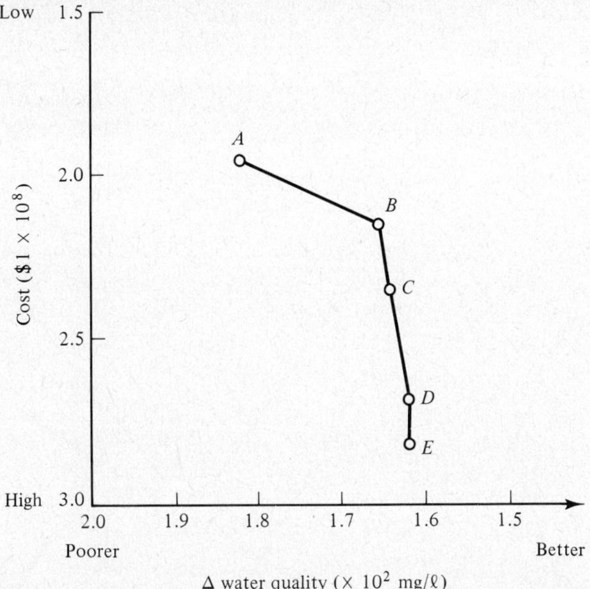

Figure 13.8 Noninferior solution set (trade-off curve for cost vs. water quality). From Figure 3, p. 49, *Journal of Water Resources Planning and Management,* "Multiobjective Water Resources Management Planning," Vol. 110, No. 1, January, 1984, by Peter Louie, William Yeh, Nien-Sheng Hsu. American Society of Civil Engineers. Used with permission.

A test case was run with three supply sources, two operational areas, each with two water use groups (municipal/industrial and agricultural), three treatment plants with various suspended solids removal efficiencies (50, 30, and 0%), and two no treatment plants used to handle agricultural return water; and six wastewater disposal sites, five within the basin, and one outside the basin. Three problems were considered, two of which will be described here.

Problem 1 considered two objectives: total cost and water quality. Optimizing each objective function separately produced the payoff table shown in Table 13.3. The total cost objective function was transformed into an incremental constraint, and the resulting model solved to generate the noninferior set shown in Figure 13.8. Figure 13.9 shows the optimal decision variable values for the extreme points of the noninferior set (points *A* and *E*). It is appropriate at this time to quote from the discussion of Problem 1 contained in the cited reference.

Observing Figure 13.8, one could almost immediately select the solution corresponding to point *B* to be the most preferred solution of the noninferior set. The reasons: By moving from point *B* through point *E,* a fairly sizable increase of cost would incur with only a slight improvement on the water quality, and moving from point *B* to point *A* would reduce the cost at the expense of deteriorating the water quality.

Through some simple calculations of the results shown in Figure 13.9, the solution corresponding to point *A* (representing the minimum cost) and indicating a greater usage of

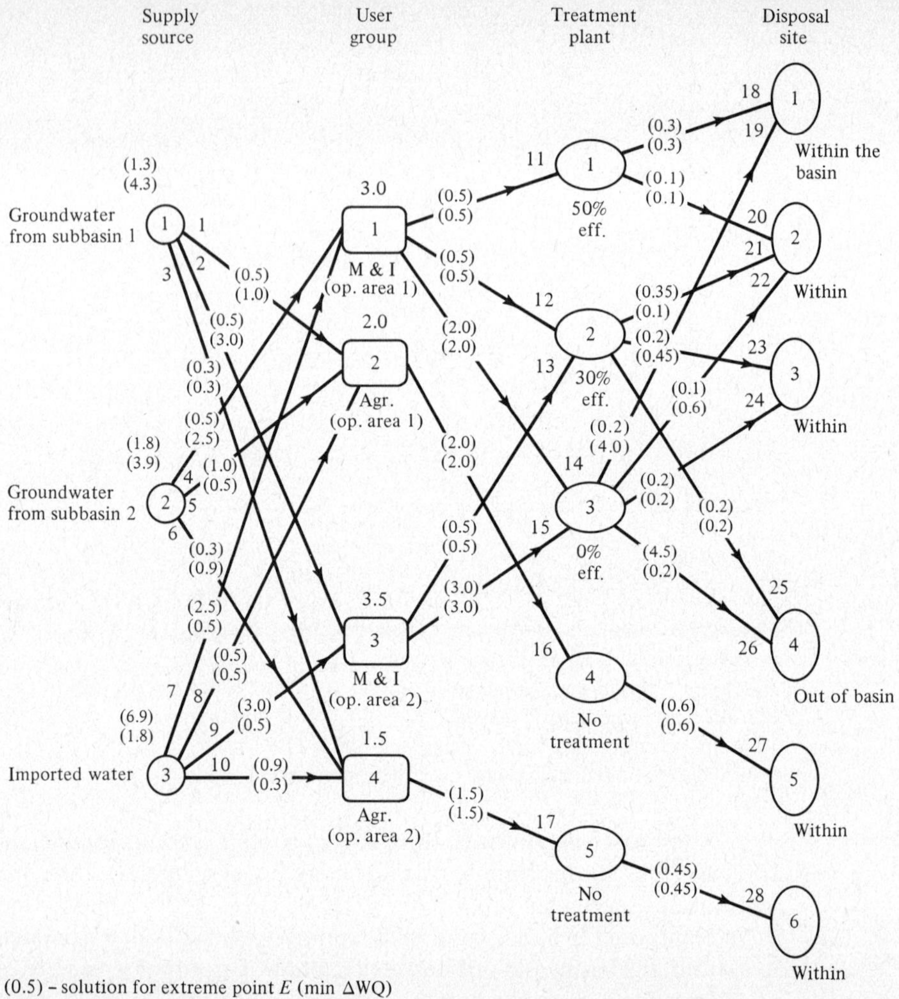

(0.5) – solution for extreme point E (min ΔWQ)
(3.0) – solution for extreme point A (min cost)

Figure 13.9 Solutions for extreme points on the noninferior solution set (trade-off curve). From Figure 4, p. 49, *Journal of Water Resources Planning and Management,* "Multiobjective Water Resources Management Planning," Vol. 110, No. 1, January 1984, by Peter Louie, William Yeh, Nien-Sheng Hsu. American Society of Civil Engineers. Used with permission.

the less expensive groundwater (82% of the total water supply), can be found. On the other hand, the solution corresponding to point E (reflecting the minimum water quality degradation) shows only a 31%. [*sic*]

It is interesting to observe that the quantities arriving at the treatment plants do not vary for the two extreme points. At first glance, one may raise the question, "shouldn't a greater amount of wastewater be transported to treatment plants with higher efficiencies in order to improve the water quality?" This reasoning is valid in general, but for the test problem, there is a unique element in the system that caused the quantities for both extreme points to be identical. Consider the following arguments:

1. To achieve the minimum cost (point A), much of the wastewater is transported to treatment plant 3 because it is the least expensive, and then the untreated wastewater is dispersed into the various disposal sites within the basins because it is less expensive than transporting it out of the basin.

2. Conversely however, to attain the minimal degradation of water quality (point E), much of the wastewater would still be transported to treatment plant 3, not just because it is the least expensive, but also because the untreated wastewater can be transported out of the basin.

For these reasons, no variation in the quantities arriving at the treatment plants can be observed for these two extreme points. However, it should also be pointed out that these reasons do not imply that there is no variation for any other points on the trade-off curve. In fact, as it can be observed from the results for Problem 2, some variabilities exist on the quantities arriving at the treatment plants.

One can further observe that a greater amount of treated (and untreated) wastewater is disposed of within the basin for point A than for point E—93.5% versus 23.6%. This is expected because it is cheaper to dispose of the wastewater within the basin (for point A), while it is better for the water quality (point E) if less is disposed of within the basin.

The second problem to be discussed included all three objective functions. The payoff table was constructed from the individual optimal solutions, as before, and is shown in Table 13.4. Note the repetition of values shown in Table 13.3. The water quality objective was chosen as the objective function, and the minimum and maximum values of the other two objective functions were used to determine through what range the associated constraint right-hand side value would be varied. The noninferior set would form a three-dimensional surface; however, to assist in visualization, a family of curves reflecting different overdraft values was plotted on a two-dimensional plane of cost versus water quality. The resulting plots are shown in Figure 13.10. Again, the authors chose two extreme points (A and B) for further analysis. Decision variable values for these two extreme points are shown in Figure 13.11.

The authors' remarks on Problem 2 are quoted below:

The set of trade-off curves shown in Figure 13.10 are parameterized by five different values of overdrafts, ranging from 5.3×10^5 acre-ft (net extraction) to -3.8×10^5 acre-ft (net recharge). It can be seen that there are only four curves on that figure because the fifth one that corresponds to a net recharge of 3.8×10^5 acre-ft cannot be obtained due to infeasibility of the constraint optimization problem at that low level of overdraft. As the "overdraft" constraint is relaxed, the solution becomes feasible; conversely, when the

TABLE 13.4 PAYOFF TABLE FOR PROBLEM 2

Optimal solution for objective function	Objective function value		
	Z_{WQ}	Z_{TC}	Z_O
WQ (water quality)	182 mg/l	1.96×10^8	1.8×10^5 acre-ft
TC (total cost)	162 mg/l	2.77×10^8	5.3×10^5 acre-ft
O (overdraft)	162 mg/l	2.60×10^8	-3.8×10^5 acre-ft

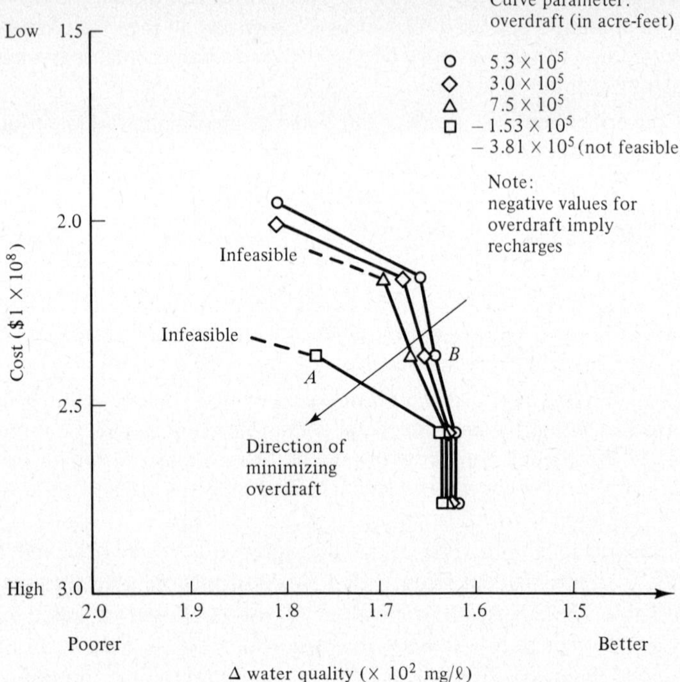

Figure 13.10 Noninferior solution set (cost vs. water quality, vs. overdraft). From Figure 7, p. 53, *Journal of Water Resources Planning and Management,* "Multiobjective Water Resources Management Planning," Vol. 110, No. 1, January 1984, by Peter Louie, William Yeh, Nien-Sheng Hsu. American Society of Civil Engineers. Used with permission.

cost constraint becomes tight, the end points for the bottom two curves cannot be obtained due again to infeasibility. If the "overdraft" constraints are relaxed enough, solutions can be obtained to complete the entire trade-off curve such as the two curves on top. This set of trade-off curves certainly provides a clear picture of the interactive behavior among all three objectives.

Further, it can be found in Figure 13.11 that for point *A* (corresponding to a smaller overdraft value), the groundwater usage is about 48% versus 51% for point *B* (corresponding to a larger overdraft value). The difference is not an appreciable amount, and the reason can be attributable to their cost values which were held to be the same; however, poorer water quality is found to be associated with point *A* because a greater amount of treated (and untreated) wastewater would be recharged into the basin to reduce overdraft.

This case study has shown a successful integration of multiobjective models with system simulation models. The procedures outlined should provide planners with a better understanding of the interaction among the planning objectives, enabling them to develop a more comprehensive plan to present to decision makers.

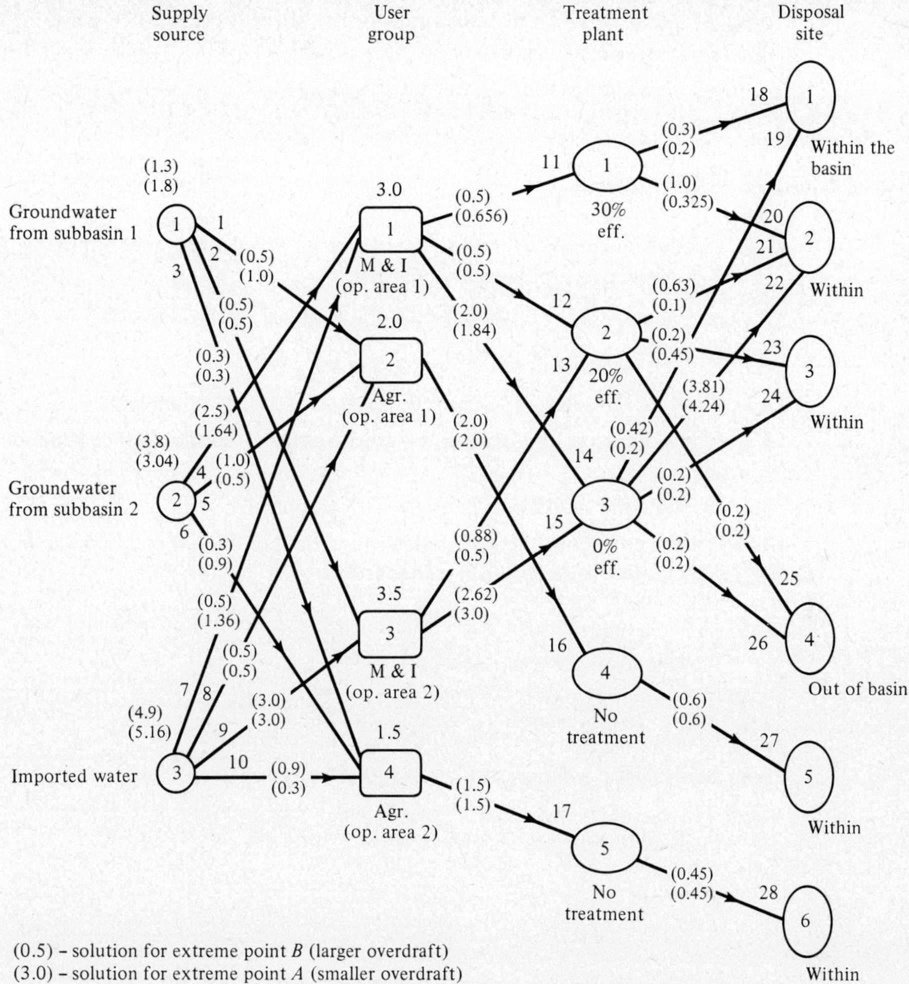

(0.5) – solution for extreme point *B* (larger overdraft)
(3.0) – solution for extreme point *A* (smaller overdraft)

Figure 13.11 Solutions for extreme points on the noninferior solution set. From Figure 8, p. 54, *Journal of Water Resources Planning and Management,* "Multiobjective Water Resources Management Planning," Vol. 110, No. 1, January 1984, by Peter Louie, William Yeh, Nien-Sheng Hsu. American Society of Civil Engineers. Used with permission.

SUMMARY

When multiple objectives must be considered during the process of determining the best operating strategy for a system, an engineer cannot rely on traditional linear or nonlinear optimization techniques to produce a solution that is optimal with respect to all of the objective functions. Two methods, the constraint and weighting techniques, are presented for producing noninferior solution sets which optimize one objective function for given values of the other objective functions. Trade-offs among the objectives can be identified, and preferences established to help determine the best operating strategy. When numerous objectives are identified, it may become computationally infeasible to find the overall optimal solution. Either some of the objectives must be reformulated as constraints, or subjective optimization must be accomplished for some or all of the objectives.

APPLICATIONS EXERCISES

13-1. Use the constraint method to determine the noninferior solution set for Equations 13.2, if the objective functions are changed to

$$Z_1 = 2X_1 + X_2 \quad \text{and} \quad Z_2 = X_1 + 3X_2$$

Plot the resulting noninferior solution set.

13-2. Illustrate the solution to Exercise 13-1 using graphical analysis.

13-3. Repeat Exercise 13-1, but use the weighting technique to produce the noninferior solution set.

13-4. Illustrate the solution to Exercise 13-3 graphically.

13-5. Use the constraint method to determine the noninferior solution set for Equations 13.4, if the objective functions are changed to

$$Z_1 = X_1 + 1.2X_2 \quad \text{and} \quad Z_2 = 5X_1 + 3X_2$$

Plot the resulting noninferior set.

13-6. Illustrate the solution to Exercise 13-5 graphically.

13-7. Repeat Exercise 13-5 using the weighting technique.

13-8. Illustrate the solution to Exercise 13-7 graphically.

APPENDIXES

Linear (Matrix) Algebra

In order to understand linear programming, and to be able to find the solutions to many mathematical models, it is important for the student to have an understanding of matrix algebra and matrix operations. A brief review will be presented here. More thorough discussion can be found in *Linear Algebra* by Hadley (1961).

Matrix algebra is important because a system of linear equations

$$2X_1 + 3X_2 + X_3 = 4$$
$$X_2 + 2X_3 = 6$$
$$4X_1 + 2X_2 = 12$$

can be represented by matrix notation as

$$\begin{bmatrix} 2 & 3 & 1 \\ 0 & 1 & 2 \\ 4 & 2 & 0 \end{bmatrix} \begin{bmatrix} X_1 \\ X_2 \\ X_3 \end{bmatrix} = \begin{bmatrix} 4 \\ 6 \\ 12 \end{bmatrix}$$

$$\underset{\text{coefficient matrix}}{}$$

or

$$\text{column} \downarrow$$

$$\text{row} \rightarrow \begin{bmatrix} a_{11} & a_{12} & a_{13} \\ a_{21} & a_{22} & a_{23} \\ a_{31} & a_{32} & a_{33} \end{bmatrix} \begin{bmatrix} X_1 \\ X_2 \\ X_3 \end{bmatrix} = \begin{bmatrix} b_1 \\ b_2 \\ b_3 \end{bmatrix}$$

$$a_{ij}$$
column identifier
row identifier

or

$$\mathbf{AX} = \mathbf{B}$$

Matrix algebra is a convenient and compact way of expressing sets of linear equations. Also, the definitions and operations given for matrix algebra will provide essential information concerning systems of linear equations, and under the proper circumstances will assist in the solution of such systems for explicit values of the independent variables.

DEFINITIONS

Square Matrix.   Has the same number of rows and columns.

Vector.   A set of n real numbers arranged either horizontally in a row or vertically in a column.

Column Vector.   All elements of the vector are arranged vertically (an $M \times 1$ matrix).

$$\mathbf{X} = \begin{bmatrix} 2 \\ 1 \\ 4 \end{bmatrix} \quad \text{is a column vector}$$

Row Vector.   All elements of the vector are arranged horizontally (a $1 \times n$ matrix).

$$\mathbf{C} = \begin{bmatrix} 1 & 6 & 8 & 2 \end{bmatrix} \quad \text{is a row vector}$$

Symmetric Matrix.   A square matrix in which corresponding off-diagonal elements are equal ($a_{ij} = a_{ji}$). The diagonal is defined as elements a_{ii} of the matrix, a_{11}, a_{22}, and a_{33} for a 3×3 matrix, **A**.

$$\mathbf{A} = \begin{bmatrix} 1 & -2 & 3 \\ -2 & 6 & 9 \\ 3 & 9 & 5 \end{bmatrix} \quad \text{is a symmetric matrix}$$

Diagonal Matrix.   All off-diagonal elements are zero.

$$\mathbf{A} = \begin{bmatrix} 1 & 0 & 0 \\ 0 & 6 & 0 \\ 0 & 0 & 5 \end{bmatrix} \quad \text{is a diagonal matrix}$$

Identity Matrix.   A diagonal matrix in which all diagonal elements are 1. The identity matrix is identified by the letter **I**.

$$\mathbf{I} = \begin{bmatrix} 1 & 0 & 0 \\ 0 & 1 & 0 \\ 0 & 0 & 1 \end{bmatrix} \quad \text{is an identity matrix}$$

Augmented Matrix.   Two or more matrices joined together. The augmented matrix helps in the solution of systems of linear equations, which will be discussed later.

$$(\mathbf{A}|\mathbf{B}) = \begin{bmatrix} 2 & 3 & 1 & 4 \\ 0 & 1 & 2 & 6 \\ 4 & 2 & 0 & 12 \end{bmatrix} \quad \text{is an augmented matrix}$$

Linear Independence.   A set of n row or column vectors $\mathbf{P}_1$, $\mathbf{P}_2$, ..., $\mathbf{P}_n$ of dimension n is linearly independent if and only if

$$\sum_{i=1}^{n} c_i \mathbf{P}_i = 0$$

only for $c_i = 0$, $i = 1, 2, \ldots, n$; otherwise, the set is linearly dependent. (The c_i are scalars). In other words, none of the vectors can be expressed as a linear combination of the others.

If $\mathbf{P}_1, \mathbf{P}_2, \ldots, \mathbf{P}_n$ is arrayed in an $n \times n$ matrix, $\mathbf{P}$, compute the determinant of $\mathbf{P}$ to find out if the set $\mathbf{P}_i$ $(i = 1, n)$ is independent.

If $|\mathbf{P}| = 0$, the set is linearly dependent and no inverse exists.

If $|\mathbf{P}| \neq 0$, the set is linearly independent.

MATRIX OPERATIONS

Addition. Corresponding elements in the two matrices are added together. The matrices must be the same size to do this.

If
$$\mathbf{A} = \begin{bmatrix} 1 & 3 \\ 2 & 4 \end{bmatrix} \quad \text{and} \quad \mathbf{B} = \begin{bmatrix} 5 & 6 \\ -2 & 1 \end{bmatrix}$$

then
$$\mathbf{A} + \mathbf{B} = \begin{bmatrix} 6 & 9 \\ 0 & 5 \end{bmatrix}$$

Subtraction. Rules for subtraction are the same as for addition.

Multiplication by a Scalar. If a matrix is multiplied by a scalar, each element of the matrix is multiplied by the scalar.

If
$$\mathbf{A} = \begin{bmatrix} 3 & 4 \\ 6 & 1 \end{bmatrix}$$

then
$$2\mathbf{A} = \begin{bmatrix} 6 & 8 \\ 12 & 2 \end{bmatrix}$$

Transpose. Rows and columns are interchanged.

$$\mathbf{A} = \begin{bmatrix} 2 & 3 & 1 \\ 0 & 1 & 2 \\ 4 & 2 & 0 \end{bmatrix} \qquad \mathbf{A}^t = \begin{bmatrix} 2 & 0 & 4 \\ 3 & 1 & 2 \\ 1 & 2 & 0 \end{bmatrix}$$

Multiplication of Two Matrices. $\mathbf{AB} = \mathbf{C}$. For matrix multiplication to be defined, the number of columns of $\mathbf{A}$ must be equal to the number of rows of $\mathbf{B}$.

$$c_{ik} = \sum_{j=1}^{n} a_{ij} b_{jk}$$

Example:

$$\begin{pmatrix} 3 & 4 & 5 \\ 1 & 2 & 3 \end{pmatrix} \begin{pmatrix} 1 & 1 \\ 0 & 1 \\ 2 & 0 \end{pmatrix} = \begin{pmatrix} c_{11} & c_{12} \\ c_{21} & c_{22} \end{pmatrix}$$

$$c_{11} = 3(1) + 4(0) + 5(2) = 13$$
$$c_{12} = 3(1) + 4(1) + 5(0) = 7$$

$$c_{21} = 1(1) + 2(0) + 3(2) = 7$$
$$c_{22} = 1(1) + 2(1) + 3(0) = 3$$

$$\mathbf{C} = \begin{pmatrix} 13 & 7 \\ 7 & 3 \end{pmatrix}$$

if $\mathbf{A}_{m \times n}$ and $\mathbf{B}_{n \times p}$ then $\mathbf{C}_{m \times p}$.

NOTE: In general, $\mathbf{AB} \neq \mathbf{BA}$, even if it is defined, that is, matrix multiplication is generally not commutative.

Matrix Division. Matrix division is not explicitly defined; however, under certain conditions there exists another matrix such that

$$\mathbf{AA}^{-1} = \mathbf{A}^{-1}\mathbf{A} = \mathbf{I}$$

This is one of the few cases where matrix multiplication is commutative.

$\mathbf{A}^{-1}$ is called the inverse of $\mathbf{A}$.

Only square matrices can have inverses.

If a matrix has an inverse, it is said to be nonsingular.

Why is it important to be able to find the inverse of a square matrix? Returning to the system of linear equations,

$$\mathbf{AX} = \mathbf{B}$$

Premultiplying by $\mathbf{A}^{-1}$

$$\mathbf{A}^{-1}\mathbf{AX} = \mathbf{A}^{-1}\mathbf{B}$$
$$\mathbf{IX} = \mathbf{A}^{-1}\mathbf{B}$$
$$\mathbf{X} = \mathbf{A}^{-1}\mathbf{B}$$

Thus, if the right-hand-side vector is premultiplied by the inverse of $\mathbf{A}$, the result will be the solution vector, $\mathbf{X}$.

$$\mathbf{X} = (X_1, X_2, \ldots, X_n)$$
$$= \text{values of indepenent variables that are to be solved}$$

The inverse of a matrix (if it is defined) can be obtained by applying elementary row operations to an augmented matrix containing the original matrix augmented with the identity matrix. These elementary row operations are:

1. Interchanging two rows.
2. Multiplying a row by a scalar.
3. Adding multiple of one row to another row.

If the original matrix is further augmented with the right-hand-side $\mathbf{B}$ vector, the solution vector can be found directly by the same elementary row operations that would be used to find the inverse. Proof of this method can be found in texts on linear algebra. An explanation by example will be sufficient here.

Given the original set of linear equations expressed in matrix form,

$$\begin{bmatrix} 2 & 3 & 1 \\ 0 & 1 & 2 \\ 4 & 2 & 0 \end{bmatrix} \begin{bmatrix} X_1 \\ X_2 \\ X_3 \end{bmatrix} = \begin{bmatrix} 4 \\ 6 \\ 12 \end{bmatrix}$$

the augmented form of the coefficient matrix is

$$\left[\begin{array}{ccc|ccc|c} 2 & 3 & 1 & 1 & 0 & 0 & 4 \\ 0 & 1 & 2 & 0 & 1 & 0 & 6 \\ 4 & 2 & 0 & 0 & 0 & 1 & 12 \end{array}\right]$$

The objective is to transform the original coefficient matrix, $\mathbf{A}$, into the identity matrix by applying elementary row operations. If the same operations are also performed on the identity matrix and the right-hand-side vector, they will become the inverse of the coefficient matrix, $\mathbf{A}^{-1}$, and the solution vector, $\mathbf{X}$, respectively. The final augmented matrix will contain the following elements:

$$[\mathbf{I}|\mathbf{A}^{-1}|\mathbf{X}]$$

Initial Matrix

$$\left[\begin{array}{ccc|ccc|c} 2 & 3 & 1 & 1 & 0 & 0 & 4 \\ 0 & 1 & 2 & 0 & 1 & 0 & 6 \\ 4 & 2 & 0 & 0 & 0 & 1 & 12 \end{array}\right]$$

Step 1. Divide row 1 by 2.

$$\left[\begin{array}{ccc|ccc|c} 1 & \frac{3}{2} & \frac{1}{2} & \frac{1}{2} & 0 & 0 & 2 \\ 0 & 1 & 2 & 0 & 1 & 0 & 6 \\ 4 & 2 & 0 & 0 & 0 & 1 & 12 \end{array}\right]$$

Step 2. Add (-4) times row 1 to row 3. NOTE: If a_{21} had not been zero at this stage, some multiple of row 1 would be added to row 2 to eliminate a_{21}.

$$\left[\begin{array}{ccc|ccc|c} 1 & \frac{3}{2} & \frac{1}{2} & \frac{1}{2} & 0 & 0 & 2 \\ 0 & 1 & 2 & 0 & 1 & 0 & 6 \\ 0 & -4 & -2 & -2 & 0 & 1 & 4 \end{array}\right]$$

Step 3. Add $(-\frac{3}{2})$ times row 2 to row 1. Add (4) times row 2 to row 3. NOTE: If a_{22} had not been 1 at this stage, an additional step would be required to divide row 2 by a_{22}.

$$\left[\begin{array}{ccc|ccc|c} 1 & 0 & -\frac{5}{2} & \frac{1}{2} & -\frac{3}{2} & 0 & -7 \\ 0 & 1 & 2 & 0 & 1 & 0 & 6 \\ 0 & 0 & 6 & -2 & 4 & 1 & 28 \end{array}\right]$$

Step 4. Divide row 3 by 6.

$$\left[\begin{array}{ccc|ccc|c} 1 & 0 & -\frac{5}{2} & \frac{1}{2} & -\frac{3}{2} & 0 & -7 \\ 0 & 1 & 2 & 0 & 1 & 0 & 6 \\ 0 & 0 & 1 & -\frac{1}{3} & \frac{2}{3} & \frac{1}{6} & \frac{14}{3} \end{array}\right]$$

Step 5. Add $(\frac{5}{2})$ times row 3 to row 1. Add (-2) times row 3 to row 2.

$$\left[\begin{array}{ccc|ccc|c} 1 & 0 & 0 & -\frac{1}{3} & \frac{1}{6} & \frac{5}{12} & \frac{14}{3} \\ 0 & 1 & 0 & \frac{2}{3} & -\frac{1}{3} & -\frac{1}{3} & -\frac{10}{3} \\ 0 & 0 & 1 & -\frac{1}{3} & \frac{2}{3} & \frac{1}{6} & \frac{14}{3} \end{array}\right]$$

From this it can be seen that

$$X_1 = \tfrac{14}{3}$$

$$X_2 = -\tfrac{10}{3}$$

$$X_3 = \tfrac{14}{3}$$

Verify this answer by multiplying

$$\mathbf{A}^{-1}\mathbf{B} = \begin{bmatrix} -\tfrac{1}{3} & \tfrac{1}{6} & \tfrac{5}{12} \\ \tfrac{2}{3} & -\tfrac{1}{3} & -\tfrac{1}{3} \\ -\tfrac{1}{3} & \tfrac{2}{3} & \tfrac{1}{6} \end{bmatrix}\begin{bmatrix} 4 \\ 6 \\ 12 \end{bmatrix}$$

Also, verify that the above matrix is in fact $\mathbf{A}^{-1}$ by multiplying $\mathbf{A}^{-1}\mathbf{A}$ to see if the product is the identity matrix. What could have been done if one of the diagonal entries on the original matrix, $\mathbf{A}$, had been zero? NOTE: By generating the inverse, $\mathbf{A}^{-1}$, solutions can be found for any right-hand-side vector by multiplying $\mathbf{A}^{-1} \cdot \mathbf{B}$. This is extremely helpful when a system of equations may be subject to a number of right-hand-side conditions. For instance, in structural analysis problems, the right-hand side may represent alternate loadings. However, if only one $\mathbf{B}$ vector is to be used, there is no need to carry out the computations for the inverse. Simply omit the middle augmentation of the original matrix.

DETERMINANT AND LINEAR INDEPENDENCE

The determinant is an important property of a square matrix. The determinant of a singular matrix is zero. Therefore, if the determinant of a coefficient matrix is zero, the matrix does not have an inverse. This generally means that the linear equations depicted by the coefficient matrix are not independent.

$$|\mathbf{A}| = \text{determinant of } \mathbf{A}$$

$$= \sum_{i=1}^{n} a_{ij}A_{ij} = \sum_{j=1}^{n} a_{ij}A_{ij}$$

$$A_{ij} \equiv \text{cofactor}$$

$$= (-1)^{(i+j)} \begin{vmatrix} \text{det} \\ \text{of} \\ \text{minor} \end{vmatrix} \begin{array}{l} \text{(a minor is the submatrix} \\ \text{formed when row } i \text{ and} \\ \text{column } j \text{ are eliminated} \\ \text{from the matrix)} \end{array}$$

Example: If $\mathbf{A} = \begin{bmatrix} 2 & 3 & 1 \\ 0 & 1 & 2 \\ 4 & 2 & 0 \end{bmatrix}$

$$|\mathbf{A}| = (1)\,(2)\left[\det\begin{pmatrix} 1 & 2 \\ 2 & 0 \end{pmatrix}\right] + (-1)\,(3)\left[\det\begin{pmatrix} 0 & 2 \\ 4 & 0 \end{pmatrix}\right]$$

$$+ (1)\,(1)\left[\det\begin{pmatrix} 0 & 1 \\ 4 & 2 \end{pmatrix}\right]$$

$$= (1)(2)(-4) + (-1)(3)(-8) + (1)(1)(-4)$$

$$= -8 + 24 - 4 = 12$$

An alternate way of finding the determinant of a 3×3 matrix is as follows:

$$|\mathbf{A}| = \begin{vmatrix} 2 & 3 & 1 & 2 & 3 \\ 0 & 1 & 2 & 0 & 1 \\ 4 & 2 & 0 & 4 & 2 \end{vmatrix}$$

Multiply the elements on the diagonal then add them according to the signs shown:

$$|\mathbf{A}| = +(2)(1)(0) + (3)(2)(4) + (1)(0)(2)$$
$$- (4)(1)(1) - (2)(2)(2) - (0)(0)(3)$$
$$= 0 + 24 + 0 - 4 - 8 - 0 = 12$$

Note: There is also a method for solving a system of equations using the determinant.

Given:
$$2X + 3Y - Z = 1$$
$$3X + 5Y + 2Z = 8$$
$$X - 2Y - 3Z = -1$$

$$\text{matrix of coefficients} = \mathbf{A} = \begin{bmatrix} 2 & 3 & -1 \\ 3 & 5 & 2 \\ 1 & -2 & -3 \end{bmatrix}$$

det $\mathbf{A} = 22$ (find by above method or method of cofactors)

$$X = \det \begin{vmatrix} 1 & 3 & -1 \\ 8 & 5 & 2 \\ -1 & -2 & -3 \end{vmatrix} = \frac{66}{22} = 3$$

det $\mathbf{A}$

Replace coefficients of the variable being solved for with the right-hand-side constants of the system of equations.

$$Y = \det \begin{vmatrix} 2 & 1 & -1 \\ 3 & 8 & 2 \\ 1 & -1 & -3 \end{vmatrix} = \frac{-22}{22} = -1 \qquad Z = \det \begin{vmatrix} 2 & 3 & 1 \\ 3 & 5 & 8 \\ 1 & -2 & -1 \end{vmatrix} = \frac{44}{22} = 2$$

det $\mathbf{A}$ det $\mathbf{A}$

This method will work for any size system of equations, provided that the determinant can be found. However, it becomes computationally difficult to find the determinant of matrices larger than 3×3.

ADDITIONAL CONCEPTS

Uniqueness of Solutions to $\mathbf{AX} = \mathbf{B}$. If $\mathbf{A}^{-1}$ exists, the system possesses a unique solution. If $\mathbf{A}^{-1}$ does not exist, there is either an infinite number of solutions or there are no solutions.

Sets of m Equations in n Unknowns, $n > m$. If values for $n - m$ variables are specified, the other m can be solved for. For example,

$$4X + 4Y + \ \ Z = 12$$
$$8X - 4Y + 2Z = 6$$
$$\text{set } Z = 0$$
$$\text{then } X = 1.5, \qquad Y = 1.5$$

This is essentially what the simplex method does to solve linear programming problems.

ADDITIONAL DEFINITIONS

Euclidian n Space (E^n). Collection of all points (vectors) in space having coordinates $(a_1, a_2, \ldots, a_n)$

or

$$\mathbf{A} = \begin{bmatrix} a_1 \\ a_2 \\ \vdots \\ a_n \end{bmatrix}$$

Spanning Set. Set of vectors $\mathbf{A}_1, \mathbf{A}_2, \ldots, \mathbf{A}_n$ from E^n is said to span or generate E^n if every vector in E^n can be written as a linear combination of those vectors, $\mathbf{A}_1, \mathbf{A}_2, \ldots, \mathbf{A}_n$.

Basis. A basis is a *linearly independent subset* of vectors from E^n which spans the space. Every basis for E^n has exactly n vectors. Given a set of basis vectors, $\mathbf{A}_1, \mathbf{A}_2, \ldots, \mathbf{A}_n$ for E^n and any other vector $\mathbf{b} \neq 0$ in E^n,

$$\mathbf{b} = \sum_{i=1}^{n} \alpha_i \mathbf{A}_i$$

If any vector $\mathbf{A}_i$, for which $\alpha_i \neq 0$, is removed and $\mathbf{b}$ is added, the new collection forms a basis, that is, the basis is not unique.

Engineering Economics

This appendix will describe the basic engineering economics formulas that will assist in economic feasibility studies. A more thorough discussion can be found in *Economic Analysis for Engineering and Managerial Decision Making,* by Barish and Kaplan (1978). Cash flow diagrams are presented to illustrate the formulas.

Interest is money that must be paid for the use of borrowed money. If you are the borrower, then you must pay the interest as a rental fee. If you are the lender, then you receive the interest as return on your investment. In basic economics problems, there are five parameters that are important; i, the interest rate; n, the number of interest compounding periods; P, a present sum of money; F, an amount of money that will accrue sometime in the future; and A, the amount of payment during any compounding period. The compounding period used for most engineering project feasibility studies is 1 year.

Four of the five parameters described above will be present in any problem, and three of the four present will be known. Also, for the basic applications, A is assumed to be a uniform series of payments (the same amount paid during each compounding period). However, as is shown in Chapter 7, the formulas can also be applied to nonuniform series of payments.

SINGLE-PAYMENT FORMULAS

Single-payment formulas involve only a single amount of money, rather than a uniform series of payments. Either some present amount of money (P) is compounded to give some estimated future amount of money (F), or some amount of money in the future (F) is discounted back to its present worth (P).

Compound Amount Factor

In compound amount factor problems, the interest rate (i), the number of compounding periods (n), and the present amount of money (P) are known and an estimate of the future amount of money (F) that would accrue after n compounding periods is desired. This is also called a compound interest problem, because the amount of interest paid each compounding period is based on the original amount (P) plus the sum of the interest paid in previous periods. For the first compounding period,

$$F_1 = P(1 + i)$$

for the second compounding period,

$$F_2 = F_1(1 + i) = P(1 + i)(1 + i) = P(1 + i)^2$$

and for the nth compounding period,

$$F_n = F = P(1 + i)^n = P(\text{CAF}, i, n)$$

The $(1 + i)^n$ is called the compound amount factor, and it is usually abbreviated as CAF. Tables have been developed for the compound amount factor and the other economic formulas that will be described. Table B.1, shown at the end of this appendix, gives these values for various interest rates.

A variation of compound interest is simple interest, in which the amount of interest paid is based only on the original amount. For this case,

$$F = P + iPn = P(1 + in)$$

A second variation is continuously compounded interest. The amount of interest paid during a compounding period can be expressed as

$$\frac{\Delta P}{\Delta t} = iP$$

If the length of the compounding period (Δt) approaches zero, this becomes a differential problem.

$$\frac{\Delta P}{\Delta t_{t \to 0}} = \frac{dP}{dt} = iP$$

$$\frac{dP}{P} = i\, dt$$

$$\int_P^F \frac{dP}{P} = \int_0^t i\, dt$$

$$\ln P \Big|_p^F = it \Big|_0^t$$

$$\ln\left(\frac{F}{P}\right) = it$$

$$\frac{F}{P} = e^{it}$$

$$F = Pe^{it}$$

The compound amount factor (CAF) is used in most engineering value projections.

Figure B.1 shows the cash flow diagram for either the single payment compound amount factor or the single payment present worth factor.

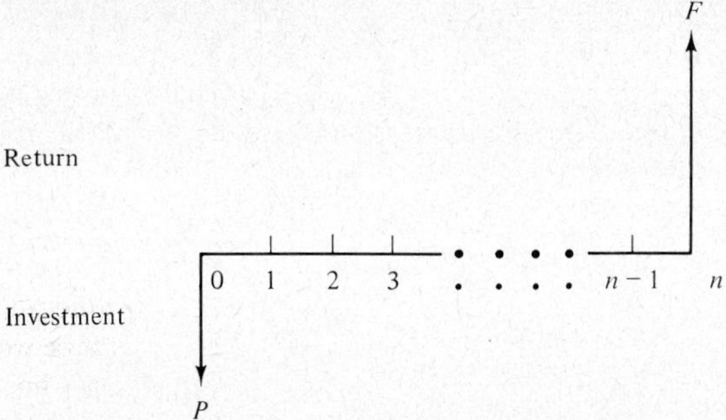

Figure B.1 Cash flow diagram: single payment compound amount factor or single payment present worth factor.

EXAMPLE B.1

You borrow $1000 for 5 yr at 5%/yr interest. At the end of 5 yr you would owe

Simple Interest

$$F = P[1 + i(n)]$$
$$= 1000[1 + 0.05(5)]$$
$$= \$1250$$

Interest Compounded Annually

$$F = P(1 + i)^n = P(\text{CAF}, 0.05, 5)$$
$$= 1000(1.05)^5 = 1000(1.276)$$
$$= \$1276.28 = \$1276$$

Interest Compounded Continuously

$$F = P(e^{it})$$
$$= 1000(e^{0.05(5)})$$
$$= 1000(1.284)$$
$$= \$1284$$

EXAMPLE B.2

What would be the effective rate of interest compounded yearly to get the $1284 found previously for continuous compounding at 5%?

$$1284 = 1000(1 + i)^5$$
$$(1.284)^{1/5} = 1 + i$$
$$i = 0.0513$$

or an effective annual rate of 5.13%.

Present Worth Factor (PWF)

The PWF is the inverse of the compound amount factor. It answers the question: if something is worth F dollars after n compounding periods at interest i, what would it be worth now? F, n, and i are known, and P is calculated. In investment terms, the PWF shows how much should be invested now to have a certain amount at some later date, if the interest rate is known.

If

$$F = P(1 + i)^n$$

then

$$P = F\left[\frac{1}{(1 + i)^n}\right] = F(\text{PWF}, i, n)$$

UNIFORM SERIES OF PAYMENT FORMULAS

Sinking Fund Factor (SFF)

The SFF indicates how much money (A) should be invested at the end of each year at $i\%/\text{yr}$ for n yr to accumulate a stipulated future sum of money (F). The first year's payment earns interest for $n - 1$ yr:

$$F_1 = A(1 + i)^{n-1}$$

Second year, $n - 2$ yr:

$$F_2 = A(1 + i)^{n-2}$$
$$\vdots$$
$$F_n = A$$
$$F = F_1 + F_2 + \cdots + F_n$$
$$F = A[(1 + i)^{n-1} + (1 + i)^{n-2} + \cdots + 1]$$
$$F(1 + i) = A[(1 + i)^n + (1 + i)^{n-1} + \cdots + (1 + i)]$$

Subtracting

$$Fi = A[(1 + i)^n - 1]$$
$$F = A\left[\frac{(1 + i)^n - 1}{i}\right]$$
$$\therefore A = F\left[\frac{i}{(1 + i)^n - 1}\right] = F(\text{SFF}, i, n)$$

Uniform Series Compound Amount Factor (USCAF)

The USCAF is the reciprocal of the sinking fund factor. If A dollars are invested at $i\%$ for n yr, how much money (F) will be accumulated?

$$F = A\left[\frac{(1 + i)^n - 1}{i}\right] = A(\text{USCAF}, i, n)$$

Figure B.2 shows the cash flow diagram for both the SFF and the USCAF.

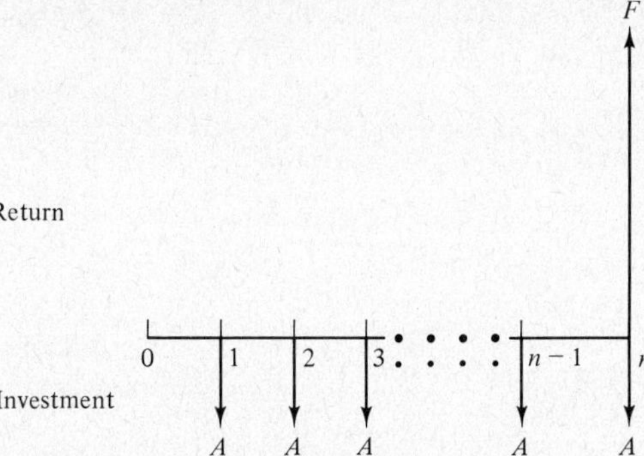

Figure B.2   Cash flow diagram: sinking fund factor
and uniform series compound amount factor.

EXAMPLE B.3

When you were born, your parents, being astute observers of the future, decided that you should
go to Union College and that your education would cost about \$7000/yr. They determined that
they could get a 6% return on their money. They wanted to have the whole \$28,000 available
when you turned 18. How much would your parents have had to set aside each year to provide
for your education?

$$n = 18 \qquad i = 0.06 \qquad F = \$28,000$$

$$A = \left[\frac{i}{(1 + i)^n - 1}\right](F) = (\text{SFF}, i, n)F$$

$$= \frac{0.06}{(1 + 0.06)^{18} - 1}\,(28,000) = 0.03236F$$

$$= 905.98 \quad \text{or about } \$906/\text{yr}$$

Capital Recovery Factor (CRF)

If P dollars are invested today at $i\%$ interest, how many dollars, A, can be secured at the end of
each year for n yr such that the initial investment, P, is just depleted?

Or, if P dollars are borrowed today at $i\%$/period, how many dollars, A, must be paid back
at the end of each period to retire the loan in n periods?

Start with P. At the end of one period,

$$F_1 = P(1 + i) - A$$

At the end of two periods,

$$F_2 = [P(1 + i) - A](1 + i) - A$$
$$= P(1 + i)^2 - A(1 + i) - A$$

At the end of three periods,

$$F_3 = P(1 + i)^3 - A(1 + i)^2 - A(1 + i) - A$$

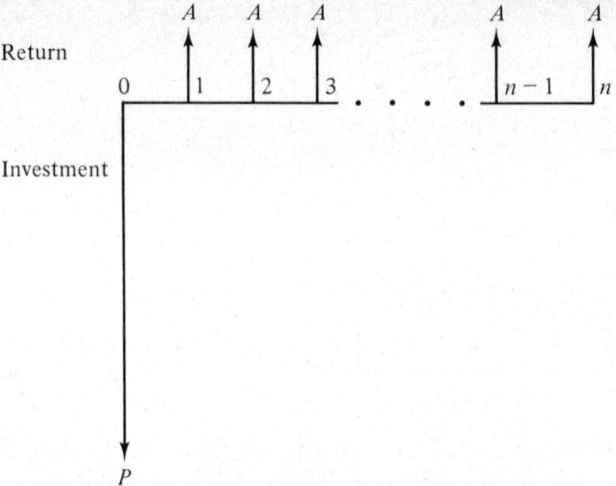

Figure B.3 Cash flow diagram: capital recovery factor and uniform series present worth factor.

After n periods,

$$F_n = 0 = P(1 + i)^n - A(1 + i)^{n-1} - \cdots - A$$
$$\therefore P(1 + i)^n = A[(1 + i)^{n-1} + (1 + i)^{n-2} + \cdots + 1]$$
$$P(1 + i)^n(1 + i) = A[(1 + i)^n + (1 + i)^{n-1} + \cdots + (1 + i)]$$

Subtracting,

$$P(1 + i)^n i = A[(1 + i)^n - 1]$$

$$A = \frac{P(1 + i)^n i}{(1 + i)^n - 1} = P(\text{CRF}, i, n)$$

Also,

$$A = P\left[\frac{i + (1 + i)^n i - i}{(1 + i)^n - 1}\right]$$

$$= P\left[\frac{i + i[(1 + i)^n - 1]}{(1 + i)^n - 1}\right]$$

$$= P\left[\frac{i}{(1 + i)^n - 1} + i\right]$$

$$= P(\text{sinking fund factor} + i)$$

Figure B.3 shows the cash flow diagram for either the CRF or the USPWF.

Uniform Series Present Worth Factor (USPWF)

The USPWF is the reciprocal of the CRF. It tells what amount P should be invested today at $i\%$ interest to recover A dollars at the end of each year for n yr.

$$P = A\left[\frac{(1 + i)^n - 1}{(1 + i)^n i}\right] = A(\text{USPWF}, i, n)$$

EXAMPLE B.4

What are the monthly payments for a $40,000 loan at $8\frac{1}{2}\%$ annual interest over 25 yr?

$$i = 0.085/12 = 0.007083 = 0.7083\%/\text{month}$$

$$n = 300 \ (25 \ \text{yr} \ @ \ 12 \ \text{payments/yr})$$

$$A = \$40,000\left(\frac{0.007083}{(1.007083)^{300} - 1} + 0.007083\right)$$

$$= \$322.10/\text{month}$$

What would the monthly payments be if the interest rate were 15%?

$$A = \$40,000\left(\text{CRF}, \frac{0.15}{12}, 300\right)$$

$$= \$40,000\left[\frac{0.0125}{(1.0125)^{300} - 1} + 0.0125\right]$$

$$= \$40,000(0.01281) = \$512.23/\text{month}$$

Notice that for large values of n, the CRF approaches the interest rate, i.

EXAMPLE B.5

An investment of $100,000 produces a return of $14,000/yr for 16 yr. What is the approximate rate of return?

$$P = 100,000$$

$$A = 14,000$$

$$\frac{P}{A} = 7.14 = \frac{(1 + i)^n - 1}{(1 + i)^n i}$$

An approximate value for i could be found using Table B.1.

$$\text{for } i = 10\% \qquad \text{PWF} = 7.824$$
$$\text{for } i = 12\% \qquad \text{PWF} = 6.974$$

By interpolation

$$i\% \approx 10\% + \left(\frac{7.82 - 7.14}{7.82 - 6.97}\right)(2\%) = 11.6\%$$

The interest rate could be found more accurately using the Newton–Raphson method. This is described in Appendix E.

EFFECTS OF INFLATION

The rate of inflation will influence the future costs of goods and services to be received or provided. If an item costs $1.00 today, and the rate of inflation is 6%/yr, the item will cost $1.06 1 yr in the future, $1.12 2 yr in the future, and $1.19 3 yr in the future. The cost at any point in the future could be predicted by the single payment CAF.

Inflation also effectively reduces the future purchasing power of an amount of money. The dollar above would buy the item now, but 3 yr from now it would pay for only a portion of the item. Effective purchasing power would then be 1/1.19, or 84% of what it is now. Therefore, it could be said that the purchasing power of the dollar would be $0.84 after 3 yr.

Example B.6 will demonstrate the effects of inflation on a cash flow diagram for a piece of equipment.

EXAMPLE B.6 ____

An item of equipment costs $100,000 today. At the time of purchase, it is expected to cost $10,000/yr to maintain the equipment over its useful life of 10 yr, and the equipment is expected to produce $30,000/yr in revenues. Presently, the salvage value of a like item of equipment is $10,000. The inflation rate is now at 6%/yr. Assuming that the inflation rate will remain approximately the same over the next 10 yr, and will affect all future amounts of money, develop the cash flow diagram that would reflect all receipts and disbursements. Consistent with other economic formulas, the base receipts and disbursements will be subjected to inflation each year. The amounts of money are tabulated in Table B.2, and the cash flow diagram is shown in Figure B.4. This cash flow diagram could be used in conjunction with the analysis techniques developed in Chapter 7.

Table B.1 gives values for the six economic factors for various interest rates and numbers of compounding periods. Table B.3 shows a compilation of the relationships among the various factors.

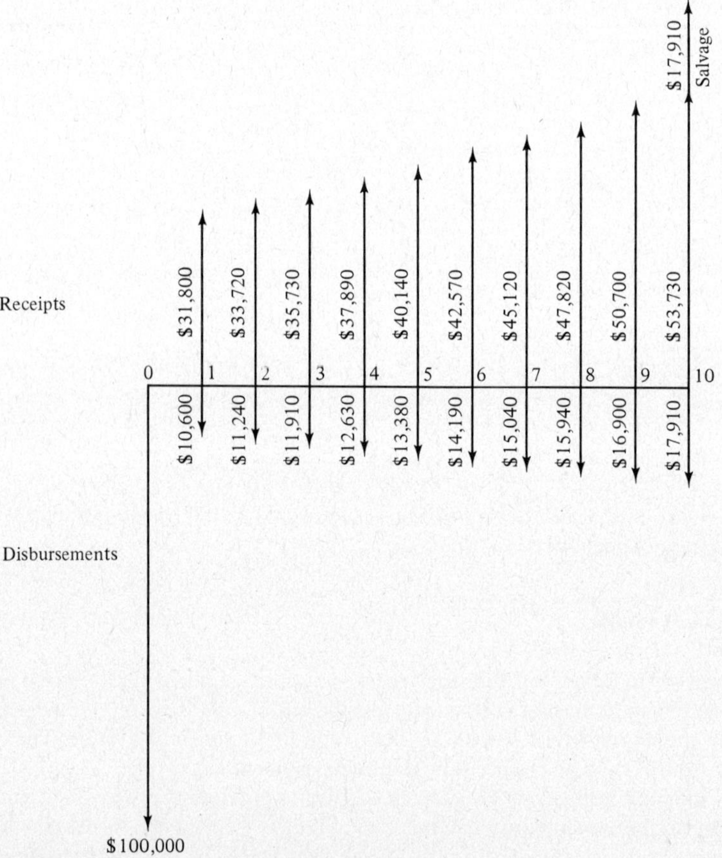

Figure B.4 Cash flow diagram for Example B.6.

                                485

TABLE B.2   INFLATION EFFECTS ON FUTURE AMOUNTS OF MONEY

Year	Compound amount factor	Maintenance ($/yr)	Revenues ($/yr)	Salvage ($)
0	1.000	10,000	30,000	
1	1.060	10,600	31,800	
2	1.124	11,240	33,720	
3	1.191	11,910	35,730	
4	1.263	12,630	37,890	
5	1.338	13,380	40,140	
6	1.419	14,190	42,570	
7	1.504	15,040	45,120	
8	1.594	15,940	47,820	
9	1.690	16,900	50,700	
10	1.791	17,910	53,730	17,910

TABLE B. 3   COMPOUND INTEREST RELATIONSHIPS

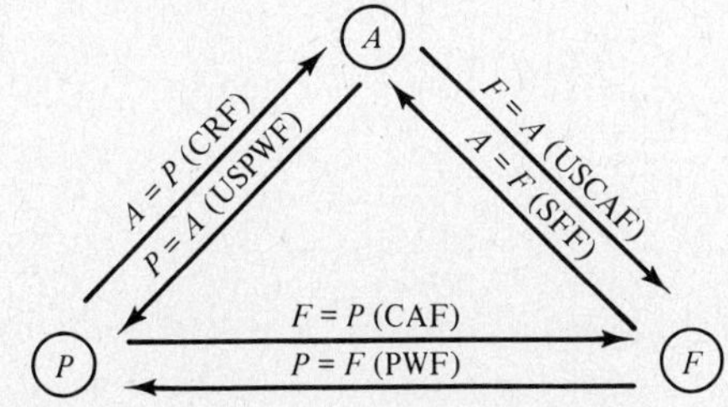

Definitions:

$$CAF = \text{single payment compound amount factor} = (1 + i)^n$$
Known: i, n, P Unknown: F

$$PWF = \text{single payment present worth factor} = 1/(1 + i)^n$$
Known: i, n, F Unknown: P

$$USCAF = \text{uniform series compound amount factor} = [(1 + i)^n - 1]/i$$
Known: A, i, n Unknown: F

$$SFF = \text{sinking fund factor} = i/[(1 + i)^n - 1]$$
Known: F, i, n Unknown: A

$$CRF = \text{capital recovery factor} = i(1 + i)^n/[(1 + i)^n - 1]$$
Known: P, i, n Unknown: A

$$USPWF = \text{uniform series present worth factor} = [(1 + i)^n - 1]/i(1 + i)^n$$
Known: A, i, n Unknown: P

Identities:

$$SFF = CRF - i \qquad CRF = 1/PWF$$
$$SFF = (CRF)(PWF) \qquad USCAF = 1/SFF$$
$$USPWF = (USCAF)(PWF)$$
$$CRF = (SFF)(CAF) \qquad CRF = 1/USPWF$$

TABLE B. 1 COMPOUND INTEREST RATES

1%

	Single payment		Uniform series				
	Compound amount factor	Present worth factor	Sinking fund factor	Capital recovery factor	Compound amount factor	Present worth factor	
n	$P \to F$	$F \to P$	$F \to A$	$P \to A$	$A \to F$	$A \to P$	n
1	1.0100	0.9901	1.00000	1.01000	1.000	0.990	1
2	1.0201	0.9803	0.49751	0.50751	2.010	1.970	2
3	1.0303	0.9706	0.33002	0.34002	3.030	2.941	3
4	1.0406	0.9610	0.24628	0.25628	4.060	3.902	4
5	1.0510	0.9515	0.19604	0.20604	5.101	4.853	5
6	1.0615	0.9420	0.16255	0.17255	6.152	5.795	6
7	1.0721	0.9327	0.13863	0.14863	7.214	6.728	7
8	1.0829	0.9235	0.12069	0.13069	8.286	7.652	8
9	1.0937	0.9143	0.10674	0.11674	9.369	8.566	9
10	1.1046	0.9053	0.09558	0.10558	10.462	9.471	10
11	1.1157	0.8963	0.08645	0.09645	11.567	10.368	11
12	1.1268	0.8874	0.07885	0.08885	12.683	11.255	12
13	1.1381	0.8787	0.07241	0.08241	13.809	12.134	13
14	1.1495	0.8700	0.06690	0.07690	14.947	13.004	14
15	1.1610	0.8613	0.06212	0.07212	16.097	13.865	15
16	1.1726	0.8528	0.05794	0.06794	17.258	14.718	16
17	1.1843	0.8444	0.05426	0.06426	18.430	15.562	17
18	1.1961	0.8360	0.05098	0.06098	19.615	16.398	18
19	1.2081	0.8277	0.04805	0.05805	20.811	17.226	19
20	1.2202	0.8195	0.04542	0.05542	22.019	18.046	20
21	1.2324	0.8114	0.04303	0.05303	23.239	18.857	21
22	1.2447	0.8034	0.04086	0.05086	24.472	19.660	22
23	1.2572	0.7954	0.03889	0.04889	25.716	20.456	23
24	1.2697	0.7876	0.03707	0.04707	26.973	21.243	24
25	1.2824	0.7798	0.03541	0.04541	28.243	22.023	25
26	1.2953	0.7720	0.03387	0.04387	29.526	22.795	26
27	1.3082	0.7644	0.03245	0.04245	30.821	23.560	27
28	1.3213	0.7568	0.03112	0.04112	32.129	24.316	28
29	1.3345	0.7493	0.02990	0.03990	33.450	25.066	29
30	1.3478	0.7419	0.02875	0.03875	34.785	25.808	30

TABLE B. 1 *(Continued)*

1%

	Single payment		Uniform series				
	Compound amount factor	Present worth factor	Sinking fund factor	Capital recovery factor	Compound amount factor	Present worth factor	
n	$P \rightarrow F$	$F \rightarrow P$	$F \rightarrow A$	$P \rightarrow A$	$A \rightarrow F$	$A \rightarrow P$	n
31	1.3613	0.7346	0.02768	0.03768	36.133	26.542	31
32	1.3749	0.7273	0.02667	0.03667	37.494	27.270	32
33	1.3887	0.7201	0.02573	0.03573	38.869	27.990	33
34	1.4026	0.7130	0.02484	0.03484	40.258	28.703	34
35	1.4166	0.7059	0.02400	0.03400	41.660	29.409	35
40	1.4889	0.6717	0.02046	0.03046	48.886	32.835	40
45	1.5648	0.6391	0.01771	0.02771	56.481	36.095	45
50	1.6446	0.6080	0.01551	0.02551	64.463	39.196	50
55	1.7285	0.5785	0.01373	0.02373	72.852	42.147	55
60	1.8167	0.5504	0.01224	0.02224	81.670	44.955	60
65	1.9094	0.5237	0.01100	0.02100	90.937	47.627	65
70	2.0068	0.4983	0.00993	0.01993	100.676	50.169	70
75	2.1091	0.4741	0.00902	0.01902	110.913	52.587	75
80	2.2167	0.4511	0.00822	0.01822	121.672	54.888	80
85	2.3298	0.4292	0.00752	0.01752	132.979	57.078	85
90	2.4486	0.4084	0.00690	0.01690	144.863	59.161	90
95	2.5735	0.3886	0.00636	0.01636	157.354	61.143	95
100	2.7048	0.3697	0.00587	0.01587	170.481	63.029	100

$1\frac{1}{4}$%

	Single payment		Uniform series				
	Compound amount factor	Present worth factor	Sinking fund factor	Capital recovery factor	Compound amount factor	Present worth factor	
n	$P \rightarrow F$	$F \rightarrow P$	$F \rightarrow A$	$P \rightarrow A$	$A \rightarrow F$	$A \rightarrow P$	n
1	1.0125	0.9877	1.00000	1.01250	1.000	0.988	1
2	1.0252	0.9755	0.49689	0.50939	2.012	1.963	2
3	1.0380	0.9634	0.32920	0.34170	3.038	2.927	3
4	1.0509	0.9515	0.24536	0.25786	4.076	3.878	4
5	1.0641	0.9398	0.19506	0.20756	5.127	4.818	5
6	1.0774	0.9282	0.16153	0.17403	6.191	5.746	6
7	1.0909	0.9167	0.13759	0.15009	7.268	6.663	7
8	1.1045	0.9054	0.11963	0.13213	8.359	7.568	8
9	1.1183	0.8942	0.10567	0.11817	9.463	8.462	9
10	1.1323	0.8832	0.09450	0.10700	10.582	9.346	10
11	1.1464	0.8723	0.08537	0.09787	11.714	10.218	11
12	1.1608	0.8615	0.07776	0.09026	12.860	11.079	12
13	1.1753	0.8509	0.07132	0.08382	14.021	11.930	13
14	1.1900	0.8404	0.06581	0.07831	15.196	12.771	14
15	1.2048	0.8300	0.06103	0.07353	16.386	13.601	15

TABLE B. 1 *(Continued)*

$1\frac{1}{4}\%$

	Single payment		Uniform series				
	Compound amount factor	Present worth factor	Sinking fund factor	Capital recovery factor	Compound amount factor	Present worth factor	
n	$P \rightarrow F$	$F \rightarrow P$	$F \rightarrow A$	$P \rightarrow A$	$A \rightarrow F$	$A \rightarrow P$	n
16	1.2199	0.8197	0.05685	0.06935	17.591	14.420	16
17	1.2351	0.8096	0.05316	0.06566	18.811	15.230	17
18	1.2506	0.7996	0.04988	0.06238	20.046	16.030	18
19	1.2662	0.7898	0.04696	0.05946	21.297	16.819	19
20	1.2820	0.7800	0.04432	0.05682	22.563	17.599	20
21	1.2981	0.7704	0.04194	0.05444	23.845	18.370	21
22	1.3143	0.7609	0.03977	0.05227	25.143	19.131	22
23	1.3307	0.7515	0.03780	0.05030	26.457	19.882	23
24	1.3474	0.7422	0.03599	0.04849	27.788	20.624	24
25	1.3642	0.7330	0.03432	0.04682	29.135	21.357	25
26	1.3812	0.7240	0.03279	0.04529	30.500	22.081	26
27	1.3985	0.7150	0.03137	0.04387	31.881	22.796	27
28	1.4160	0.7062	0.03005	0.04255	33.279	23.503	28
29	1.4337	0.6975	0.02882	0.04132	34.695	24.200	29
30	1.4516	0.6889	0.02768	0.04018	36.129	24.889	30
31	1.4698	0.6804	0.02661	0.03911	37.581	25.569	31
32	1.4881	0.6720	0.02561	0.03811	39.050	26.241	32
33	1.5067	0.6637	0.02467	0.03717	40.539	26.905	33
34	1.5256	0.6555	0.02378	0.03628	42.045	27.560	34
35	1.5446	0.6474	0.02295	0.03545	43.571	28.208	35
40	1.6436	0.6084	0.01942	0.03192	51.490	31.327	40
45	1.7489	0.5718	0.01669	0.02919	59.916	34.258	45
50	1.8610	0.5373	0.01452	0.02702	68.882	37.013	50
55	1.9803	0.5050	0.01275	0.02525	78.422	39.602	55
60	2.1072	0.4746	0.01129	0.02379	88.575	42.035	60
65	2.2422	0.4460	0.01006	0.02256	99.377	44.321	65
70	2.3859	0.4191	0.00902	0.02152	110.872	46.470	70
75	2.5388	0.3939	0.00812	0.02062	123.103	48.489	75
80	2.7015	0.3702	0.00735	0.01985	136.119	50.387	80
85	2.8746	0.3479	0.00667	0.01917	149.968	52.170	85
90	3.0588	0.3269	0.00607	0.01857	164.705	53.846	90
95	3.2548	0.3072	0.00554	0.01804	180.386	55.421	95
100	3.4634	0.2887	0.00507	0.01757	197.072	56.901	100

$1\frac{1}{2}\%$

	Single payment		Uniform series				
	Compound amount factor	Present worth factor	Sinking fund factor	Capital recovery factor	Compound amount factor	Present worth factor	
n	$P \rightarrow F$	$F \rightarrow P$	$F \rightarrow A$	$P \rightarrow A$	$A \rightarrow F$	$A \rightarrow P$	n
1	1.0150	0.9852	1.00000	1.01500	1.000	0.985	1
2	1.0302	0.9707	0.49628	0.51128	2.015	1.956	2

TABLE B. 1 *(Continued)*

$1\frac{1}{2}\%$

	Single payment		Uniform series				
	Compound amount factor	Present worth factor	Sinking fund factor	Capital recovery factor	Compound amount factor	Present worth factor	
n	$P \rightarrow F$	$F \rightarrow P$	$F \rightarrow A$	$P \rightarrow A$	$A \rightarrow F$	$A \rightarrow P$	n
3	1.0457	0.9563	0.32838	0.34338	3.045	2.912	3
4	1.0614	0.9422	0.24444	0.25944	4.091	3.854	4
5	1.0773	0.9283	0.19409	0.20909	5.152	4.783	5
6	1.0934	0.9145	0.16053	0.17553	6.230	5.697	6
7	1.1098	0.9010	0.13656	0.15156	7.323	6.598	7
8	1.1265	0.8877	0.11858	0.13358	8.433	7.486	8
9	1.1434	0.8746	0.10461	0.11961	9.559	8.361	9
10	1.1605	0.8617	0.09343	0.10843	10.703	9.222	10
11	1.1779	0.8489	0.08429	0.09929	11.863	10.071	11
12	1.1956	0.8364	0.07668	0.09168	13.041	10.908	12
13	1.2136	0.8240	0.07024	0.08524	14.237	11.732	13
14	1.2318	0.8118	0.06472	0.07972	15.450	12.543	14
15	1.2502	0.7999	0.05994	0.07494	16.682	13.343	15
16	1.2690	0.7880	0.05577	0.07077	17.932	14.131	16
17	1.2880	0.7764	0.05208	0.06708	19.201	14.908	17
18	1.3073	0.7649	0.04881	0.06381	20.489	15.673	18
19	1.3270	0.7536	0.04588	0.06088	21.797	16.426	19
20	1.3469	0.7425	0.04325	0.05825	23.124	17.169	20
21	1.3671	0.7315	0.04087	0.05587	24.470	17.900	21
22	1.3876	0.7207	0.03870	0.05370	25.838	18.621	22
23	1.4084	0.7100	0.03673	0.05173	27.225	19.331	23
24	1.4300	0.6995	0.03492	0.04992	28.634	20.030	24
25	1.4509	0.6892	0.03326	0.04826	30.063	20.720	25
26	1.4727	0.6790	0.03173	0.04673	31.514	21.399	26
27	1.4948	0.6690	0.03032	0.04532	32.987	22.068	27
28	1.5172	0.6591	0.02900	0.04400	34.481	22.727	28
29	1.5400	0.6494	0.02778	0.04278	35.999	23.376	29
30	1.5631	0.6398	0.02664	0.04164	37.539	24.016	30
31	1.5865	0.6303	0.02557	0.04057	39.102	24.646	31
32	1.6103	0.6210	0.02458	0.03958	40.688	25.267	32
33	1.6345	0.6118	0.02364	0.03864	42.299	25.879	33
34	1.6590	0.6028	0.02276	0.03776	43.933	26.482	34
35	1.6839	0.5939	0.02193	0.03693	45.592	27.076	35
40	1.8140	0.5513	0.01843	0.03343	54.268	29.916	40
45	1.9542	0.5117	0.01572	0.03072	63.614	32.552	45
50	2.1052	0.4750	0.01357	0.02857	73.683	35.000	50
55	2.2679	0.4409	0.01183	0.02683	84.530	37.271	55
60	2.4432	0.4093	0.01039	0.02539	96.215	39.380	60
65	2.6320	0.3799	0.00919	0.02419	108.803	41.338	65
70	2.8355	0.3527	0.00817	0.02317	122.364	43.155	70
75	3.0546	0.3274	0.00730	0.02230	136.973	44.842	75
80	3.2907	0.3039	0.00655	0.02155	152.711	46.407	80
85	3.5450	0.2821	0.00589	0.02089	169.665	47.861	85

TABLE B. 1 *(Continued)*

$1\frac{1}{2}\%$

	Single payment		Uniform series				
	Compound amount factor	Present worth factor	Sinking fund factor	Capital recovery factor	Compound amount factor	Present worth factor	
n	$P \rightarrow F$	$F \rightarrow P$	$F \rightarrow A$	$P \rightarrow A$	$A \rightarrow F$	$A \rightarrow P$	n
90	3.8189	0.2619	0.00532	0.02022	187.930	49.210	90
95	4.1141	0.2431	0.00482	0.01982	207.606	50.462	95
100	4.4320	0.2256	0.00437	0.01937	228.803	51.625	100

$1\frac{3}{4}\%$

	Single payment		Uniform series				
	Compound amount factor	Present worth factor	Sinking fund factor	Capital recovery factor	Compound amount factor	Present worth factor	
n	$P \rightarrow F$	$F \rightarrow P$	$F \rightarrow A$	$P \rightarrow A$	$A \rightarrow F$	$A \rightarrow P$	n
1	1.0175	0.9828	1.00000	1.01750	1.000	0.983	1
2	1.0353	0.9695	0.49566	0.51316	2.018	1.949	2
3	1.0534	0.9493	0.32757	0.34507	3.053	2.898	3
4	1.0719	0.9330	0.24353	0.26103	4.106	3.831	4
5	1.0906	0.9169	0.19312	0.21062	5.178	4.748	5
6	1.1097	0.9011	0.15952	0.17702	6.269	5.649	6
7	1.1291	0.8856	0.13553	0.15303	7.378	6.535	7
8	1.1489	0.8704	0.11754	0.13504	8.508	7.405	8
9	1.1690	0.8554	0.10356	0.12106	9.656	8.260	9
10	1.1894	0.8407	0.09238	0.10988	10.825	9.101	10
11	1.2103	0.8263	0.08323	0.10073	12.015	9.927	11
12	1.2314	0.8121	0.07561	0.09311	13.225	10.740	12
13	1.2530	0.7981	0.06917	0.08667	14.457	11.538	13
14	1.2749	0.7844	0.06366	0.08116	15.710	12.322	14
15	1.2972	0.7709	0.05888	0.07638	16.984	13.093	15
16	1.3199	0.7576	0.05470	0.07220	18.282	13.850	16
17	1.3430	0.7446	0.05102	0.06852	19.602	14.595	17
18	1.3665	0.7318	0.04774	0.06524	20.945	15.327	18
19	1.3904	0.7192	0.04482	0.06232	22.311	16.046	19
20	1.4148	0.7068	0.04219	0.05969	23.702	16.753	20
21	1.4395	0.6947	0.03981	0.05731	25.116	17.448	21
22	1.4647	0.6827	0.03766	0.05516	26.556	18.130	22
23	1.4904	0.6710	0.03569	0.05319	28.021	18.801	23
24	1.5164	0.6594	0.03389	0.05139	29.511	19.461	24
25	1.5430	0.6481	0.03223	0.04973	31.027	20.109	25
26	1.5700	0.6369	0.03070	0.04820	32.570	20.746	26
27	1.5975	0.6260	0.02929	0.04679	34.140	21.372	27
28	1.6254	0.6152	0.02798	0.04548	35.738	21.987	28
29	1.6539	0.6046	0.02676	0.04426	37.363	22.592	29
30	1.6828	0.5942	0.02563	0.04313	39.017	23.186	30

TABLE B. 1 *(Continued)*

$1\frac{3}{4}\%$

| | Single payment | | Uniform series | | | | |
| | Compound amount factor | Present worth factor | Sinking fund factor | Capital recovery factor | Compound amount factor | Present worth factor | |
n	$P \rightarrow F$	$F \rightarrow P$	$F \rightarrow A$	$P \rightarrow A$	$A \rightarrow F$	$A \rightarrow P$	n
31	1.7122	0.5840	0.02457	0.04207	40.700	23.770	31
32	1.7422	0.5740	0.02358	0.04108	42.412	24.344	32
33	1.7727	0.5641	0.02265	0.04015	44.154	24.908	33
34	1.8037	0.5544	0.02177	0.03927	45.927	25.462	34
35	1.8353	0.5449	0.02095	0.03845	47.731	26.007	35
40	2.0016	0.4996	0.01747	0.03497	57.234	28.594	40
45	2.1830	0.4581	0.01479	0.03229	67.599	30.966	45
50	2.3808	0.4200	0.01267	0.03017	78.902	33.141	50
55	2.5965	0.3851	0.01096	0.02846	91.230	35.135	55
60	2.8318	0.3531	0.00955	0.02705	104.675	36.964	60
65	3.0884	0.3238	0.00838	0.02588	119.339	38.641	65
70	3.3683	0.2969	0.00739	0.02489	135.331	40.178	70
75	3.6735	0.2722	0.00655	0.02405	152.772	41.587	75
80	4.0064	0.2496	0.00582	0.02332	171.794	42.880	80
85	4.3694	0.2289	0.00519	0.02269	192.539	44.065	85
90	4.7654	0.2098	0.00465	0.02215	215.165	45.152	90
95	5.1972	0.1924	0.00417	0.02167	239.840	46.148	95
100	5.6682	0.1764	0.00375	0.02125	266.752	47.061	100

2%

| | Single payment | | Uniform series | | | | |
| | Compound amount factor | Present worth factor | Sinking fund factor | Capital recovery factor | Compound amount factor | Present worth factor | |
n	$P \rightarrow F$	$F \rightarrow P$	$F \rightarrow A$	$P \rightarrow A$	$A \rightarrow F$	$A \rightarrow P$	n
1	1.0200	0.9804	1.00000	1.02000	1.000	0.980	1
2	1.0404	0.9612	0.49505	0.51505	2.020	1.942	2
3	1.0612	0.9423	0.32675	0.34675	3.060	2.884	3
4	1.0824	0.9238	0.24262	0.26262	4.122	3.808	4
5	1.1041	0.9057	0.19216	0.21216	5.204	4.713	5
6	1.1262	0.8880	0.15853	0.17853	6.308	5.601	6
7	1.1487	0.8706	0.13451	0.15451	7.434	6.472	7
8	1.1717	0.8535	0.11651	0.13651	8.583	7.325	8
9	1.1951	0.8368	0.10252	0.12252	9.755	8.162	9
10	1.2190	0.8203	0.09133	0.11133	10.950	8.983	10
11	1.2434	0.8043	0.08218	0.10218	12.169	9.787	11
12	1.2682	0.7885	0.07456	0.09456	13.412	10.575	12
13	1.2936	0.7730	0.06812	0.08812	14.680	11.348	13
14	1.3195	0.7579	0.06260	0.08260	15.974	12.106	14
15	1.3459	0.7430	0.05783	0.07783	17.293	12.849	15

TABLE B. 1 *(Continued)*

2%

	Single payment		Uniform series				
	Compound amount factor	Present worth factor	Sinking fund factor	Capital recovery factor	Compound amount factor	Present worth factor	
n	$P \rightarrow F$	$F \rightarrow P$	$F \rightarrow A$	$P \rightarrow A$	$A \rightarrow F$	$A \rightarrow P$	n
16	1.3728	0.7284	0.05365	0.07365	18.639	13.578	16
17	1.4002	0.7142	0.04997	0.06997	20.012	14.292	17
18	1.4282	0.7002	0.04670	0.06670	11.412	14.992	18
19	1.4568	0.6864	0.04378	0.06378	22.841	15.678	19
20	1.4859	0.6730	0.04116	0.06116	24.297	16.351	20
21	1.5157	0.6598	0.03878	0.05878	25.783	17.011	21
22	1.5460	0.6468	0.03663	0.05663	27.299	17.658	22
23	1.5769	0.6342	0.03467	0.05467	28.845	18.292	23
24	1.6084	0.6217	0.03287	0.05287	30.422	18.914	24
25	1.6406	0.6095	0.03122	0.05122	32.030	19.523	25
26	1.6734	0.5976	0.02970	0.04970	33.671	20.121	26
27	1.7069	0.5859	0.02829	0.04829	35.344	20.707	27
28	1.7410	0.5744	0.02699	0.04699	37.051	21.281	28
29	1.7758	0.5631	0.02578	0.04578	38.792	21.844	29
30	1.8114	0.5521	0.02465	0.04465	40.568	22.396	30
31	1.8476	0.5412	0.02360	0.04360	42.379	22.938	31
32	1.8845	0.5306	0.02261	0.04261	44.227	23.468	32
33	1.9222	0.5202	0.02169	0.04169	46.112	23.989	33
34	1.9607	0.5100	0.02082	0.04082	48.034	24.499	34
35	1.9999	0.5000	0.02000	0.04000	49.994	24.999	35
40	2.2080	0.4529	0.01656	0.03656	60.402	27.355	40
45	2.4379	0.4102	0.01391	0.03391	71.893	29.490	45
50	2.6916	0.3715	0.01182	0.03182	84.579	31.424	50
55	2.9717	0.3365	0.01014	0.03014	98.587	33.175	55
60	3.2810	0.3048	0.00877	0.02877	114.052	34.761	60
65	3.6225	0.2761	0.00763	0.02763	131.126	36.197	65
70	3.9996	0.2500	0.00667	0.02667	149.978	37.499	70
75	4.4158	0.2265	0.00586	0.02586	170.792	38.677	75
80	4.8754	0.2051	0.00516	0.02516	193.772	39.745	80
85	5.3829	0.1858	0.00456	0.02456	219.144	40.711	85
90	5.9431	0.1683	0.00405	0.02405	247.157	41.587	90
95	6.5617	0.1524	0.00360	0.02360	278.085	42.380	95
100	7.2446	0.1380	0.00320	0.02320	312.232	43.098	100

$2\frac{1}{2}\%$

	Single payment		Uniform series				
	Compound amount factor	Present worth factor	Sinking fund factor	Capital recovery factor	Compound amount factor	Present worth factor	
n	$P \rightarrow F$	$F \rightarrow P$	$F \rightarrow A$	$P \rightarrow A$	$A \rightarrow F$	$A \rightarrow P$	n
1	1.0250	0.9756	1.00000	0.02500	1.000	0.976	1
2	1.0506	0.9518	0.49383	0.51883	2.025	1.927	2

TABLE B. 1 *(Continued)*

$2\frac{1}{2}\%$

	Single payment		Uniform series				
	Compound amount factor	Present worth factor	Sinking fund factor	Capital recovery factor	Compound amount factor	Present worth factor	
n	$P \to F$	$F \to P$	$F \to A$	$P \to A$	$A \to F$	$A \to P$	n
3	1.0769	0.9286	0.32514	0.35014	3.076	2.856	3
4	1.1038	0.9060	0.24082	0.26582	4.153	3.762	4
5	1.1314	0.8839	0.19025	0.21525	5.256	4.646	5
6	1.1597	0.8623	0.15655	0.18155	6.388	5.508	6
7	1.1887	0.8413	0.13250	0.15750	7.547	6.349	7
8	1.2184	0.8207	0.11447	0.13947	8.736	7.170	8
9	1.2489	0.8007	0.10046	0.12546	9.955	7.971	9
10	1.2801	0.7812	0.08926	0.11426	11.203	8.752	10
11	1.3121	0.7621	0.08011	0.10511	12.483	9.514	11
12	1.3449	0.7436	0.07249	0.09749	13.796	10.258	12
13	1.3785	0.7254	0.06605	0.09105	15.140	10.983	13
14	1.4130	0.7077	0.06054	0.08554	16.519	11.691	14
15	1.4483	0.6905	0.05577	0.08077	17.932	12.381	15
16	1.4845	0.6736	0.05160	0.07660	19.380	13.055	16
17	1.5216	0.6572	0.04793	0.07293	20.865	13.712	17
18	1.5597	0.6412	0.04467	0.06967	22.386	14.353	18
19	1.5987	0.6255	0.04176	0.06676	23.946	14.979	19
20	1.6386	0.6103	0.03915	0.06415	25.545	15.589	20
21	1.6796	0.5954	0.03679	0.06179	27.183	16.185	21
22	1.7216	0.5809	0.03465	0.05965	28.863	16.765	22
23	1.7646	0.5667	0.03270	0.05770	30.584	17.332	23
24	1.8087	0.5529	0.03091	0.05591	32.349	17.885	24
25	1.8539	0.5394	0.02928	0.05428	34.158	18.424	25
26	1.9003	0.5262	0.02777	0.05277	36.012	18.951	26
27	1.9478	0.5134	0.02638	0.05138	37.912	19.464	27
28	1.9965	0.5009	0.02509	0.05009	39.860	19.965	28
29	2.0464	0.4887	0.02389	0.04889	41.856	20.454	29
30	2.0976	0.4767	0.02278	0.04778	43.903	20.930	30
31	2.1500	0.4651	0.02174	0.04674	46.000	21.395	31
32	2.2038	0.4538	0.02077	0.04577	48.150	21.849	32
33	2.2589	0.4427	0.01986	0.04486	50.354	22.292	33
34	2.3153	0.4319	0.01901	0.04401	52.613	22.724	34
35	2.3732	0.4214	0.01821	0.04321	54.928	23.145	35
40	2.6851	0.3724	0.01484	0.03984	67.403	25.103	40
45	3.0379	0.3292	0.01227	0.03727	81.516	26.883	45
50	3.4371	0.2909	0.01026	0.03526	97.484	28.362	50
55	3.8888	0.2572	0.00865	0.03365	115.551	29.714	55
60	4.3998	0.2273	0.00735	0.03235	135.992	30.909	60
65	4.9780	0.2009	0.00628	0.03128	159.118	31.965	65
70	5.6321	0.1776	0.00540	0.03040	185.284	32.898	70
75	6.3722	0.1569	0.00465	0.02965	214.888	33.723	75
80	7.2100	0.1387	0.00403	0.02903	248.383	34.452	80
85	8.1570	0.1226	0.00349	0.02849	286.279	35.096	85

TABLE B. 1 *(Continued)*

$2\frac{1}{2}\%$

| | Single payment | | Uniform series | | | | |
| | Compound amount factor | Present worth factor | Sinking fund factor | Capital recovery factor | Compound amount factor | Present worth factor | |
n	$P \rightarrow F$	$F \rightarrow P$	$F \rightarrow A$	$P \rightarrow A$	$A \rightarrow F$	$A \rightarrow P$	n
90	9.2289	0.1084	0.00304	0.02804	329.154	35.666	90
95	10.4416	0.0958	0.00265	0.02765	377.664	36.169	95
100	11.8137	0.0846	0.00231	0.02731	432.549	36.614	100

3%

| | Single payment | | Uniform series | | | | |
| | Compound amount factor | Present worth factor | Sinking fund factor | Capital recovery factor | Compound amount factor | Present worth factor | |
n	$P \rightarrow F$	$F \rightarrow P$	$F \rightarrow A$	$P \rightarrow A$	$A \rightarrow F$	$A \rightarrow P$	n
1	1.0300	0.9709	1.00000	1.03000	1.000	0.971	1
2	1.0609	0.9426	0.49261	0.52261	2.030	1.913	2
3	1.0927	0.9151	0.32353	0.35353	3.091	2.829	3
4	1.1255	0.8885	0.23903	0.26903	4.184	3.717	4
5	1.1593	0.8626	0.18835	0.21835	5.309	4.580	5
6	1.1941	0.8375	0.15460	0.18460	6.468	5.417	6
7	1.2299	0.8131	0.13051	0.16051	7.662	6.230	7
8	1.2668	0.7894	0.11246	0.14246	8.892	7.020	8
9	1.3048	0.7664	0.09843	0.12843	10.159	7.786	9
10	1.3439	0.7441	0.08723	0.11723	11.464	8.530	10
11	1.3842	0.7224	0.07808	0.10808	12.808	9.253	11
12	1.4258	0.7014	0.07046	0.10046	14.192	9.954	12
13	1.4685	0.6810	0.06403	0.09403	15.618	10.635	13
14	1.5126	0.6611	0.05853	0.08853	17.086	11.296	14
15	1.5580	0.6419	0.05377	0.08377	18.599	11.938	15
16	1.6047	0.6232	0.04961	0.07961	20.157	12.561	16
17	1.6528	0.6050	0.04595	0.07595	21.762	13.166	17
18	1.7024	0.5874	0.04271	0.07271	23.414	13.754	18
19	1.7535	0.5703	0.03981	0.06981	25.117	14.324	29
20	1.8061	0.5537	0.03722	0.06722	26.870	14.877	20
21	1.8603	0.5375	0.03487	0.06487	28.676	15.415	21
22	1.9161	0.5219	0.03275	0.06275	30.537	15.937	22
23	1.9736	0.5067	0.03081	0.06081	32.453	16.444	23
24	2.0238	0.4919	0.02905	0.05905	34.426	16.936	24
25	2.0938	0.4776	0.02743	0.05743	36.459	17.413	25
26	2.1566	0.4637	0.02594	0.05594	38.553	17.877	26
27	2.2213	0.4502	0.02456	0.05456	40.710	18.327	27
28	2.2879	0.4371	0.02329	0.05329	42.931	18.764	28
29	2.3566	0.4243	0.02211	0.05211	45.219	19.188	29
30	2.4273	0.4120	0.02102	0.05102	47.575	19.600	30

TABLE B. 1 *(Continued)*

3%

| | Single payment | | Uniform series | | | | |
| | Compound amount factor | Present worth factor | Sinking fund factor | Capital recovery factor | Compound amount factor | Present worth factor | |
n	$P \rightarrow F$	$F \rightarrow P$	$F \rightarrow A$	$P \rightarrow A$	$A \rightarrow F$	$A \rightarrow P$	n
31	2.5001	0.4000	0.02000	0.05000	50.003	20.000	31
32	2.5751	0.3883	0.01905	0.04905	52.503	20.389	32
33	2.6523	0.3770	0.01816	0.04816	55.078	20.766	33
34	2.7319	0.3660	0.01732	0.04732	57.730	21.132	34
35	2.8139	0.3554	0.01654	0.04654	60.462	21.487	35
40	3.2620	0.3066	0.01326	0.04326	75.401	23.115	40
45	3.7816	0.2644	0.01079	0.04079	92.720	24.519	45
50	4.3839	0.2281	0.00887	0.03887	112.797	25.730	50
55	5.0821	0.1968	0.00735	0.03735	136.072	26.774	55
60	5.8916	0.1697	0.00613	0.03613	163.053	27.676	60
65	6.8300	0.1464	0.00515	0.03515	194.333	28.453	65
70	7.9178	0.1283	0.00434	0.03434	230.594	29.123	70
75	9.1789	0.1089	0.00367	0.03367	272.631	29.702	75
80	10.6409	0.0940	0.00311	0.03311	321.363	30.201	80
85	12.3357	0.0811	0.00265	0.03265	377.857	30.631	85
90	14.3005	0.0699	0.00226	0.03226	443.349	31.002	90
95	16.5782	0.0603	0.00193	0.03193	519.272	31.323	95
100	19.2186	0.0520	0.00165	0.03165	607.288	31.599	100

$3\frac{1}{2}\%$

| | Single payment | | Uniform series | | | | |
| | Compound amount factor | Present worth factor | Sinking fund factor | Capital recovery factor | Compound amount factor | Present worth factor | |
n	$P \rightarrow F$	$F \rightarrow P$	$F \rightarrow A$	$P \rightarrow A$	$A \rightarrow F$	$A \rightarrow P$	n
1	1.0350	0.9662	1.00000	1.03500	1.000	0.966	1
2	1.0712	0.9335	0.49140	0.52640	2.035	1.900	2
3	1.1087	0.9019	0.32193	0.35693	3.106	2.802	3
4	1.1475	0.8714	0.23725	0.27225	4.215	3.673	4
5	1.1877	0.8420	0.18648	0.22148	5.362	4.515	5
6	1.2293	0.8135	0.15267	0.18767	6.550	5.329	6
7	1.2723	0.7860	0.12854	0.16354	7.779	6.115	7
8	1.3168	0.7594	0.11048	0.14548	9.052	6.874	8
9	1.3629	0.7337	0.09645	0.13145	10.368	7.608	9
10	1.4106	0.7089	0.08524	0.12024	11.731	8.317	10
11	1.4600	0.6849	0.07609	0.11109	13.142	9.002	11
12	1.5111	0.6618	0.06848	0.10348	14.602	9.663	12
13	1.5640	0.6394	0.06206	0.09706	16.113	10.303	13
14	1.6187	0.6178	0.05657	0.09157	17.677	10.921	14
15	1.6753	0.5969	0.05183	0.08683	19.296	11.517	15

TABLE B. 1 *(Continued)*

$3\frac{1}{2}\%$

| | Single payment | | Uniform series | | | | |
| | Compound amount factor | Present worth factor | Sinking fund factor | Capital recovery factor | Compound amount factor | Present worth factor | |
n	$P \rightarrow F$	$F \rightarrow P$	$F \rightarrow A$	$P \rightarrow A$	$A \rightarrow F$	$A \rightarrow P$	n
16	1.7340	0.5767	0.04768	0.08268	20.971	12.094	16
17	1.7947	0.5572	0.04404	0.07904	22.705	12.651	17
18	1.8575	0.5384	0.04082	0.07582	24.500	13.190	18
19	1.9225	0.5202	0.03794	0.07294	26.357	13.710	19
20	1.9898	0.5026	0.03536	0.07036	28.280	14.212	20
21	2.0594	0.4856	0.03304	0.06804	30.269	14.698	21
22	2.1315	0.4692	0.03093	0.06593	32.329	15.167	22
23	2.2061	0.4533	0.02902	0.06402	34.460	15.620	23
24	2.2833	0.4380	0.02727	0.06227	36.667	16.058	24
25	2.3632	0.4231	0.02567	0.06067	38.950	16.482	25
26	2.4460	0.4088	0.02421	0.05921	41.313	16.890	26
27	2.5316	0.3950	0.02285	0.05785	43.759	17.285	27
28	2.6202	0.3817	0.02160	0.05660	46.291	17.667	28
29	2.7119	0.3687	0.02045	0.05545	48.911	18.036	29
30	2.8068	0.3563	0.01937	0.05437	51.623	18.392	30
31	2.9050	0.3442	0.01837	0.05337	54.429	18.736	31
32	3.0067	0.3326	0.01744	0.05244	57.335	19.069	32
33	3.1119	0.3213	0.01657	0.05157	60.341	19.390	33
34	3.2209	0.3105	0.01576	0.05076	63.453	19.701	34
35	3.3336	0.3000	0.01500	0.05000	66.674	20.001	35
40	3.9593	0.2526	0.01183	0.04683	84.550	21.355	40
45	4.7024	0.2127	0.00945	0.04445	106.782	22.495	45
50	5.5849	0.1791	0.00763	0.04263	130.998	23.456	50
55	6.6331	0.1508	0.00621	0.04121	160.947	24.264	55
60	7.8781	0.1269	0.00509	0.04009	196.517	24.945	60
65	9.3567	0.1069	0.00419	0.03919	238.763	25.518	65
70	11.1128	0.0900	0.00346	0.03846	288.938	26.000	70
75	13.1986	0.0758	0.00287	0.03787	348.530	26.407	75
80	15.6757	0.0638	0.00238	0.03738	419.307	26.749	80
85	18.6179	0.0537	0.00199	0.03699	503.367	27.037	85
90	22.1122	0.0452	0.00166	0.03666	603.205	27.279	90
95	26.2623	0.0381	0.00139	0.03639	721.781	27.484	95
100	31.1914	0.0321	0.00116	0.03616	862.612	27.655	100

4%

| | Single payment | | Uniform series | | | | |
| | Compound amount factor | Present worth factor | Sinking fund factor | Capital recovery factor | Compound amount factor | Present worth factor | |
n	$P \rightarrow F$	$F \rightarrow P$	$F \rightarrow A$	$P \rightarrow A$	$A \rightarrow F$	$A \rightarrow P$	n
1	1.0400	0.9615	1.00000	1.04000	1.000	0.962	1
2	1.0816	0.9246	0.49020	0.53020	2.040	1.886	2

TABLE B. 1 (Continued)

4%

	Single payment		Uniform series				
	Compound amount factor	Present worth factor	Sinking fund factor	Capital recovery factor	Compound amount factor	Present worth factor	
n	$P \rightarrow F$	$F \rightarrow P$	$F \rightarrow A$	$P \rightarrow A$	$A \rightarrow F$	$A \rightarrow P$	n
3	1.1249	0.8890	0.32035	0.36035	3.122	2.775	3
4	1.1699	0.8548	0.23549	0.27549	4.246	3.630	4
5	1.2167	0.8219	0.18463	0.22463	5.416	4.452	5
6	1.2653	0.7903	0.15076	0.19076	6.633	5.242	6
7	1.3159	0.7599	0.12661	0.16661	7.898	6.002	7
8	1.3686	0.7307	0.10853	0.14853	9.214	6.733	8
9	1.4233	0.7026	0.09449	0.13449	10.583	7.435	9
10	1.4802	0.6756	0.08329	0.12329	12.006	8.111	10
11	1.5395	0.6496	0.07415	0.11415	13.486	8.760	11
12	1.6010	0.6246	0.06655	0.10655	15.026	9.385	12
13	1.6651	0.6006	0.06014	0.10014	16.627	9.986	13
14	1.7317	0.5775	0.05467	0.09467	18.292	10.563	14
15	1.8009	0.5553	0.04994	0.08994	20.024	11.118	15
16	1.8730	0.5339	0.04582	0.08582	21.825	11.652	16
17	1.9479	0.5134	0.04220	0.08220	23.698	12.166	17
18	2.0258	0.4936	0.03899	0.07899	25.645	12.659	18
19	2.1068	0.4746	0.03614	0.07614	27.671	13.134	19
20	2.1911	0.4564	0.03358	0.07358	29.778	13.590	20
21	2.2788	0.4388	0.03128	0.07128	31.969	14.029	21
22	2.3699	0.4220	0.02920	0.06920	34.248	14.451	22
23	2.4647	0.4057	0.02731	0.06731	36.618	14.857	23
24	2.5633	0.3901	0.02559	0.06559	39.083	15.247	24
25	2.6658	0.3751	0.02401	0.06401	41.646	15.622	25
26	2.7725	0.3607	0.02257	0.06257	44.312	15.983	26
27	2.8834	0.3468	0.02124	0.06124	47.084	16.330	27
28	2.9987	0.3335	0.02001	0.06001	49.968	16.663	28
29	3.1187	0.3207	0.01888	0.05888	52.966	16.984	29
30	3.2434	0.3083	0.01783	0.05783	56.085	17.292	30
31	3.3731	0.2965	0.01686	0.05686	59.328	17.588	31
32	3.5081	0.2851	0.01595	0.05595	62.701	17.874	32
33	3.6484	0.2741	0.01510	0.05510	66.210	18.148	33
34	3.7943	0.2636	0.01431	0.05431	69.858	18.411	34
35	3.9461	0.2534	0.01358	0.05358	73.652	18.665	35
40	4.8010	0.2083	0.01052	0.05052	95.026	19.793	40
45	5.8412	0.1712	0.00826	0.04826	121.029	20.720	45
50	7.1067	0.1407	0.00655	0.04655	152.667	21.482	50
55	8.6464	0.1157	0.00523	0.04523	191.159	22.109	55
60	10.5196	0.0951	0.00420	0.04420	237.991	22.623	60
65	12.7987	0.0781	0.00339	0.04339	294.968	23.047	65
70	15.5716	0.0642	0.00275	0.04275	364.290	23.395	70
75	18.9453	0.0528	0.00223	0.04223	448.631	23.680	75
80	23.0500	0.0434	0.00181	0.04181	551.245	23.915	80
85	28.0436	0.0357	0.00148	0.04148	676.090	24.109	85

TABLE B. 1 *(Continued)*

4%

	Single payment		Uniform series				
	Compound amount factor	Present worth factor	Sinking fund factor	Capital recovery factor	Compound amount factor	Present worth factor	
n	$P \rightarrow F$	$F \rightarrow P$	$F \rightarrow A$	$P \rightarrow A$	$A \rightarrow F$	$A \rightarrow P$	n
90	34.1193	0.0293	0.00121	0.04121	827.983	24.267	90
95	41.5114	0.0241	0.00099	0.04099	1,012.785	24.398	95
100	50.5049	0.0198	0.00081	0.04081	1,237.624	24.505	100

$4\frac{1}{2}\%$

	Single payment		Uniform series				
	Compound amount factor	Present worth factor	Sinking fund factor	Capital recovery factor	Compound amount factor	Present worth factor	
n	$P \rightarrow F$	$F \rightarrow P$	$F \rightarrow A$	$P \rightarrow A$	$A \rightarrow F$	$A \rightarrow P$	n
1	1.0450	0.9569	1.00000	1.04500	1.000	0.957	1
2	1.0920	0.9157	0.48900	0.53400	2.045	1.873	2
3	1.1412	0.8763	0.31877	0.36377	3.137	2.749	3
4	1.1925	0.8386	0.23374	0.27874	4.278	3.588	4
5	1.2462	0.8025	0.18279	0.22779	5.471	4.390	5
6	1.3023	0.7679	0.14888	0.19388	6.717	5.158	6
7	1.3609	0.7348	0.12470	0.16970	8.019	5.893	7
8	1.4221	0.7032	0.10661	0.15161	9.380	6.596	8
9	1.4861	0.6729	0.09257	0.13757	10.802	7.269	9
10	1.5530	0.6439	0.08138	0.12638	12.288	7.913	10
11	1.6229	0.6162	0.07225	0.11725	13.841	8.529	11
12	1.6959	0.5897	0.06467	0.10967	15.464	9.119	12
13	1.7722	0.5643	0.05828	0.10328	17.160	9.683	13
14	1.8519	0.5400	0.05282	0.09782	18.932	10.223	14
15	1.9353	0.5167	0.04811	0.09311	20.784	10.740	15
16	2.0224	0.4945	0.04402	0.08902	22.719	11.234	16
17	2.1134	0.4732	0.04042	0.08542	24.742	11.707	17
18	2.2085	0.4528	0.03724	0.08224	26.855	12.160	18
19	2.3079	0.4333	0.03441	0.07941	29.064	12.593	19
20	2.4117	0.4146	0.03188	0.07688	31.371	13.008	20
21	2.5202	0.3968	0.02960	0.07460	33.783	13.405	21
22	2.6337	0.3797	0.02755	0.07255	36.303	13.784	22
23	2.7522	0.3634	0.02568	0.07068	38.937	14.148	23
24	2.8760	0.3477	0.02399	0.06899	41.689	14.495	24
25	3.0054	0.3327	0.02244	0.06744	44.565	14.828	25
26	3.1407	0.3184	0.02102	0.06602	47.571	15.147	26
27	3.2820	0.3047	0.01972	0.06472	50.711	15.451	27
28	3.4397	0.2916	0.01852	0.06352	53.993	15.743	28
29	3.5840	0.2790	0.01741	0.06241	57.423	16.022	29
30	3.7453	0.2670	0.01639	0.06139	61.007	16.289	30

499

TABLE B. 1 *(Continued)*

$4\frac{1}{2}\%$

	Single payment		Uniform series				
	Compound amount factor	Present worth factor	Sinking fund factor	Capital recovery factor	Compound amount factor	Present worth factor	
n	$P \rightarrow F$	$F \rightarrow P$	$F \rightarrow A$	$P \rightarrow A$	$A \rightarrow F$	$A \rightarrow P$	n
31	3.9139	0.2555	0.01544	0.06044	64.752	16.544	31
32	4.0900	0.2445	0.01456	0.05956	68.666	16.789	32
33	4.2740	0.2340	0.01374	0.05874	72.756	17.023	33
34	4.4664	0.2239	0.01298	0.05798	77.030	17.247	34
35	4.6673	0.2143	0.01277	0.05727	81.497	17.461	35
40	5.8164	0.1719	0.00934	0.05434	107.030	18.402	40
45	7.2482	0.1380	0.00720	0.05220	138.850	19.156	45
50	9.0326	0.1107	0.00560	0.05060	178.503	19.762	50
55	11.2563	0.0888	0.00439	0.04939	227.918	20.248	55
60	14.0274	0.0713	0.00345	0.04845	289.498	20.638	60
65	17.4807	0.0572	0.00273	0.04773	366.238	20.951	65
70	21.7841	0.0459	0.00217	0.04717	461.870	21.202	70
75	27.1470	0.0368	0.00172	0.04672	581.044	21.404	75
80	33.8301	0.0296	0.00137	0.04637	729.558	21.565	80
85	42.1585	0.0237	0.00109	0.04609	914.632	21.695	85
90	52.5371	0.0190	0.00087	0.04587	1,145.269	21.799	90
95	65.4708	0.0153	0.00070	0.04570	1,432.684	21.883	95
100	81.5885	0.0123	0.00056	0.04556	1,790.856	21.950	100

5%

	Single payment		Uniform series				
	Compound amount factor	Present worth factor	Sinking fund factor	Capital recovery factor	Compound amount factor	Present worth factor	
n	$P \rightarrow F$	$F \rightarrow P$	$F \rightarrow A$	$P \rightarrow A$	$A \rightarrow F$	$A \rightarrow P$	n
1	1.0500	0.9524	0.00000	1.05000	1.000	0.952	1
2	1.1025	0.9070	0.48780	0.53780	2.050	1.859	2
3	1.1576	0.8638	0.31721	0.36721	3.153	2.723	3
4	1.2155	0.8227	0.23201	0.28201	4.310	3.546	4
5	1.2763	0.7835	0.18097	0.23097	5.526	4.329	5
6	1.3401	0.7462	0.14702	0.19702	6.802	5.076	6
7	1.4071	0.7107	0.12282	0.17282	8.142	5.786	7
8	1.4775	0.6768	0.10472	0.15472	9.549	6.463	8
9	1.5513	0.6446	0.09069	0.14069	11.027	7.108	9
10	1.6289	0.6139	0.07950	0.12950	12.578	7.722	10
11	1.7103	0.5847	0.07039	0.12039	14.207	8.306	11
12	1.7959	0.5568	0.06283	0.11283	15.917	8.863	12
13	1.8856	0.5303	0.05646	0.10646	17.713	9.394	13
14	1.9800	0.5051	0.05102	0.10102	19.599	9.899	14
15	2.0789	0.4810	0.04634	0.09634	21.579	10.380	15

TABLE B. 1 *(Continued)*

5%

	Single payment		Uniform series				
	Compound amount factor	Present worth factor	Sinking fund factor	Capital recovery factor	Compound amount factor	Present worth factor	
n	$P \to F$	$F \to P$	$F \to A$	$P \to A$	$A \to F$	$A \to P$	n
16	2.1829	0.4581	0.04227	0.09227	23.657	10.838	16
17	2.2920	0.4363	0.03870	0.08870	25.840	11.274	17
18	2.4066	0.4155	0.03555	0.08555	28.132	11.690	18
19	2.5270	0.3957	0.03275	0.08275	30.539	12.085	19
20	2.6533	0.3769	0.03024	0.08024	33.066	12.462	20
21	2.7860	0.3589	0.02800	0.07800	35.719	12.821	21
22	2.9253	0.3418	0.02597	0.07597	38.505	13.163	22
23	3.0715	0.3256	0.02414	0.07414	41.430	13.489	23
24	3.2251	0.3101	0.02247	0.07247	44.502	13.799	24
25	3.3864	0.2953	0.02095	0.07095	47.727	14.094	25
26	3.5557	0.2812	0.01956	0.06956	51.113	14.375	26
27	3.7335	0.2678	0.01829	0.06829	54.669	14.643	27
28	3.9201	0.2551	0.01712	0.06712	58.403	14.898	28
29	4.1161	0.2429	0.01605	0.06605	62.323	15.141	29
30	4.3219	0.2314	0.01505	0.06505	66.439	15.372	30
31	4.5380	0.2204	0.01413	0.06413	70.761	15.593	31
32	4.7649	0.2099	0.01328	0.06328	75.299	15.803	32
33	5.0032	0.1999	0.01249	0.06249	80.064	16.003	33
34	5.2533	0.1904	0.01176	0.06176	85.067	16.193	34
35	5.5160	0.1813	0.01107	0.06107	90.320	16.374	35
40	7.0400	0.1420	0.00828	0.05828	120.800	17.159	40
45	8.9850	0.1113	0.00626	0.05626	159.700	17.774	45
50	11.4674	0.0872	0.00478	0.05478	209.348	18.256	50
55	14.6356	0.0683	0.00367	0.05367	272.713	18.633	55
60	18.6792	0.0535	0.00283	0.05283	353.584	18.929	60
65	23.8399	0.0419	0.00219	0.05219	456.798	19.161	65
70	30.4264	0.0329	0.00170	0.05170	588.529	19.343	70
75	38.8327	0.0258	0.00132	0.05132	756.654	19.485	75
80	49.5614	0.0202	0.00103	0.05103	971.229	19.596	80
85	63.2544	0.0158	0.00080	0.05080	1,245.087	19.684	85
90	80.7304	0.0124	0.00063	0.05063	1,594.607	19.752	90
95	103.0357	0.0097	0.00049	0.05049	2,040.694	19.806	95
100	131.5013	0.0076	0.00038	0.05038	2,610.025	19.848	100

$5\frac{1}{2}\%$

	Single payment		Uniform series				
	Compound amount factor	Present worth factor	Sinking fund factor	Capital recovery factor	Compound amount factor	Present worth factor	
n	$P \to F$	$F \to P$	$F \to A$	$P \to A$	$A \to F$	$A \to P$	n
1	1.0550	0.9479	1.00000	1.05500	1.000	0.948	1
2	1.1130	0.8985	0.48662	0.54162	2.055	1.846	2

TABLE B. 1 *(Continued)*

$5\frac{1}{2}\%$

	Single payment		Uniform series				
	Compound amount factor	Present worth factor	Sinking fund factor	Capital recovery factor	Compound amount factor	Present worth factor	
n	$P \rightarrow F$	$F \rightarrow P$	$F \rightarrow A$	$P \rightarrow A$	$A \rightarrow F$	$A \rightarrow P$	n
3	1.1742	0.8516	0.31565	0.37065	3.168	2.698	3
4	1.2388	0.8072	0.23029	0.28529	4.342	3.505	4
5	1.3070	0.7651	0.17918	0.23418	5.581	4.270	5
6	1.3788	0.7252	0.14518	0.20018	6.888	4.996	6
7	1.4547	0.6874	0.12096	0.17596	8.267	5.683	7
8	1.5347	0.6516	0.10286	0.15786	9.722	6.335	8
9	1.6191	0.6176	0.08884	0.14384	11.256	6.952	9
10	1.7081	0.5854	0.07767	0.13267	12.875	7.538	10
11	1.8021	0.5549	0.06857	0.12357	14.583	8.093	11
12	1.9012	0.5260	0.06103	0.11603	16.386	8.619	12
13	2.0058	0.4986	0.05468	0.10968	18.287	9.117	13
14	2.1161	0.4726	0.04928	0.10428	20.293	9.590	14
15	2.2325	0.4479	0.04463	0.09963	22.409	10.038	15
16	2.3553	0.4246	0.04058	0.09558	24.641	10.462	16
17	2.4848	0.4024	0.03704	0.09204	26.996	10.865	17
18	2.6215	0.3815	0.03392	0.08892	29.481	11.246	18
19	2.7656	0.3616	0.03115	0.08615	32.103	11.608	19
20	2.9178	0.3427	0.02868	0.08368	34.868	11.950	20
21	3.0782	0.3249	0.02646	0.08146	37.786	12.275	21
22	3.2475	0.3079	0.02447	0.07947	40.864	12.583	22
23	3.4262	0.2919	0.02267	0.07767	44.112	12.875	23
24	3.6146	0.2767	0.02104	0.07604	47.538	13.152	24
25	3.8134	0.2622	0.01955	0.07455	51.153	13.414	25
26	4.0231	0.2486	0.01819	0.07319	54.966	13.662	26
27	4.2444	0.2356	0.01695	0.07195	58.989	13.898	27
28	4.4778	0.2233	0.01581	0.07081	63.234	14.121	28
29	4.7241	0.2117	0.01477	0.06977	67.711	14.333	29
30	4.9840	0.2006	0.01381	0.06881	72.435	14.534	30
31	5.2581	0.1902	0.01292	0.06792	77.419	14.724	31
32	5.5473	0.1803	0.01210	0.06710	82.677	14.904	32
33	5.8524	0.1709	0.01133	0.06633	88.225	15.075	33
34	6.1742	0.1620	0.01063	0.06563	94.077	15.237	34
35	6.5138	0.1535	0.00997	0.06497	100.251	15.391	35
40	8.5133	0.1175	0.00732	0.06232	136.606	16.046	40
45	11.1266	0.0899	0.00543	0.06043	184.119	16.548	45
50	14.5420	0.0688	0.00406	0.05906	246.217	16.932	50
55	19.0058	0.0526	0.00305	0.05805	327.377	17.225	55
60	24.8398	0.0403	0.00231	0.05731	433.450	17.450	60
65	32.4646	0.0308	0.00175	0.05675	572.083	17.622	65
70	42.4299	0.0236	0.00133	0.05633	753.271	17.753	70
75	55.4542	0.0180	0.00101	0.05601	990.076	17.854	75
80	72.4764	0.0138	0.00077	0.05577	1,299.571	17.931	80
85	94.7238	0.0106	0.00059	0.05559	1,704.069	17.990	85

TABLE B. 1 *(Continued)*

$5\frac{1}{2}\%$

	Single payment		Uniform series				
	Compound amount factor	Present worth factor	Sinking fund factor	Capital recovery factor	Compound amount factor	Present worth factor	
n	$P \to F$	$F \to P$	$F \to A$	$P \to A$	$A \to F$	$A \to P$	n
90	123.8002	0.0081	0.00045	0.05545	2,232.731	18.035	90
95	161.8019	0.0062	0.00034	0.05334	2,923.671	18.069	95
100	211.4686	0.0047	0.00026	0.05526	3,826.702	18.096	100

6%

	Single payment		Uniform series				
	Compound amount factor	Present worth factor	Sinking fund factor	Capital recovery factor	Compound amount factor	Present worth factor	
n	$P \to F$	$F \to P$	$F \to A$	$P \to A$	$A \to F$	$A \to P$	n
1	1.0600	0.9434	1.00000	1.06000	1.000	0.943	1
2	1.1236	0.8900	0.48544	0.54544	2.060	1.833	2
3	1.1910	0.8396	0.31411	0.37411	3.184	2.673	3
4	1.2625	0.7921	0.22859	0.28859	4.375	3.465	4
5	1.3382	0.7473	0.17740	0.23740	5.637	4.212	5
6	1.4185	0.7050	0.14336	0.20336	6.975	4.917	6
7	1.5036	0.6651	0.11914	0.17914	8.394	5.582	7
8	1.5938	0.6274	0.10104	0.16104	9.897	6.210	8
9	1.6895	0.5919	0.08702	0.14702	11.491	6.802	9
10	1.7908	0.5584	0.07587	0.13587	13.181	7.360	10
11	1.8983	0.5268	0.06679	0.12679	14.972	7.887	11
12	2.0122	0.4970	0.05928	0.11928	16.870	8.384	12
13	2.1329	0.4688	0.05296	0.11296	18.882	8.853	13
14	2.2609	0.4423	0.04758	0.10758	21.015	9.295	14
15	2.3966	0.4173	0.04296	0.10296	23.276	9.712	15
16	2.5404	0.3936	0.03895	0.09895	25.673	10.106	16
17	2.6928	0.3714	0.03544	0.09544	28.213	10.477	17
18	2.8543	0.3503	0.03236	0.09236	30.906	10.828	18
19	3.0256	0.3305	0.02962	0.08962	33.760	11.158	19
20	3.2071	0.3118	0.02718	0.08718	36.786	11.470	20
21	3.3996	0.2942	0.02500	0.08500	39.993	11.764	21
22	3.6035	0.2775	0.02305	0.08305	43.392	12.042	22
23	3.8197	0.2618	0.02128	0.08128	46.996	12.303	23
24	4.0489	0.2470	0.01968	0.07968	50.816	12.550	24
25	4.2919	0.2330	0.01823	0.07823	54.865	12.783	25
26	4.5494	0.2198	0.01690	0.07690	59.156	13.003	26
27	4.8223	0.2074	0.01570	0.07570	63.706	13.211	27
28	5.1117	0.1956	0.01459	0.07459	68.528	13.406	28
29	5.4184	0.1846	0.01358	0.07358	73.640	13.591	29
30	5.7435	0.1741	0.01265	0.07265	79.058	13.765	30

TABLE B. 1 *(Continued)*

	6%						
	Single payment		Uniform series				
	Compound amount factor	Present worth factor	Sinking fund factor	Capital recovery factor	Compound amount factor	Present worth factor	
n	$P \rightarrow F$	$F \rightarrow P$	$F \rightarrow A$	$P \rightarrow A$	$A \rightarrow F$	$A \rightarrow P$	n
31	6.0881	0.1643	0.01179	0.07179	84.802	13.929	31
32	6.4534	0.1550	0.01100	0.07100	90.890	14.084	32
33	6.8406	0.1462	0.01027	0.07027	97.343	14.230	33
34	7.2510	0.1379	0.00960	0.06960	104.184	14.368	34
35	7.6861	0.1301	0.00897	0.06897	111.435	14.498	35
40	10.2857	0.0972	0.00646	0.06646	154.762	15.046	40
45	13.7646	0.0727	0.00470	0.06470	212.744	15.456	45
50	18.4202	0.0543	0.00344	0.06344	290.336	15.762	50
55	24.6503	0.0406	0.00254	0.06254	394.172	15.991	55
60	32.9877	0.0303	0.00188	0.06188	533.128	16.161	60
65	44.1450	0.0227	0.00139	0.06139	719.083	16.289	65
70	59.0759	0.0169	0.00103	0.06103	967.932	16.385	70
75	79.0569	0.0126	0.00077	0.06077	1,300.949	16.456	75
80	105.7960	0.0095	0.00057	0.06057	1,746.600	16.509	80
85	141.5789	0.0071	0.00043	0.06043	2,342.982	16.549	85
90	189.4645	0.0053	0.00032	0.06032	3,141.075	16.579	90
95	253.5463	0.0039	0.00024	0.06024	4,209.104	16.601	95
100	339.3021	0.0029	0.00018	0.06018	5,638.368	16.618	100

	7%						
	Single payment		Uniform series				
	Compound amount factor	Present worth factor	Sinking fund factor	Capital recovery factor	Compound amount factor	Present worth factor	
n	$P \rightarrow F$	$F \rightarrow P$	$F \rightarrow A$	$P \rightarrow A$	$A \rightarrow F$	$A \rightarrow P$	n
1	1.0700	0.9346	1.00000	1.07000	1.000	0.935	1
2	1.1449	0.8734	0.48309	0.55309	2.070	1.808	2
3	1.2250	0.8163	0.31105	0.38105	3.215	2.624	3
4	1.3108	0.7629	0.22523	0.29523	4.440	3.387	4
5	1.4026	0.7130	0.17389	0.24389	5.751	4.100	5
6	1.5007	0.6663	0.13980	0.20980	7.153	4.767	6
7	1.6058	0.6227	0.11555	0.18555	8.654	5.389	7
8	1.7182	0.5820	0.09747	0.16747	10.260	5.971	8
9	1.8385	0.5439	0.08349	0.15349	11.978	6.515	9
10	1.9672	0.5083	0.07238	0.14238	13.816	7.024	10
11	2.1049	0.4751	0.06336	0.13336	15.784	7.499	11
12	2.2522	0.4440	0.05590	0.12590	17.888	7.943	12
13	2.4098	0.4150	0.04965	0.11965	20.141	8.358	13
14	2.5785	0.3878	0.04434	0.11434	22.550	8.745	14
15	2.7590	0.3624	0.03979	0.10979	25.129	9.108	15

TABLE B. 1 *(Continued)*

	7%						
	Single payment		Uniform series				
n	Compound amount factor $P \rightarrow F$	Present worth factor $F \rightarrow P$	Sinking fund factor $F \rightarrow A$	Capital recovery factor $P \rightarrow A$	Compound amount factor $A \rightarrow F$	Present worth factor $A \rightarrow P$	n
16	2.9522	0.3387	0.03586	0.10586	27.888	9.447	16
17	3.1588	0.3166	0.03243	0.10243	30.840	9.763	17
18	3.3799	0.2959	0.02941	0.09941	33.999	10.059	18
19	3.6165	0.2765	0.02675	0.09675	37.379	10.336	19
20	3.8697	0.2584	0.02439	0.09439	40.995	10.594	20
21	4.1406	0.2415	0.02229	0.09229	44.865	10.836	21
22	4.4304	0.2257	0.02041	0.09041	49.006	11.061	22
23	4.7405	0.2109	0.01871	0.08871	53.436	11.272	23
24	5.0724	0.1971	0.01719	0.08719	58.177	11.469	24
25	5.4274	0.1842	0.01581	0.08581	63.249	11.654	25
26	5.8074	0.1722	0.01456	0.08456	68.676	11.826	26
27	6.2139	0.1609	0.01343	0.08343	74.484	11.987	27
28	6.6488	0.1504	0.01239	0.08239	80.698	12.137	28
29	7.1143	0.1406	0.01145	0.08145	87.347	12.278	29
30	7.6123	0.1314	0.01059	0.08059	94.461	12.409	30
31	8.1451	0.1228	0.00980	0.07980	102.073	12.532	31
32	8.7153	0.1147	0.00907	0.07907	110.218	12.647	32
33	9.3253	0.1072	0.00841	0.07841	118.933	12.754	33
34	9.9781	0.1002	0.00780	0.07780	128.259	12.854	34
35	10.6766	0.0937	0.00723	0.07723	138.237	12.948	35
40	14.9745	0.0668	0.00501	0.07501	199.635	13.332	40
45	21.0025	0.0476	0.00350	0.07350	285.749	13.606	45
50	29.4570	0.0339	0.00246	0.07246	406.529	13.801	50
55	41.3150	0.0242	0.00174	0.07174	575.929	13.940	55
60	57.9464	0.0173	0.00123	0.07123	813.520	14.039	60
65	81.2729	0.0123	0.00087	0.07087	1,146.755	14.110	65
70	113.9894	0.0088	0.00062	0.07062	1,614.134	14.160	70
75	159.8760	0.0063	0.00044	0.07044	2,269.657	14.196	75
80	224.2344	0.0045	0.00031	0.00031	3,189.063	14.222	80
85	314.5003	0.0032	0.00022	0.07022	4,478.576	14.240	85
90	441.1030	0.0023	0.00016	0.07016	6,287.185	14.253	90
95	618.6697	0.0016	0.00011	0.07011	8,823.854	14.263	95
100	867.7163	0.0012	0.00008	0.07008	12,381.662	14.269	100

	8%						
	Single payment		Uniform series				
n	Compound amount factor $P \rightarrow F$	Present worth factor $F \rightarrow P$	Sinking fund factor $F \rightarrow A$	Capital recovery factor $P \rightarrow A$	Compound amount factor $A \rightarrow F$	Present worth factor $A \rightarrow P$	n
1	1.0800	0.9259	1.00000	1.08000	1.000	0.926	1
2	1.1664	0.8573	0.48077	0.56077	2.080	1.783	2

TABLE B. 1 *(Continued)*

	8%						
	Single payment		Uniform series				
	Compound amount factor	Present worth factor	Sinking fund factor	Capital recovery factor	Compound amount factor	Present worth factor	
n	$P \rightarrow F$	$F \rightarrow P$	$F \rightarrow A$	$P \rightarrow A$	$A \rightarrow F$	$A \rightarrow P$	*n*
3	1.2597	0.7938	0.30803	0.38803	3.246	2.577	3
4	1.3605	0.7350	0.22192	0.30192	4.506	3.312	4
5	1.4693	0.6806	0.17046	0.25046	5.867	3.993	5
6	1.5869	0.6302	0.13632	0.21632	7.336	4.623	6
7	1.7138	0.5835	0.11207	0.19207	8.923	5.206	7
8	1.8509	0.5403	0.09401	0.17401	10.637	5.747	8
9	1.9990	0.5002	0.08008	0.16008	12.488	6.247	9
10	2.1589	0.4632	0.06903	0.14903	14.487	6.710	10
11	2.3316	0.4289	0.06008	0.14008	16.645	7.139	11
12	2.5182	0.3971	0.05270	0.13270	18.977	7.536	12
13	2.7196	0.3677	0.04652	0.12652	21.495	7.904	13
14	2.9372	0.3405	0.04130	0.12130	24.215	8.244	14
15	3.1722	0.3152	0.03683	0.11683	27.152	8.559	15
16	3.4259	0.2919	0.03298	0.11298	30.324	8.851	16
17	3.7000	0.2703	0.02963	0.10963	33.750	9.122	17
18	3.9960	0.2502	0.02670	0.10670	37.450	9.372	18
19	4.3157	0.2317	0.02413	0.10413	41.446	9.604	19
20	4.6610	0.2145	0.02185	0.10185	45.762	9.818	20
21	5.0338	0.1987	0.01983	0.09983	50.423	10.017	21
22	5.4365	0.1839	0.01803	0.09803	55.457	10.201	22
23	5.8715	0.1703	0.01642	0.09642	60.893	10.371	23
24	6.3412	0.1577	0.01498	0.09498	66.765	10.529	24
25	6.8485	0.1460	0.01368	0.09368	73.106	10.675	25
26	7.3964	0.1352	0.01251	0.09251	79.954	10.810	26
27	7.9881	0.1252	0.01145	0.09145	87.351	10.935	27
28	8.6271	0.1159	0.01049	0.09049	95.339	11.051	28
29	9.3173	0.1073	0.00962	0.08962	103.966	11.158	29
30	10.0627	0.0994	0.00883	0.08883	113.283	11.258	30
31	10.8677	0.0920	0.00811	0.08811	123.346	11.350	31
32	11.7371	0.0852	0.00745	0.08745	134.214	11.435	32
33	12.6760	0.0789	0.00685	0.08685	145.951	11.514	33
34	13.6901	0.0730	0.00630	0.08630	158.627	11.587	34
35	14.7853	0.0676	0.00580	0.08580	172.317	11.655	35
40	21.7245	0.0460	0.00386	0.08386	259.057	11.925	40
45	31.9204	0.0313	0.00259	0.08259	386.506	12.108	45
50	46.9016	0.0213	0.00174	0.08174	573.770	12.233	50
55	68.9139	0.0145	0.00118	0.08118	848.923	12.319	55
60	101.2571	0.0099	0.00080	0.08080	1,253.213	12.377	60
65	148.7798	0.0067	0.00054	0.08054	1,847.248	12.416	65
70	218.6064	0.0046	0.00037	0.08037	2,720.080	12.443	70
75	321.2045	0.0031	0.00025	0.08025	4,002.557	12.461	75
80	471.9548	0.0021	0.00017	0.08017	5,886.935	12.474	80
85	693.4565	0.0014	0.00012	0.08012	8,655.706	12.482	85

TABLE B. 1 *(Continued)*

8%

	Single payment		Uniform series				
	Compound amount factor	Present worth factor	Sinking fund factor	Capital recovery factor	Compound amount factor	Present worth factor	
n	$P \to F$	$F \to P$	$F \to A$	$P \to A$	$A \to F$	$A \to P$	n
90	1,018.9151	0.0010	0.00008	0.08008	12,723.939	12.488	90
95	1,497.1205	0.0007	0.00005	0.08005	18,701.507	12.492	95
100	2,199.7613	0.0005	0.00004	0.08004	27,484.516	12.494	100

10%

	Single payment		Uniform series				
	Compound amount factor	Present worth factor	Sinking fund factor	Capital recovery factor	Compound amount factor	Present worth factor	
n	$P \to F$	$F \to P$	$F \to A$	$P \to A$	$A \to F$	$A \to P$	n
1	1.1000	0.9091	1.00000	1.10000	1.000	0.909	1
2	1.2100	0.8264	0.47619	0.57619	2.100	1.736	2
3	1.3310	0.7513	0.30211	0.40211	3.310	2.487	3
4	1.4641	0.6830	0.21547	0.31547	4.641	3.170	4
5	1.6105	0.6209	0.16380	0.26380	6.105	3.791	5
6	1.7716	0.5645	0.12961	0.22961	7.716	4.355	6
7	1.9487	0.5132	0.10541	0.20541	9.487	4.868	7
8	2.1436	0.4665	0.08744	0.18744	11.436	5.335	8
9	2.3579	0.4241	0.07364	0.17364	13.579	5.759	9
10	2.5937	0.3855	0.06275	0.16275	15.937	6.144	10
11	2.8531	0.3505	0.05396	0.15396	18.531	6.495	11
12	3.1384	0.3186	0.04676	0.14676	21.384	6.814	12
13	3.4523	0.2897	0.04078	0.14078	24.523	7.103	13
14	3.7975	0.2633	0.03575	0.13575	27.975	7.367	14
15	4.1772	0.2394	0.03147	0.13147	31.772	7.606	15
16	4.5950	0.2176	0.02782	0.12782	35.950	7.824	16
17	5.0545	0.1978	0.02466	0.12466	40.545	8.022	17
18	5.5599	0.1799	0.02193	0.12193	45.599	8.201	18
19	6.1159	0.1635	0.01955	0.11955	51.159	8.365	19
20	6.7275	0.1486	0.01746	0.11746	57.275	8.514	20
21	7.4002	0.1351	0.01562	0.11562	64.002	8.649	21
22	8.1403	0.1228	0.01401	0.11401	71.403	8.772	22
23	8.9543	0.1117	0.01257	0.11257	79.543	8.883	23
24	9.8497	0.1015	0.01130	0.11130	88.497	8.985	24
25	10.8347	0.0923	0.01017	0.11017	98.347	9.077	25
26	11.9182	0.0839	0.00916	0.10916	109.182	9.161	26
27	13.1100	0.0763	0.00826	0.10826	121.100	9.237	27
28	14.4210	0.0693	0.00745	0.10745	134.210	9.307	28
29	15.8631	0.0630	0.00673	0.10673	148.631	9.370	29
30	17.4494	0.0573	0.00608	0.10608	164.494	9.427	30

TABLE B. 1 *(Continued)*

10%

	Single payment		Uniform series				
	Compound amount factor	Present worth factor	Sinking fund factor	Capital recovery factor	Compound amount factor	Present worth factor	
n	$P \rightarrow F$	$F \rightarrow P$	$F \rightarrow A$	$P \rightarrow A$	$A \rightarrow F$	$A \rightarrow P$	n
31	19.1943	0.0521	0.00550	0.10550	181.943	9.479	31
32	21.1138	0.0474	0.00497	0.10497	201.138	9.526	32
33	23.2252	0.0431	0.00450	0.10450	222.252	9.569	33
34	25.5477	0.0391	0.00407	0.10407	245.477	9.609	34
35	28.1024	0.0356	0.00369	0.10369	271.024	9.644	35
40	45.2593	0.0221	0.00226	0.10226	442.593	9.779	40
45	72.8905	0.0137	0.00139	0.10139	718.905	9.863	45
50	117.3909	0.0085	0.00086	0.10086	1,163.909	9.915	50
55	189.0591	0.0053	0.00053	0.10053	1,880.591	9.947	55
60	304.4816	0.0033	0.00033	0.10033	3,034.816	9.967	60
65	490.3707	0.0020	0.00020	0.10020	4,893.707	9.980	65
70	789.7470	0.0013	0.00013	0.10013	7,887.470	9.987	70
75	1,271.8952	0.0008	0.00008	0.10008	12,708.954	9.992	75
80	2,048.4002	0.0005	0.00005	0.10005	20,474.002	9.995	80
85	3,298.9690	0.0003	0.00003	0.10003	32,979.690	9.997	85
90	5,313.0226	0.0002	0.00002	0.10002	53,120.226	9.998	90
95	8,556.6760	0.0001	0.00001	0.10001	85,556.760	9.999	95
100	13,780.6123	0.0001	0.00001	0.10001	137,796.123	9.999	100

12%

	Single payment		Uniform series				
	Compound amount factor	Present worth factor	Sinking fund factor	Capital recovery factor	Compound amount factor	Present worth factor	
n	$P \rightarrow F$	$F \rightarrow P$	$F \rightarrow A$	$P \rightarrow A$	$A \rightarrow F$	$A \rightarrow P$	n
1	1.1200	0.8929	1.00000	1.12000	1.000	0.893	1
2	1.2544	0.7972	0.47170	0.59170	2.120	1.690	2
3	1.4049	0.7118	0.29635	0.41635	3.374	2.402	3
4	1.5735	0.6355	0.20923	0.32923	4.779	3.037	4
5	1.7623	0.5674	0.15741	0.27741	6.353	3.605	5
6	1.9738	0.5066	0.12323	0.24323	8.115	4.111	6
7	2.2107	0.4523	0.09912	0.21912	10.089	4.564	7
8	2.4760	0.4039	0.08130	0.20130	12.300	4.968	8
9	2.7731	0.3606	0.06768	0.18768	14.776	5.328	9
10	3.1058	0.3220	0.05698	0.17698	17.549	5.650	10
11	3.4785	0.2875	0.04842	0.16842	20.655	5.938	11
12	3.8960	0.2567	0.04144	0.16144	24.133	6.194	12
13	4.3635	0.2292	0.03568	0.15568	28.029	6.424	13
14	4.8871	0.2046	0.03087	0.15087	32.393	6.628	14
15	5.4736	0.1827	0.02682	0.14682	37.280	6.811	15

TABLE B. 1 *(Continued)*

12%

	Single payment		Uniform series				
	Compound amount factor	Present worth factor	Sinking fund factor	Capital recovery factor	Compound amount factor	Present worth factor	
n	$P \rightarrow F$	$F \rightarrow P$	$F \rightarrow A$	$P \rightarrow A$	$A \rightarrow F$	$A \rightarrow P$	n
16	6.1304	0.1631	0.02339	0.14339	42.753	6.974	16
17	6.8660	0.1456	0.02046	0.14046	48.884	7.120	17
18	7.6900	0.1300	0.01794	0.13794	55.750	7.250	18
19	8.6128	0.1161	0.01576	0.13576	63.440	7.366	19
20	9.6463	0.1037	0.01388	0.13388	72.052	7.469	20
21	10.8038	0.0926	0.01224	0.13224	81.699	7.562	21
22	12.1003	0.0826	0.01081	0.13081	92.503	7.645	22
23	13.5523	0.0738	0.00956	0.12956	104.603	7.718	23
24	15.1786	0.0659	0.00846	0.12846	118.155	7.784	24
25	17.0001	0.0588	0.00750	0.12750	133.334	7.843	25
26	19.0401	0.0525	0.00665	0.12665	150.334	7.896	26
27	21.3249	0.0469	0.00590	0.12590	169.374	7.943	27
28	23.8839	0.0419	0.00524	0.12524	190.699	7.984	28
29	26.7499	0.0374	0.00466	0.12466	214.583	8.022	29
30	29.9599	0.0334	0.00414	0.12414	241.333	8.055	30
31	33.5551	0.0298	0.00369	0.12369	271.292	8.085	31
32	37.5817	0.0266	0.00328	0.12328	304.847	8.112	32
33	42.0915	0.0238	0.00292	0.12292	342.429	8.135	33
34	47.1425	0.0212	0.00260	0.12260	384.520	8.157	34
35	52.7996	0.0189	0.00232	0.12232	431.663	8.176	35
40	93.0510	0.0107	0.00130	0.12130	767.091	8.244	40
45	163.9876	0.0061	0.00074	0.12074	1,358.230	8.283	45
50	289.0022	0.0035	0.00042	0.12042	2,400.018	8.305	50
∞				0.12000		8.333	∞

15%

	Single payment		Uniform series				
	Compound amount factor	Present worth factor	Sinking fund factor	Capital recovery factor	Compound amount factor	Present worth factor	
n	$P \rightarrow F$	$F \rightarrow P$	$F \rightarrow A$	$P \rightarrow A$	$A \rightarrow F$	$A \rightarrow P$	n
1	1.1500	0.8696	1.00000	1.15000	1.000	0.870	1
2	1.3225	0.7561	0.46512	0.61512	2.150	1.626	2
3	1.5209	0.6575	0.28798	0.43798	3.472	2.283	3
4	1.7490	0.5718	0.20026	0.35027	4.993	2.855	4
5	2.0114	0.4972	0.14822	0.29832	6.742	3.352	5
6	2.3131	0.4323	0.11424	0.26424	8.754	3.784	6
7	2.6600	0.3759	0.09036	0.24036	11.067	4.160	7
8	3.0590	0.3269	0.07285	0.22285	13.727	4.487	8

TABLE B. 1 *(Continued)*

15%

	Single payment		Uniform series				
	Compound amount factor	Present worth factor	Sinking fund factor	Capital recovery factor	Compound amount factor	Present worth factor	
n	$P \rightarrow F$	$F \rightarrow P$	$F \rightarrow A$	$P \rightarrow A$	$A \rightarrow F$	$A \rightarrow P$	n
9	3.5179	0.2843	0.05957	0.20957	16.786	4.772	9
10	4.0456	0.2472	0.04925	0.19925	20.304	5.019	10
11	4.6524	0.2149	0.04107	0.19107	24.349	5.234	11
12	5.3503	0.1869	0.03448	0.18448	29.002	5.421	12
13	6.1528	0.1625	0.02911	0.17911	34.352	5.583	13
14	7.0757	0.1413	0.02469	0.17469	40.505	5.724	14
15	8.1371	0.1229	0.02102	0.17102	47.580	5.847	15
16	9.3576	0.1069	0.01795	0.16795	55.717	5.954	16
17	10.7613	0.0929	0.01537	0.16537	65.075	6.047	17
18	12.3755	0.0808	0.01319	0.16319	75.836	6.128	18
19	14.2318	0.0703	0.01134	0.16134	88.212	6.198	19
20	16.3665	0.0611	0.00976	0.15976	102.444	6.259	20
21	18.8215	0.0531	0.00842	0.15842	118.810	6.312	21
22	21.6447	0.0462	0.00727	0.15727	137.632	6.359	22
23	24.8915	0.0402	0.00628	0.15628	159.276	6.399	23
24	28.6252	0.0349	0.00543	0.15543	184.168	6.434	24
25	32.9190	0.0304	0.00470	0.15470	212.793	6.464	25
26	37.8568	0.0264	0.00407	0.15407	245.712	6.491	26
27	43.5353	0.0230	0.00353	0.15353	283.569	6.514	27
28	50.0656	0.0200	0.00306	0.15306	327.104	6.534	28
29	57.5755	0.0174	0.00265	0.15265	377.170	6.551	29
30	66.2118	0.0151	0.00230	0.15230	434.745	6.566	30
31	76.1435	0.0131	0.00200	0.15200	500.957	6.579	31
32	87.5651	0.0114	0.00173	0.15173	577.100	6.591	32
33	100.6998	0.0099	0.00150	0.15150	664.666	6.600	33
34	115.8048	0.0086	0.00131	0.15131	765.365	6.609	34
35	133.1755	0.0075	0.00113	0.15113	881.170	6.617	35
40	267.8635	0.0037	0.00056	0.15056	1,779.090	6.642	40
45	538.7693	0.0019	0.00028	0.15028	3,585.128	6.654	45
50	1,083.6574	0.0009	0.00014	0.15014	7,217.716	6.661	50
∞				0.15000		6.667	∞

20%

	Single payment		Uniform series				
	Compound amount factor	Present worth factor	Sinking fund factor	Capital recovery factor	Compound amount factor	Present worth factor	
n	$P \rightarrow F$	$F \rightarrow P$	$F \rightarrow A$	$P \rightarrow A$	$A \rightarrow F$	$A \rightarrow P$	n
1	1.2000	0.8333	1.00000	1.20000	1.000	0.833	1
2	1.4400	0.6944	0.45455	0.65455	2.200	1.528	2

TABLE B. 1 *(Continued)*

	Single payment		Uniform series				
	Compound amount factor	Present worth factor	Sinking fund factor	Capital recovery factor	Compound amount factor	Present worth factor	
n	$P \rightarrow F$	$F \rightarrow P$	$F \rightarrow A$	$P \rightarrow A$	$A \rightarrow F$	$A \rightarrow P$	n
3	1.7280	0.5787	0.27473	0.47473	3.640	2.106	3
4	2.0736	0.4823	0.18629	0.38629	5.368	2.589	4
5	2.4883	0.4019	0.13438	0.33438	7.442	2.991	5
6	2.9860	0.3349	0.10071	0.30071	9.930	3.326	6
7	3.5832	0.2791	0.07742	0.27742	12.916	3.605	7
8	4.2998	0.2326	0.06061	0.26061	16.499	3.837	8
9	5.1598	0.1938	0.04808	0.24808	20.799	4.031	9
10	6.1917	0.1615	0.03852	0.23852	25.959	4.192	10
11	7.4301	0.1346	0.03110	0.23110	32.150	4.327	11
12	8.9161	0.1122	0.02526	0.22526	39.581	4.439	12
13	10.6993	0.0935	0.02062	0.22062	48.497	4.533	13
14	12.8392	0.0779	0.01689	0.21689	59.196	4.611	14
15	15.4070	0.0649	0.01388	0.21388	72.035	4.675	15
16	18.4884	0.0541	0.01144	0.21144	87.442	4.730	16
17	22.1861	0.0451	0.00944	0.20944	105.931	4.775	17
18	26.6233	0.0376	0.00781	0.20781	128.117	4.812	18
19	31.9480	0.0313	0.00646	0.20646	154.740	4.844	19
20	38.3376	0.0261	0.00536	0.20536	186.688	4.870	20
21	46.0051	0.0217	0.00444	0.20444	225.026	4.891	21
22	55.2061	0.0181	0.00369	0.20369	271.031	4.909	22
23	66.2474	0.0151	0.00307	0.20307	326.237	4.925	23
24	79.4968	0.0126	0.00255	0.20255	392.484	4.937	24
25	95.3962	0.0105	0.00212	0.20212	471.981	4.948	25
26	114.4755	0.0087	0.00176	0.20176	567.377	4.956	26
27	137.3706	0.0073	0.00147	0.20147	681.853	4.964	27
28	164.8447	0.0061	0.00122	0.20122	819.223	4.970	28
29	197.8136	0.0051	0.00102	0.20102	984.068	4.975	29
30	237.3763	0.0042	0.00085	0.20085	1,181.882	4.979	30
31	284.8516	0.0035	0.00070	0.20070	1,419.258	4.982	31
32	341.8219	0.0029	0.00059	0.20059	1,704.109	4.985	32
33	410.1863	0.0024	0.00049	0.20049	2,045.931	4.988	33
34	492.2235	0.0020	0.00041	0.20041	2,456.118	4.990	34
35	590.6682	0.0017	0.00034	0.20034	2,948.341	4.992	35
40	1,469.7716	0.0007	0.00014	0.20014	7,343.858	4.997	40
45	3,657.2620	0.0003	0.00005	0.20005	18,281.310	4.999	45
50	9,100.4382	0.0001	0.00002	0.20002	45,497.191	4.999	50
∞				0.20000		5.000	∞

TABLE B. 1 *(Continued)*

25%

	Single payment		Uniform series				
	Compound amount factor	Present worth factor	Sinking fund factor	Capital recovery factor	Compound amount factor	Present worth factor	
n	$P \rightarrow F$	$F \rightarrow P$	$F \rightarrow A$	$P \rightarrow A$	$A \rightarrow F$	$A \rightarrow P$	n
1	1.2500	0.8000	1.00000	1.25000	1.000	0.800	1
2	1.5625	0.6400	0.44444	0.69444	2.250	1.440	2
3	1.9531	0.5120	0.26230	0.51230	3.813	1.952	3
4	2.4414	0.4096	0.17344	0.42344	5.766	2.362	4
5	3.0518	0.3277	0.12185	0.37185	8.207	2.689	5
6	3.8147	0.2621	0.08882	0.33882	11.259	2.951	6
7	4.7684	0.2097	0.06634	0.31634	15.073	3.161	7
8	5.9605	0.1678	0.05040	0.30040	19.842	3.329	8
9	7.4506	0.1342	0.03876	0.28876	25.802	3.463	9
10	9.3132	0.1074	0.03007	0.28007	33.253	3.571	10
11	11.6415	0.0859	0.02349	0.27349	42.566	3.656	11
12	14.5519	0.0687	0.01845	0.26845	54.208	3.725	12
13	18.1899	0.0550	0.01454	0.26454	68.760	3.780	13
14	22.7374	0.0440	0.01150	0.26150	86.949	3.824	14
15	28.4217	0.0352	0.00912	0.25912	109.687	3.859	15
16	35.5271	0.0281	0.00724	0.25724	138.109	3.887	16
17	44.4089	0.0225	0.00576	0.25576	173.636	3.910	17
18	55.5112	0.0180	0.00459	0.25459	218.045	3.928	18
19	69.3889	0.0144	0.00366	0.25366	273.556	3.942	19
20	86.7362	0.0115	0.00292	0.25292	342.945	3.954	20
21	108.4202	0.0092	0.00233	0.25233	429.681	3.963	21
22	135.5253	0.0074	0.00186	0.25186	538.101	3.970	22
23	169.4066	0.0059	0.00148	0.25148	673.626	3.976	23
24	211.7582	0.0047	0.00119	0.25119	843.033	3.981	24
25	264.6978	0.0038	0.00095	0.25095	1,054.791	3.985	25
26	330.8722	0.0030	0.00076	0.25076	1,319.489	3.988	26
27	413.5903	0.0024	0.00061	0.25061	1,650.361	3.990	27
28	516.9879	0.0019	0.00048	0.25048	2,063.952	3.992	28
29	646.2349	0.0015	0.00039	0.25039	2,580.939	3.994	29
30	807.7936	0.0012	0.00031	0.25031	3,227.174	3.995	30
31	1,009.7420	0.0010	0.00025	0.25025	4,034.968	3.996	31
32	1,262.1774	0.0008	0.00020	0.25020	5,044.710	3.997	32
33	1,577.7218	0.0006	0.00016	0.25016	6,306.887	3.997	33
34	1,972.1523	0.0005	0.00013	0.25013	7,884.609	3.998	34
35	2,465.1903	0.0004	0.00010	0.25010	9,856.761	3.998	35
40	7,523.1638	0.0001	0.00003	0.25003	30,088.655	3.999	40
45	22,958.8740	0.0001	0.00001	0.25001	91,831.496	4.000	45
50	70.064.9232	0.0000	0.00000	0.25000	280,255.693	4.000	50
∞				0.25000		4.000	∞

TABLE B. 1 *(Continued)*

	30%						
	Single payment		Uniform series				
	Compound amount factor	Present worth factor	Sinking fund factor	Capital recovery factor	Compound amount factor	Present worth factor	
n	$P \to F$	$F \to P$	$F \to A$	$P \to A$	$A \to F$	$A \to P$	n
1	1.3000	0.7692	1.00000	1.30000	1.000	0.769	1
2	1.6900	0.5917	0.43478	0.73478	2.300	1.361	2
3	2.1970	0.4552	0.25063	0.55063	3.990	1.816	3
4	2.8561	0.3501	0.16163	0.46163	6.187	2.166	4
5	3.7129	0.2693	0.11058	0.41058	9.043	2.436	5
6	4.8268	0.2072	0.07839	0.37839	12.756	2.643	6
7	6.2749	0.1594	0.05687	0.35687	17.583	2.802	7
8	8.1573	0.1226	0.04192	0.34192	23.858	2.925	8
9	10.6045	0.0943	0.03124	0.33124	32.015	3.019	9
10	13.7858	0.0725	0.02346	0.32346	42.619	3.092	10
11	17.9216	0.0558	0.01773	0.31773	56.405	3.147	11
12	23.2981	0.0429	0.01345	0.31345	74.327	3.190	12
13	30.2875	0.0330	0.01024	0.31024	97.625	3.223	13
14	39.3738	0.0254	0.00782	0.30782	127.913	3.249	14
15	51.1859	0.0195	0.00598	0.30598	167.286	3.268	15
16	66.5417	0.0150	0.00458	0.30458	218.472	3.283	16
17	86.5042	0.0116	0.00351	0.30351	285.014	3.295	17
18	112.4554	0.0089	0.00269	0.30269	371.518	3.304	18
19	146.1920	0.0068	0.00207	0.30207	483.973	3.311	19
20	190.0496	0.0053	0.00159	0.30159	630.165	3.316	20
21	247.0645	0.0040	0.00122	0.30122	820.215	3.320	21
22	321.1839	0.0031	0.00094	0.30094	1,067.280	3.323	22
23	417.5391	0.0024	0.00072	0.30072	1,388.464	3.325	23
24	542.8008	0.0018	0.00055	0.30055	1,806.003	3.327	24
25	705.6410	0.0014	0.00043	0.30043	2,348.803	3.329	25
26	917.3333	0.0011	0.00033	0.30033	3,054.444	3.330	26
27	1,192.5333	0.0008	0.00025	0.30025	3,971.778	3.331	27
28	1,550.2933	0.0006	0.00019	0.30019	5,164.311	3.331	28
29	2,015.3813	0.0005	0.00015	0.30015	6,714.604	3.332	29
30	2,619.9956	0.0004	0.00011	0.30011	8,729.985	3.332	30
31	3,405.9943	0.0003	0.00009	0.30009	11,349.981	3.332	31
32	4,427.7926	0.0002	0.00007	0.30007	14,755.975	3.333	32
33	5,756.1304	0.0002	0.00005	0.30005	19,183.768	3.333	33
34	7,482.9696	0.0001	0.00004	0.30004	24,939.899	3.333	34
35	9.727.8604	0.0001	0.00003	0.30003	32,422.868	3.333	35
∞				0.30000		3.333	∞

TABLE B. 1 *(Continued)*

| | Single payment | | Uniform series | | | | |
| | Compound amount factor | Present worth factor | Sinking fund factor | Capital recovery factor | Compound amount factor | Present worth factor | |
n	$P \to F$	$F \to P$	$F \to A$	$P \to A$	$A \to F$	$A \to P$	n
1	1.3500	0.7407	1.00000	1.35000	1.000	0.741	1
2	1.8225	0.5487	0.42553	0.77553	2.350	1.289	2
3	2.4604	0.4064	0.23966	0.58966	4.172	1.696	3
4	3.3215	0.3011	0.15076	0.50076	6.633	1.997	4
5	4.4840	0.2230	0.10046	0.45046	9.954	2.220	5
6	6.0534	0.1652	0.06926	0.41926	14.438	2.385	6
7	8.1722	0.1224	0.04880	0.39880	20.492	2.507	7
8	11.0324	0.0906	0.03489	0.38489	28.664	2.598	8
9	14.8937	0.0671	0.02519	0.37519	39.696	2.665	9
10	20.1066	0.0497	0.01832	0.36832	54.590	2.715	10
11	27.1439	0.0368	0.01339	0.36339	74.697	2.752	11
12	36.6442	0.0273	0.00982	0.35982	101.841	2.779	12
13	49.4697	0.0202	0.00722	0.35722	138.485	2.799	13
14	66.7841	0.0150	0.00532	0.35532	187.954	2.814	14
15	90.1585	0.0111	0.00393	0.35393	254.738	2.825	15
16	121.7139	0.0082	0.00290	0.35290	344.897	2.834	16
17	164.3138	0.0061	0.00214	0.35214	466.611	2.840	17
18	221.8236	0.0045	0.00159	0.35158	630.925	2.844	18
19	299.4619	0.0033	0.00117	0.35117	852.748	2.848	19
20	404.2736	0.0025	0.00087	0.35087	1,152.210	2.850	20
21	545.7693	0.0018	0.00064	0.35064	1,556.484	2.852	21
22	736.7886	0.0014	0.00048	0.35048	2,102.253	2.853	22
23	994.6646	0.0010	0.00035	0.35035	2,839.042	2.854	23
24	1,342.7973	0.0007	0.00026	0.35026	3,833.706	2.855	24
25	1,812.7763	0.0006	0.00019	0.35019	5,176.504	2.856	25
26	2,447.2480	0.0004	0.00014	0.35014	6,989.280	2.856	26
27	3,303.7848	0.0003	0.00011	0.35011	9,436.528	2.856	27
28	4,460.1095	0.0002	0.00008	0.35008	12,740.313	2.857	28
29	6,021.1478	0.0002	0.00006	0.35006	17,200.422	2.857	29
30	8,128.5495	0.0001	0.00004	0.35004	23,221.570	2.857	30
31	10,973.5418	0.0001	0.00003	0.35003	31,350.120	2.857	31
32	14,814.2815	0.0001	0.00002	0.35002	42,323.661	2.857	32
33	19,999.2800	0.0001	0.00002	0.35002	57,137.943	2.857	33
34	26,999.0280	0.0000	0.00001	0.35001	77,137.223	2.857	34
35	36,448.6678		0.00001	0.35001	104,136.251	2.857	35
∞				0.35000		2.857	∞

TABLE B. 1 *(Continued)*

	Single payment		Uniform series				
	Compound amount factor	Present worth factor	Sinking fund factor	Capital recovery factor	Compound amount factor	Present worth factor	
n	$P \rightarrow F$	$F \rightarrow P$	$F \rightarrow A$	$P \rightarrow A$	$A \rightarrow F$	$A \rightarrow P$	n
				40%			
1	1.4000	0.7143	1.00000	1.40000	1.000	0.714	1
2	1.9600	0.5102	0.41667	0.81667	2.400	1.224	2
3	2.7440	0.3644	0.22936	0.62936	4.360	1.589	3
4	3.8416	0.2603	0.14077	0.54077	7.104	1.849	4
5	5.3782	0.1859	0.09136	0.49136	10.946	2.035	5
6	7.5295	0.1328	0.06126	0.46126	16.324	2.168	6
7	10.5414	0.0949	0.04192	0.44192	23.853	2.263	7
8	14.7579	0.0678	0.02907	0.42907	34.395	2.331	8
9	20.6610	0.0484	0.02034	0.42034	49.153	2.379	9
10	28.9255	0.0346	0.01432	0.41432	69.814	2.414	10
11	40.4957	0.0247	0.01013	0.41013	98.739	2.438	11
12	56.6939	0.0176	0.00718	0.40718	139.235	2.456	12
13	79.3715	0.0126	0.00510	0.40510	195.929	2.469	13
14	111.1201	0.0090	0.00363	0.40363	275.300	2.478	14
15	155.5681	0.0064	0.00259	0.40259	386.420	2.484	15
16	217.7953	0.0046	0.00185	0.40185	541.988	2.489	16
17	304.9135	0.0033	0.00132	0.40132	759.784	2.492	17
18	426.8789	0.0023	0.00094	0.40094	1,064.697	2.494	18
19	597.6304	0.0017	0.00067	0.40067	1,491.576	2.496	19
20	836.6826	0.0012	0.00048	0.40048	2,089.206	2.497	20
21	1,171.3554	0.0009	0.00034	0.40034	2,925.889	2.498	21
22	1,639.8976	0.0006	0.00024	0.40024	4,097.245	2.498	22
23	2,295.8569	0.0004	0.00317	0.40017	5,737.142	2.499	23
24	3,214.1997	0.0003	0.00012	0.40012	8,032.999	2.499	24
25	4,499.8796	0.0002	0.00009	0.40009	11,247.199	2.499	25
26	6,299.8314	0.0002	0.00006	0.40006	15,747.079	2.500	26
27	8,819.7640	0.0001	0.00005	0.40005	22,046.910	2.500	27
28	12,347.6696	0.0001	0.00003	0.40003	30,866.674	2.500	28
29	17,286.7374	0.0001	0.00002	0.40002	43,214.343	2.500	29
30	24,201.4324	0.0000	0.00001	0.40002	60,501.081	2.500	30
31	33,882.0053		0.00001	0.40001	84,702.513	2.500	31
32	47,434.8074		0.00001	0.40001	118,584.519	2.500	32
33	66,408.7304		0.00001	0.40001	166,019.326	2.500	33
34	92,972.2225		0.00000	0.40000	232,428.056	2.500	34
35	130,161.1116			0.40000	325,400.279	2.500	35
∞				0.40000		2.500	∞

TABLE B. 1 *(Continued)*

45%

	Single payment		Uniform series				
	Compound amount factor	Present worth factor	Sinking fund factor	Capital recovery factor	Compound amount factor	Present worth factor	
n	$P \to F$	$F \to P$	$F \to A$	$P \to A$	$A \to F$	$A \to P$	n
1	1.4500	0.6897	1.00000	1.45000	1.000	0.690	1
2	2.1025	0.4756	0.40816	0.85816	2.450	1.165	2
3	3.0486	0.3280	0.21966	0.66966	4.552	1.493	3
4	4.4205	0.2262	0.13156	0.58156	7.601	1.720	4
5	6.4097	0.1560	0.08318	0.53318	12.022	1.876	5
6	9.2941	0.1076	0.05426	0.50426	18.431	1.983	6
7	13.4765	0.0742	0.03607	0.48607	27.725	2.057	7
8	19.5409	0.0512	0.02427	0.47427	41.202	2.109	8
9	28.3343	0.0353	0.01646	0.46646	60.743	2.144	9
10	41.0847	0.0243	0.01123	0.46123	89.077	2.168	10
11	59.5728	0.0168	0.00768	0.45768	130.162	2.185	11
12	86.3806	0.0116	0.00527	0.45527	189.735	2.196	12
13	125.2518	0.0080	0.00362	0.45362	276.115	2.204	13
14	181.6151	0.0055	0.00249	0.45249	401.367	2.210	14
15	263.3419	0.0038	0.00172	0.45172	582.982	2.214	15
16	381.8458	0.0026	0.00118	0.45118	846.324	2.216	16
17	553.6764	0.0018	0.00081	0.45081	1,228.170	2.218	17
18	802.8308	0.0012	0.00056	0.45056	1,781.846	2.219	18
19	1,164.1047	0.0009	0.00039	0.45039	2,584.677	2.220	19
20	1,687.9518	0.0006	0.00027	0.45027	3,748.782	2.221	20
21	2,447.5301	0.0004	0.00018	0.45018	5,436.734	2.221	21
22	3,548.9187	0.0003	0.00013	0.45013	7,884.264	2.222	22
23	5,145.9321	0.0002	0.00009	0.45009	11,433.182	2.222	23
24	7,461.6015	0.0001	0.00006	0.45006	16,579.115	2.222	24
25	10,819.3222	0.0001	0.00004	0.45004	24,040.716	2.222	25
26	15,688.0173	0.0001	0.00003	0.45003	34,860.038	2.222	26
27	22,747.6250	0.0000	0.00002	0.45002	50,548.056	2.222	27
28	32,984.0563		0.00001	0.45001	73,295.681	2.222	28
29	47,826.8816		0.00001	0.45001	106,279,737	2.222	29
30	69,348.9783		0.00001	0.45001	154,106.618	2.222	30
∞				0.45000		2.222	∞

50%

	Single payment		Uniform series				
	Compound amount factor	Present worth factor	Sinking fund factor	Capital recovery factor	Compound amount factor	Present worth factor	
n	$P \to F$	$F \to P$	$F \to A$	$P \to A$	$A \to F$	$A \to P$	n
1	1.5000	0.6667	1.00000	1.50000	1.000	0.667	1
2	2.2500	0.4444	0.40000	0.90000	2.500	1.111	2

TABLE B. 1 *(Continued)*

50%

| | Single payment | | Uniform series | | | | |
| | Compound amount factor | Present worth factor | Sinking fund factor | Capital recovery factor | Compound amount factor | Present worth factor | |
n	$P \rightarrow F$	$F \rightarrow P$	$F \rightarrow A$	$P \rightarrow A$	$A \rightarrow F$	$A \rightarrow P$	n
3	3.3750	0.2963	0.21053	0.71053	4.750	1.407	3
4	5.0625	0.1975	0.12308	0.62308	8.125	1.605	4
5	7.5938	0.1317	0.07583	0.57583	13.188	1.737	5
6	11.3906	0.0878	0.04812	0.54812	20.781	1.824	6
7	17.0859	0.0585	0.03108	0.53108	32.172	1.883	7
8	25.6289	0.0390	0.02030	0.52030	49.258	1.922	8
9	38.4434	0.0260	0.01335	0.51335	74.887	1.948	9
10	57.6650	0.0173	0.00882	0.50882	113.330	1.965	10
11	86.4976	0.0116	0.00585	0.50585	170.995	1.977	11
12	129.7463	0.0077	0.00388	0.50388	257.493	1.985	12
13	194.6195	0.0051	0.00258	0.50258	387.239	1.990	13
14	291.9293	0.0034	0.00172	0.50172	581.859	1.993	14
15	437.8939	0.0023	0.00114	0.50114	873.788	1.995	15
16	656.8408	0.0015	0.00076	0.50076	1,311.682	1.997	16
17	985.2613	0.0010	0.00051	0.50051	1,968.523	1.998	17
18	1,477.8919	0.0007	0.00034	0.50034	2,953.784	1.999	18
19	2,216.8378	0.0005	0.00023	0.50023	4,431.676	1.999	19
20	3,325.2567	0.0003	0.00015	0.50015	6,648.513	1.999	20
21	4,987.8851	0.0002	0.00010	0.50010	9,973.770	2.000	21
22	7.481.8276	0.0001	0.00007	0.50007	14,961.655	2.000	22
23	11,222.7415	0.0001	0.00004	0.50004	22,443.483	2.000	23
24	16,834.1122	0.0001	0.00003	0.50003	33,666.224	2.000	24
25	25,251.1683	0.0000	0.00002	0.50002	50,500.337	2.000	25
∞				0.50000		2.000	∞

Source: From the text of Rudolofo J. Aguilar, *Systems Analysis and Design in Engineering, Architecture, Construction, and Planning,* © 1973, pp. 374–398. Reprinted by permission of Prentice-Hall, Inc., Englewood Cliffs, N.J.

Probability and Statistics

This appendix contains background information on probability and statistics that will be helpful in understanding the stochastic concepts presented in the text.

POPULATION AND SAMPLE

The basic element of statistics is the observation of some characteristic of interest. Each observation provides a data point, or value, for the characteristic of interest. The entire set of all possible observations forms the population for the characteristic. There is no set size for a population. For example, the education level of all citizens in the United States would form one population, while the weight of all students in a 30-member class would form another population. A collection of observations representing a portion of a population is called a sample.

It would be possible to consider the whole population of the 30-member class; however, it would generally not be practical to consider the entire population of the United States when analyzing educational level. Therefore, a sample is chosen, and inferences made concerning the whole population based on the evaluation of the sample.

Sampling theory, which is beyond the scope of this text, is designed to make sure that a representative sample of the population is observed. Correct choice of the sample size is especially important when making observations requires consumption of or destruction of the item being observed. This would be the case, for example, when failure ages for electronic components are being examined.

NORMAL DISTRIBUTION

When repetitive measurements are made of some measurand that should theoretically be constant, the results are not generally identical. The reason for this discrepancy is generally that

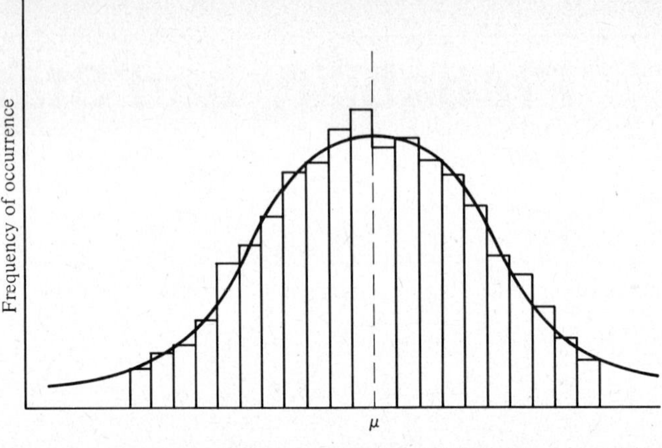

Figure C.1 Normal distribution.

the measurements are not exactly precise, but vary over some range. However, when the frequency of occurrence of each measurand value is plotted versus the corresponding measured characteristic value, a bell-shaped frequency distribution often results. Figure C.1 shows how one such distribution might look. The frequency of occurrence values cluster around a central, or expected, value for the distribution. This type of symmetrical distribution is called normal or Gaussian. If it can be assumed that the distribution is normal (there are statistical methods for checking this assumption), then the expected value is taken as the mean value of all the data points used to generate the frequency distribution. This is an estimate of the population mean, μ.

If the estimate of the population standard deviation, σ, is calculated (calculation of this statistic was described in Chapter 3), it forms a measure of the scatter of the data about the mean. A large standard deviation would indicate a lot of scatter and a wide, low bell-shaped curve [Figure C.2(a)]. A small standard deviation would indicate a tightly packed distribution and a narrow, high bell-shaped curve [Figure C.2(b)].

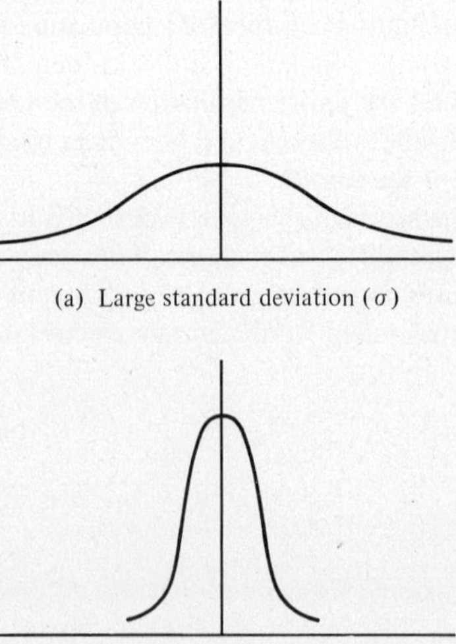

(a) Large standard deviation (σ)

(b) Small standard deviation (σ)

Figure C.2 Influence of data scatter.

Also, the standard deviation for the normal distribution is useful in estimating what percentage of the values of the measured characteristic should lie within a certain distance from the mean: 68% of the values should lie within $\pm 1\sigma$, 95% of the values within $\pm 2\sigma$, and 99.7% of the values within $\pm 3\sigma$. This property is useful in generating tables such as Table C.1, which can be used to make probability estimates.

TABLE C.1 AREAS UNDER THE NORMAL CURVE

(Proportion of total area under the curve from $-\infty$ to designated Z value)

Z^a	0.09	0.08	0.07	0.06	0.05	0.04	0.03	0.02	0.01	0.00
−3.5	0.00017	0.00017	0.00018	0.00019	0.00019	0.00020	0.00021	0.00022	0.00022	0.00023
−3.4	0.00024	0.00025	0.00026	0.00027	0.00028	0.00029	0.00030	0.00031	0.00033	0.00034
−3.3	0.00035	0.00036	0.00038	0.00039	0.00040	0.00042	0.00043	0.00045	0.00017	0.00048
−3.2	0.00050	0.00052	0.00054	0.00056	0.00058	0.00060	0.00062	0.00064	0.00066	0.00069
−3.1	0.00071	0.00074	0.00076	0.00079	0.00082	0.00085	0.00087	0.00090	0.00094	0.00097
−3.0	0.00100	0.00104	0.00107	0.00111	0.00114	0.00118	0.00122	0.00126	0.00131	0.00135
−2.9	0.0014	0.0014	0.0015	0.0015	0.0016	0.0016	0.0017	0.0017	0.0018	0.0019
−2.8	0.0019	0.0020	0.0020	0.0021	0.0022	0.0023	0.0023	0.0024	0.0025	0.0026
−2.7	0.0026	0.0027	0.0028	0.0029	0.0030	0.0031	0.0032	0.0033	0.0034	0.0035
−2.6	0.0036	0.0037	0.0038	0.0039	0.0040	0.0041	0.0043	0.0044	0.0045	0.0047
−2.5	0.0048	0.0049	0.0051	0.0052	0.0054	0.0055	0.0057	0.0059	0.0060	0.0062
−2.4	0.0064	0.0066	0.0068	0.0069	0.0071	0.0073	0.0075	0.0078	0.0080	0.0082
−2.3	0.0084	0.0087	0.0089	0.0091	0.0094	0.0096	0.0099	0.0102	0.0104	0.0107
−2.2	0.0110	0.0113	0.0116	0.0119	0.0122	0.0125	0.0129	0.0132	0.0136	0.0139
−2.1	0.0143	0.0146	0.0150	0.0154	0.0158	0.0162	0.0166	0.0170	0.0174	0.0179
−2.0	0.0183	0.0188	0.0192	0.0197	0.0202	0.0207	0.0212	0.0217	0.0222	0.0227
−1.9	0.0233	0.0239	0.0244	0.0250	0.0256	0.0262	0.0268	0.0274	0.0281	0.0287
−1.8	0.0294	0.0301	0.0307	0.0314	0.0322	0.0329	0.0336	0.0344	0.0351	0.0359
−1.7	0.0367	0.0375	0.0384	0.0392	0.0401	0.0409	0.0418	0.0427	0.0436	0.0446
−1.6	0.0455	0.0465	0.0475	0.0485	0.0495	0.0505	0.0516	0.0526	0.0537	0.0548
−1.5	0.0559	0.0571	0.0582	0.0594	0.0606	0.0618	0.0630	0.0643	0.0655	0.0668
−1.4	0.0681	0.0694	0.0703	0.0721	0.0735	0.0749	0.0764	0.0778	0.0793	0.0808
−1.3	0.0823	0.0838	0.0853	0.0869	0.0885	0.0901	0.0918	0.0934	0.0951	0.0968
−1.2	0.0985	0.1003	0.1020	0.1038	0.1057	0.1075	0.1093	0.1112	0.1131	0.1151
−1.1	0.1170	0.1190	0.1210	0.1230	0.1251	0.1271	0.1292	0.1314	0.1335	0.1357
−1.0	0.1379	0.1401	0.1423	0.1446	0.1469	0.1492	0.1515	0.1539	0.1562	0.1587
−0.9	0.1611	0.1635	0.1660	0.1685	0.1711	0.1736	0.1762	0.1788	0.1814	0.1841
−0.8	0.1867	0.1894	0.1922	0.1949	0.1977	0.2005	0.2033	0.2061	0.2000	0.2119
−0.7	0.2148	0.2177	0.2207	0.2236	0.2266	0.2297	0.2327	0.2358	0.2389	0.2420
−0.6	0.2451	0.2483	0.2514	0.2546	0.2578	0.2611	0.2643	0.2676	0.2709	0.2743
−0.5	0.2776	0.2810	0.2843	0.2877	0.2912	0.2946	0.2981	0.3015	0.3050	0.3085
−0.4	0.3121	0.3156	0.3192	0.3228	0.3264	0.3300	0.3336	0.3372	0.3409	0.3446
−0.3	0.3483	0.3520	0.3557	0.3594	0.3632	0.3669	0.3707	0.3745	0.3783	0.3821
−0.2	0.3859	0.3897	0.3936	0.3974	0.4013	0.4052	0.4090	0.4129	0.4168	0.4207
−0.1	0.4247	0.4286	0.4325	0.4364	0.4404	0.4443	0.4483	0.4522	0.4562	0.4602
−0.0	0.4641	0.4681	0.4721	0.4761	0.4801	0.4840	0.4880	0.4920	0.4960	0.5000

(*continued overleaf*)

TABLE C.1 *(Continued)*

(Proportion of total area under the curve from $-\infty$ to designated Z value)

Z	0.00	0.01	0.02	0.03	0.04	0.05	0.06	0.07	0.08	0.09
+0.0	0.5000	0.5040	0.5080	0.5120	0.5160	0.5199	0.5239	0.5279	0.5319	0.5359
+0.1	0.5398	0.5438	0.5478	0.5517	0.5557	0.5596	0.5636	0.5675	0.5714	0.5753
+0.2	0.5793	0.5832	0.5871	0.5910	0.5948	0.5987	0.6026	0.6064	0.6103	0.6141
+0.3	0.6179	0.6217	0.6255	0.6293	0.6331	0.6368	0.6406	0.6443	0.6480	0.6517
+0.4	0.6554	0.6591	0.6628	0.6664	0.6700	0.6736	0.6772	0.6808	0.6844	0.6879
+0.5	0.6915	0.6950	0.6985	0.7019	0.7054	0.7088	0.7123	0.7157	0.7190	0.7224
+0.6	0.7257	0.7291	0.7324	0.7357	0.7389	0.7422	0.7454	0.7486	0.7517	0.7549
+0.7	0.7580	0.7611	0.7642	0.7673	0.7704	0.7734	0.7764	0.7794	0.7823	0.7852
+0.8	0.7881	0.7910	0.7939	0.7967	0.7995	0.8023	0.8051	0.8079	0.8106	0.8133
+0.9	0.8159	0.8186	0.8212	0.8238	0.8264	0.8289	0.8315	0.8340	0.8365	0.8389
+1.0	0.8413	0.8438	0.8461	0.8485	0.8508	0.8531	0.8554	0.8577	0.8599	0.8621
+1.1	0.8643	0.8665	0.8686	0.8708	0.8729	0.8749	0.8770	0.8790	0.8810	0.8830
+1.2	0.8849	0.8869	0.8888	0.8907	0.8925	0.8944	0.8962	0.8980	0.8997	0.9015
+1.3	0.9032	0.9049	0.9066	0.9082	0.9099	0.9115	0.9131	0.9147	0.9162	0.9177
+1.4	0.9192	0.9207	0.9222	0.9236	0.9251	0.9265	0.9279	0.9292	0.9306	0.9319
+1.5	0.9332	0.9345	0.9357	0.9370	0.9382	0.9394	0.9406	0.9418	0.9429	0.9441
+1.6	0.9452	0.9463	0.9474	0.9484	0.9495	0.9505	0.9515	0.9525	0.9535	0.9545
+1.7	0.9554	0.9564	0.9573	0.9582	0.9591	0.9599	0.9608	0.9616	0.9625	0.9633
+1.8	0.9641	0.9649	0.9656	0.9664	0.9671	0.9678	0.9686	0.9693	0.9699	0.9706
+1.9	0.9713	0.9719	0.9726	0.9732	0.9738	0.9744	0.9750	0.9756	0.9761	0.9767
+2.0	0.9773	0.9778	0.9783	0.9788	0.9793	0.9798	0.9803	0.9808	0.9812	0.9817
+2.1	0.9821	0.9826	0.9830	0.9834	0.9838	0.9842	0.9846	0.9850	0.9854	0.9857
+2.2	0.9861	0.9864	0.9868	0.9871	0.9875	0.9878	0.9881	0.9884	0.9887	0.9890
+2.3	0.9893	0.9896	0.9898	0.9901	0.9904	0.9906	0.9909	0.9911	0.9913	0.9916
+2.4	0.9918	0.9920	0.9922	0.9925	0.9927	0.9929	0.9931	0.9932	0.9934	0.9936
+2.5	0.9938	0.9940	0.9941	0.9943	0.9945	0.9946	0.9948	0.9949	0.9951	0.9952
+2.6	0.9953	0.9955	0.9956	0.9957	0.9959	0.9960	0.9961	0.9962	0.9963	0.9964
+2.7	0.9965	0.9966	0.9967	0.9968	0.9969	0.9970	0.9971	0.9972	0.9973	0.9974
+2.8	0.9974	0.9975	0.9976	0.9977	0.9977	0.9978	0.9979	0.9979	0.9980	0.9981
+2.9	0.9981	0.9982	0.9983	0.9983	0.9984	0.9984	0.9985	0.9985	0.9986	0.9986
+3.0	0.99865	0.99869	0.99874	0.99878	0.99882	0.99886	0.99889	0.99893	0.99896	0.99900
+3.1	0.99903	0.99906	0.99910	0.99913	0.99915	0.99918	0.99921	0.99924	0.99926	0.99929
+3.2	0.99931	0.99934	0.99936	0.99938	0.99940	0.99942	0.99944	0.99946	0.99948	0.99950
+3.3	0.99952	0.99953	0.99955	0.99957	0.99958	0.99960	0.99961	0.99962	0.99964	0.99965
+3.4	0.99966	0.99967	0.99969	0.99970	0.99971	0.99972	0.99973	0.99974	0.99975	0.99976
+3.5	0.99977	0.99978	0.99978	0.99979	0.99980	0.99981	0.99981	0.99982	0.99983	0.99983

[a] $Z = [x_i - E(x_i)]/\sigma;$ $E(x_i)$ = expected value for x_i, x_i = actual value.

Source: From the text of Rodolfo J. Aguilar, *Systems Analysis and Design in Engineering, Architecture, Construction, and Planning,* © 1973, pp. 400, 401. Reprinted by permission of Prentice-Hall, Inc., Englewood Cliffs, N.J.

CONDITIONAL PROBABILITY

The probability that outcome A will occur is called an unconditional probability and is represented as $P(A)$. If two outcomes are of concern, and it is desired to know the probability of outcome A occurring given that outcome B occurs, a conditional probability must be estimated. $P(A/B)$ is the conditional probability that A will occur, given that B has occurred.

Outcomes A and B each have their own probability of occurring. The probability of both A and B occurring is called the joint probability,

$$P(A \text{ and } B) = P(A) \times P(B/A)$$
$$= P(B) \times P(A/B) \tag{C.1}$$

If outcomes A and B are independent, that is, one outcome has no influence over the other, then

$$P(A \text{ and } B) = P(A) \times P(B) \tag{C.2}$$

Conversely, it may be desirable to know the probability of either A or B occurring. This inclusive probability includes the probability of either A or B and the probability of both A and B occurring. Then,

$$P(A \text{ or } B) = P(A) + P(B) - P(A \text{ and } B) \tag{C.3}$$

However, if A and B are mutually exclusive, they cannot both happen, $P(A \text{ and } B) = 0$, and

$$P(A \text{ or } B) = P(A) + P(B) \tag{C.4}$$

EXAMPLE C.1

Consider two independent events: the state of rainfall (rain or no rain) and the state of operability of storm sewer overflow regulators. The former is dependent on meteorological conditions, while the latter is dependent on the age of the regulator and past maintenance practices.

Suppose that the probability of having rain on a given day is 0.3, and the probability of a given overflow regulator being inoperative on that day is 0.1; then, there are four possible joint outcomes:

$$P(\text{rain}) \cdot P(\text{reg. inop.}) = (0.30)(0.10) = 0.03$$
$$P(\text{rain}) \cdot P(\text{reg. oper.}) = (0.30)(0.90) = 0.27$$
$$P(\text{no rain}) \cdot P(\text{reg. inop.}) = (0.70)(0.10) = 0.07$$
$$P(\text{no rain}) \cdot P(\text{reg. oper.}) = (0.70)(0.90) = \underline{0.63}$$
$$\text{Total} \qquad 1.00$$

Note that the sum of the joint probabilities for all possible joint outcomes has to add up to 1.0, just as the sum of probabilities for outcomes of an individual event is 1.0.

Computer Usage

INTRODUCTION

Computers have rapidly become essential tools in our society. The cost of computations has gone down by at least two orders of magnitude in the past 20 yr, while the amount of computational ability per given size of computer has increased tremendously. There are no indications that this trend will end.

Computers have allowed the widespread application of many sophisticated mathematical techniques that were impractical to implement by hand. A computer can store a vast amount of information in its memory, and can analyze and manipulate data quickly (roughly 10^6 times as fast as the human brain) without making errors. However, it will only do these things if correctly programmed to do so by humans. The computer does not think for itself: it will carry out instructions even if they are totally erroneous and nonsensical. The computer is a tool that, if properly used, will greatly increase the productivity of an engineer.

While computer competency is not necessary to understand the theory of systems analysis and the systems approach, it is necessary if many of the concepts are to be applied to real-world problems. The development of computer technology has gone hand in hand with advances in applied systems theory. Early application of many of the tools of systems analysis was hampered by the lack of computational capabilities. Large-scale practical applications were too computation intensive to be undertaken by hand. More advanced techniques were developed as a result of the need demonstrated by increased computer capabilities. This parallel development continues.

Similar comments could be made concerning the relationship between computers and numerical analysis. The objective of numerical analysis is to approximate solutions to mathematical functions for which an analytical solution is either impossible, or not practical. Since these are approximation methods, iteration is generally required, a task that is most efficiently done by the computer.

Background

Early computers were not practical computational machines for general engineering practice. They were behemoths, filling rooms with hot vacuum tubes, and requiring huge air conditioning systems to keep the tubes from burning out. The memory available in these machines was less than what is available in today's home microcomputers. Lack of a high-level programming language made these machines cumbersome to program. Nevertheless, they were a start and showed that computers could do repetitive calculations orders of magnitude faster than people could. The introduction of transistors, and then integrated circuits, allowed computer memory to take up less space, and thus, machines with much larger memories could be produced. Also, the speed of computations was significantly improved. Costs of computers and computation went down to the point that both could be afforded by even small consulting firms. The time is fast approaching when computer capability will be a necessity for all engineering firms that want to remain competitive.

Since the mid-1960s, civil engineering students have been introduced in varying degrees to computer programming during their undergraduate education. Many of them did not become dedicated users, however, until they were graduate students or entered practice. However, a revolution is also taking place in how and when students are introduced to computers, and computer programming. With the introduction of low-priced microcomputers, every level of formal education, including the elementary school level, is introducing some form of computer instruction. In the near future, the microcomputer will replace the calculator as the primary analysis tool for the civil engineering student, and undergraduate programs will operate under the assumption that students have a working level of computer competence when they enter as freshmen.

Paralleling the development of computer hardware has been the development of numerous comprehensive software packages designed to assist civil engineers in their analysis and design work. Some of the better known of these packages include COGO, ICES, STRESS, STRUDL, SWMM, HEC1, and HEC2. Civil engineers who have not actually used these, or similar packages, are at least cognizant of their existence through reading the literature, reviewing submitted reports, or attending engineering conferences. The best of these packages are under continual updating and development, have larger user support groups associated with them, and have generally been written so they will operate on several different brands of computer hardware. It is important for engineers to understand the capabilities and limitations of these applications packages.

Computer hardware and software will continue to evolve, and there will be a real need for personnel who can transfer the technical expertise of civil engineers into easily verifiable software programs which, in conjunction with modern computing machines, can be used to solve many engineering problems. A worthwhile approach might be for an engineering firm to own or rent access to a library of computer program modules. Each module would have a fairly specific purpose, but the modules would be written in such a manner that they could be easily linked together for solving more complex or larger problems. Each module could be independently verified. If new modules were needed for analyzing segments of problems not already programmed, they could be written with compatible input and output formats and incorporated into the library.

To implement such a plan would require technically competent engineers who are also competent computer programmers. They would have to be able to define the engineering problem, decide which parts of the problem lend themselves to computer-aided analysis or design and which do not, choose the modules that would perform the analyses required by the problem, develop any new modules required, and link the modules together in the proper sequence. The person or persons responsible for this program development should also be held responsible for verification of modules and communications of this verification to all colleagues who might have use for the module. Since each module would be for a specific purpose, it should not have to be verified unless it was modified in some way. In preparing a report for a client, the engineer responsible for program development would present the results in such a manner that the reader could determine how the results were obtained and could ascertain that adequate program verification was accomplished to ensure accuracy of output data.

Even in the absence of an idealized situation as just described, there will still be a great need for engineers with both technical and computer competence. Only someone with technical competence can interpret computer results and spot potential inconsistencies or errors in output data. After doing manual calculations to confirm that there is a problem, that same person, or someone of equal combined competency, would be the best one to track down the error in the program and correct it. A dually competent engineer would also be the best person to make modifications or additions to existing software packages and reverify the program accuracy.

The outlook is for computer hardware and computer usage to continue to develop, probably at an increasing rate, thus, it is of increasing importance that there be computer competent engineers who are able to accomplish the integration of hardware, software, and users. Computer hardware is one of the few commodities that has gone down in price during the same time that capabilities were being increased. Development of very large-scale integrated circuits (VLSI) will further reduce computer size and enhance their capabilities. Even if computer hardware costs do not continue to go down, some type of computer will be within the means of every civil engineer or civil engineering firm. Even home microcomputers have progressed to the stage that they are valuable analysis tools, not just computer game processors. The hardware will be available, but it is people who will have to develop the user-oriented software to assist in technical analysis and design.

PROGRAMMING

Civil engineers do not have to be trained as computer scientists; however, they must be familiar with computer operating systems and have the confidence and ability to access the computer when required. Some engineers will logically become expert programmers and work on transforming civil engineering theory into practical software to be used in anaysis and design. Other engineers should have at least enough programming expertise to interpret programs, to track down bugs in programs, and to customize programs for unique applications. Engineers must be able to flowchart problems for computer programming, as this is nothing more than charting the logical sequence of a problem solution.

There is no single high-level computer language that all engineers should learn. FORTRAN is probably the most widely used, but BASIC is experiencing a strong resurgence because of microcomputer and graphics applications. PASCAL is also gaining wide acceptance as a scientific and engineering language. Numerous other specialized languages exist, and others are continually being developed. The engineer should be competent in at least one language, and at least familiar with the capabilities and limitations of the others.

When debugging programs, the engineer must be able to recognize logical blocks of computer code, even if they are poorly identified within the program. Interim output statements should be inserted into the program to print out computations as they occur. Comparing this output with results that are known to be correct can pinpoint where within the program there is a problem. Then the logic can be checked and corrected where necessary. Also, any program should be constructed so that all input data are output for checking.

Errors occurring in computer programming can be divided into three general categories: system, syntax, and logic errors. System errors involve mistakes in the instructions necessary to run a program or manipulate files on a given computer system. These instructions are system dependent, so they must be ascertained from the technical manuals for the system or from consultants familiar with its operation. Since all operating systems perform essentially the same tasks, but in different ways, a general familiarity with them can be extremely helpful. System errors will normally result in an aborted run, with a descriptive error message printed out.

Syntax errors refer to the type of error that occurs when the rules of the particular high-level programming language being used are violated. Again, these errors generally result in an aborted run. Most computer systems will identify what type of error is involved and where within the program it occurs. Logic errors are the most insidious of the three, and generally take the most time to identify and correct. They involve either errors in the theoretical development of the algorithms being utilized in the programs, or errors in transcribing the algorithms to computer code. Logic errors are difficult to identify because they do not result in any error message being printed out, unless they also violate some syntax rule. Often, the only indication is erroneous program output, which may not be evident itself unless correct answers are known for comparison. This is the reason for it being imperative that all computer programs be checked against a known correct solution for the problem being modeled. The best way to isolate logic errors is through the judicious use of intermediate write statements, as was previously discussed.

PROGRAM DOCUMENTATION

Any program or applications package is only as good as the documentation that accompanies it. Documentation is the way of instructing users other than the developer how to access and run the program. A general description of good program documentation is described here. In some instances, some elements of program documentation can be left off if users will still be able to efficiently utilize the program.

Elements of Program Documentation

1. *Description of Problem to Be Solved.* This should be a brief abstract of the general problem addressed by the program, so anyone searching for the appropriate program for a certain application will know whether or not to investigate this program more thoroughly.

2. *Capabilities and Limitations of the Program.* Size, theoretical, and other program limitations should be concisely described so reviewers can assess its applicability.

3. *Theoretical Development upon Which the Program Is Based.* This is especially important when more than one method is available to solve a given problem.

4. *Description of Program Structure with Flowchart.* This is essential if users are to understand the logical flow of the program. It should include a definition of the variable names used in the program.

5. *Special Machine Requirements.* Any unusual computer memory and/or input/output device requirements should be identified.

6. *User Data Input Requirements and Instructions.* This should include detailed instructions as to how to access, input data to, and execute the available options for the program.

7. *Sample Problem.* This should include a problem statement, sample input data, and sample program output for as many program options as necessary to provide a complete sample.

8. *Program Listing.* A program listing should be provided for all but proprietary programs. This will facilitate program verification, debugging, and modification.

USE OF APPLICATIONS PACKAGES

Applications packages can be powerful tools, but engineers must always remember that they have final responsibility for any decisions that are made based on computer analysis. Therefore, it is imperative that an engineer verify and become thoroughly familiar with an applications package before utilizing it in design or planning work. The engineer must be able to interpret the results generated by the computer and stake his or her professional reputation on the answers given.

Numerical Analysis

INTRODUCTION

Many problems cannot be solved explicitly, so the engineer has to resort to a numerical technique to get an approximate answer. The accuracy of these approximate answers can vary widely, depending on the characteristics of the particular problem and the skill of the person doing the analysis. Anyone using numerical analysis techniques should be knowledgeable as to the limitations and special requirements of the technique to be used. This appendix will present the essential concepts for three numerical analysis techniques that are useful in the application of systems analysis. They are the Newton–Raphson method for finding real roots of a polynomial equation, and two climbing techniques, gradient search and pattern search, which are useful for finding minimum or maximum solutions for nonlinear optimization problems. Additional information about these and other numerical analysis techniques can be found in *Introduction to Numerical Analysis* by Hildebrand (1974).

NEWTON–RAPHSON METHOD

The Newton–Raphson method is an efficient method for finding real roots of nearly any polynomial equation that would be of interest to civil engineers. As with most numerical approximation techniques, it requires an initial estimate of the root desired. The more accurate the estimate is, the quicker the method will converge to the solution. If the initial estimate is too far off, there is a chance that the method could converge to a different root than the one desired.

When the root has been determined correctly, the value of the function will be zero. Since this is an approximation technique, the value of the function will never be exactly zero; but, as will be shown, it is possible to estimate the root accurately to any number of decimal places desired.

Assume that X is the dependent variable, and $f(X)$ is some polynomial function of X. To apply the Newton–Raphson method, any constants have to be on the same side of the equation as the dependent variable, and

$$f(X) = 0 \tag{E.1}$$

is the polynomial equation. The estimation algorithm can be written

$$X_{n+1} = X_n - \frac{f(X_n)}{f'(X_n)} \tag{E.2}$$

where
X_{n+1} = next estimate of the root,
X_n = present estimate of the root,
$f(X_n)$ = equation $[f(X)]$ evaluated at X_n, and
$f'(X_n)$ = derivative of $f(X)$ evaluated at X_n

The initial estimate of the root (X_0) can be estimated by a rough graph or by knowledge of the approximate value expected.

It can also be shown that the accuracy of the method is directly related to the absolute value of the difference between successive approximations of X.

If
$$|X_{n+1} - X_n| \le 10^{-a}$$

then the method is accurate to a decimal places. This provides a mechanism for knowing when enough iterations have been accomplished. The desired decimal place accuracy is determined beforehand, then iterations are performed until the above stated criterion is met. The estimate of X, accurate to a decimal places, is X_{n+1}.

A graphical example will effectively demonstrate the Newton–Raphson method. Figure E.1 shows a typical polynomial function. At X_n, the height of the curve above the X axis is $f(X_n)$. At the point that the vertical dashed line through X_n contacts the curve, the slope of the curve is equal to the derivative of the function evaluated at X_n, $f'(X_n)$. Dividing the height of

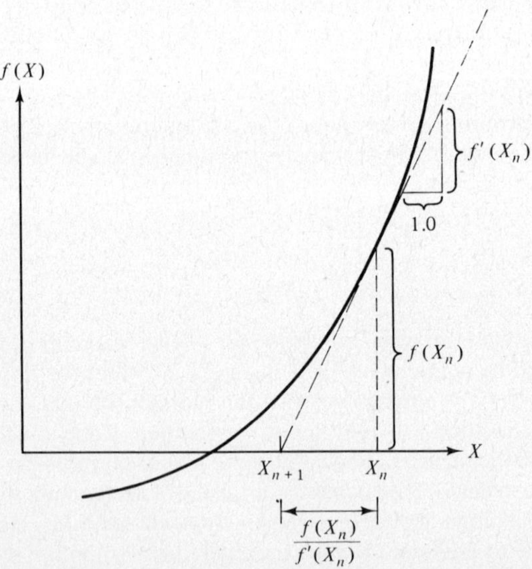

Figure E.1 Graphical interpretation of Newton–Raphson method for finding real roots of a polynomial equation.

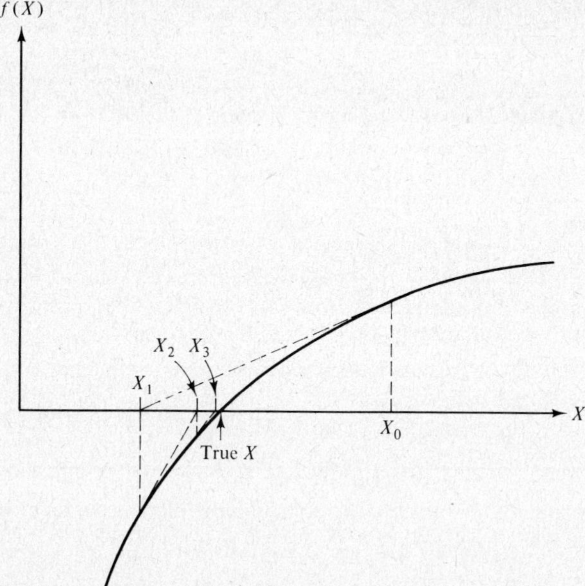

Figure E.2 Successive approximations for Newton–Raphson method illustrating switching of X_n from $> X_{\text{root}}$ to $< X_{\text{root}}$.

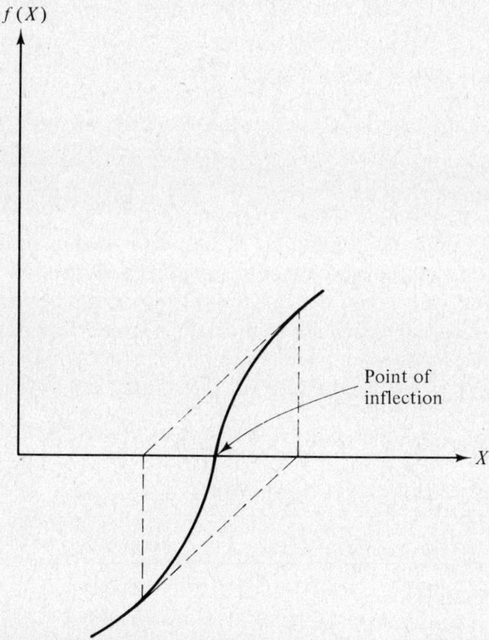

Figure E.3 Oscillation of successive approximations about a point of inflection.

the right side of the right triangle, $f(X_n)$, by the slope of the hypotenuse, $f'(X_n)$, gives the distance along the X axis between X_n and X_{n+1}, which is the correction factor. Successive approximations would be made in the same manner, using X_{n+1} as the starting value.

Figure E.2 shows another case in which the shape of the function is concave downward in the vicinity of X_{root}. If the initial estimate of X is greater than the true root, the derivative will cause the next estimate to be less then X_{root}. Successive iterations would converge normally toward the true root.

Problems can occur if the initial estimate of X is in the vicinity of a minimum or maximum point for $f(X)$. This would cause $f'(X)$ to be small and $f(X)/f'(X)$ large, which might throw the next estimate of X out of the vicinity of the root that was desired. This problem can be avoided by judicious choice of the initial estimate of X. Also, convergence can be slow, or nonexistent, if a point of inflection in $f(X)$ occurs near the desired root, as illustrated in Figure E.3. This does not occur often in practice.

EXAMPLE E.1 ____

Solve for the approximate rate of return (to four-decimal place accuracy) for the problem developed in Example B.5.

$$7.14 = \frac{(1 + i)^{16} - 1}{(1 + i)^{16}i} \tag{E.3}$$

This is actually a polynomial equation, even though it is not in the normal form. If the denominator is cleared,

$$7.14(1 + i)^{16}i = (1 + i)^{16} - 1 \tag{E.4}$$

and all terms brought over to the right-hand side,

$$f(i) = 0 = 7.14(i)(1 + i)^{16} - (1 + i)^{16} + 1 \tag{E.5}$$

which is the standard form for application of the Newton–Raphson method. The derivative of $f(i)$ is

$$f'(i) = 7.14(16)(i)(1 + i)^{15} + 7.14(1 + i)^{16} - 16(1 + i)^{15} \tag{E.6}$$

The approximate answer of 0.116 found in Example B.5 could be used as an initial estimate of the root; however, in this case an initial estimate of 0.11 will be used to show how quickly the method converges. When doing the method with a hand calculator, it is best to set up a tabular format similar to Table E.1. A computer spread sheet program, such as VISICALC, is an excellent way of streamlining this tabular analysis, or a short computer program could be written. Note that $|f(i_n)/f'(i_n)|$ is the same as $|X_{n+1} - X_n|$; therefore, the last column can be used as the accuracy check.

TABLE E.1 NEWTON–RAPHSON SOLUTION FOR EXAMPLE B.5

Iteration	i_n	$f(i_n)$	$f'(i_n)$	$f(i_n)/f'(i_n)$
0	0.11	−0.1397	21.4914	−0.0065
1	0.1165	0.0193	27.5790	0.0007
2	0.1158	0.0002	26.8814	0.0000
3	0.1158			

For this problem, the estimate of the rate of return (accurate to four decimal places) is 0.1158 or 11.58%, which, to three significant figures, is the same as the estimate made from Table B.1. Note also that the accuracy used here is mathematical accuracy. That does not mean that it is physical accuracy. It is up to the user to determine how many significant figures are meaningful to be reported.

CLIMBING TECHNIQUES

Climbing techniques use previously determined information to locate solution points that will improve the objective function value. Climbing techniques generally have three distinct phases:

1. An opening exploration to set the stage for climbing techniques. This is the determination of the initial estimate of the solution. The better this initial estimate is, the more likely that climbing techniques will converge to the global optimum solution.
2. A series of middle steps that push to improve the objective function value. These are the climbing steps that are designed to converge toward an optimum solution.
3. A final exploratory stage to confirm that the stationary point (point to which the climbing steps converge) found is actually the optimum being searched for. Example E.2 will demonstrate a stationary point that is not an optimum. This illustrates the importance of carefully examining the final solution and the region around it.

Two climbing techniques, pattern search and gradient search (method of steepest descent), will be presented here. Each technique will be developed in the multidimensional environment, then applied to a two-dimensional nonlinear optimization problem. Pattern search will be applied to a constrained optimization example, while gradient search will be applied to an unconstrained optimization example.

Pattern Search

Pattern search involves starting at a solution point, then sequentially changing (perturbing) each independent variable above and below its present value, in an effort to find a better solution. After all variables are perturbed, the new solution point and old solution point are used to predict a new point that has a good chance of being an even better solution. This procedure establishes the pattern that will be used to step toward the final solution. When the solution approaches the vicinity of the optimum, smaller and smaller perturbations are made until the solution converges to the optimum.

There is no guarantee that the optimum solution found will be a global optimum. Starting the search procedure at two or more points some distance apart will improve the chance of finding the global optimum.

The method is best described as a series of steps that will be referred back to during the numerical example.

Step 1. Choose the initial starting point, $\mathbf{b}_1 = (b_1, b_2, \ldots, b_n)$ for the n variable problem. The choice of this starting point is not altogether arbitrary. A rough estimate of the optimum solution will greatly improve the initial solution point, and should speed convergence toward the optimum. Also, prudent choice of the initial solution point may circumvent computational problems that can sometimes arise during application of the technique.

Step 2.   Choose the step size for each independent variable, $\delta_i = (0, \ldots, 0, \delta_i, 0, \ldots, 0)$. Too large a step size may obscure subtle changes in the objective function surface, while excessively small steps will necessitate extra computation time. Choice of the appropriate step size is a combination of good judgment and trial-and-error adjustments.

Step 3.   Measure the objective function value at the original base starting point, $\mathbf{b}_1$. This point, $\mathbf{b}_1$, will represent the first temporary solution, $\mathbf{t}_{10}$. The first subscript (i) of $\mathbf{t}_{ij}$ represents the pattern (iteration) number, while the second subscript (j) identifies which variable is being perturbed. Steps 4–7 will describe the local explorations (perturbations) of the solution about $\mathbf{b}_i$.

Step 4.   Check the value of the objective function at $\mathbf{t}_{i0} + \delta_1$. If $Z(\mathbf{t}_{i0} + \delta_1)$ is better than $Z(\mathbf{t}_{i0})$, then the solution $\mathbf{t}_{i0} + \delta_1$ becomes the new temporary solution point, t_{i1}, and control should shift to step 6. If not, go to step 5.

Step 5.   Check the value of the objective function at $\mathbf{t}_{i0} - \delta_1$. If $Z(\mathbf{t}_{i0} - \delta_1)$ is better than $Z(\mathbf{t}_{i0})$, then the solution $\mathbf{t}_{i0} - \delta_1$ becomes the new temporary solution point, t_{i1}. If not, then $t_{i1} = \mathbf{t}_{i0}$.

Step 6.   Repeat steps 4 and 5 for all δ_j.

Step 7.   The last temporary solution point, $\mathbf{t}_{in}$, becomes the new base solution, $\mathbf{b}_{i+1}$, that is,

$$\mathbf{b}_{i+1} = \mathbf{t}_{in} \tag{E.7}$$

The latest two base solutions, $\mathbf{b}_i$ and $\mathbf{b}_{i+1}$, will be used to establish, continue, or expand the search pattern.

Step 8.   In establishing or continuing the search pattern, the reasoning is used that if the new base solution is an improvement over the old one, that is, $Z(\mathbf{b}_{i+1})$ is better than $Z(\mathbf{b}_i)$, then there is a good chance that a projection of the solution beyond $\mathbf{b}_{i+1}$ may further improve the objective function value. Therefore the first temporary solution for the next pattern is found by the relationship

$$\mathbf{t}_{(i+1)0} = \mathbf{b}_i + 2(\mathbf{b}_{i+1} - \mathbf{b}_i)$$

$(i + 1)$st pattern
no variables perturbed yet

$$= 2\mathbf{b}_{i+1} - \mathbf{b}_i \tag{E.8}$$

Step 9.   Local explorations are undertaken about the new temporary solution point, $\mathbf{t}_{(i+1)0}$ as per steps 4 through 7. The last temporary solution point becomes the new base solution.

Step 10.   If $Z(\mathbf{b}_{i+1})$ is an improvement over $Z(\mathbf{t}_{i0})$, then $\mathbf{t}_{(i+1)0}$ is projected as per step 8 (Equation E.8), and the length of the step will increase. Control is recycled to step 9.

Step 11.   If the solution $\mathbf{b}_{i+1}$ does not improve the temporary solution point $\mathbf{t}_{i0}$, but the solution $\mathbf{t}_{i0}$ is better than $\mathbf{b}_i$, then the pattern will maintain its direction, but the length of the step will not increase. A new temporary solution, $\mathbf{t}_{(i+1)0}$, is found using Equation E.8, and control recycles to step 9.

Step 12.   If $\mathbf{b}_{i+1}$ does not improve $\mathbf{b}_i$, then the pattern is destroyed and a new exploration (starting at step 4) must be undertaken to try to establish a new search pattern. For this pattern exploration, the new temporary solution, $\mathbf{t}_{i0}$, is set equal to $\mathbf{b}_i$.

1. If the local explorations find a better solution, then a new pattern is established and the explorations can continue.
2. If the local explorations do not find a better solution, then the step sizes are reduced and the explorations repeated. Control is recycled to step 4.

Step 13. The search is terminated when the step sizes fall below preselected minimum values.

EXAMPLE E.2

Example 6.9 presents an optimization model for estimating the minimum cost method of using water from two wells to supply a municipality. The cost of water from each well varies as a quadratic function of the flowrate from that well. If Z_1 and Z_2 represent the cost per unit flow (Q_1 and Q_2 in millions of gallons per day) from each of the sources, then

$$\left.\begin{array}{l} Z_1 + 13 + 4Q_1 + Q_1^2 \\ Z_2 + 26 + 8Q_2 + 4Q_2^2 \end{array}\right\} \tag{6.52}$$

and the overall objective function is

$$Z = Z_1 + Z_2 = 39 + 4Q_1 + Q_1^2 + 8Q_2 + 4Q_2^2 \tag{E.9}$$

in which Z = total supply costs (\$/day).

Two million gallons of water per day are required. Well 1 can supply up to 5 million gal/day, while well 2 can supply up to 10 million gal/day. Therefore, the following constraints apply to the solution values for Q_1 and Q_2 (in million gal/day):

$$\left.\begin{array}{l} Q_1 + Q_2 \geq 2 \\ Q_1 \leq 5 \\ Q_2 \leq 10 \end{array}\right\} \tag{6.54}$$

The method of Lagrange multipliers was used to find the minimum cost of \$52/day, with Q_1 = 2 million gal/day and Q_2 = 0. The climbing technique of pattern search will now be used to estimate the optimum operating policy.

The base vector, $\mathbf{b}_i$, will be replaced by $\mathbf{Q}_i$, with

$$\mathbf{Q}_i = (Q_{1i}, Q_{2i}). \tag{E.10}$$

Other variables will be the same as those used in the description of the method. To accommodate the constraints imposed on the problem, the objective function value for any solution vector, $\mathbf{Q}$, that does not satisfy the constraints will have an artificially high value of M added to it.

An initial base solution within the interior of the feasible region will be chosen. This initial solution will more than meet the constraint of at least 2 million gal/day, since the pattern search method will fail to generate a pattern if the initial solution is chosen with $Q_1 + Q_2 = 2.0$. Let

$$\left.\begin{array}{l} \mathbf{Q}_1 = (Q_{11}, Q_{21}) = (0.5, 2.8) \\ \delta_1 = (0.1, 0) \\ \delta_2 = (0, 0.1) \end{array}\right\} \tag{E.11}$$

Steps referenced in the following computations refer back to the steps of the pattern search method description.

The objective function value for the initial base solution is

$$Z(\mathbf{Q}_1) = Z_1(0.5, 2.8) = 95.01$$

From step 4, $Z(\mathbf{Q}_1 + \delta_1) = Z(0.6, 2.8) = 95.52$, which is greater than $Z(\mathbf{Q}_1)$, so $\mathbf{t}_{11}$ remains at $(0.5, 2.8)$. Applying step 5, $Z(\mathbf{Q}_1 - \delta_1) = Z(0.4, 2.8) = 94.52$, which is less than 95.01, so the temporary solution point, $\mathbf{t}_{11}$, becomes $(0.4, 2.8)$.

Step 6 is used to explore the perturbations of Q_2, and to find the temporary solution point, $\mathbf{t}_{12}$.

$$Z(\mathbf{Q}_1 + \delta_2) = Z(0.4, 2.9) = 97.60 > 94.52$$
$$Z(\mathbf{Q}_1 - \delta_2) = Z(0.4, 2.7) = 91.52 < 94.52$$

Therefore, $\mathbf{t}_{12} = (0.4, 2.7) = \mathbf{Q}_2$, by step 7. $\mathbf{Q}_1 = (0.5, 2.8)$ and $\mathbf{Q}_2 = (0.4, 2.7)$ establishes the first search pattern, and $\mathbf{t}_{20} = 2\mathbf{Q}_2 - \mathbf{Q}_1 = (0.3, 2.6)$. Local exploration about this temporary solution point shows the following:

$$Z(\mathbf{t}_{20}) = Z(0.3, 2.6) = 88.13 < 91.52$$
$$Z(\mathbf{t}_{20}, + \delta) = Z(0.4, 2.6) = 88.60 > 88.13$$
$$Z(\mathbf{t}_{20}, \delta_1) = Z(0.2, 2.6) = 87.68 < 88.13$$

Therefore $\mathbf{t}_{21} = (0.2, 2.6)$,

$$Z(\mathbf{t}_{21} + \delta_2) = Z(0.2, 2.7) = 90.60 > 87.68$$
$$Z(\mathbf{t}_{21} - \delta_2) = Z(0.2, 2.5) = 84.84 < 87.68$$
$$\mathbf{t}_{22} = (0.2, 2.5) = \mathbf{Q}_3$$

which is an improvement over $\mathbf{Q}_2$. Applying step 10,

$$\mathbf{t}_{30} = 2\mathbf{Q}_3 - \mathbf{Q}_2 = 2(0.2, 2.5) - (0.4, 2.7) = (0.0, 2.3)$$

which demonstrates the acceleration of steps.

Local explorations about the third temporary solution point follow:

$$Z(\mathbf{t}_{30}) = Z(0.0, 2.3) = 78.56 < 84.84$$
$$Z(\mathbf{t}_{30} + \delta_1) = Z(0.1, 2.3) = 78.97 > 78.56$$
$$Z(\mathbf{t}_{30} - \delta_1) = Z(-0.1, 2.3) = M \left(\begin{array}{c} \text{violates} \\ \text{constraint} \end{array}\right) > 78.56$$
$$\mathbf{t}_{31} = (0.0, 2.3)$$
$$Z(\mathbf{t}_{31} + \delta_2) = Z(0.0, 2.4) = 81.24 > 78.56$$
$$Z(\mathbf{t}_{31} - \delta_2) = Z(0.0, 2.2) = 75.96 < 78.56$$
$$\mathbf{t}_{32} = (0.0, 2.2) = \mathbf{Q}_4$$

The fourth temporary solution point, $\mathbf{t}_{40}$, will be

$$\mathbf{t}_{40} = 2\mathbf{Q}_4 - \mathbf{Q}_3 = (-0.2, 1.9)$$

which violates the constraints for the problem. Furthermore, when Q_1 and Q_2 are perturbed for explorations around $\mathbf{t}_{40}$, all temporary solution points will violate the constraints. Therefore, no improved solutions have been found, and by step 12, the pattern has been destroyed. $\mathbf{Q}_4$ is used as a new base point to try to reestablish a search pattern. Local explorations about $\mathbf{Q}_4$ produce

$$\mathbf{t}_{40} = (0.0, 2.2) \qquad Z(\mathbf{t}_{40}) = 75.96$$
$$Z(\mathbf{t}_{40} + \delta_1) = Z(0.1, 2.2) = 76.37 > 75.96$$
$$Z(\mathbf{t}_{40} - \delta_1) = Z(-0.1, 2.2) = M > 75.96$$
$$\mathbf{t}_{41} = (0.0, 2.2)$$
$$Z(\mathbf{t}_{41} + \delta_2) = Z(0.0, 2.3) = 78.56 > 75.96$$
$$Z(\mathbf{t}_{41} - \delta_2) = Z(0.0, 2.1) = 73.44 < 75.96$$
$$\mathbf{t}_{42} = (0.0, 2.1) = \mathbf{Q}_5$$

The next temporary solution point will be determined by step 8,

$$\mathbf{t}_{50} = 2\mathbf{Q}_5 - \mathbf{Q}_4 = (0.0,\ 2.0)$$

Local explorations show that

$$Z(\mathbf{t}_{50}) = 71.00 < 73.44$$
$$Z(\mathbf{t}_{50} + \delta_1) = Z(0.1,\ 2.0) = 71.41 > 71.00$$
$$Z(\mathbf{t}_{50} - \delta_1) = Z(-0.1,\ 2.0) = M > 71.00$$
$$\mathbf{t}_{51} = (0.0,\ 2.0)$$
$$Z(\mathbf{t}_{51} + \delta_2) = Z(0.0,\ 2.1) = 73.44 > 71.00$$
$$Z(\mathbf{t}_{51} - \delta_2) = Z(0.0,\ 1.9) = M > 71.00$$
$$\mathbf{t}_{52} = (0.0,\ 2.0) = \mathbf{Q}_6$$

None of the local explorations produce a better solution than $\mathbf{t}_{50}$, but $Z(\mathbf{t}_{50})$ is better than $Z(\mathbf{Q}_5)$, so the pattern is continued, without growth, according to step 11.

$$\mathbf{t}_{60} = 2\mathbf{Q}_6 - \mathbf{Q}_5 = (0.0,\ 1.9)$$
$$Z(\mathbf{t}_{60}) = Z(0.0,\ 1.9) = 68.64 + M > 71.00$$
$$Z(\mathbf{t}_{60} + \delta_1) = Z(0.1,\ 1.9) = 69.05 < 71.00$$
$$Z(\mathbf{t}_{60} - \delta_1) = Z(-0.1,\ 1.9) = M > 69.05$$
$$\mathbf{t}_{61} = (0.1,\ 1.9)$$
$$Z(\mathbf{t}_{61} + \delta_2) = Z(0.1,\ 2.0) = 71.41 > 69.05$$
$$Z(\mathbf{t}_{61} - \delta_2) = Z(0.1,\ 1.8) = 66.77 + M > 69.05$$
$$\mathbf{t}_{62} = (0.1,\ 1.9) = \mathbf{Q}_7$$
$$\mathbf{t}_{70} = 2\mathbf{Q}_7 - \mathbf{Q}_6 = (0.2,\ 1.8)$$
$$Z(\mathbf{t}_{70}) = 67.20 < 69.05$$
$$Z(\mathbf{t}_{70} + \delta_1) = Z(0.3,\ 1.8) = 67.65 > 67.20$$
$$Z(\mathbf{t}_{70} - \delta_1) = Z(0.1,\ 1.8) = 66.77 + M > 67.20$$
$$\mathbf{t}_{71} = (0.2,\ 1.8)$$
$$Z(\mathbf{t}_{71} + \delta_2) = Z(0.2,\ 1.9) = 69.48 > 67.20$$
$$Z(\mathbf{t}_{71} - \delta_2) = Z(0.2,\ 1.7) = 65.00 + M > 67.20$$
$$\mathbf{t}_{72} = (0.2,\ 1.8) = \mathbf{Q}_8$$
$$\mathbf{t}_{80} = 2\mathbf{Q}_8 - \mathbf{Q}_7 = (0.3,\ 1.7)$$
$$Z(\mathbf{t}_{80}) = 65.45 < 67.20$$

Succeeding patterns would mirror the previous one, since the solution is stepping along the constraint $Q_1 + Q_2 \geq 2$. Values for $\mathbf{Q}_i$ and $Z(\mathbf{Q}_i)$ are given in the following table.

Pattern (i)	$\mathbf{Q}_i$	$Z(\mathbf{Q}_i)$
8	(0.2, 1.8)	67.20
9	(0.3, 1.7)	65.45
10	(0.4, 1.6)	63.80
11	(0.5, 1.5)	62.25
12	(0.6, 1.4)	60.80
13	(0.7, 1.3)	59.45
14	(0.8, 1.2)	58.20
15	(0.9, 1.1)	57.05
16	(1.0, 1.0)	56.00
17	(1.1, 0.9)	55.05

Pattern (i)	$\mathbf{Q}_i$	$Z(\mathbf{Q}_i)$	(Continued)
18	(1.2, 0.8)	54.20	
19	(1.3, 0.7)	53.45	
20	(1.4, 0.6)	52.80	
21	(1.5, 0.5)	52.25	
22	(1.6, 0.4)	51.80	
23	(1.7, 0.3)	51.45	
24	(1.8, 0.2)	51.20	
25	(1.9, 0.1)	51.05	
26	(2.0, 0.0)	51.00	

The next solution, $\mathbf{Q}_{27} = (2.1, -0.1)$, would violate the constraints and would not produce any improved solutions during local explorations. Therefore, the pattern would be destroyed, and $\mathbf{Q}_{26}$ would become the new base. Local explorations around $\mathbf{Q}_{26}$ will not establish a pattern, no matter how small δ_1 and δ_2 are made. Therefore, $\mathbf{Q}_{26} = (2.0, 0.0)$ represents the optimal solution, with $Z(\mathbf{Q}_{26}) = 51.00$, which is the same optimum solution that was determined by Lagrange multipliers. Figure E.4 shows the progression of solutions and local explorations undertaken to find $\mathbf{Q}_{26}$.

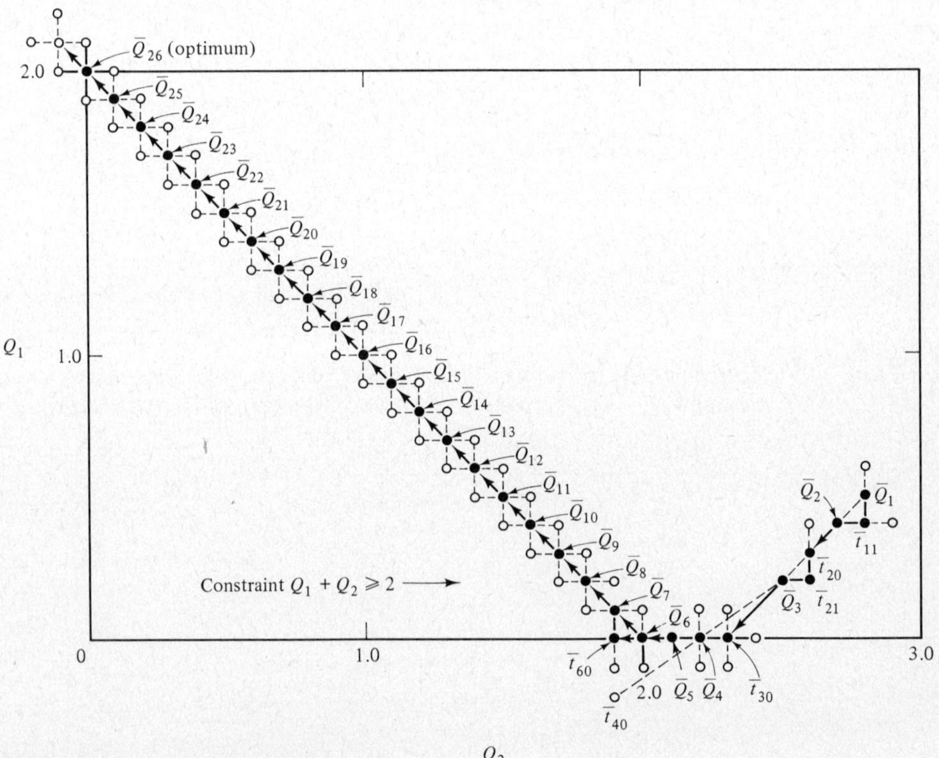

Figure E.4 Pattern search solution.

The search technique starts at an interior point, develops a pattern with Q_1 and Q_2 both decreasing, and continues this pattern until the $Q_1 \geq 0$ constraint is reached. The new pattern turns and follows this constraint, with $Q_1 = 0$ and Q_2 decreasing, until the $Q_1 + Q_2 \geq 2$ constraint is reached. Another new pattern is developed that follows this constraint, until the optimum solution of $Q_1 = 2$, $Q_2 = 0$, $Z = 51.00$ is reached. No further patterns can be developed, since this is the optimum point. The reasons for the patterns of the search can be ascertained by examining the contours of the objective function given in Figure 6.11(c).

The pattern search technique successfully found the optimum solution from the starting point given. However, if a point nearer the $Q_1 + Q_2 \geq 2$ constraint had been chosen, problems would have arisen. Suppose the initial solution had been $\mathbf{Q}_1 = (1.4, 1.4)$. Successive patterns would have identified the following values for $\mathbf{Q}_i$:

$\mathbf{Q}_i$	(Q_1, Q_2)	$Z(\mathbf{Q}_i)$
$\mathbf{Q}_1$	(1.4, 1.4)	65.60
$\mathbf{Q}_2$	(1.3, 1.3)	63.05
$\mathbf{Q}_3$	(1.1, 1.1)	58.25
$\mathbf{Q}_4$	(1.0, 1.0) [new pattern]	56.00

Steps leading to these solutions can be verified by the student.

Attempts to initiate a new pattern from solution $\mathbf{Q}_4$ will prove fruitless, as illustrated in Figure E.5. The standard cruciform pattern is initiated; however, two of the points produce infeasible solutions (1.0, 0.9) and (0.9, 1.0), while the other two points produce feasible solutions with Z values greater than the base solution at (1.0, 1.0). Making the step sizes smaller will not solve the problem, because no matter how small δ_1 and δ_2 are made, the same outcome will

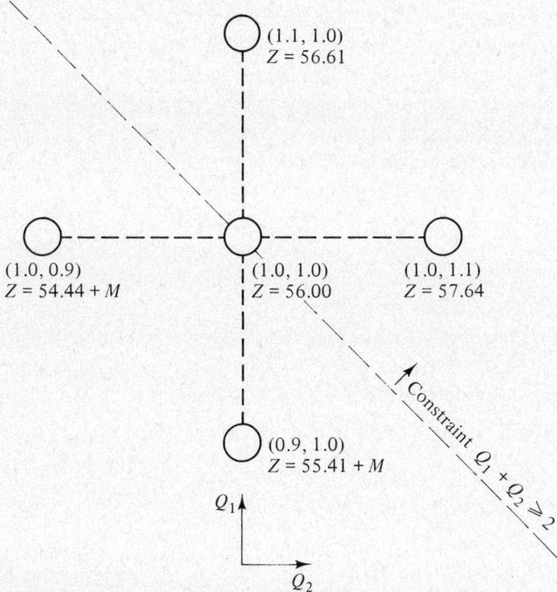

Figure E.5 Cruciform search pattern in vicinity of $Q_1 + Q_2 \geq 2$.

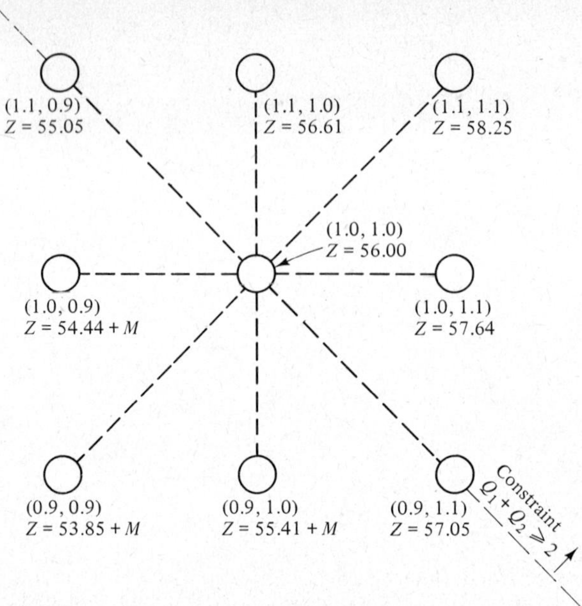

Figure E.6   Box pattern to resolve pattern.

result. Thus, the indication is that an optimum solution has been found, when it is not optimum at all. It is not even a local optimum.

The pattern search technique is subject to this type of problem when approaching a ridge or valley from the side. The constraint $Q_1 + Q_2 \leq 2$ does represent a valley, because on one side the objective function value is decreasing as the constraint is approached. On the other side, the solutions are infeasible, and the Z values are increased by the arbitrarily high value M. One method for turning the solution procedure up the ridge or down the valley incorporates a quadratic strategy to detect the trend in the ridge or valley from the information provided by the cruciform pattern (Wilde and Beightler, 1967). This method works well for unconstrained problems, but will not reorient the patterns for this search because of the way the constraints are treated.

Another method is to extend the local exploration to a box pattern, instead of the cruciform pattern. Figure E.6 shows this pattern, which would identify the improved solution at $\mathbf{Q} = (1.1, 0.9)$, and turn the new pattern in this direction.

Engineers applying numerical techniques to help find problem solutions must expect to encounter problems such as the one previously described. This is why it is extremely important that the engineer thoroughly understand both the characteristics of the problem being analyzed and the characteristics of the numerical technique being applied.

Gradient Search

Gradient search is a widely used climbing technique that is based on the gradient of an n-dimensional function, which is defined mathematically as

$$\nabla Z = \frac{\partial Z}{\partial X_1}, \frac{\partial Z}{\partial X_2}, \ldots, \frac{\partial Z}{\partial X_n} \qquad (E.12)$$

Using the gradient will point the solution in the direction of steepest descent or ascent. The steps for implementing the technique follow.

Step 1. Determine the starting solution vector,

$$\mathbf{X}_i = X_1, X_2, \ldots, X_n \tag{E.13}$$

which will either be the initial starting point, or the solution point estimated from the previous iteration of the technique.

Step 2. Determine the gradient of the objective function at the solution point $\mathbf{X}_i$.

$$\nabla Z(\mathbf{X}_i) = \nabla Z(X_{1i}, X_{2i}, \ldots, X_{ni})$$

$$= \left.\frac{\partial Z}{\partial X_1}\right|_{X_{1i}}, \left.\frac{\partial Z}{\partial X_2}\right|_{X_{2i}}, \ldots, \left.\frac{\partial Z}{\partial X_n}\right|_{X_{ni}} \tag{E.14}$$

Step 3. Set up a relationship to estimate the amount that each independent variable should be changed to start the next iteration.

$$\Delta\mathbf{X}_i = X_{1(i+1)} - X_{1i}, X_{2(i+1)} - X_{2i}, \ldots, X_{n(i+1)} - X_{ni} = \rho_i \nabla Z(\mathbf{X}_i) \tag{E.15}$$

The coefficient ρ_i is a weighting factor to project each independent variable from its present value, along the gradient, to the next value for that independent variable. There is no guarantee that the value for each independent variable will move closer to its optimum value on each iteration, but the overall scheme is to move the solution point, $\mathbf{X}_{i+1}$, closer to the optimum. At this point, ρ_i is a variable that will be solved for.

Step 4. Determine the value for ρ_i to be used to project the solution.
 a. Substitute $\mathbf{X}_i + \Delta\mathbf{X}_i$ into the objective function and evaluate it in terms of ρ_i.

$$Z(\mathbf{X}_i + \Delta\mathbf{X}_i) = Z[\mathbf{X}_i + \rho_i \nabla Z(\mathbf{X}_i)]$$

$$= Z\left(X_{1i} + \rho_i \left.\frac{\partial Z}{\partial X_1}\right|_{X_{1i}}, X_{2i} + \rho_i \left.\frac{\partial Z}{\partial X_2}\right|_{X_{2i}}, \ldots, X_{ni} + \rho_i \left.\frac{\partial Z}{\partial X_n}\right|_{X_{ni}}\right) \tag{E.16}$$

 b. If this new solution point is actually an optimum point, then the partial derivatives evaluated at that point will be zero. Therefore, the partial derivative of the objective function from step 4(a) can be taken with respect to ρ_i and set equal to zero. the resulting equation can be used to provide an estimate of ρ_i.

$$\frac{\partial Z[\mathbf{X}_i + \rho_i \nabla Z(\mathbf{X}_i)]}{\partial \rho_i} = 0 \tag{E.17}$$

 c. Solve Equation E.17 to find an estimate of ρ_i.

Step 5. Use the estimate of ρ_i to determine the new estimate of $\mathbf{X}$.

$$\mathbf{X}_{i+1} = \mathbf{X}_i + \rho_i \nabla Z(\mathbf{X}_i) \tag{E.18}$$

Step 6. Iterate through steps 1–5 until the differences between successive solution estimates are smaller than an acceptable margin of error.

EXAMPLE E.3 ____

Given the function

$$Z = 5 + 4X_1 + 3X_1^2 - 2X_1X_2 - 2X_2 + 2X_2^2 \tag{E.19}$$

Use the gradient search method to estimate the minimum point for this two variable function, accurate to three decimal places.

For Equation E.19, the partial derivatives are

$$\frac{\partial Z}{\partial X_1} = 4 + 6X_1 - 2X_2$$

$$\frac{\partial Z}{\partial X_2} = -2X_1 - 2 + 4X_2$$

(E.20)

If Equations E.20 are set equal to zero, values of X_1 and X_2 that minimize the function can be found by the usual means. This will give

$$X_1 = -0.60 \qquad X_2 = 0.20, \qquad \text{and} \quad Z = 3.60$$

Second partial derivatives will confirm that this is a minimum.

Three iterations of the gradient search technique will be carried through to demonstrate the method.

Iteration 1

1. Choose $\mathbf{X}_1 = (0, 0)$.

2. $\nabla Z(\mathbf{X}_1) = \left.\frac{\partial Z}{\partial X_1}\right|_0, \quad \left.\frac{\partial Z}{\partial X_2}\right|_0 = (4, -2)$

3. $\Delta\mathbf{X}_1 = \rho_1 \nabla Z(\mathbf{X}_1) = (4\rho_1, -2\rho_1)$

4. a. $Z[\mathbf{X}_1 + \rho_1 \nabla Z(\mathbf{X}_1)]$

$$= Z(0 + 4\rho_1, 0 - 2\rho_1)$$

$$= 5 + 20\rho_1 + 72\rho_1^2$$

b. $\dfrac{\partial Z[\mathbf{X}_1 + \rho_1 \nabla Z(\mathbf{X}_1)]}{\partial \rho_1}$

$$= \frac{\partial(5 + 20\rho_1 + 72\rho_1^2)}{\partial \rho_1}$$

$$= 20 + 144\rho_1 = 0$$

c. $\rho_1 = -0.139$

5. $\mathbf{X}_2 = \mathbf{X}_1 + \rho_1 \nabla Z(\mathbf{X}_1)$

$$= (0, 0) + (-0.139)(4, -2)$$

Iteration 2

1. $\mathbf{X}_2 = (-0.556, 0.278)$

2. $\nabla Z(\mathbf{X}_2) = (0.108, 0.224)$

3. $\Delta\mathbf{X}_2 = (0.108\rho_2, 0.224\rho_2)$

4. a. $Z[\mathbf{X}_2 + \rho_2 \nabla Z(\mathbf{X}_2)]$

$$= Z[-0.556 + 0.108\rho_1, 0.278 + 0.224\rho_1]$$

$$= 3.61 + 0.0618\rho_2 + 0.0870\rho_2^2$$

b. $\dfrac{\partial[3.61 + 0.0618\rho_2 + 0.0870\rho_2^2]}{\partial \rho_2}$

$$= 0.0618 + 0.174\rho_2$$

c. $\rho_2 = -0.356$

5. $\mathbf{X}_3 = \mathbf{X}_2 + \rho_2 \nabla Z(\mathbf{X}_2)$

$$= (-0.556, 0.278) + (-0.356)(0.108, 0.224)$$

Iteration 3

1. $\mathbf{X}_3 = (-0.594, 0.198)$

2. $\nabla Z(\mathbf{X}_3) = (0.040, -0.020)$

3. $\Delta \mathbf{X}_3 = (0.040\rho_3, -0.020\rho_3)$

4. a. $Z[\mathbf{X}_3 + \rho_3 \nabla Z(\mathbf{X}_3)]$

$$= Z[-0.594 + 0.040\rho_3, 0.198 - 0.020\rho_3]$$

$$= 3.600 + 0.002\rho_3 + 0.0072\rho_3^2$$

b. $\dfrac{\partial(3.600 + 0.002\rho_3 - 0.0072\rho_3^2)}{\partial\rho_3}$

$$= 0.002 - 0.0144\rho_3$$

c. $\rho_3 = -0.139$

5. $\mathbf{X}_4 = \mathbf{X}_3 + \rho_3 \nabla Z(\mathbf{X}_3)$

$$= (-0.594, 0.198) + (-0.139)(0.040, -0.020)$$

$$= (-0.600, 0.201)$$

Iteration 4 would produce a new estimate, $\mathbf{X}_5 = (-0.600, 0.200)$, which is accurate to three decimal places.

Another starting point will be used to verify the minimum solution. A summary of the results is presented below:

Iteration	X_1	X_2	$Z(X_1, X_2)$
1	1.0	1.0	10.0
2	−0.333	1.0	4.667
3	−0.333	0.333	3.778
4	−0.556	0.333	3.630
5	−0.556	0.222	3.605
6	−0.593	0.222	3.601
7	−0.593	0.204	3.600
8	−0.600	0.204	3.600
9	−0.600	0.201	3.600
10	−0.600	0.200	3.600

This is the same answer as before, and is the global minimum for this function.

REFERENCES

Accreditation Board for Engineering and Technology, *Criteria for Accrediting Programs in Engineering in the United States* (1980).

Agin, N., "Optimum Seeking with Branch and Bound," *Management Science,* Vol. 13, No. 4, pp. B176–185 (1966).

Barish, N. N., and Kaplan, S., *Economic Analysis for Engineering and Managerial Decision Making,* 2nd Ed., McGraw-Hill, New York (1978).

Bellman, R., *Dynamic Programming,* Princeton University Press, Princeton, NJ (1957).

Biswas, A. K., Ed., *Systems Approach to Water Management,* McGraw-Hill, New York (1976).

Block, H. D., and Lynn, W. R., *Systems Analysis Applications,* Cornell University Water Resources Center, Ithaca, NY (1963).

Bradley, S. P., Hax, A. C., and Magnanti, T. L., *Applied Mathematical Programming,* Addison-Wesley, Reading, MA (1977).

Carmichael, D. G., *Structural Modelling and Optimization: A General Methodology for Engineering and Control,* Ellis Horwood Limited, Chichester, UK (1981).

Churchman, C. W., *The Systems Approach and Its Enemies,* Basic Books, New York (1979).

Clough, R. H., and Sears, G. A., *Construction Project Management,* 2nd Ed., Wiley, New York (1979).

Cohon, J. L., *Multiobjective Programming and Planning,* Academic Press, New York (1978).

Dixon, W. J., Ed., *Biomedical Computer Programs, BMDP,* University of California Press, Berkeley, CA (1977).

Dreyfus, S. E., and Law, A. M., *The Art and Theory of Dynamic Programming,* Academic Press, New York (1977).

Gallagher, R. H., and Zienkiewicz, O. C., *Optimum Structural Design, Theory and Applications,* Wiley, New York (1973).

Garfinkel, R. S., and Nemhauser, G. L., *Integer Programming,* Wiley, New York (1972).

Gass, S. I., *Linear Programming, Methods and Applications,* 4th Ed., McGraw-Hill, New York (1975).

Gates, M., "Bidding Strategies and Probabilities," *Proceedings, ASCE,* Vol. 93, No. C02, pp. 75–107 (March, 1967).

Greenberg, H., *Integer Programming,* Academic Press, New York (1971).

Hadley, G., *Linear Algebra,* Addison-Wesley, Reading, MA (1961).

Hadley, G., *Linear Programming,* Addison-Wesley, Reading, MA (1962).

Hall, W. A., and Dracup, J. A., *Water Resources Systems Engineering,* McGraw-Hill, New York (1970).

Haug, E. J., and Arora, J. S., *Applied Optimal Design: Mechanical and Structural Systems,* Wiley, New York (1979).

Hildebrand, F. B., *Introduction to Numerical Analysis,* 2nd Ed., McGraw-Hill, New York (1974).

Hill, L. A. Jr., "Seven Instructional Objectives that Characterize a Design Course," *Proceedings, 1981 American Society for Engineering Education Annual Conference,* pp. 155–158 (1981).

Jewell, T. K., Adrian, D. D., and Hosmer, D. W., "Analysis of Stormwater Pollutant Washoff Estimation Techniques," *Proceedings of the International Symposium on Urban Storm Runoff,* University of Kentucky at Lexington, pp. 135–141 (July, 1980).

Kirsch, U., *Optimal Structural Design,* McGraw-Hill, New York (1981).

Loucks, D. P., Stedinger, J. R., and Haith, D. A., *Water Resource Systems Planning and Analysis,* Prentice-Hall, Englewood Cliffs, NJ (1981).

Louie, P. W., Yeh, W. G., and Hsu, N. S., "Multiobjective Water Resources Management Planning," *Journal of Water Resources Planning and Management,* American Society of Civil Engineers, Vol. 110, No. 1 (January, 1984).

Major, D. C., "Multiobjective Water Resource Planning," Water Resources Monograph No. 4, American Geophysical Union, Washington, D.C. (1977).

Major, D. C., and Lenton, R. L., *Applied Water Resources Systems Planning,* Prentice-Hall, Englewood Cliffs, NJ (1979).

Mayer, R. H. Jr., Stark, R. M., and Fitzgerald, W., *Unbalanced Bidding Models-Applications,* University of Delaware Technical Report (March, 1969).

McCormick, G. P., *Nonlinear Programming,* Wiley, New York (1983).

Meredith, D. D., Wong, K. W., Woodhead, R. W., and Wortman, R. H., *Design and Planning of Engineering Systems,* Prentice-Hall, Englewood Cliffs, NJ (1973).

Meta Systems, Inc., *Systems Analysis in Water Resources Planning,* Water Information Center, Inc., Port Washington, NY (1975).

Nie, N., Hull, C., Jenkins, J., Steinbrenner K., and Bent, D., *Statistical Package for the Social Sciences (SPSS),* 2nd Ed., McGraw-Hill, New York (1975).

Phillips, K. J., Beltrami, E. J., Carroll, T. O., and Kellogg, S. R., "Optimization of Areawide Wastewater Management," *Journal of the Water Pollution Control Federation,* Vol. 54, No. 1, pp. 87–93 (January, 1982).

Schmit, L. A. Jr., "Structural Design by Systematic Synthesis," *Proceedings of Second Conference on Electronic Computation,* ASCE, Pittsburgh, PA, pp. 105–132 (1960).

Shelley, P. E., and Kirkpatrick, G. A., *An Assessment of Automatic Sewer Flow Samplers— 1975,* Report EPA-600/2-75-065, U. S. Environmental Protection Agency, Cincinnati, OH (1975).

Standard Methods for the Examination of Water and Wastewater, 15th Ed., American Public Health Association, Washington, D.C. (1980).

Stark, R. M., "Unbalanced Bidding Models—Theory," *Proceedings, ASCE,* Vol. 94, No. C02, pp. 197–209 (October, 1968).

Stark, R. M., and Nicholls, R. L., *Mathematical Foundations for Design: Civil Engineering Systems,* McGraw-Hill, New York (1972).

Thomann, R. V., *Systems Analysis and Water Quality Management,* McGraw-Hill, New York (1972).

Whipple, W. Jr., Hunter, J. V., and Yu, S. L., "Effects of Storm Frequency on Pollution from Urban Runoff," *Journal of the Water Pollution Control Federation,* Vol. 49, No. 11, pp. 2243–2254 (November, 1977).

White, J. A., Agee, M. H., and Case, K. E., *Principles of Engineering Economic Analysis,* Wiley, New York (1977).

Wilde, I. J., and Beightler, C. S., *Foundations of Optimization,* Prentice-Hall, Englewood Cliffs, NJ (1967).

Wismer, D., and Chattergy, R., *Introduction to Nonlinear Optimization,* North-Holland, New York (1978).

BIBLIOGRAPHY

Afifi, A. A., and Azen, S. P., *Statistical Analysis, A Computer Oriented Approach,* Academic Press, New York (1972).

Aguilar, R. J., *Systems Analysis and Design in Engineering, Architecture, Construction, and Planning,* Prentice-Hall, Englewood Cliffs, NJ (1973).

Au, T., and Stelson, T. E., *Introduction to Systems Engineering: Deterministic Models,* Addison-Wesley, Reading, MA (1969).

de Neufville, R., and Marks, D., *Systems Planning and Design, Case Studies in Modeling, Optimization, and Evaluation,* Prentice-Hall, Englewood Cliffs, NJ (1974).

de Neufville, R., and Stafford, J. H., *Systems Analysis for Engineers and Managers,* McGraw-Hill, New York (1971).

Draper, N., and Smith, H., *Applied Regression Analysis,* Wiley, New York (1966).

Dunn, O. J., and Clark, V. A., *Applied Statistics: Analysis of Variance and Regression,* Wiley, New York (1974).

Dym, C. L., and Ivey, E. S., *Principles of Mathematical Modeling,* Academic Press, New York (1980).

Haith, D. A., *Environmental Systems Optimization,* Wiley, New York (1982).

Hajek, V. G., *Management of Engineering Projects,* McGraw-Hill, New York (1977).

Harris, R. B., *Precedence and Arrow Networking Techniques for Construction,* Wiley, New York (1978).

Henderson, J. M., "Yes Design Can be Taught," *Engineering Education,* American Society for Engineering Education, pp. 302–304 (January, 1980).

James, M. L., Smith, G. M., and Wolford, J. C., *Applied Numerical Methods for Digital Computation with Fortran and CSMP,* 2nd Ed., Harper & Row, New York (1977).

Jewell, T. K., "The Role of Systems Analysis in Teaching Design," *Proceedings,* American Society for Engineering Education Annual Conference, pp. 159–164 (1981).

Jewell, T. K., "Urban Stormwater Pollutant Loadings," Ph.D. Dissertation presented to Department of Civil Engineering, University of Massachusetts, Amherst (February, 1980).

Jewell, T. K., Adrian, D. D., and DiGiano, F. A., *Urban Stormwater Pollutant Loadings,* Publication No. 113, Water Resources Research Center, University of Massachusetts, Amherst (1980).

Jewell, T. K., Nunno, T. J., and Adrian, D. D., "Methodology for Calibrating Stormwater Management Models," *Journal of the Environmental Engineering Division,* Proceedings of the American Society of Civil Engineers, Vol. 104, No. EE3 (June, 1978).

Kane, S., "Introducing Systems Concepts to all University Students," *Engineering Education,* American Society for Engineering Education, pp. 427–429 (February, 1980).

Kerzner, H., *Project Management, A Systems Approach to Planning, Scheduling, and Controlling,* Van Nostrand Reinhold, New York (1979).

Lapin, L. L., *Probability and Statistics for Modern Engineering,* Brooks/Cole Publishing Company, Monterey, CA (1983).

Middlebrooks, E. J., *Statistical Calculations,* Ann Arbor Science Publishers, Ann Arbor, MI (1976).

Miles, R. F. Jr., Ed., *Systems Concepts, Lectures on Contemporary Approaches to Systems,* Wiley, New York (1973).

Ossenbruggen, P. J., *Systems Analysis for Civil Engineers,* Wiley, New York (1984).

Ostrofsky, B., *Design, Planning, and Development Methodology,* Prentice-Hall, Englewood Cliffs, NJ (1977).

Pilcher, R., *Principles of Construction Management,* McGraw-Hill, London (1966).

Pinnel, S. S., "Critical Path Scheduling: An Overview and a Practical Alternative," *Civil Engineering Magazine,* American Society of Civil Engineers, New York (July, 1980).

Pistrang, J., "A Senior Course in Civil Engineering Systems," presented at American Society for Engineering Education Annual Meeting, University of California, Los Angeles (June 17–20, 1968).

Pollock, S. M., "Mathematical Modeling: Applying the Principles of the Art Studio," *Engineering Education,* American Society for Engineering Education, pp. 167–171 (November, 1976).

Rice, R. G., "Experiences with Systems Engineering in the University of Toronto Undergraduate Curriculum," presented at Conference, Systems Theory for the Civil Engineer, University of Calgary, Calgary, Alberta, Canada (October, 1979).

Shelley, P. E., and Kirkpatrick, G. A., *Sewer Flow Measurement—A State-of-the-Art Assessment,* Report EPA-600/2-75-027, U. S. Environmental Protection Agency, Cincinnati, OH (1975).

Shoup, T. E., Fletcher, L. S., and Mochel, E. V., *Introduction to Engineering Design with Graphics and Design Projects,* Prentice-Hall, Englewood Cliffs, NJ (1981).

Volk, W., *Applied Statistics for Engineers,* 2nd Ed., McGraw-Hill, New York (1969).

Whipple, W. Jr., Hunter, J. V., and Yu, S. L., Author's response to discussion of paper "Effects of Storm Frequency on Pollution from Urban Runoff," *Journal of the Water Pollution Control Federation,* Vol. 50, No. 5, pp. 979–982 (May, 1978).

Index

Index